Crystal Symmetry, Lattice Vibrations, and Optical Spectroscopy of Solids

A Group Theoretical Approach

Crystal Symmetry, Lattice Vibrations, and Optical Spectroscopy of Solids

A Group Theoretical Approach

Baldassare Di Bartolo
Boston College, USA

Richard C. Powell
University of Arizona, USA

World Scientific

NEW JERSEY • LONDON • SINGAPORE • BEIJING • SHANGHAI • HONG KONG • TAIPEI • CHENNAI

Published by

World Scientific Publishing Co. Pte. Ltd.

5 Toh Tuck Link, Singapore 596224

USA office: 27 Warren Street, Suite 401-402, Hackensack, NJ 07601

UK office: 57 Shelton Street, Covent Garden, London WC2H 9HE

Library of Congress Cataloging-in-Publication Data

Di Bartolo, Baldassare, author.

Crystal symmetry, lattice vibrations and optical spectroscopy of solids : a group theoretical approach / Baldassare Di Bartolo (Boston College, USA) & Richard C. Powell (University of Arizona, USA).

pages cm

Includes bibliographical references and index.

ISBN 978-9814579209 (hardcover : alk. paper)

1. Lattice dynamics. 2. Phonons. 3. Symmetry (Physics) 4. Optical spectroscopy. 5. Solid state physics. I. Powell, Richard C. (Richard Conger), 1939– author. II. Title.

QC176.8.L3D53 2014

530.4'11--dc23

2013047941

British Library Cataloguing-in-Publication Data

A catalogue record for this book is available from the British Library.

In-house Editor: Song Yu

Typeset by Stallion Press
Email: enquiries@stallionpress.com

Printed in Singapore by World Scientific Printers.

To Rita and Gwen

Preface

The field of *solid state physics* consists of many different topics, each addressing a specific physical property of the material, e.g. electrical, mechanical, optical, or magnetic properties. However, there are two fundamental aspects of a solid that affect all of its physical properties. The first is the microscopic arrangement of the atomic constituents of the material. This can be amorphous leading to a glass material or have a specific order leading to a crystalline material. The basic structure of crystals and glasses is a major factor in determining all of their physical properties. The second is the motion of the atomic constituents about their equilibrium positions. These atomic vibrations modulate the physical properties of the material. This book focuses on these two fundamental aspects of solids and thus is appropriate to use as a textbook for an introductory class in solid state physics within a curriculum which includes additional classes on specific properties of solids. It assumes the student has taken a one year course in quantum mechanics.

The major mathematical tool used to describe the symmetry properties of crystal lattices and their thermal vibrations is *group theory*. The techniques of group theory have important applications in almost every area of physics. This book covers the principles of group theory starting at a fundamental level and providing specific examples of how to use group theory in studying crystalline solids. It assumes no prior knowledge of the subject. Because of the extensive treatment of the topic, it would be possible to use this book as a textbook for a one semester course on group theory.

The book is divided into three parts:

I. The first part focuses on the symmetry of crystal structures. The Hamiltonian of a crystalline solid is introduced and the adiabatic approximation is described. The importance of symmetry for this Hamiltonian is discussed. The basic concepts of group theory are reviewed and applied to the treatment of spatial symmetries and quantum mechanical processes. Different types of crystal lattices are described and the experimental measurement of crystal lattices by x-ray diffraction is discussed.
II. The second part of the book deals with the lattice vibrations of crystals (i.e. phonons). The Hamiltonian of the crystal developed in Part I is modified to describe phonons and different types of phonon modes are discussed. The thermodynamics of lattice vibrations is described as well as the experimental measurements of lattice vibrations by neutron scattering and the effects of thermal vibrations on x-ray diffraction. Again symmetry and group theory play important roles in describing different types of lattice vibrations.
III. In the third part of the book the optical spectroscopy of crystals is described with a special emphasis on the effects of lattice symmetry and thermal vibrations on the spectroscopic properties. The energy levels of a solid are in general determined by the electrons on its atomic constituents or by its lattice vibrations. These energy levels can be either delocalized throughout the material or localized at a specific point in the lattice. Transitions between energy levels can produce spectral lines throughout the ultraviolet, visible, and infrared regions of the spectra. Measurement techniques such as absorption, emission, and scattering of light are used to determine the properties of these energy levels. This section describes in detail the important case of electronic transitions between the energy levels of an impurity ion in a crystal with special emphasis on the effects of thermal vibrations on these transitions. In addition, the measurement of phonon energies through infrared absorption and inelastic light scattering is described. The important application of these concepts to solid state lasers is summarized in the final chapter. This ties

together how crystal symmetry, lattice vibrations, and optical spectroscopic properties determine the operational parameters of a laser.

This book provides an example of the importance of symmetry and group theory techniques in describing fundamental physical properties of solids. As well as its potential use as a textbook, it should be of interest to researchers studying lattice vibrations of crystals or spectroscopists whose results are affected by lattice vibrations. It should be noted that there are other thermal effects in solids (such as phonon transport properties) that are not treated here since they are not closely related to spectral data and have been treated extensively elsewhere.

Much of the basic material in this book was originally published in 1976 by John Wiley & Sons under the title *Phonons and Resonances in Solids* which is now out of print. We found the book to be a useful text for advanced courses in solid state physics but felt that the organization of the book was somewhat cumbersome and the connection between the various chapters was not transparent. Therefore we have revised the book by reorganizing the chapters and adding new material. In addition, numerous typographical errors in the original book have been corrected and the title has been changed to reflect the new structure of the text. These changes should significantly increase the usefulness of the book as a basic text on group theory and solid state physics.

Baldassare Di Bartolo
Chestnut Hill, Massachusetts
August 2013

Richard C. Powell
Tucson, Arizona

Contents

Part I

Symmetry of Crystals

The first part of this book deals with the symmetry of an ordered array of atoms. Group Theory is introduced starting from fundamental principles and is applied to the study of the symmetry of crystalline solids. The symmetry of solids is intimately related to their physical properties and a study of their symmetry is essential for an understanding of their behaviour.

Chapter 1

Introduction

This chapter sets the stage for the entire book by treating the Hamiltonian of a crystalline solid, introducing the adiabatic approximation, and considering the consequences of this approximation on the role played by symmetry.

1.1. The Hamiltonian of a Crystalline Solid

A crystalline solid is an ordered array of atoms bound together. The Hamiltonian of such a system that includes n electrons and N nuclei is given by

$$H = \sum_{i=1}^{n} \frac{\underset{\sim}{p}_i^2}{2m} + \sum_{\alpha=1}^{N} \frac{\underset{\sim}{P}_\alpha^2}{2M_\alpha} + V(\underset{\sim}{r}_i, \underset{\sim}{R}_\alpha), \tag{1.1.1}$$

where m is the mass of the electron, M_α is the mass of the αth nucleus, $\underset{\sim}{r}_i$ is the position coordinate of the ith electron, $\underset{\sim}{p}_i$ is the linear momentum of the ith electron, $\underset{\sim}{R}_\alpha$ is the position coordinate of the αth nucleus, and $\underset{\sim}{P}_\alpha$ is the linear momentum of the αth nucleus. Also

$$V(\underset{\sim}{r}_i, \underset{\sim}{R}_\alpha) = V_{ee} + V_{nn} + V_{ne}, \tag{1.1.2}$$

where

$$V_{ee} = \frac{1}{2} \sum_{i=1}^{n} \sum_{\substack{j=1 \\ i \neq j}}^{n} \frac{e^2}{|\underset{\sim}{r}_i - \underset{\sim}{r}_j|},$$

$$V_{nn} = \frac{1}{2}\sum_{\alpha=1}^{N}\sum_{\substack{\beta=1\\ \alpha\neq\beta}}^{N}\frac{e^2 Z_\alpha Z_\beta}{|\underset{\sim}{R}_\alpha - \underset{\sim}{R}_\beta|},$$

$$V_{ne} = -\sum_{i=1}^{n}\sum_{\alpha=1}^{N}\frac{e^2 Z_\alpha}{|\underset{\sim}{R}_\alpha - \underset{\sim}{r}_i|},$$

The Schrödinger equation of the system is given by

$$H\Psi(\underset{\sim}{r}_i, \underset{\sim}{R}_\alpha) = E\Psi(\underset{\sim}{r}_i, \underset{\sim}{R}_\alpha), \tag{1.1.3}$$

where the Hamiltonian operator H is expressed as follows:

$$H = -\frac{\hbar^2}{2m}\sum_{i=1}^{n}\nabla_i^2 - \frac{\hbar^2}{2}\sum_{\alpha=1}^{N}\frac{\nabla_\alpha^2}{M_\alpha} + V(\underset{\sim}{r}_i, \underset{\sim}{R}_\alpha). \tag{1.1.4}$$

Therefore the eigenfunctions and the eigenvalues of the system are given by

$$-\frac{\hbar^2}{2m}\sum_{i=1}^{n}\nabla_i^2\Psi - \frac{\hbar^2}{2}\sum_{\alpha=1}^{N}\frac{1}{M_\alpha}\nabla_\alpha^2\Psi + V\Psi = E\Psi. \tag{1.1.5}$$

In order to solve the above Schrödinger equation we seek solutions of the type

$$\Psi(\underset{\sim}{r}_i, \underset{\sim}{R}_\alpha) = \phi(\underset{\sim}{R}_\alpha)\psi(\underset{\sim}{r}_i, \underset{\sim}{R}_\alpha). \tag{1.1.6}$$

Using (1.1.6) in (1.1.5) we obtain, dropping the subscripts i and α,

$$\begin{aligned} -\frac{\hbar^2}{2m}\phi(\underset{\sim}{R})\sum_{i=1}^{n}\nabla_i^2\psi(\underset{\sim}{r}, \underset{\sim}{R}) - \frac{\hbar^2}{2}\sum_{\alpha=1}^{N}\frac{\nabla_\alpha^2}{M_\alpha}\phi(\underset{\sim}{R})\psi(\underset{\sim}{r}, \underset{\sim}{R}) \\ + V(\underset{\sim}{r}, \underset{\sim}{R})\phi(\underset{\sim}{R})\psi(\underset{\sim}{r}, \underset{\sim}{R}) = E\phi(\underset{\sim}{R})\psi(\underset{\sim}{r}, \underset{\sim}{R}). \end{aligned} \tag{1.1.7}$$

But

$$\begin{aligned} \nabla_\alpha^2\phi(\underset{\sim}{R})\psi(\underset{\sim}{r}, \underset{\sim}{R}) = \phi(\underset{\sim}{R})\nabla_\alpha^2\psi(\underset{\sim}{r}, \underset{\sim}{R}) + \psi(\underset{\sim}{r}, \underset{\sim}{R})\nabla_\alpha^2\phi(\underset{\sim}{R}) \\ + 2\underset{\sim}{\nabla}_\alpha\phi(\underset{\sim}{R})\cdot\underset{\sim}{\nabla}_\alpha\psi(\underset{\sim}{r}, \underset{\sim}{R}). \end{aligned} \tag{1.1.8}$$

Then

$$
\begin{aligned}
&-\sum_{\alpha=1}^{N}\frac{\hbar^2}{M_\alpha}\underset{\sim}{\nabla}_\alpha\phi(\underset{\sim}{R})\cdot\underset{\sim}{\nabla}_\alpha\psi(\underset{\sim}{r},\underset{\sim}{R})-\sum_{\alpha=1}^{N}\frac{\hbar^2}{2M_\alpha}\phi(\underset{\sim}{R})\nabla_\alpha^2\psi(\underset{\sim}{r},\underset{\sim}{R})\\
&-\psi(\underset{\sim}{r},\underset{\sim}{R})\sum_{\alpha=1}^{N}\frac{\hbar^2}{2M_\alpha}\underset{\sim}{\nabla}_\alpha^2\phi(\underset{\sim}{R})-\phi(\underset{\sim}{R})\sum_{i=1}^{n}\frac{\hbar^2}{2m_\alpha}\nabla_i^2\psi(\underset{\sim}{r},\underset{\sim}{R})\\
&\quad+V(\underset{\sim}{r},\underset{\sim}{R})\phi(\underset{\sim}{R})\psi(\underset{\sim}{r},\underset{\sim}{R})=E\phi(\underset{\sim}{R})\psi(\underset{\sim}{r},\underset{\sim}{R}).
\end{aligned}
\tag{1.1.9}
$$

We will assume that

$$
\left|-\sum_{\alpha=1}^{N}\frac{\hbar^2}{M_\alpha}\left[\underset{\sim}{\nabla}_\alpha\phi\cdot\underset{\sim}{\nabla}_\alpha\psi+\frac{1}{2}\phi\nabla_\alpha^2\psi\right]\right|\ll\left|-\psi\sum_{\alpha=1}^{N}\frac{\hbar^2}{2M_\alpha}\nabla_\alpha^2\phi\right|.
\tag{1.1.10}
$$

We shall return to the validity of this assumption later. Because of (1.1.10), (1.1.9) becomes

$$
-\frac{\hbar^2}{2}\frac{\psi}{\phi}\sum_{\alpha=1}^{N}\frac{1}{M_\alpha}\nabla_\alpha^2\phi+\left[-\frac{\hbar^2}{2m}\sum_{i=1}^{n}\nabla_i^2+V(\underset{\sim}{r},\underset{\sim}{R})\right]\psi=E\psi.
\tag{1.1.11}
$$

The operator in the square brackets on the left-hand side of (1.1.11) represents the Hamiltonian for the system if the nuclei are assumed to be fixed in space. We call this Hamiltonian H_e:

$$
H_e=-\frac{\hbar^2}{2m}\sum_{i=1}^{n}\nabla_i^2+V(\underset{\sim}{r},\underset{\sim}{R}).
\tag{1.1.12}
$$

The corresponding eigenfunctions and eigenvalues are given by the Schrödinger equation,

$$
-\frac{\hbar^2}{2m}\sum_{i=1}^{n}\nabla_i^2\psi(\underset{\sim}{r},\underset{\sim}{R})+V(\underset{\sim}{r},\underset{\sim}{R})\psi(\underset{\sim}{r},\underset{\sim}{R})=\varepsilon(\underset{\sim}{R})\psi(\underset{\sim}{r},\underset{\sim}{R}).
\tag{1.1.13}
$$

Replacing (1.1.13) in (1.1.11) we obtain

$$
-\frac{\hbar^2}{2}\sum_{\alpha=1}^{N}\frac{1}{M_\alpha}\nabla_\alpha^2\phi(\underset{\sim}{R})+\varepsilon(\underset{\sim}{R})\phi(\underset{\sim}{R})=E\phi(\underset{\sim}{R}).
\tag{1.1.14}
$$

The solution of the Schrödinger equation (1.1.3) reduces then to the solution of the two equations (1.1.13) and (1.1.14); we now rewrite this solution with the proper subscripts as follows:

$$-\frac{\hbar^2}{2m}\sum_{i=1}^{n}\nabla_i^2\psi_k(\underset{\sim}{r},\underset{\sim}{R})+V(\underset{\sim}{r},\underset{\sim}{R})\psi_k(\underset{\sim}{r},\underset{\sim}{R})=\varepsilon_k(\underset{\sim}{R})\psi_k(\underset{\sim}{r},\underset{\sim}{R}), \quad (1.1.15)$$

$$-\frac{\hbar^2}{2}\sum_{\alpha=1}^{N}\frac{1}{M_\alpha}\nabla_\alpha^2\phi_{kl}(\underset{\sim}{R})+\varepsilon_k(\underset{\sim}{R})\phi_{kl}(\underset{\sim}{R})=E_{kl}\phi_{kl}(\underset{\sim}{R}). \quad (1.1.16)$$

A stationary state of the system will be represented by the eigenfunction

$$\Psi_n(\underset{\sim}{r},\underset{\sim}{R})=\psi_k(\underset{\sim}{r},\underset{\sim}{R})\phi_{kl}(\underset{\sim}{R}), \quad (1.1.17)$$

where $\psi_k(\underset{\sim}{r},\underset{\sim}{R})$ and $\phi_{kl}(\underset{\sim}{R})$ are eigenfunctions of the Hamiltonians

$$H_e=-\frac{\hbar^2}{2m}\sum_{i=1}^{n}\nabla_i^2+V(\underset{\sim}{r},\underset{\sim}{R}), \quad (1.1.18)$$

and

$$H_v=-\frac{\hbar^2}{2}\sum_{\alpha=1}^{N}\frac{\nabla_\alpha^2}{M_\alpha}+\varepsilon_k(\underset{\sim}{R}), \quad (1.1.19)$$

respectively.

1.2. The Adiabatic Approximation

Let us consider now in detail equations (1.1.15) and (1.1.16). Equation (1.1.15) is an eigenvalue equation whose eigenfunctions represent the motion of the electrons in the crystal when the nuclei are kept fixed in space. The energy eigenvalues of (1.1.15) depend parametrically on the nuclear coordinates. Equation (1.1.16) is an eigenvalue equation whose eigenfunctions represent the motion of the nuclei in the crystal. In the Hamiltonian (1.1.19) for nuclear motion, the energy $\varepsilon_k(\underset{\sim}{R})$, which is a function of the nuclear coordinates regarded as parameters, plays the role of the potential energy for nuclear motion. This potential energy $\varepsilon_k(\underset{\sim}{R})$ is an eigenvalue of (1.1.15) and, as such, depends on the quantum number k. The eigenfunction $\phi(\underset{\sim}{R})$ also depends on k; however, k does not play the role of a quantum number for $\phi(\underset{\sim}{R})$ even if it is used as one of its subscripts. Therefore the functions $\phi_{kl}(\underset{\sim}{R})$ and $\phi_{k'l}(\underset{\sim}{R})$ with $k' \neq k$ are *not* mutually orthogonal.

We now return to the approximation (1.1.10) and discuss its physical meaning. The implication of (1.1.10) is that the function $\psi(\underset{\sim}{r}, \underset{\sim}{R})$, which represents the motion of the electrons, is a function that varies slowly with the nuclear coordinates, so that $|\underset{\sim}{\nabla}_\alpha \psi(\underset{\sim}{r}, \underset{\sim}{R})|$ is much smaller than $|\underset{\sim}{\nabla}_\alpha \phi(\underset{\sim}{R})|$. In pictorial terms we may say that this is the case since the electrons, having much smaller masses than the nuclei, go through their orbits many times before the nuclei have shifted from their equilibrium position by any considerable distance.

In the light of this fact the approximation (1.1.10), which has allowed us to express the eigenstates of the system in the product form (1.1.6), is called the *adiabatic approximation*. It is also called sometimes the *Born–Oppenheimer approximation.*[1]

The implications of the adiabatic approximation are far reaching. We are now in the position of treating the electrons and the nuclei independently; we will, however, keep in mind that we are allowed to do so only within the limits of validity of the adiabatic approximation.

1.3. The Role of Symmetry

The quantum-mechanical treatment of a physical system implies generally the solution of a Schrödinger equation. This solution gives the energy eigenvalues and the eigenfunctions of the Hamiltonian. In general the eigenfunctions are degenerate; that is, several of them correspond to the same energy eigenvalue. The degeneracy and the transformation properties of the eigenfunctions are closely related to the symmetry properties of the Hamiltonian; indeed, both degeneracy and transformation properties can be derived from the knowledge of symmetries.

Before considering these symmetries for the case of a crystalline solid, it is worthwhile to review some basic concepts regarding coordinate transformations. In particular we are concerned with those transformations that leave the distance between two points unchanged. The most general transformation of this type can be expressed by the symbol $\{R|\underset{\sim}{t}\}$ and involves a rotational operation R followed by a translation $\underset{\sim}{t}$. A position vector $\underset{\sim}{x}$, when acted upon by $\{R|\underset{\sim}{t}\}$, becomes

$$\underset{\sim}{x}' = \underset{\sim}{R}\,\underset{\sim}{x} + \underset{\sim}{t}, \tag{1.3.1}$$

with

$$\begin{aligned} x_1' &= R_{11}x_1 + R_{12}x_2 + R_{13}x_3 + t_1, \\ x_2' &= R_{21}x_1 + R_{22}x_2 + R_{23}x_3 + t_2, \\ x_3' &= R_{31}x_1 + R_{32}x_2 + R_{33}x_3 + t_3. \end{aligned} \tag{1.3.2}$$

$\underset{\sim}{R}$ is a 3×3 real orthogonal matrix: if its determinant is $+1$, the rotation is called *proper*; if its determinant is -1, the rotation is called *improper*.

A pure rotation is indicated by $\{R|\underset{\sim}{0}\}$ and pure translation by $\{E|\underset{\sim}{t}\}$. The identity operation is represented by $\{E|\underset{\sim}{0}\}$.

If two operations $\{R_1|\underset{\sim}{t}'\}$ and $\{R_2|\underset{\sim}{t}''\}$ act in succession upon a vector $\underset{\sim}{x}$, the result is

$$x' = \underset{\sim}{R}_1\underset{\sim}{x} + \underset{\sim}{t}', \tag{1.3.3}$$

and

$$\begin{aligned} \underset{\sim}{x}'' &= \underset{\sim}{R}_2\underset{\sim}{x}' + \underset{\sim}{t}'' = \underset{\sim}{R}_2(\underset{\sim}{R}_1\underset{\sim}{x} + \underset{\sim}{t}') + \underset{\sim}{t}'' \\ &= \underset{\sim}{R}_2\underset{\sim}{R}_1\underset{\sim}{x} + \underset{\sim}{R}_2\underset{\sim}{t}' + \underset{\sim}{t}''. \end{aligned} \tag{1.3.4}$$

Therefore, the product of the two operations $\{R_2|\underset{\sim}{t}''\}$ and $\{R_1|\underset{\sim}{t}'\}$ is given by

$$\{R_2|\underset{\sim}{t}''\}\{R_1|\underset{\sim}{t}'\} = \{R_2R_1|R_2\underset{\sim}{t}' + \underset{\sim}{t}''\}. \tag{1.3.5}$$

The inverse of an operation $\{R|\underset{\sim}{t}\}$ is given by

$$\{R|\underset{\sim}{t}\}^{-1} = \{R^{-1}| - R^{-1}\underset{\sim}{t}\}. \tag{1.3.6}$$

In fact, applying (1.3.5) we find

$$\begin{aligned} \{R^{-1}| - R^{-1}\underset{\sim}{t}\}\{R|\underset{\sim}{t}\} &= \{R^{-1}R|R^{-1}\underset{\sim}{t} - R^{-1}\underset{\sim}{t}\} \\ &= \{E|\underset{\sim}{0}\}. \end{aligned} \tag{1.3.7}$$

1.4. The Symmetries of the Hamiltonian

Let us consider first the "exact" Hamiltonian (1.1.4) given by

$$H = -\sum_{i=1}^{n} \frac{\hbar^2}{2m}\nabla_i^2 - \sum_{\alpha=1}^{N} \frac{\hbar^2}{2M_\alpha}\nabla_\alpha^2 + \frac{1}{2}\sum_{i=1}^{n}\sum_{\substack{j=1\\ i\neq j}}^{n} \frac{e^2}{|\underset{\sim}{r}_i - \underset{\sim}{r}_j|} + \frac{1}{2}\sum_{\alpha=1}^{N}\sum_{\substack{\beta=1\\ \alpha\neq\beta}}^{N} \frac{e^2 Z_\alpha Z_\beta}{|\underset{\sim}{R}_\alpha - \underset{\sim}{R}_\beta|} - \sum_{i=1}^{n}\sum_{\alpha=1}^{N} \frac{e^2 Z_\alpha}{|\underset{\sim}{R}_\alpha - \underset{\sim}{r}_i|}. \tag{1.4.1}$$

It is clear that if we translate each coordinate by an amount $\underset{\sim}{d}$, that is, if we replace the electronic and nuclear coordinates as follows,

$$\begin{aligned} \underset{\sim}{R}'_\alpha &= \underset{\sim}{R}_\alpha + \underset{\sim}{d}, \\ \underset{\sim}{r}'_i &= \underset{\sim}{r}_i + \underset{\sim}{d}, \end{aligned} \tag{1.4.2}$$

the Hamiltonian does not change. This invariance can be eliminated from our consideration if we replace the $3(n+N)$ coordinates by means of the three coordinates of the center of mass of the system and $3(n+N-1)$ coordinates depending only on the relative vectorial distances between particles. In the new coordinate system the Hamiltonian can be separated into the sum

$$H = -\frac{\hbar^2}{2M}\nabla^2 + H', \tag{1.4.3}$$

where M is the total mass of the crystal and the operator ∇^2 is defined with respect to the coordinates of the center of mass. In (1.4.3) the first term is the kinetic energy of the center of mass and the second term is the Hamiltonian of the motion relative to the center of mass. H' depends only on the relative coordinates and is therefore translationally invariant.

The eigenfunctions of the Hamiltonian H can be written as

$$\psi = e^{i\underset{\sim}{k}\cdot\underset{\sim}{U}}\psi', \tag{1.4.4}$$

and the energy is given by

$$E = E' + \frac{\hbar^2}{2M}|\underset{\sim}{k}|^2, \tag{1.4.5}$$

where $\underset{\sim}{U}$ is the coordinate of the center of mass, ψ is an eigenfunction of H with energy E, and ψ' is an eigenfunction of H' with energy E', and $\underset{\sim}{k}$ is an arbitrary vector. The problem is then restricted to finding the energy eigenvalues E' and H' and the eigenfunctions ψ' of H'.

Since the rotational symmetry properties of H are the same as those of H', we can now restrict ourselves to the consideration of the rotational symmetries of H'.

Let us consider now the effect of a real orthogonal transformation $\underset{\sim}{R}$ on the Hamiltonian H'. This transformation does not change the distances between particles, and therefore the terms in the Hamiltonian representing Coulomb interactions remain invariant. The Laplacian operator ∇^2 also remains invariant:

$$\begin{aligned} \nabla'^2 &= \sum_{i=1}^{3} \frac{\partial^2}{\partial x_i'^2} = \sum_{i,jk} R_{ij} R_{ik} \frac{\partial}{\partial x_j}\frac{\partial}{\partial x_k} \\ &= \sum_{jk} \delta_{jk} \frac{\partial}{\partial x_j}\frac{\partial}{\partial x_k} = \nabla^2. \end{aligned} \tag{1.4.6}$$

Therefore the "exact" Hamiltonian H is invariant under *all* spatial transformations, which include translations, rotations (proper and improper), and combinations of the two. It is evident that under these nonrestrictive conditions no relevant information on the eigenfunctions and eigenvalues of H can be gained by the consideration of the symmetry properties of H.

The situation becomes completely different when we consider the Hamiltonian

$$H_e = -\frac{\hbar^2}{2m}\sum_{i=1}^{n} \nabla_i^2 + V(\underset{\sim}{r}, \underset{\sim}{R}). \tag{1.4.7}$$

The Hamiltonian H_e represents the motion of the n electrons in the field of the nuclei, which are assumed to be at fixed positions. Considering the symmetry properties of H_e, the nuclear coordinates are not to be operated upon, because they appear only as *parameters* in the Hamiltonian; we can only perform symmetry operations on the electronic coordinates. Let us perform the most general symmetry operation $\{R|\underset{\sim}{t}\}$ on the electronic coordinates. If we call

$$\underset{\sim}{r}'_i = \{\underset{\sim}{R}|\underset{\sim}{t}\}\underset{\sim}{r}_i, \tag{1.4.8}$$

then the Hamiltonian H_e in the new coordinates is given by

$$-\frac{\hbar^2}{2m}\sum_{i=1}^{n}\nabla_i'^2 + \frac{1}{2}\sum_{\substack{i=1 \\ i\neq j}}^{n}\sum_{j=1}^{n}\frac{e^2}{|\underset{\sim}{r}_i' - \underset{\sim}{r}_j'|} + \frac{1}{2}\sum_{\substack{\alpha=1 \\ \alpha\neq\beta}}^{N}\sum_{\beta=1}^{N}\frac{e^2 Z_\alpha Z_\beta}{|\underset{\sim}{R}_\alpha - \underset{\sim}{R}_\beta|}$$

$$-\sum_{i=1}^{n}\sum_{\alpha=1}^{N}\frac{e^2 Z_\alpha}{|\underset{\sim}{R}_\alpha - \underset{\sim}{r}_i'|}. \tag{1.4.9}$$

The only term that has changed is the last sum. But, if the operation $\{\underset{\sim}{R}|\underset{\sim}{t}\}$ is such that

$$|\underset{\sim}{R}_\alpha - \underset{\sim}{r}_i'| = |\underset{\sim}{R}_\beta - \underset{\sim}{r}_i|, \tag{1.4.10}$$

where $\underset{\sim}{R}_\beta$ indicates the position of a nucleus identical to the α nucleus, then the last sum in (1.4.9) will go into itself. This fact implies that $\{\underset{\sim}{R}|\underset{\sim}{t}\}$ must be such that

$$\{\underset{\sim}{R}|\underset{\sim}{t}\}^{-1}\underset{\sim}{R}_\alpha = \underset{\sim}{R}_\beta, \tag{1.4.11}$$

or

$$\{\underset{\sim}{R}|\underset{\sim}{t}\}\underset{\sim}{R}_\beta = \underset{\sim}{R}_\alpha, \tag{1.4.12}$$

for all identical nuclei. Therefore the Hamiltonian H_e is invariant under all those operations on the electronic coordinates which, when used on the nuclear coordinate parameters, send identical nuclei into one another.

The Hamiltonian H_v represents the nuclear motion. The eigenvalues of H_e are used as the potential in which the nuclei experience their motion; this Hamiltonian depends *only* on the nuclear coordinates. The symmetry operations are now to be performed on the nuclear coordinates, and it is clear that H_v is invariant under all those operations which send identical nuclei into one another.

We have gone through this long discussion to show how symmetry considerations are relevant to the quantum-mechanical treatment of a crystalline solid. We have found that in the adiabatic approximation the relevant symmetries are those related to the nuclear coordinates. We learn how to consider these symmetries in a formal way in the next three chapters.

Reference

1. M. Born and J. R. Oppenheimer, *Ann. Phys.* 84, 457 (1927).

Chapter 2

Concepts of Group Theory

Group theory studies the algebraic structures known as groups. This chapter defines a group and presents its basic properties, most important among them the representations. It then examines the connections between group theory and quantum mechanics.

2.1. Properties of a Group

An ensemble of elements forms a *group* if the elements have the following properties:

a. Closure: the product of two elements is an element of the ensemble.
b. The ensemble contains the element E (identity).
c. The multiplication of elements is associative: A(BC) = (AB)C.
d. Every element has a reciprocal element contained in the ensemble: Given an element A, the ensemble contains an element B such that AB = BA = E.

In particular, the elements of a group may be symmetry operations that transform a geometrical body of a certain shape into itself.

Example. Equilateral triangle (see Fig. 2.1). The operations that transform the equilateral triangle into itself are the following:

1. E: identity,
2. C_3: 120° clockwise rotation about the axis Z,
3. C_3^2: 240° clockwise rotation about the axis Z,
4. σ_1: reflection through the plane containing the axis Z and axis 11,

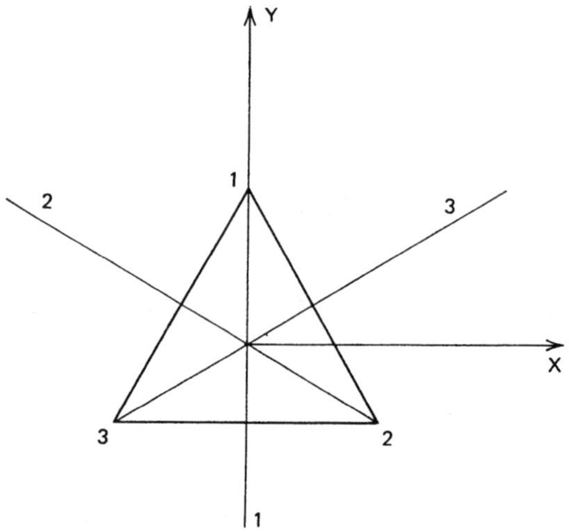

Fig. 2.1. Equilateral triangle with related axes. The axis Z is perpendicular to the X and Y axes.

5. σ_2: reflection through the plane containing the axis Z and axis 22,
6. σ_3: reflection through the plane containing the axis Z and axis 33.

The group of these operations is called C_{3v}.

The number of elements in a group is called the *order of the group.*

The elements of a group can be arranged in a *multiplication table.* In the example of the equilateral triangle the table is given by

	E	C_3	C_3^2	σ_1	σ_2	σ_3
E	E	C_3	C_3^2	σ_1	σ_2	σ_3
C_3	C_3	C_3^2	E	σ_3	σ_1	σ_2
C_3^2	C_3^2	E	C_3	σ_2	σ_3	σ_1
σ_1	σ_1	σ_2	σ_3	E	C_3	C_3^2
σ_2	σ_2	σ_3	σ_1	C_3^2	E	C_3
σ_3	σ_3	σ_1	σ_2	C_3	C_3^2	E

In general, if A and B are two elements of a group, then $AB \neq BA$. If $AB = BA$, A and B being any two elements of the group, the group is said to be *Abelian.*

2.2. Subgroups, Cosets, and Classes

Given a group G of order g, any subset of elements of G that is itself a group is called a *subgroup*. The group C_{3v} then has the following subgroups:

$$\{E\}, \quad \{E, \sigma_1\}, \quad \{E, \sigma_2\}, \quad \{E, \sigma_3\}, \quad \{E, C_3, C_3^2\}. \tag{2.2.1}$$

Let us consider now the subgroup $\{E, \sigma_1\}$ of C_{3v}. We can multiply all the elements of $\{E, \sigma_1\}$ by all the other elements of C_{3v} not in $\{E, \sigma_1\}$. Multiplying from the left, we obtain the following *left cosets*:

$$\begin{aligned} C_3\{E, \sigma_1\} &= \{C_3, \sigma_3\} \\ C_3^2\{E, \sigma_1\} &= \{C_3^2, \sigma_2\} \\ \sigma_2\{E, \sigma_1\} &= \{\sigma_2, C_3^2\} \\ \sigma_3\{E, \sigma_1\} &= \{\sigma_3, C_3\} \end{aligned} \tag{2.2.2}$$

Multiplying from the right, we obtain the *right cosets*:

$$\begin{aligned} \{E, \sigma_1\}C_3 &= \{C_3, \sigma_2\} \\ \{E, \sigma_1\}C_3^2 &= \{C_3^2, \sigma_3\} \\ \{E, \sigma_1\}\sigma_2 &= \{\sigma_2, C_3\} \\ \{E, \sigma_1\}\sigma_3 &= \{\sigma_3, C_3^2\} \end{aligned} \tag{2.2.3}$$

For each case we have two distinct cosets.

We notice here that the group C_{3v} can be expressed in terms of the subgroup $\{E, \sigma_1\}$ and the corresponding left or right cosets:

$$\begin{aligned} C_{3v}\colon \{E, \sigma_1\} &+ C_3\{E, \sigma_1\} + C_3^2\{E, \sigma_1\} \\ &= \{E, \sigma_1\} + \{C_3, \sigma_3\} + \{C_3^2, \sigma_2\}, \end{aligned} \tag{2.2.4}$$

$$\begin{aligned} C_{3v}\colon \{E, \sigma_1\} &+ \{E, \sigma_1\}C_3 + \{E, \sigma_1\}C_3^2 \\ &= \{E, \sigma_1\} + \{C_3, \sigma_2\} + \{C_3^2, \sigma_3\}. \end{aligned} \tag{2.2.5}$$

This property is of general nature, in the sense that a group may be decomposed into right or left cosets with respect to a definite subgroup.

If A and B are two elements of a group, the element

$$C = B^{-1}AB \tag{2.2.6}$$

is called *conjugate of A with respect to B.*

If a group G contains a certain subgroup H such that, for any element X in G

$$X^{-1}HX = H, \tag{2.2.7}$$

then the subgroup is said to be *invariant* or *self-conjugate.* A subgroup of order g/2 of a group G of order g is an invariant subgroup of G. The proof of this last property is left to the reader.

The right and left cosets of a group with respect to an invariant subgroup are the same, because HX = XH.

Given a certain group G with an invariant subgroup H, the collection of elements that consists of H and of all the cosets with respect to H forms a group that is called the *factor group* and is indicated by G/H. In the case of the group C_{3v} an invariant subgroup is given by $H = \{E, C_3, C_3^2\}$. The factor group C_{3v}/H is then given by

$$\{E, C_3, C_3^2\}, \quad \{\sigma_1, \sigma_2, \sigma_3\}$$

and is a group of order 2.

A collection of elements conjugate to each other form a *class.*

Example. Group C_{3v}. The elements conjugate of C_3 and C_3^2 can be found from the following operations:

$$\begin{aligned}
EC_3E &= C_3 & EC_3^2E &= C_3^2\\
C_3^2C_3C_3 &= C_3 & C_3^2C_3^2C_3 &= C_3^2\\
C_3C_3C_3^2 &= C_3 & C_3C_3^2C_3^2 &= C_3^2\\
\sigma_1C_3\sigma_1 &= C_3^2 & \sigma_1C_3^2\sigma_1 &= C_3\\
\sigma_2C_3\sigma_2 &= C_3^2 & \sigma_2C_3^2\sigma_2 &= C_3\\
\sigma_3C_3\sigma_3 &= C_3^2 & \sigma_3C_3^2\sigma_3 &= C_3
\end{aligned}$$

Therefore the elements C_3 and C_3^2 belong to the same class.

It can be shown easily that the identity E forms a class by itself (a general property) and that the elements σ_1, σ_2, and σ_3 also form a class.

The group C_{3v} consists therefore of the following three classes:

$$\begin{aligned} \mathcal{C}_1 &: \mathrm{E} \\ \mathcal{C}_2 &: \mathrm{C}_3, \mathrm{C}_3^2 \\ \mathcal{C}_3 &: \sigma_1, \sigma_2, \sigma_3. \end{aligned}$$

Any group can be decomposed into *non-overlapping* classes.

The product of two classes can be expressed in general as a linear combination of all the classes of the group:

$$\mathcal{C}_i \mathcal{C}_j = \sum_{\ell=1}^{h_i} \sum_{k=1}^{h_j} R_\ell^i R_k^j = \sum_{m=1}^{r} c_{ijm} \mathcal{C}_m \tag{2.2.8}$$

where

$$\begin{aligned} h_i &= \text{number of elements in } \mathcal{C}_i \\ h_j &= \text{number of elements in } \mathcal{C}_j \\ R_\ell^i &= \text{generic element of } \mathcal{C}_i \\ R_k^j &= \text{generic element of } \mathcal{C}_j \\ r &= \text{number of classes in the group} \\ c_{ijm} &= \text{positive integer or zero.} \end{aligned}$$

In the case of group C_{3v}, for example,

$$\begin{aligned} \mathcal{C}_2 \mathcal{C}_3 &= (C_3, C_3^2) \times (\sigma_1, \sigma_2, \sigma_3) \\ &= C_3\sigma_1, C_3\sigma_2, C_3\sigma_3, C_3^2\sigma_1, C_3^2\sigma_2, C_3^2\sigma_3 \\ &= \sigma_3, \sigma_1, \sigma_2, \sigma_2, \sigma_3, \sigma_1 = 2\mathcal{C}_3. \end{aligned} \tag{2.2.9}$$

If a group is Abelian, then

$$B^{-1}AB = AB^{-1}B = A, \tag{2.2.10}$$

B and A being any two elements of the group. Therefore in an Abelian group every element forms a class by itself.

2.3. Theory of Representations

Given a group G, any set of elements that multiply according to the multiplication table of G is said to be a *representation* of G. If the elements of the representation are matrices the representation is called a *matrix representation.* A restriction on these matrices is that they must be *nonsingular*; that is, each matrix must posses an inverse. A necessary and sufficient condition for a matrix to be nonsingular is that its determinant must not vanish.

Example. Three matrix representations of the group C_{3v} are given in the following table.

C_{3v}	E	C_3	C_3^2
Γ_1	1	1	1
Γ_2	1	1	1
Γ_3	$\begin{pmatrix} 1 & 0 \\ 0 & 1 \end{pmatrix}$	$\begin{pmatrix} -\frac{1}{2} & \frac{\sqrt{3}}{2} \\ -\frac{\sqrt{3}}{2} & -\frac{1}{2} \end{pmatrix}$	$\begin{pmatrix} -\frac{1}{2} & -\frac{\sqrt{3}}{2} \\ \frac{\sqrt{3}}{2} & -\frac{1}{2} \end{pmatrix}$
C_{3v}	σ_1	σ_2	σ_3
Γ_1	1	1	1
Γ_2	-1	-1	-1
Γ_3	$\begin{pmatrix} -1 & 0 \\ 0 & 1 \end{pmatrix}$	$\begin{pmatrix} \frac{1}{2} & -\frac{\sqrt{3}}{2} \\ -\frac{\sqrt{3}}{2} & -\frac{1}{2} \end{pmatrix}$	$\begin{pmatrix} \frac{1}{2} & \frac{\sqrt{3}}{2} \\ \frac{\sqrt{3}}{2} & -\frac{1}{2} \end{pmatrix}$

It is easy to see that the matrices multiply as the elements they represent. For example $C_3 \times \sigma_1 = \sigma_3$ corresponds to

$$\begin{pmatrix} -\frac{1}{2} & \frac{\sqrt{3}}{2} \\ -\frac{\sqrt{3}}{2} & -\frac{1}{2} \end{pmatrix} \begin{pmatrix} -1 & 0 \\ 0 & 1 \end{pmatrix} = \begin{pmatrix} \frac{1}{2} & \frac{\sqrt{3}}{2} \\ \frac{\sqrt{3}}{2} & -\frac{1}{2} \end{pmatrix}$$

Let us assume that

$$\underset{\sim}{a}, \underset{\sim}{b}, \underset{\sim}{c}, \underset{\sim}{d}, \underset{\sim}{e}, \underset{\sim}{f}, \ldots \tag{2.3.1}$$

are the matrices of a representation of a group. Let us transform each matrix by using a certain matrix $\underset{\sim}{\beta}$ in a similarity transformation as follows:

$$\underset{\sim}{a}' = \underset{\sim}{\beta}^{-1}\underset{\sim}{a}\underset{\sim}{\beta}, \quad \underset{\sim}{b}' = \underset{\sim}{\beta}^{-1}\underset{\sim}{b}\underset{\sim}{\beta}, \quad \underset{\sim}{c}' = \underset{\sim}{\beta}^{-1}\underset{\sim}{c}\underset{\sim}{\beta}. \tag{2.3.2}$$

The new matrices $\underset{\sim}{a}', \underset{\sim}{b}', \underset{\sim}{c}', \ldots$ form a representation of the group. In fact, if

$$\underset{\sim}{a}\underset{\sim}{b} = \underset{\sim}{d}, \tag{2.3.3}$$

then

$$\underset{\sim}{\beta}^{-1}\underset{\sim}{a}\underset{\sim}{b}\underset{\sim}{\beta} = \underset{\sim}{\beta}^{-1}\underset{\sim}{a}\underset{\sim}{\beta}\underset{\sim}{\beta}^{-1}\underset{\sim}{b}\underset{\sim}{\beta} = \underset{\sim}{\beta}^{-1}\underset{\sim}{d}\underset{\sim}{\beta}, \tag{2.3.4}$$

that is

$$\underset{\sim}{a}'\underset{\sim}{b}' = \underset{\sim}{d}'. \tag{2.3.5}$$

A representation Γ with matrices $\underset{\sim}{a}, \underset{\sim}{b}, \underset{\sim}{c}, \underset{\sim}{d}, \underset{\sim}{e}, \underset{\sim}{f}, \ldots$ is said to be *reducible* if it is possible to find a similarity transformations which transform each matrix of Γ as follows:

$$\underset{\sim}{a}' = \underset{\sim}{\beta}^{-1}\underset{\sim}{a}\underset{\sim}{\beta} = \begin{pmatrix} \boxed{\underset{\sim}{a}'_1} & 0 & 0 & 0 \\ 0 & \boxed{\underset{\sim}{a}'_2} & 0 & 0 \\ 0 & 0 & \boxed{\underset{\sim}{a}'_3} & 0 \\ 0 & 0 & 0 & \boxed{\underset{\sim}{a}'_4} \end{pmatrix}. \tag{2.3.6}$$

The matrices $\underset{\sim}{a}'_1$ are square matrices. Correspondingly the matrices $\underset{\sim}{b}', \underset{\sim}{c}', \ldots$ in their reduced form are blocked out in the same way. Since

$$\underset{\sim}{a}'\underset{\sim}{b}' = \underset{\sim}{d}', \tag{2.3.7}$$

it is also true that

$$\begin{aligned} \underset{\sim}{a}'_1\underset{\sim}{b}'_1 &= \underset{\sim}{d}'_1 \\ \underset{\sim}{a}'_2\underset{\sim}{b}'_2 &= \underset{\sim}{d}'_2 \\ \underset{\sim}{a}'_3\underset{\sim}{b}'_3 &= \underset{\sim}{d}'_3 \\ &\cdots \end{aligned} \tag{2.3.8}$$

Therefore the matrices

$$\underset{\sim}{a}'_1, \underset{\sim}{b}'_1, \underset{\sim}{c}'_1, \underset{\sim}{d}'_1, \underset{\sim}{e}'_1, \underset{\sim}{f}'_1, \ldots \tag{2.3.9}$$

form a representation of the group. Other representations are given by

$$\underset{\sim}{a}'_2, \underset{\sim}{b}'_2, \underset{\sim}{c}'_2, \underset{\sim}{d}'_2, \underset{\sim}{e}'_2, \underset{\sim}{f}'_2, \ldots \tag{2.3.10}$$

$$\underset{\sim}{a}'_3, \underset{\sim}{b}'_3, \underset{\sim}{c}'_3, \underset{\sim}{d}'_3, \underset{\sim}{e}'_3, \underset{\sim}{f}'_3, \ldots \tag{2.3.11}$$

If we indicate the representations (2.3.9), (2.3.10), (2.3.11),... by $\Gamma_1, \Gamma_2, \Gamma_3, \ldots$, then the reduction of the matrix representation is indicated as follows:

$$\Gamma = \Gamma_1 + \Gamma_2 + \Gamma_3 + \cdots \tag{2.3.12}$$

Given a matrix representation Γ, if it is not possible to find a similarity transformation that reduces it, the representation Γ is said to be *irreducible.*

Two irreducible representations that differ only by a similarity transformation are said to be *equivalent.*

The representations of the group C_{3v} given in the example are inequivalent and irreducible; this statement is given here without proof, but will be checked later.

2.4. Orthogonality Relations

We now examine some important relations that apply to irreducible inequivalent matrix representations.

We consider in particular the representations of the group C_{3v} reported in Section 2.3. If we call R the generic symmetry operation we find

$$\sum_R \Gamma_1(R)_{11}\Gamma_1(R)_{11} = 1+1+1+1+1+1 = 6$$

$$\sum_R \Gamma_2(R)_{11}\Gamma_2(R)_{11} = 1+1+1+1+1+1 = 6$$

$$\sum_R \Gamma_3(R)_{11}\Gamma_3(R)_{11} = 1+\frac{1}{4}+\frac{1}{4}+1+\frac{1}{4}+\frac{1}{4} = 3$$

$$\sum_R \Gamma_3(R)_{12}\Gamma_3(R)_{12} = \frac{3}{4}+\frac{3}{4}+\frac{3}{4}+\frac{3}{4} = 3$$

$$\sum_R \Gamma_3(R)_{21}\Gamma_3(R)_{21} = \frac{3}{4} + \frac{3}{4} + \frac{3}{4} + \frac{3}{4} = 3$$

$$\sum_R \Gamma_3(R)_{22}\Gamma_3(R)_{22} = 1 + \frac{1}{4} + \frac{1}{4} + 1 + \frac{1}{4} + \frac{1}{4} = 3$$

Also, we find

$$\begin{aligned}
\sum_R \Gamma_1(R)_{11}\Gamma_2(R)_{11} &= 1 + 1 + 1 - 1 - 1 - 1 = 0 \\
\sum_R \Gamma_1(R)_{11}\Gamma_3(R)_{11} &= 1 - \frac{1}{2} - \frac{1}{2} + 1 - \frac{1}{2} - \frac{1}{2} = 0 \\
\sum_R \Gamma_2(R)_{11}\Gamma_3(R)_{22} &= 1 - \frac{1}{2} - \frac{1}{2} - 1 + \frac{1}{2} + \frac{1}{2} = 0 \\
\sum_R \Gamma_3(R)_{11}\Gamma_3(R)_{22} &= 1 + \frac{1}{4} + \frac{1}{4} - 1 - \frac{1}{4} - \frac{1}{4} = 0 \\
&\cdots
\end{aligned}$$

In summary we obtain

$$\sum_R \Gamma_\alpha(R)_{ik}\Gamma_\beta(R)_{\ell j} = 0, \quad \text{if } \alpha \neq \beta, \text{ or } i \neq \ell, \text{ or } k \neq j, \tag{2.4.1}$$

and

$$\sum_R \Gamma_\alpha(R)_{ik}\Gamma_\alpha(R)_{ik} = \frac{g}{n_\alpha}, \tag{2.4.2}$$

where g is the order of the group and n_α is the dimension of the representation Γ_α.

The relations (2.4.1) and (2.4.2) reflect a general property of the matrix elements of irreducible inequivalent representations that can be expressed as follows*:

$$\sum_R \Gamma_\alpha(R^{-1})_{ki}\Gamma_\beta(R)_{\ell j} = \delta_{\alpha\beta}\delta_{i\ell}\delta_{kj}\frac{g}{n_\alpha}. \tag{2.4.3}$$

*The generic element of a matrix may be a complex number.

These relations are called *orthogonality relations* and apply to inequivalent irreducible representations. If the matrices are unitary the relation above can be expressed as follows:

$$\sum_R \Gamma_\alpha(R)^*_{ik}\Gamma_\beta(R)_{\ell j} = \delta_{\alpha\beta}\delta_{i\ell}\delta_{kj}\frac{g}{n_\alpha}. \tag{2.4.4}$$

The g numbers

$$\Gamma_\alpha(R_1)_{ij}, \Gamma_\alpha(R_2)_{ij}, \Gamma_\alpha(R_3)_{ij}, \ldots \tag{2.4.5}$$

may be considered to be the components of a g-dimensional vector orthogonal to any other vector of the same type, obtained by a different choice of the subscripts i, j, or α. If the dimensions of the representations are $n_\alpha, n_\beta, n_\gamma, \ldots$ then the total number of these vectors is

$$n_\alpha^2 + n_\beta^2 + n_\gamma^2 + \cdots \tag{2.4.6}$$

This number must be equal to g, because it is possible to construct only g orthogonal vectors in a g-dimensional space:

$$n_\alpha^2 + n_\beta^2 + n_\gamma^2 + \cdots = g, \tag{2.4.7}$$

or

$$\sum_{\alpha=1}^{r} n_\alpha^2 = g, \tag{2.4.8}$$

where r is the number of irreducible inequivalent representations. In the example of the group C_{3v},

$$\begin{aligned} n_1 &= 1 \\ n_2 &= 1 \\ n_3 &= 2 \end{aligned}$$

and the number of elements in the group is

$$\sum_{\alpha=1}^{3} n_\alpha^2 = 1 + 1 + 4 = 6.$$

Example. If g = 4, since we always have a one-dimensional representation with all matrices equal to 1, it must also be

$$\sum_\alpha n_\alpha^2 = 1^2 + 1^2 + 1^2 + 1^2 = 4.$$

Therefore if g = 4, the group has four one-dimensional irreducible representations.

If g = 10

$$\sum_\alpha n_\alpha^2 = 1^2 + 1^2 + 1^2 + 1^2 + 1^2 + 1^2 + 1^2 + 1^2 + 1^2 + 1^2,$$

or

$$1^2 + 1^2 + 1^2 + 1^2 + 1^2 + 1^2 + 2^2,$$

or

$$1^2 + 1^2 + 2^2 + 2^2,$$

or

$$1^2 + 3^2.$$

That is, the group has either ten one-dimensional or six one-dimensional and one two-dimensional or two one-dimensional and two two-dimensional or one one-dimensional and one three-dimensional irreducible representations.

2.5. Characters of a Matrix Representation

Given a certain matrix representation, the traces of the matrices are called *characters* of the representation. The character of a representation Γ_α corresponding to the generic operation R is given by

$$\chi_\alpha(R) = \sum_m \Gamma_\alpha(R)_{mm}. \tag{2.5.1}$$

The characters of the irreducible representations of the group C_{3v} can be easily derived by looking at the table of these representations in Section 2.3:

C_{3v}	E	C_3	C_3^2	σ_1	σ_2	σ_3
Γ_1	1	1	1	1	1	1
Γ_2	1	1	1	−1	−1	−1
Γ_3	2	−1	−1	0	0	0

We notice here that the character corresponding to the operation E (identity) is equal to the dimension of the representation, namely

$$\chi_\alpha(E) = n_\alpha. \tag{2.5.2}$$

The character of a matrix is unchanged by a similarity transformation. In fact, if the character of a matrix A is given by

$$\chi_A = \sum_j A_{jj}, \tag{2.5.3}$$

and if

$$B = C^{-1} A\, C, \tag{2.5.4}$$

then

$$\begin{aligned}\chi_B &= \sum_i B_{ii} = \sum_i \left(\sum_j \sum_k C_{ij}^{-1} A_{jk} C_{ki} \right) \\ &= \sum_j \sum_k \left(\sum_i C_{ij}^{-1} C_{ki} \right) A_{jk} = \sum_j \sum_k \delta_{jk} A_{jk} \\ &= \sum_j A_{jj} = \chi_A.\end{aligned} \tag{2.5.5}$$

From (2.5.4) and (2.5.5) we can deduce that if two operations belong to the same class, the corresponding matrices for a given representation have the same character.

Also, two nonequivalent, irreducible representations have different character systems and two irreducible representations with the same character system are equivalent.

Let us consider again the orthogonality relations (2.4.4) and let us put $i = k, \ell = j$. We obtain

$$\sum_R \Gamma_\alpha(R)_{ii}^* \Gamma_\beta(R)_{\ell\ell} = \delta_{\alpha\beta} \delta_{i\ell} \frac{g}{n_\alpha}. \tag{2.5.6}$$

Summing over i from 1 to n_α and over ℓ form 1 to n_β

$$\sum_R \left[\sum_{i=1}^{n_\alpha} \Gamma_\alpha(R)^*_{ii} \sum_{\ell=1}^{n_\beta} \Gamma_\beta(R)_{\ell\ell} \right]$$

$$= \sum_R \chi_\alpha(R)^* \chi_\beta(R) = \frac{g}{n_\alpha} \delta_{\alpha\beta} \sum_{i=1}^{n_\alpha} \sum_{\ell=1}^{n_\beta} \delta_{i\ell} = g\delta_{\alpha\beta}. \tag{2.5.7}$$

Therefore we find the following orthogonality relation for the characters:

$$\sum_R \chi_\alpha(R)^* \chi_\beta(R) = g\,\delta_{\alpha\beta}, \tag{2.5.8}$$

and we can say that the characters of the irreducible representations of a group of order g form a set of orthogonal g-dimensional vectors.

An important property can be derived from (2.5.8). Since for a certain representation the characters corresponding to operations of the same class are equal, we may write (2.5.8) as

$$\sum_{\rho=1}^{r} \chi_\alpha(R_\rho)^* \chi_\beta(R_\rho) g_\rho = g\delta_{\alpha\beta}, \tag{2.5.9}$$

or

$$\sum_{\rho=1}^{r} \left[\chi_\alpha(R_\rho)^* \sqrt{\frac{g_\rho}{g}} \right] \left[\chi_\beta(R_\rho) \sqrt{\frac{g_\rho}{g}} \right] = \delta_{\alpha\beta}, \tag{2.5.10}$$

where g_ρ is the number of elements in the ρth class, R_ρ is the generic operation in the ρth class, and r is the number of classes. The normalized characters $\chi_\alpha(R_\rho)\sqrt{g_\rho/g}$ are the components of a set of orthogonal vectors in an r-dimensional space. Since we can construct only r such vectors, we see that the number of irreducible inequivalent representations of a group must be equal to the number of classes.

2.6. Reduction of a Reducible Representation

Any reducible representation can be reduced in terms of its irreducible representations by a proper similarity transformation.

This transformation leaves the character of the representation unchanged. Therefore, if we designate by $\chi(R)$ and $\chi_j(R)$ the character of the reducible representation and of the generic irreducible representation, respectively, we can write

$$\chi(R) = \sum_{j=1}^{r} c_j \chi_j(R), \tag{2.6.1}$$

where r is the number of irreducible representations and c_j is the number of times the jth irreducible representation is contained in the reducible representation.

Let us multiply both members of (2.6.1) by $\chi_i(R)^*$ and sum over R:

$$\begin{aligned}
\sum_R \chi(R)\chi_i(R)^* &= \sum_R \sum_j c_j \chi_j(R)\chi_i(R)^* \\
&= \sum_j c_j \sum_R \chi_j(R)\chi_i(R)^* \\
&= \sum_j c_j g \delta_{ij} \\
&= c_i g,
\end{aligned} \tag{2.6.2}$$

or

$$c_i = \frac{1}{g} \sum_R \chi(R)\chi_i(R)^*. \tag{2.6.3}$$

It is also true that

$$\begin{aligned}
\sum_R |\chi(R)|^2 &= \sum_R \chi(R)^* \chi(R) \\
&= \sum_R \left\{ \left[\sum_i c_i \chi_i(R)^* \right] \left[\sum_j c_j \chi_j(R) \right] \right\} \\
&= \sum_i \sum_j c_i c_j \sum_R \chi_i(R)^* \chi_j(R) = g \sum_i c_i^2,
\end{aligned} \tag{2.6.4}$$

where we have made use of (2.5.8).

Example. Group C_{3v}. A reducible representation $\Gamma$of this group has the following character system:

	E	C_3	C_3^2	σ_1	σ_2	σ_3
Γ	5	-1	-1	1	1	1

The representation will reduce as

$$\Gamma = c_1\Gamma_1 + c_2\Gamma_2 + c_3\Gamma_3,$$

where

$$c_1 = \frac{1}{6}\sum_R \chi(R)\chi_1(R) = \frac{1}{6}[5 - 1 - 1 + 1 + 1 + 1] = 1,$$

$$c_2 = \frac{1}{6}\sum_R \chi(R)\chi_2(R) = \frac{1}{6}[5 - 1 - 1 - 1 - 1 - 1] = 0,$$

$$c_3 = \frac{1}{6}\sum_R \chi(R)\chi_3(R) = \frac{1}{6}[10 + 1 + 1] = 2.$$

Therefore

$$\Gamma = \Gamma_1 + 2\Gamma_3.$$

Also

$$\sum_R |\chi(R)|^2 = 30 = g\sum_i c_i^2.$$

2.7. Basis Functions for Irreducible Representations

Let us consider an irreducible representation Γ_α of dimension n_α and let

$$u_1^\alpha, u_2^\alpha, \ldots, u_{n_\alpha}^\alpha \tag{2.7.1}$$

be a set of n_α functions. These functions are said to form a *basis* for the representation Γ_α if when applying an operation R of the group to a function u_i^α of the set we obtain

$$Ru_i^\alpha = \sum_{j=1}^{n_\alpha} \Gamma_\alpha(R)_{ji} u_j^\alpha. \tag{2.7.2}$$

If a set of functions form a basis for a representation Γ_α it is also said that they *transform irreducibly* according to Γ_α. In particular it is said that u_i^α transforms according to the i-th column of the matrix representation Γ_α.

Example. Group C_{3v}. The representation Γ_3 of the group C_{3v} is given by

E	C_3	C_3^2
$\begin{pmatrix} 1 & 0 \\ 0 & 1 \end{pmatrix}$	$\begin{pmatrix} -\frac{1}{2} & \frac{\sqrt{3}}{2} \\ -\frac{\sqrt{3}}{2} & -\frac{1}{2} \end{pmatrix}$	$\begin{pmatrix} -\frac{1}{2} & -\frac{\sqrt{3}}{2} \\ \frac{\sqrt{3}}{2} & -\frac{1}{2} \end{pmatrix}$
σ_1	σ_2	σ_3
$\begin{pmatrix} -1 & 0 \\ 0 & 1 \end{pmatrix}$	$\begin{pmatrix} \frac{1}{2} & -\frac{\sqrt{3}}{2} \\ -\frac{\sqrt{3}}{2} & -\frac{1}{2} \end{pmatrix}$	$\begin{pmatrix} \frac{1}{2} & \frac{\sqrt{3}}{2} \\ -\frac{\sqrt{3}}{2} & -\frac{1}{2} \end{pmatrix}$

A set of functions that form a basis for this representation is given by x and y. Applying (2.7.2) to these two functions, we obtain, for example, in the case $R = C_3$

$$C_3 x = -\frac{1}{2}x - \frac{\sqrt{3}}{2}y,$$

$$C_3 y = \frac{\sqrt{3}}{2}x - \frac{1}{2}y.$$

The operation R, being in general a *symmetry operation*, may be considered to be linear and unitary. Therefore

$$(Ru_i^\alpha, Ru_j^\alpha) = (u_i^\alpha, u_j^\alpha), \tag{2.7.3}$$

where the parentheses indicate the *inner product* of two functions. Let us assume that the functions of the set (2.7.1) are orthonormal:

$$(u_i^\alpha, u_j^\alpha) = \delta_{ij}. \tag{2.7.4}$$

Then, since

$$Ru_i^\alpha = \sum_{\ell=1}^{n_\alpha} \Gamma_\alpha(R)_{\ell i} u_\ell^\alpha, \tag{2.7.5}$$

$$Ru_j^\alpha = \sum_{m=1}^{n_\alpha} \Gamma_\alpha(R)_{mj} u_m^\alpha, \tag{2.7.6}$$

(2.7.3) becomes

$$\begin{aligned}(Ru_i^\alpha, Ru_j^\alpha) &= \sum_{\ell=1}^{n_\alpha}\sum_{m=1}^{n_\alpha} \Gamma_\alpha(R)_{\ell i}^* \Gamma_\alpha(R)_{mj} (u_\ell^\alpha, u_m^\alpha) \\ &= \sum_{\ell=1}^{n_\alpha}\sum_{m=1}^{n_\alpha} \Gamma_\alpha(R)_{\ell i}^* \Gamma_\alpha(R)_{mj} \delta_{\ell m} \\ &= \sum_{\ell=1}^{n_\alpha} \Gamma_\alpha(R)_{\ell i}^* \Gamma_\alpha(R)_{\ell j} = \delta_{ij},\end{aligned} \tag{2.7.7}$$

or

$$\underset{\sim}{\Gamma}_\alpha^+(R) \underset{\sim}{\Gamma}_\alpha(R) = \underset{\sim}{1}. \tag{2.7.8}$$

Thus we can say:

If the functions forming a basis for an irreducible representation of a group of linear unitary operations are orthonormal, the matrices representing the operations are unitary.

2.8. Direct Product Representations

Let us assume that

$$\begin{aligned}&u_1^\alpha, u_2^\alpha, \ldots, u_{n_\alpha}^\alpha \\ &v_1^\beta, v_2^\beta, \ldots, v_{n_\beta}^\beta\end{aligned} \tag{2.8.1}$$

are two sets of functions that form the basis for two representations of the same group Γ_α and Γ_β, respectively. The products $u_i^\alpha v_k^\beta$ form a set of functions that is called the *direct product* of the two sets of functions. The direct product set forms a basis for a representation that has dimension $n_\alpha \times n_\beta$ and is in general reducible.

If R is an operation of the group, then

$$Ru_i^\alpha = \sum_{j=1}^{n_\alpha} \Gamma_\alpha(R)_{ji} u_j^\alpha,$$

$$Rv_k^\beta = \sum_{\ell=1}^{n_\beta} \Gamma_\beta(R)_{\ell k} v_\ell^\beta, \tag{2.8.2}$$

and

$$R(u_i^\alpha v_k^\beta) = \sum_{j=1}^{n_\alpha}\sum_{\ell=1}^{n_\beta} \Gamma_\alpha(R)_{ji}\Gamma_\beta(R)_{\ell k} u_j^\alpha v_\ell^\beta$$
$$= \sum_{j\ell} \Gamma(R)_{j\ell,ik} u_j^\alpha v_\ell^\beta. \tag{2.8.3}$$

The character of the direct product representation Γ is equal to the product of the characters of the individual representations:

$$\chi(R) = \sum_{j\ell} \Gamma(R)_{j\ell,j\ell}$$
$$= \sum_{j=1}^{n_\alpha}\sum_{\ell=1}^{n_\beta} \Gamma_\alpha(R)_{jj}\Gamma_\beta(R)_{\ell\ell} = \chi_\alpha(R)\chi_\beta(R). \tag{2.8.4}$$

A direct product representation can, in general, be reduced in terms of the irreducible representation of the group.

Example. Group C_{3v}. We obtain

$$\Gamma_1\Gamma_1 = \Gamma_1; \quad \Gamma_2\Gamma_2 = \Gamma_1$$
$$\Gamma_3\Gamma_3 = \Gamma_1 + \Gamma_2 + \Gamma_3; \quad \Gamma_1\Gamma_2 = \Gamma_2$$
$$\Gamma_1\Gamma_3 = \Gamma_3; \quad \Gamma_2\Gamma_3 = \Gamma_3$$

2.9. The Fundamental Theorem for Functions Transforming Irreducibly

A set of n_α functions u_i^α are said to form a basis for an irreducible representation Γ_α or to transform irreducibly according to representation Γ_α of a symmetry group G if

$$Ru_i^\alpha = \sum_{j=1}^{n_\alpha} \Gamma_\alpha(R)_{ji} u_j^\alpha, \tag{2.9.1}$$

where R is any operation of G. The above relation can be rewritten in the following form:

$$R(u_1^\alpha, u_2^\alpha, \ldots, u_{n_\alpha}^\alpha)$$
$$= (u_1^\alpha, u_2^\alpha, \ldots, u_{n_\alpha}^\alpha) \begin{pmatrix} \Gamma_\alpha(R)_{11}\Gamma_\alpha(R)_{12} \ldots \Gamma_\alpha(R)_{1n_\alpha} \\ \Gamma_\alpha(R)_{21}\Gamma_\alpha(R)_{22} \ldots \Gamma_\alpha(R)_{2n_\alpha} \\ \Gamma_\alpha(R)_{n_\alpha 1}\Gamma_\alpha(R)_{n_\alpha 2} \ldots \Gamma_\alpha(R)_{n_\alpha n_\alpha} \end{pmatrix}, \tag{2.9.2}$$

or, more concisely,

$$R\,\underset{\sim}{u}^\alpha = \underset{\sim}{u}^\alpha \Gamma_\alpha(R). \tag{2.9.3}$$

In (2.9.1) the i-th function u_i^α is said to transform according to the ith column of the Γ_α representation. We assume that the basis functions are orthonormal, and therefore the representations are unitary.

A set of n_α operators Q_i^α are said to *transform irreducibly* according to a representation Γ_α of a symmetry group G if

$$R\,Q_i^\alpha R^{-1} = \sum_j Q_j^\alpha \Gamma_\alpha(R)_{ji}, \tag{2.9.4}$$

where R is any operation of G.

The case of an invariant operator is a particular one in which the operator transforms according to the identity representation Γ_1:

$$RQ(x)R^{-1} = Q(x). \tag{2.9.5}$$

Let us consider a set of orthonormal functions u_i^α transforming irreducibly according to a representation Γ_α and a set of orthonormal functions v_j^β transforming irreducibly according to a representation Γ_β. The product functions $u_i^\alpha v_j^\beta$ form a basis for the representation $\Gamma_\alpha \times \Gamma_\beta$ that is, in general, reducible:

$$\Gamma_\alpha \times \Gamma_\beta = \sum_\gamma g_{\alpha\beta\gamma}\Gamma_\gamma. \tag{2.9.6}$$

The representation $\Gamma_\alpha \times \Gamma_\beta$ is brought into a reduced from by a unitary matrix $\underset{\sim}{T}$:

$$\underset{\sim}{T}^{\alpha\beta+}[\underset{\sim}{\Gamma}_\alpha(R) \times \underset{\sim}{\Gamma}_\beta(R)]\underset{\sim}{T}^{\alpha\beta} = \underset{\sim}{\Gamma}'(R). \tag{2.9.7}$$

The generic element of $\underset{\sim}{T}$ may be indicated by $T_{ij,n}^{\alpha\beta,\gamma q}$ where i and j individuate two basis functions, respectively, in the Γ_α and Γ_β manifold, Γ_γ is the generic irreducible representation in (2.9.6), and n individuates a basis function in the Γ_γ manifold. Finally, since Γ_γ may be contained more than once in $\Gamma_\alpha \times \Gamma_\beta$ ($g_{\alpha\beta\gamma}$ may be greater than 1), the index q distinguishes among the different Γ_γ representations.

We may consider the function

$$w_n^{\gamma,q} = \sum_{i,j} T_{ij,n}^{\alpha\beta,\gamma q}\, u_i^\alpha\, v_j^\beta. \tag{2.9.8}$$

Then

$$u_i^\alpha v_j^\beta = \sum_{\gamma' n' q'} [T_{ij,n'}^{\alpha\beta,\gamma' q'}]^*\, w_{n'}^{\gamma',q'}, \tag{2.9.9}$$

where $[T_{ij,n}^{\alpha\beta,\gamma q}]^*$ is the generic element of $\underset{\sim}{T}^+$.

Let us consider now an integral of the type

$$(u_i^\alpha, Qv_j^\beta), \tag{2.9.10}$$

where Q is an invariant operator. If the representations Γ_α and Γ_β are unitary, besides being irreducible, then

$$\begin{aligned}(u_i^\alpha, Qv_j^\beta) &= (Ru_i^\alpha, RQv_j^\beta)\\ &= (Ru_i^\alpha, QRv_j^\beta) = \frac{1}{g}\sum_R (Ru_i^\alpha, QRv_j^\beta)\\ &= \frac{1}{g}\sum_R\sum_k\sum_m \Gamma_\alpha(R)_{ki}^*\Gamma_\beta(R)_{mj}(u_k^\alpha, Qv_m^\beta)\\ &= \frac{1}{n_\alpha}\delta_{\alpha\beta}\delta_{ij}\sum_k (u_k^\alpha, Qv_k^\beta).\end{aligned} \tag{2.9.11}$$

We see that the matrix element is different from zero if $\alpha = \beta$, $i = j$ and also is independent of the subscript i. A particular case is given by

$$(u_i^\alpha, v_j^\beta) = 0, \quad \text{unless } \alpha = \beta, \quad i = j, \tag{2.9.12}$$

as expected.

We may now enunciate the first part of the *Fundamental Theorem*:

I. Matrix elements of invariant operators taken between two given functions vanish unless the two functions transform according to the same column of the same irreducible representation. If it does not vanish, the matrix element is independent of the column according to which the two functions transform.

Let us consider, further, an integral of the following type:

$$(f_n^\gamma, Q_i^\alpha v_j^\beta), \tag{2.9.13}$$

where Q_i^α indicates the i component of a set of operators that transform irreducibly according to the Γ_α representation. Following (2.9.8) we may write

$$Q_i^\alpha v_j^\beta = \sum_{\gamma' n' q'} [T_{ij,n'}^{\alpha\beta,\gamma' q'}]^* w_{n'}^{\gamma' q'}, \tag{2.9.14}$$

where the $w_{n'}^{\gamma' q'}$ function transforms irreducibly according to the n'th column of the $\Gamma_{\gamma'}$ representation.

The integral (2.9.13) is then given by

$$(f_n^\gamma, Q_i^\alpha v_j^\beta) = \sum_{q=1}^{g_{\alpha\beta\gamma}} c_q [T_{ij,n}^{\alpha\beta,\gamma q}]^*, \tag{2.9.15}$$

where the coefficient c_q are independent of the indices i, j, and n and depend only on the operators in question and on the indices α, β, γ, and q.

If $g_{\alpha\beta\gamma} = 1$, then

$$(f_n^\gamma, Q_i^\alpha v_j^\beta) = [T_{ij,n}^{\alpha\beta,\gamma}]^* \times \text{const}, \tag{2.9.16}$$

where the constant is independent of the subscripts.

If P_i^α is the i component of another set of operators that transform irreducibly according to Γ_α we get also

$$(f_n^\gamma, P_i^\alpha v_j^\beta) = [T_{ij,n}^{\alpha\beta,\gamma}]^* \times \text{const}, \tag{2.9.17}$$

the constant being different from the one in (2.9.16), but still independent of subscripts. From (2.9.16) and (2.9.17) we get

$$(f_n^\gamma, Q_i^\alpha v_j^\beta) = \text{const} \times (f_n^\gamma, P_i^\alpha v_j^\beta),$$

the constant being independent of the subscript.

We can now enunciate the second part of the *Fundamental Theorem:*

II. Matrix elements of operators Q_i^α transforming irreducibly according to a representation Γ_α, taken between two functions f_n^γ and v_j^β that transform irreducibly according to columns n and j, respectively, of representation Γ_γ and Γ_β, vanish unless the representation $\Gamma_\alpha \times \Gamma_\beta$ contains the representation Γ_γ. If Γ_γ is contained q times in $\Gamma_\alpha \times \Gamma_\beta$, then the value of the matrix element is uniquely determined by the symmetry property of f_n^γ, Q_i^α, and v_j^β apart from q constants. These constants do not depend on the subscripts n, i, and j.

2.10. Product Groups and Their Representations

Suppose we are given two groups,

$$\begin{aligned} g &= e, a, b, c, \ldots \\ G &= E, A, B, C, \ldots \end{aligned} \tag{2.10.1}$$

and suppose that all products between the elements of the two groups are commutative:

$$rR = Rr, \tag{2.10.2}$$

where r is any element of g and R is any element of G.

If the groups g and G are, respectively, of order m and M and if $\underset{\sim}{\gamma}(r)$ and $\underset{\sim}{\Gamma}(R)$ are the matrices representing g and G, respectively, it is possible to order the $(m \times M)^2$ numbers $\gamma(r)_{ij} \times \Gamma(R)_{k\ell}$ in such a way that they form a mM × mM matrix. For example, the element appearing in the ik-th row and jℓ-th column may be

$$\gamma(r)_{ij} \times \Gamma(R)_{k\ell}. \tag{2.10.3}$$

This matrix is called the *product group matrix.*

The character $\chi(rR)$ of $\underset{\sim}{\gamma}(r) \times \underset{\sim}{\Gamma}(R)$ is given by

$$\chi(rR) = \chi(r)\chi(R). \tag{2.10.4}$$

All the irreducible representations of g × G can be obtained by forming the direct product representations that have as components the respective irreducible representations of the component groups.

If g has s irreducible representations and G has S irreducible representations, g × G has sS irreducible representations.

2.11. Connection of Quantum Mechanics with Group Theory

Let us consider a quantum-mechanical system that is represented by the eigenfunctions of the equation

$$H\psi = E\psi. \tag{2.11.1}$$

Let us also assume that the Hamiltonian is invariant under certain operations

$$R\,H(\underset{\sim}{x})R^{-1} = H(\underset{\sim}{x}). \tag{2.11.2}$$

It is easily seen that these operations form a group.

1. The operation that does not change the change the coordinates corresponds to the identity element of the group.
2. If R and S are two generic operations that leave H invariant, that is, if

$$\begin{aligned} RHR^{-1} &= H, \\ SHS^{-1} &= H, \end{aligned} \tag{2.11.3}$$

then

$$\begin{aligned} (SR)H(SR)^{-1} &= (SR)H(R^{-1}S^{-1}) = S(RHR^{-1})S^{-1} \\ &= SHS^{-1} = H. \end{aligned} \tag{2.11.4}$$

3. The associative law also holds.
4. Moreover, from (2.11.2), multiplying by R^{-1} on the left and R on the right we obtain

$$H = R^{-1} H R,$$

that is, the inverse of any operation R that leaves H invariant is also an operation that leaves H invariant.

Therefore the operations that leave the Hamiltonian invariant form a group. We call this group the *group of the Schrödinger equation* and designate it by the symbol G.

Let us now apply an operation R of G to both sides of (2.11.1):

$$R H \psi = E R \psi. \tag{2.11.5}$$

Because of the invariance of H,

$$R H = H R, \tag{2.11.6}$$

and

$$H(R\psi) = E(R\psi). \tag{2.11.7}$$

$R\psi$ is a solution of the eigenvalue equation belonging to the eigenvalue E.

If E is not degenerate, it must be

$$R\psi = c\psi, \tag{2.11.8}$$

with c a constant.

If E is an m-fold degenerate eigenvalue, then

$$R \psi_\ell = \sum_{i=1}^{m} a_{i\ell} \psi_i, \tag{2.11.9}$$

where ψ_i are the degenerate eigenfunctions corresponding to the eigenvalue E. If we apply another operation S of G to ψ_ℓ we get

$$S\psi_\ell = \sum_{j=1}^{m} b_{j\ell}\psi_j. \tag{2.11.10}$$

If we apply the operation $T = SR$ to ψ_ℓ we obtain

$$(SR)\psi_\ell = \sum_{j=1}^{m} c_{j\ell}\psi_j. \tag{2.11.11}$$

Since the operation SR consists of operation R followed by operation S, we have also from (2.11.9)

$$\begin{aligned} SR\psi_\ell = S(R\psi_\ell) &= \sum_{i=1}^{m} a_{i\ell} S\psi_i \\ &= \sum_{i=1}^{m} a_{i\ell} \sum_{j=1}^{m} b_{ji}\psi_j \\ &= \sum_{i=1}^{m}\sum_{j=1}^{m} a_{i\ell} b_{ji}\psi_j. \end{aligned} \tag{2.11.12}$$

Then from (2.11.11) and (2.11.12)

$$c_{j\ell} = \sum_{i=1}^{m} b_{ji}\, a_{i\ell}. \tag{2.11.13}$$

The matrix of the coefficients $c_{j\ell}$ is the product of the matrix of the coefficients b_{ji} and of the matrix of the coefficients $a_{i\ell}$:

$$\underset{\sim}{C} = \underset{\sim}{B} \times \underset{\sim}{A}. \tag{2.11.14}$$

We note that if the eigenfunctions are orthonormal these matrices are unitary. The above relation mirrors the relation

$$T = SR. \tag{2.11.15}$$

By applying all the operations of G to ψ_ℓ we find a matrix for every element of the group. These matrices form a representation of the group G of the Schrödinger equation and the set of degenerate eigenfunctions $\psi_\ell(\ell = 1, 2, \ldots, m)$ are said to form a basis for a representation of the group G.

The representation is irreducible. In fact, if it were not so, it could be possible to form subsets of linear combinations of the ψ_ℓ functions

$$\psi'_1, \psi'_2, \ldots; \quad \psi'_i, \psi'_{i+1}, \ldots; \quad \psi'_j, \psi'_{j+1}, \ldots \psi'_m, \tag{2.11.16}$$

such that each operation of G transforms a function of a subset into a linear combination of functions of the same subset. But, if this were the case, the eigenvalues corresponding to the different subsets could be different, contrary to the assumption that all the m eigenfunctions ψ_ℓ have the same eigenvalue.

Therefore we can state the following:

Eigenfunctions belonging to the same eigenvalue of the energy of a system form a basis for an irreducible representation of the group of operations that leave the Hamiltonian invariant. The dimension of this irreducible representation is equal to the degree of degeneracy of the eigenvalue.

Example. Let us assume that the system under consideration has a Hamiltonian that is left invariant by all the symmetry operations of the group C_{3v}.

An eigenfunction representing the system is of one of the following types:

a. An eigenfunction that forms a basis for the representation Γ_1 of C_{3v}; that is, it remains unchanged if acted upon by any operation of C_{3v}.
b. An eigenfunction that forms a basis for the irreducible representation Γ_2 of C_{3v}; that is, it remains unchanged if acted upon by the operations E, C_3, C_3^2, but changes sign if subjected to the operations σ_1, σ_2, and σ_3 of C_{3v}.
c. One of two functions that form a basis for the representation Γ_3 of C_{3v} and are bound by the relation

$$R(\psi_1\psi_2) = (\psi_1\psi_2)\begin{pmatrix} \Gamma_3(R)_{11} & \Gamma_3(R)_{12} \\ \Gamma_3(R)_{21} & \Gamma_3(R)_{22} \end{pmatrix}, \qquad (2.11.17)$$

or

$$\begin{aligned} R\psi_1 &= \Gamma_3(R)_{11}\psi_1 + \Gamma_3(R)_{21}\psi_2, \\ R\psi_2 &= \Gamma_3(R)_{12}\psi_1 + \Gamma_3(R)_{22}\psi_2, \end{aligned} \qquad (2.11.18)$$

where R is the generic operation of C_{3v}.

The rule just enunciated has the following exceptions:

1. Accidental Degeneracy

The eigenvalues of two eigenfunctions may be equal by fortuitous coincidence. The eigenfunctions do not belong to the same set of basis function for a representation of the group of the Schrödinger equation. Examples of systems that present this type of degeneracy are the following:

a. An atomic system under the action of a magnetic field. It may happen that two Zeeman levels corresponding to two different orientations of the magnetic moment of the atom in the magnetic field may have the same energy in correspondence to a certain value of the magnetic field. If we change the intensity of the magnetic field this degeneracy is lifted.
b. Atomic hydrogen presents degeneracy between $\mathrm{s, p, d, f, \ldots}$ levels with the same n number. If the Coulomb field is replaced by a non-Coulombic central field (as in hydrogen-like atoms, for example, Na), the degeneracy is lifted.

2. Excess Degeneracy

This type of degeneracy derives from the fact that the group of the Schrödinger equation under consideration does not contain all the operation that leave the Hamiltonian invariant. An example of this type of degeneracy is the *Kramers' degeneracy* due to time-reversal symmetry.

In summary, we can say, that, apart from the two cases described above,

1. All the possible eigenfunctions of a Hamiltonian must form a basis for some irreducible representation of the group of the operations that leave the Hamiltonian invariant.
2. By knowing the irreducible representations of the group of operations that leave the Hamiltonian invariant, we know the possible degrees of degeneracy of the eigenfunctions and the transformation properties of the eigenfunctions under the operations of the group.

Chapter 3

Crystal Symmetries

This chapter deals with the basic concepts underlying the symmetry of an ordered array of atoms. The notions of crystal systems and Bravais lattices are introduced.

3.1. Unit Cells and Space Lattices

A crystal is a periodic array of atoms whose basic repeating unit is called *unit cell.* The smallest unit cell is called *primitive unit cell.*

The three linearly independent vectors $\underset{\sim}{a}$, $\underset{\sim}{b}$, and $\underset{\sim}{c}$ that define the primitive unit cell are called *basic primitive translations.* The array of points generated by the vectors

$$\underset{\sim}{T}_n = n_1\underset{\sim}{a} + n_2\underset{\sim}{b} + n_3\underset{\sim}{c}, \tag{3.1.1}$$

where n_1, n_2, and n_3 are integer numbers, is called a *lattice.*

In a crystal lattice, each lattice site has the same surroundings and is at a corner of a primitive unit cell. Therefore the primitive unit cell contains one lattice site. Unit cells other than primitive are sometimes used; these cells contain more than one lattice site.

Every site that occurs on the surface or edge or corner of a unit cell must also occur at *equivalent sites* on the same surface or edge or corner, respectively, of every other unit cell (see Fig. 3.1).

Any point in or on the unit cell may be indicated by three coordinates with respect to the axes $\underset{\sim}{a}$, $\underset{\sim}{b}$, and $\underset{\sim}{c}$. In order to identify a position that repeats itself at equivalent locations, it is sufficient to give the coordinate of one location. In general, the coordinates

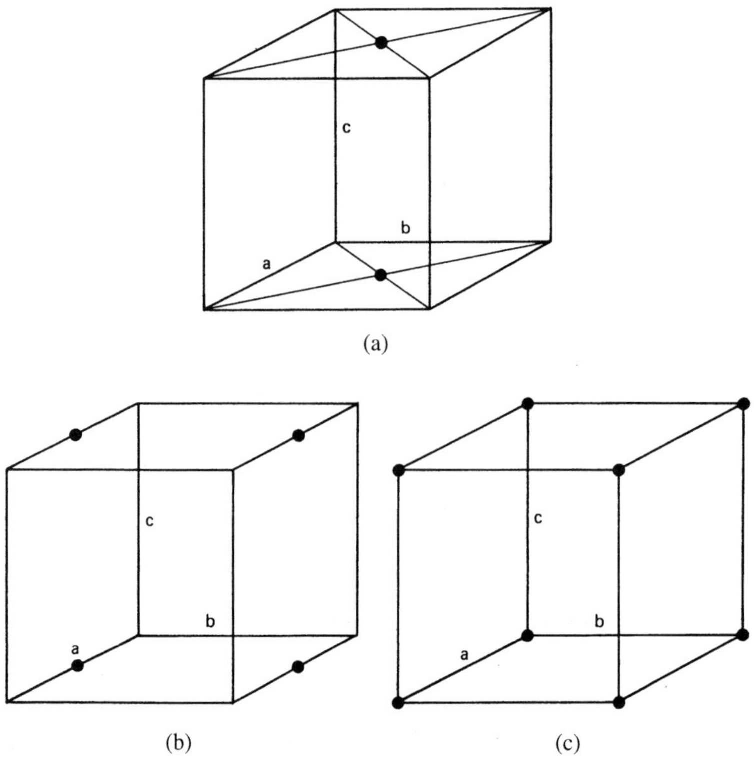

Fig. 3.1. Equivalent sites on a (a) surface, (b) edge, and (c) corner of a unit cell.

involving as many zeros as possible are used. In Fig. 3.1 the equivalent positions are identified by the following coordinates:

$$\text{a: } \frac{1}{2}, \frac{1}{2}, 0$$

$$\text{b: } \frac{1}{2}, 0, 0$$

$$\text{c: } 0, 0, 0$$

Atomic positions in the unit cell may or may not coincide with lattice sites. In general, only the inequivalent atomic sites in the unit cell are given. The unit cell of Fig. 3.2 has one atomic site at $(0, 0, 0)$. The unit cell of Fig. 3.3 has two atomic sites at $(\frac{1}{2}, 0, 0)$ and $(0, \frac{1}{2}, 0)$. The unit cell of Fig. 3.4 has three atomic sites at $(\frac{1}{2}, \frac{1}{2}, 0)$, $(\frac{1}{2}, 0, \frac{1}{2})$, and $(0, \frac{1}{2}, \frac{1}{2})$.

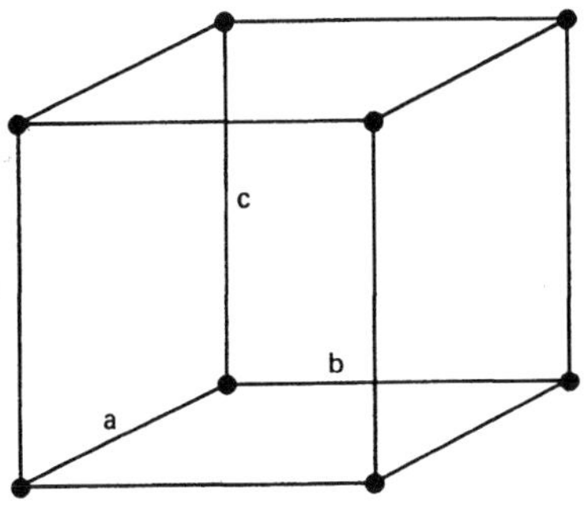

Fig. 3.2. Unit cell with one atomic site at 000.

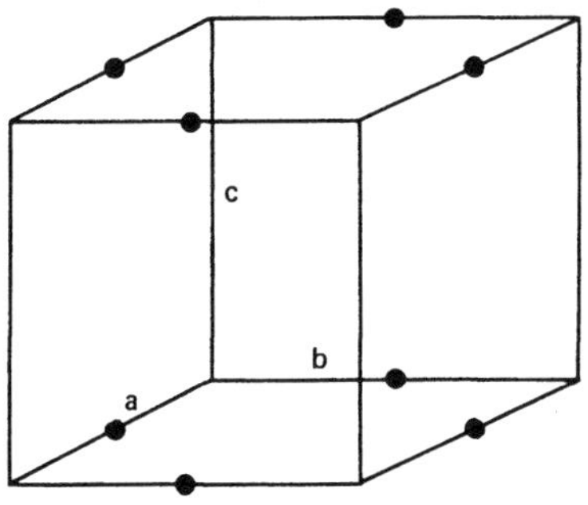

Fig. 3.3. Unit cell with two atomic sites at $\frac{1}{2}\,00$ and $0\,\frac{1}{2}\,0$.

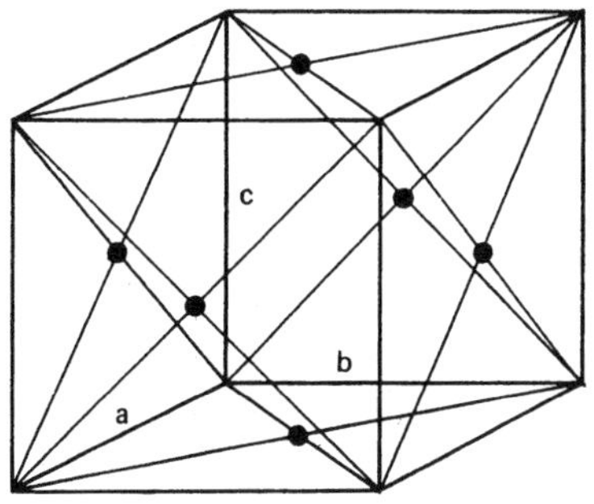

Fig. 3.4. Unit cell with three atomic sites at $\frac{1}{2}\,\frac{1}{2}\,0$, $\frac{1}{2}\,0\,\frac{1}{2}$, $0\,\frac{1}{2}\,\frac{1}{2}$.

There are, in general, several ways of positioning and choosing the unit cell with respect to the atomic sites. However, it should be clear that the lattice associated with a certain crystal structure is unique.

3.2. Miller Indices

A specific direction in a crystal may be identified by considering a vector $\underset{\sim}{T}_n$ parallel to this direction and giving its three components in the basic directions of the unit cell. The vector is so chosen that the three components n_1, n_2, n_3 are the lowest possible integers defining the direction. These three numbers are called *Miller indices* of the direction and are presented within square brackets as follows: $[n_1n_2n_3]$. In Fig. 3.5 the directions $\underset{\sim}{s}$ and $\underset{\sim}{t}$ have indices [121] and [111], respectively. The directions of the basic vectors $\underset{\sim}{a}$, $\underset{\sim}{b}$, and $\underset{\sim}{c}$ are designated by [100], [010], and [001], respectively. If any of the three indices is negative a minus sign is put on top of the negative index. The directions opposite to $\underset{\sim}{s}$ in Fig. 3.5 is indicated by $[\bar{1}\bar{2}\bar{1}]$.

A plane in a crystal is identified by three *Miller indices* that are determined from the intersection of the plane with the three coordinate axes. The intercepts are expressed in terms of the basic vectors $\underset{\sim}{a}$, $\underset{\sim}{b}$, and $\underset{\sim}{c}$. We illustrate the procedure to be followed in order to find the Miller indices of a plane with the following example.

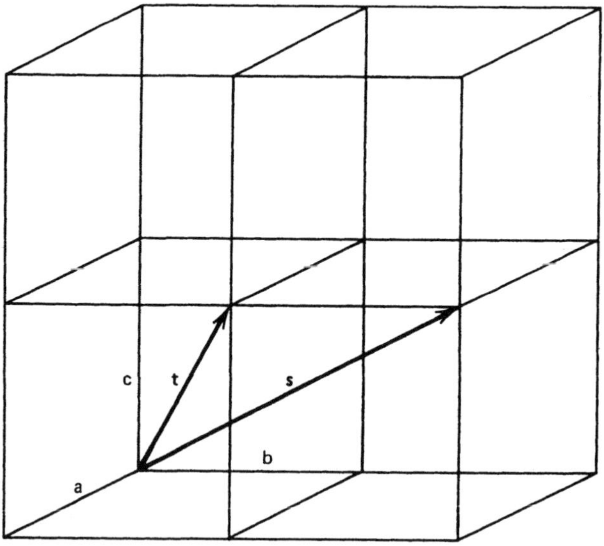

Fig. 3.5. Directions in a crystal: $\underset{\sim}{s} = [121], \underset{\sim}{t} \equiv [111]$.

Example. Let us assume that a plane has the following intercepts, with the axes $\underset{\sim}{a}$, $\underset{\sim}{b}$, and $\underset{\sim}{c}$, respectively: $2, 1, 3$. We proceed as follows:

1. We take the reciprocal of the intercepts:

$$\frac{1}{2}, 1, \frac{1}{3}$$

2. We given the above fractions the minimum common denominator:

$$\frac{3}{6}, \frac{6}{6}, \frac{2}{6}$$

3. We take the numerators of the fractions above as indices:

$$3, 6, 2$$

4. We enclose the indices in parentheses:

$$(362)$$

If the point of interception of a plane has a negative coordinate, the correspondent Miller index is given with a negative sign on top of it. Notice that two planes (h k l) and $(\bar{h}\ \bar{k}\ \bar{l})$ are parallel.

If a plane is parallel to one of the axes, the point of intersection between the plane and the axis is at infinity; the corresponding Miller index is, in this case, zero. In particular we notice the following planes:

Plane	Miller indices
$\underset{\sim}{a}\underset{\sim}{b}$	(001)
$\underset{\sim}{b}\underset{\sim}{c}$	(100)
$\underset{\sim}{a}\underset{\sim}{c}$	(010)

In a cubic lattice a plane (h k l) and a direction [h k l] are mutually perpendicular.

Also, in a cubic lattice the directions of fourfold rotation

$$[100],\quad [010],\quad [001],\quad [\bar{1}00],\quad [0\bar{1}0],\quad [00\bar{1}]$$

are considered *equivalent* and collectively designated by $\langle 100\rangle$. Similarly, the directions of threefold rotation

$$[111],\quad [\bar{1}11],\quad [1\bar{1}1],\quad [11\bar{1}],\quad [\bar{1}\bar{1}1],\quad [\bar{1}1\bar{1}],\quad [1\bar{1}\bar{1}],\quad [\bar{1}\bar{1}\bar{1}]$$

are equivalent and are designated by $\langle 111\rangle$. The six twofold rotation axes are of the form $\langle 110\rangle$.

In a cubic lattice the six faces are considered to be equivalent and collectively designated by $\{100\}$. The six dodecahedral planes are designated by $\{110\}$ and the eight octahedral planes by $\{111\}$.

Consider now in Fig. 3.6 the unit cell of a hexagonal lattice defined by the basic vectors $\underset{\sim}{a}_1$, $\underset{\sim}{a}_2$, and $\underset{\sim}{c}$. The six planes

$$(100),\quad (010),\quad (\bar{1}10),\quad (\bar{1}00),\quad (0\bar{1}0),\quad (1\bar{1}0)$$

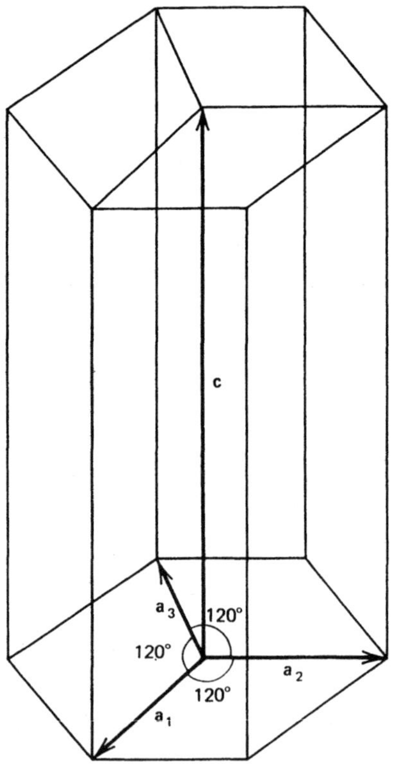

Fig. 3.6. Unit cell of a hexagonal lattice.

are equivalent in the hexagonal lattice, but do not have similar Miller indices. This inconvenience is obviated by using a four-index notation, referred to the four basic vectors $\underset{\sim}{a}_1$, $\underset{\sim}{a}_2$, $\underset{\sim}{a}_3$, and $\underset{\sim}{c}$. In this notation the six planes become

$$(10\bar{1}0),\quad (01\bar{1}0),\quad (\bar{1}100),\quad (\bar{1}010),\quad (0\bar{1}10),\quad (1\bar{1}00)$$

A consequence of choosing the three basic vectors $\underset{\sim}{a}_1$, $\underset{\sim}{a}_2$, and $\underset{\sim}{a}_3$ with 120° between each other is the fact that in a four-index notation the sum of the first three indices is always zero. Therefore one index can always be determined if the other three are known and the redundant index may be replaced by a dot. For example, plane $(11\bar{2}0)$ may be indicated by (11.0).

If a direction is given in terms of the unit cell defined by the vectors $\underset{\sim}{a}_1$, $\underset{\sim}{a}_2$, and $\underset{\sim}{c}$ in Fig. 3.6 by means of the indices [MNQ], it is possible to express it in terms of the unit cell defined by the four vectors $\underset{\sim}{a}_1$, $\underset{\sim}{a}_2$, $\underset{\sim}{a}_3$, and $\underset{\sim}{c}$ by means of four indices [m n p q]. These indices can be obtained by using the following relations

$$\begin{aligned} m &= \frac{1}{3}(2M - N), \\ n &= \frac{1}{3}(2N - M), \\ p &= -(m + n), \\ q &= Q, \end{aligned} \tag{3.2.1}$$

and reducing the new indices to the lowest integers. Thus, if [MNQ] = [110], then

$$m = \frac{1}{3}, \quad n = \frac{1}{3},$$
$$p = -\frac{2}{3}, \quad q = 0,$$

and [m n p q] = $[11\bar{2}0]$.

3.3. The Crystal Systems

The unit cells of a crystal structure can take only a certain number of distinct shapes. This number is four for two-dimensional crystals

Table 3.1. The four two-dimensional crystal systems.

System	Axial relationship	Angular relationship[a]	Characteristic symmetry[b]	C
Oblique	$a \neq b$	$\gamma = 90°$	No symmetry	C_2
Rectangular	$a \neq b$	$\gamma = 90°$	1⬯	C_{2v}
Square	$a = b$	$\gamma = 90°$	1□	C_{4v}
Hexagonal	$a = b$	$\gamma = 120°$	1⬡	C_{6v}

[a]*The convention for the angle is*

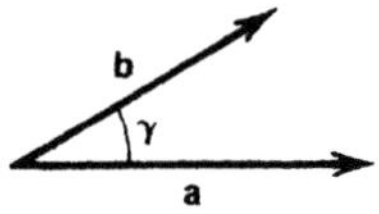

[b]⬯, *2-fold symmetry axis*
□, *4-fold symmetry axis*
⬡, *6-fold symmetry axis*
[c]*The meaning of the notation used in this column is explained in Chapter 4.*

and seven for four-dimensional crystals. Correspondingly there are four *crystal systems* in two dimensions and seven *crystal systems* in three dimensions.

3.3.1. *The Four Two-Dimensional Crystal Systems*

The crystal systems in two dimensions are reported together with the axial and angular relationships of their unit cells in Table 3.1. Each system has also a characteristic symmetry. The square lattice has, for example, fourfold rotational symmetry; that is, it will reproduce itself four times in a complete 360° rotation.

In general, a figure with an n-fold axis reproduces itself every 360/n degrees in a rotation about such an axis.

3.3.2. *The Seven Three-Dimensional Crystal Systems*

The crystal systems in three dimensions are reported in Table 3.2 together with the axial and angular relationships of their unit cells.

The only rotation axes that can occur in crystals are 1-, 2-, 3-, 4-, and 6-fold. In a rotation about an $n - 1$ fold axis, a body with such an axis reproduces itself every 360/n degrees.

Table 3.2. The seven three-dimensional crystal systems.

System	Axial relationship	Angular relationship[a]	Characteristic symmetry[b]	C
Triclinic	$a \neq b \neq c$	$\alpha \neq \beta \neq \gamma$	No symmetry	S_2
Monoclinic	$a \neq b \neq c$	$\alpha = \gamma = 90° \neq \beta$	1 0 in b direction	C_{2h}
Orthorhombic	$a \neq b \neq c$	$\alpha = \beta = \gamma = 90°$	3 0 mutually ⊥	D_{2h}
Tetragonal	$a = b \neq c$	$\alpha = \beta = \gamma = 90°$	1□ in c direction	D_{4h}
Cubic	$a = b = c$	$\alpha = \beta = \gamma = 90°$	4△⊥ to octahedral planes 3□⊥ to cube faces	O_h
Trigonal	$a = b = c$	$\alpha = \beta = \gamma < 120° \neq 90°$	1△	D_{3d}
Hexagonal	$a = b \neq c$	$\alpha = \beta = 90°\ \gamma = 120°$	1⬡	D_{6h}

[a]*The convention for the angles is*

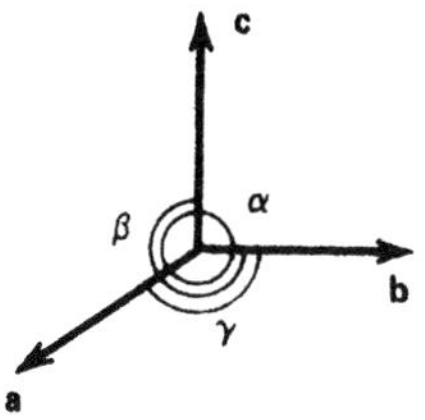

[b]0, *two-fold symmetry axis*
△, *three-fold symmetry axis*
□, *four-fold symmetry axis*
⬡, *six-fold symmetry axis*
[c]*The meaning of the notation used in this column is explained in Chapter 4.*

Each system has a characteristic symmetry. Given a certain crystal the characteristic symmetry can be used to individuate the system to which the structure belongs.

For the cubic systems, either the presence of the four 3-fold axes (perpendicular to the octahedral {111} planes) or the presence of the three fourfold axes (perpendicular to the cube faces {100} are sufficient to identify the crystal system.

The characteristic symmetries of trigonal, tetragonal, and hexagonal systems are one 3-fold, one 4-fold and one 6-fold axis, respectively.

The presence of three mutually orthogonal 2-fold axes is sufficient to identify the orthorhombic system.

The monoclinic system is identifiable by the presence of one 2-fold axis in the direction b.

Finally the triclinic system, having a unit cell with the three basis vectors and the three angles different from each other, has no symmetry.

We notice that, in general, a crystal structure has more symmetry elements than the ones strictly necessary to identify it. For example, the cubic system has also 2-fold axes perpendicular to the dodecahedral planes $\{110\}$.

3.4. The Bravais Lattices

A crystal structure is represented by an array of points called a lattice. Each lattice site is at a corner of a primitive cell and has surroundings identical to those of any other site. We have shown that there are four distinct shapes of primitive cells (and crystal systems) in two dimensions and seven distinct shapes of primitive cells (and crystal system) in three dimensions. The corresponding lattices are called *primitive Bravais lattices.*

There are other lattices that have additional sites not at the corners of their unit cells but with surroundings identical to those at the corner. These lattices are called *nonprimitive Bravais lattices.* There is one nonprimitive Bravais lattice in two dimensions and there are seven nonprimitive Bravais lattices in three dimensions.

Therefore we have a total of five Bravais lattices in two dimensions and fourteen Bravais lattices in three dimensions.

3.4.1. *The Five Bravais Lattices in Two Dimensions*

The five Bravais lattices in two dimensions are given in Table 3.3 and Fig. 3.7. Of these, four (the oblique, the rectangular, the square and the hexagonal) are primitive. The internally centered rectangular is the only nonprimitive Bravais lattice in two dimensions.

We shall illustrate now the reason for the presence of the nonprimitive Bravais lattices. The following considerations apply to both the two-dimensional and the three-dimensional cases.

Any crystal can be represented by one of the primitive Bravais lattices; that is, it may have a primitive cell in one of the four (seven)

Table 3.3. The five Bravais lattices in two dimensions.

System	Bravais lattice[a]
Oblique	P
Rectangular	P, I
Square	P
Hexagonal	P

[a]*P, primitive; I, internally centered.*

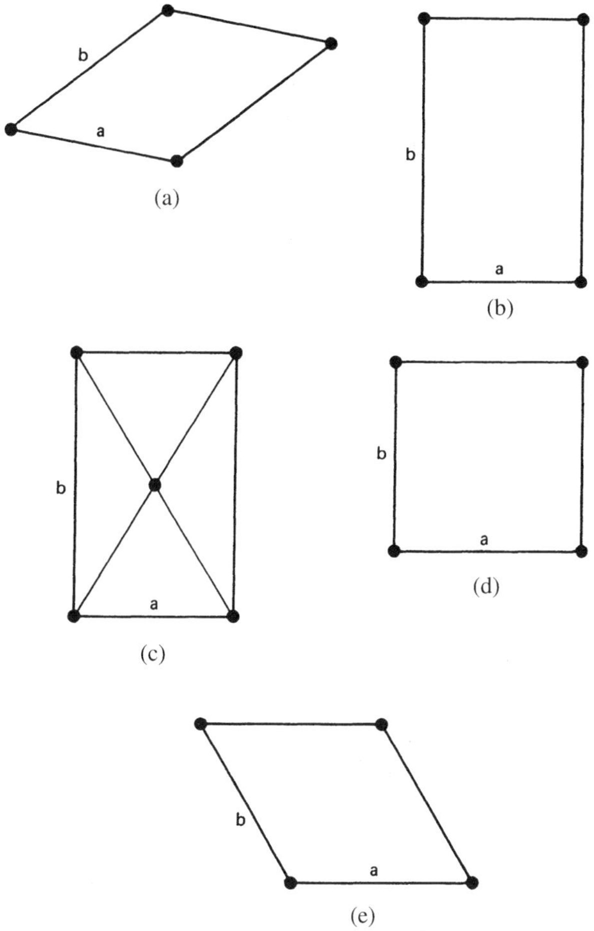

Fig. 3.7. The five Bravais lattices in two dimensions: (a) oblique P, (b) rectangular P, (c) rectangular I, (d) square P, (e) hexagonal P.

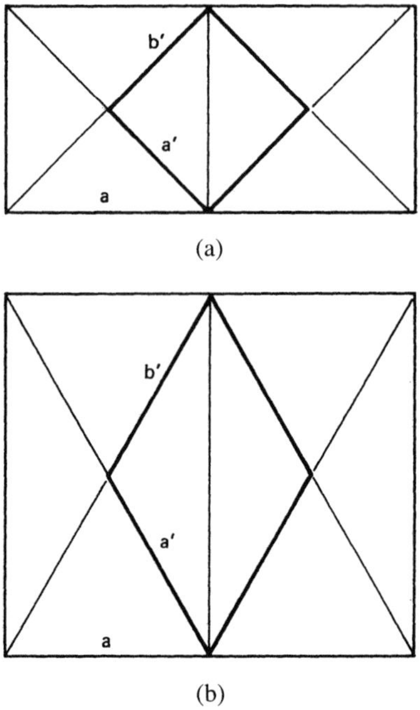

Fig. 3.8. Unit cells of (a) square lattices, (b) rectangular lattice.

crystal systems. Therefore, a primitive unit cell can always be drawn for any lattice, including a nonprimitive Bravais lattice. This primitive cell contains the smallest possible number of atomic sites.

We illustrate these considerations with an example. Consider the square lattice in Fig. 3.8(a) with a nonprimitive unit cell of basis vectors $\underset{\sim}{a}$ and $\underset{\sim}{b}$ in the form of an internally centered square. A primitive cell can be found for this lattice; this is the cell with basis vectors $\underset{\sim}{a}'$ and $\underset{\sim}{b}'$. This primitive cell belongs to the square crystal system.

The Bravais unit cell is always chosen as the smallest unit cell that presents the symmetry of the structure. Therefore, since the primitive unit cell (with basis vectors $\underset{\sim}{a}'$ and $\underset{\sim}{b}'$) belongs to the same crystal system of and is smaller than the unit cell with basis vector $\underset{\sim}{a}$ and $\underset{\sim}{b}$, the square P is the Bravais lattice of the structure in Fig. 3.8(a) and the internally centered square is *not* a Bravais lattice.

Consider now the rectangular I lattice in Fig. 3.8(b). A primitive cell can also be found for this lattice; this is the cell with basis vectors

$\underset{\sim}{a}'$ and $\underset{\sim}{b}'$. This cell, however, does not belong to the rectangular but to the oblique system, and therefore it cannot be the Bravais lattice for this structure. Therefore the internally centered rectangle is a Bravais lattice.

3.4.2. *The Fourteen Bravais Lattices in Three Dimensions*

The fourteen Bravais lattices in three dimensions are given in Table 3.4 and Figs. 3.9–3.15. Of these fourteen lattices seven are primitive and seven are nonprimitive.

The primitive Bravais lattices have one lattice site at 000.

Table 3.4. The fourteen Bravais lattices in three dimensions.

System	Bravais lattice[a]
Triclinic	P
Monoclinic	P, C
Orthorhombic	P, C, F, I
Tetragonal	P, I
Cubic	P, F, I
Trigonal	R(P)
Hexagonal	P

[a] *P, primitive; C, two faces centered; F, all faces centered; I, internally centered.*

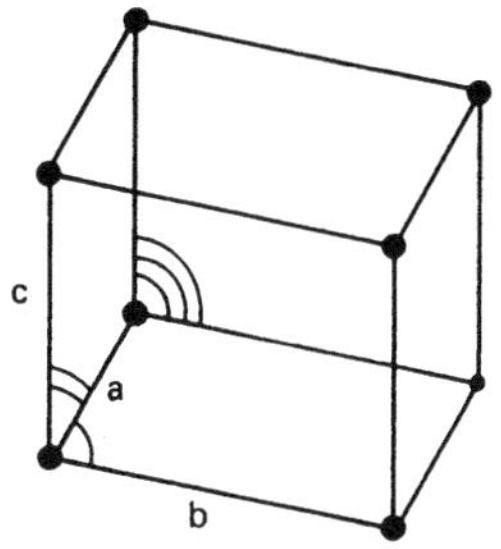

Fig. 3.9. Triclinic Bravais lattice. Vector lengths are indicated by letters a, b, and c. Angles different from 90° are indicated by ∢, ∢, and ∢.

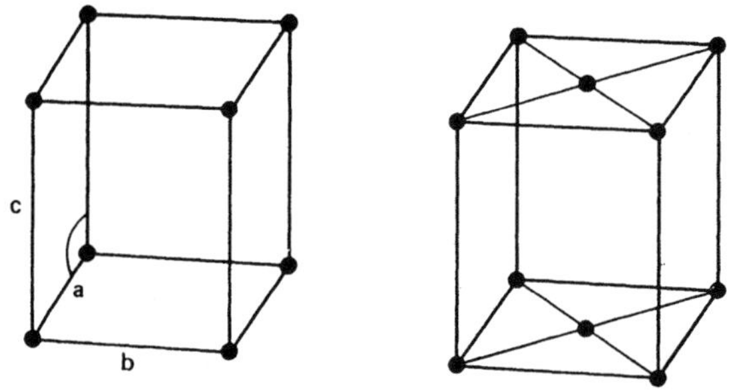

(a) (b)

Fig. 3.10. Bravais lattices of the monoclinic system: (a) simple monoclinic, (b) two face-centered monoclinic.

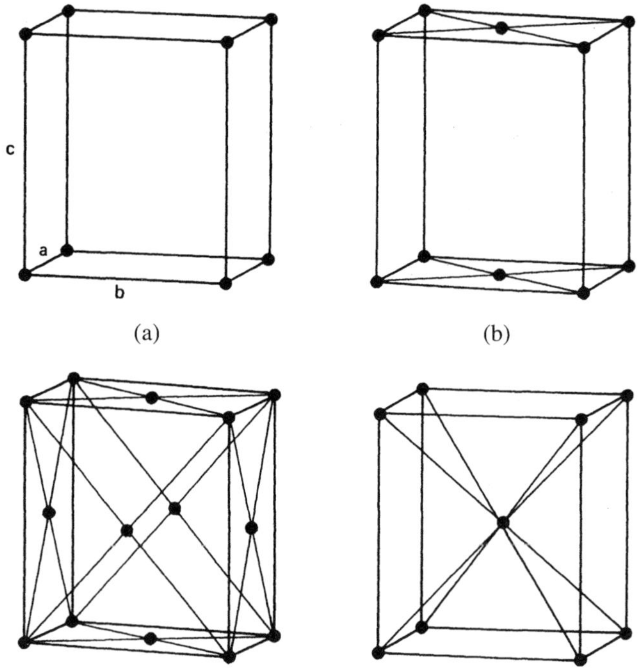

(a) (b)

(c) (d)

Fig. 3.11. Bravais lattices of the orthorhombic system: (a) simple orthorhombic, (b) two-face-centered orthorhombic, (c) all face-centered orthorhombic, (d) body-centered orthorhombic.

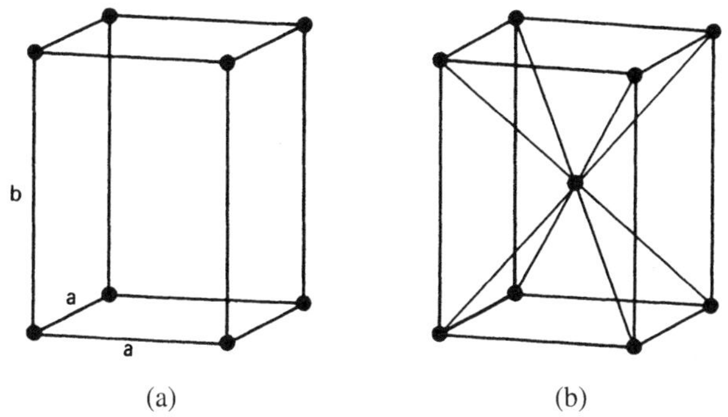

Fig. 3.12. Bravais lattices of the tetragonal system: (a) simple tetragonal, (b) body-centered tetragonal.

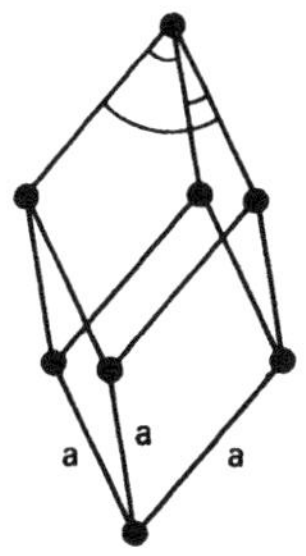

Fig. 3.13. Bravais lattices of the trigonal system.

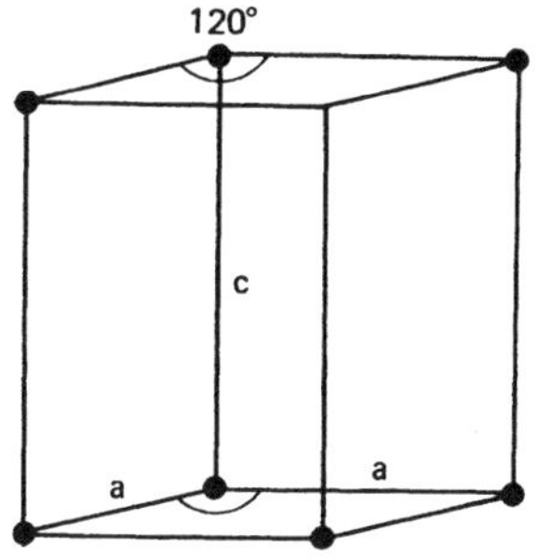

Fig. 3.14. Bravais lattices of the hexagonal system.

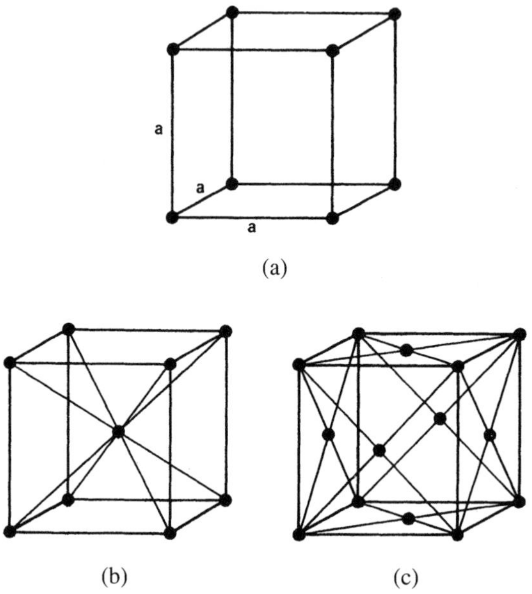

Fig. 3.15. Bravais lattices of the cubic system: (a) simple cubic, (b) body-centered cubic, (c) face-centered cubic.

The nonprimitive Bravais lattices have the following sites:

C two face centered:	000,	$\frac{1}{2}\frac{1}{2}0$
I internally centered:	000,	$\frac{1}{2}\frac{1}{2}\frac{1}{2}$
F all faces centered:	000,	$\frac{1}{2}\frac{1}{2}0,\quad \frac{1}{2}0\frac{1}{2},\quad 0\frac{1}{2}\frac{1}{2}$

As in the two-dimensional case, the Bravais unit cell is always chosen as the smallest unit cell that presents the symmetry of the structure. This is true also for nonprimitive Bravais lattices for which no primitive cell can be found that is in the same crystal system as the nonprimitive unit cell.

Chapter 4

Group Theoretical Treatment of Crystal Symmetries

This chapter considers the crystal symmetries in light of group theory and introduces the concepts of space groups and crystallographic point groups.

4.1. Space Groups

The complete symmetry of a crystal structure can be described in terms of all the symmetry operations that leave the structure invariant. These operations form a group that is called the *space group* of the crystal.

The most general operation of a space group can be expressed by the symbol $\{R \mid \underset{\sim}{t}\}$ and involves a rotational operation R followed by a translation $\underset{\sim}{t}$. A position vector $\underset{\sim}{x}$, when acted upon by $\{R \mid \underset{\sim}{t}\}$, becomes

$$\underset{\sim}{x}' = R\underset{\sim}{x} + \underset{\sim}{t}, \tag{4.1.1}$$

with

$$\begin{aligned} x_1' &= R_{11}x_1 + R_{12}x_2 + R_{13}x_3 + t_1, \\ x_2' &= R_{21}x_1 + R_{22}x_2 + R_{23}x_3 + t_2, \\ x_3' &= R_{31}x_1 + R_{32}x_2 + R_{33}x_3 + t_3. \end{aligned} \tag{4.1.2}$$

R is a 3×3 real orthogonal matrix; if its determinant is $+1$, the rotation is called *proper*, if it is -1, the rotation is called *improper*.

A pure rotation is indicated by $\{R \mid \underset{\sim}{0}\}$ and a pure translation by $\{E \mid \underset{\sim}{t}\}$. The identity operation is represented by $\{E \mid \underset{\sim}{0}\}$.

If two operations $\{R_1 \mid \underset{\sim}{t}'\}$ and $\{R_2 \mid \underset{\sim}{t}''\}$ act in succession upon a vector $\underset{\sim}{x}$, the result is

$$\underset{\sim}{x}' = R_1\underset{\sim}{x}_1 + \underset{\sim}{t}', \tag{4.1.3}$$

and

$$\begin{aligned}\underset{\sim}{x}'' &= R_2\underset{\sim}{x}' + \underset{\sim}{t}'' = R_2(R_1\underset{\sim}{x}_1 + \underset{\sim}{t}') + \underset{\sim}{t}'' \\ &= R_2R_1\underset{\sim}{x}_1 + R_2\underset{\sim}{t}' + \underset{\sim}{t}''.\end{aligned} \tag{4.1.4}$$

Therefore, the product of the two operations $\{R_2 \,|\, \underset{\sim}{t}''\}$ and $\{R_1 \,|\, \underset{\sim}{t}'\}$ is given by

$$\{R_2 \,|\, \underset{\sim}{t}''\}\{R_1 \,|\, \underset{\sim}{t}'\} = \{R_2R_1 \,|\, R_2\underset{\sim}{t}' + \underset{\sim}{t}''\}. \tag{4.1.5}$$

The inverse of an operation $\{R \,|\, \underset{\sim}{t}\}$ is given by

$$\{R \,|\, \underset{\sim}{t}\}^{-1} = \{R^{-1} \,|\, -R^{-1}\underset{\sim}{t}\}. \tag{4.1.6}$$

In fact, applying (4.1.5) we find

$$\begin{aligned}\{R^{-1} \,|\, -R^{-1}\underset{\sim}{t}\}\{R \,|\, \underset{\sim}{t}\} &= \{R^{-1}R \,|\, R^{-1}\underset{\sim}{t} - R^{-1}\underset{\sim}{t}\} \\ &= \{E \,|\, \underset{\sim}{0}\}.\end{aligned} \tag{4.1.7}$$

All the rotations R refer to axes passing through a specific point in the crystal. Consider in particular a rotation R through an angle α about an axis passing through a point A (see Fig. 4.1). It is evident that this rotation is equivalent to a rotation through the same angle about an axis parallel to the original axis and passing through $B(AB \equiv \underset{\sim}{t}')$ followed by a translation. This can be expressed as follows:

$$\begin{aligned}\{E \,|\underset{\sim}{t}'\}^{-1}\{R \,|\, \underset{\sim}{0}\}\{E \,|\, \underset{\sim}{t}'\} &= \{E| - \underset{\sim}{t}'\}\{R \,|\, R\underset{\sim}{t}'\} \\ &= \{R \,|\, R\underset{\sim}{t}' - t'\} = \{R \,|\, \underset{\sim}{t}\},\end{aligned} \tag{4.1.8}$$

where $\underset{\sim}{t} = R\underset{\sim}{t}' - \underset{\sim}{t}'$. Because of (4.1.8), the point of the crystal through which all the rotation axes pass can be chosen arbitrarily. In general, a point of the highest local symmetry is chosen, so that as many operations as possible appear in the form $\{R \,|\, \underset{\sim}{0}\}$. This does not, however, mean that all the rotation-translation operations can be put in the

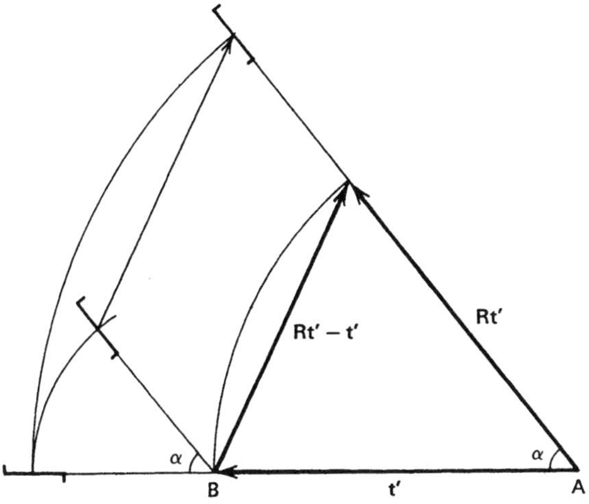

Fig. 4.1. Diagram illustrating (4.1.8).

form $\{R\,|\,\underset{\sim}{0}\}$. Two rotation-translation operations cannot be put in this form: the *screw* operation, which consists of a rotation about some axis followed by a translation along the same axis, and the *glide* operation, which consists of a reflection about a plane, followed by a translation in a direction parallel to the plane. For these operations the rotation R is associated with a nonprimitive translation $\underset{\sim}{t}$ in $\{R\,|\,\underset{\sim}{t}\}$.

Space groups that contain no glide or screw operations are called *symmorphic*, while space groups that do contain glide and/or screw operations are called *asymmorphic*. There are 13 symmorphic and 4 asymmorphic space groups in two dimensions. In three dimensions there are 73 symmorphic and 157 asymmorphic space groups.

The purely translational operations of a space group form an invariant subgroup of the space group. This can be shown as follows. If $\{E\,|\,\underset{\sim}{T}_n\}$ is a translational operation, and $\{R\,|\,\underset{\sim}{t}\}$ is a generic operation of the space group, the similarity transformation

$$\begin{aligned}\{R\,|\,\underset{\sim}{t}\}^{-1}\{E\,|\,\underset{\sim}{T}_n\}\{R\,|\,\underset{\sim}{t}\} &= \{R^{-1}\,|\,-R^{-1}\underset{\sim}{t}\}\{R\,|\,\underset{\sim}{t}+\underset{\sim}{T}_n\}\\ &= \{E\,|\,R^{-1}\underset{\sim}{T}_n\},\end{aligned} \tag{4.1.9}$$

gives also a translational operation.

The rotational operations R that appear in the symmetry operation of the space group form also a group that is called the *crystallographic point group.*

The entire crystallographic point group is contained as a subgroup in a symmorphic space group. This is not the case if the space group is asymmorphic.

4.2. The Crystallographic Point Groups

In a periodic structure, rotational and translational operations must be compatible. This fact restricts the possible rotation axes to 1-, 2-, 3-, 4- and 6-fold axes. A simple argument can be made for the exclusion of, say, 5-fold axes of rotation: it is impossible to cover a floor with pentagonal tiles!

4.2.1. *Two-Dimensional Crystallographic Point Groups*

The possible symmetry operations in two dimensions are the proper rotations

$$E, C_2, C_4, C_4^3, C_3, C_3^2, C_6, C_6^5$$

and the improper rotations

$$\sigma, \sigma C_2, \sigma C_4, \sigma C_4^3, \sigma C_3, \sigma C_3^2, \sigma C_6, \sigma C_6^5$$

where C_n represents clockwise rotation by an angle $2\pi/n$ and σ indicates reflection through an axis.

The following ten point groups can be formed:

a. Groups containing only proper rotations:

C_1: E

C_2: E, C_2

C_4: $E, C_4, C_2 = C_4^2, C_4^3$

C_3: E, C_3, C_3^2

C_6: $E, C_6, C_3 = C_6^2, C_2 = C_6^3, C_3^2 = C_6^4, C_6^5$

b. Groups containing as many proper as improper rotations:

C_{1v}: E, σ

C_{2v}: $E, C_2, \sigma, \sigma C_2$

C_{4v}: $E, C_4, C_2, C_4^3, \sigma, \sigma C_4, \sigma C_2, \sigma C_4^3$

C_{3v}: $E, C_3, C_3^2, \sigma, \sigma C_3, \sigma C_3^2$

C_{6v}: $E, C_6, C_3, C_2, C_3^2, C_6^5, \sigma, \sigma C_6, \sigma C_3, \sigma C_2, \sigma C_3^2, \sigma C_6^5$

All these groups can be represented by the planar figures they leave invariant as in Figs. 4.2 and 4.3.

4.2.2. *Three-Dimensional Crystallographic Point Groups*

The possible symmetry operations in three dimensions are:

$E =$ identity; leaves each particle in its original position

$C_n =$ clockwise rotation about an axis of symmetry by an angle $2\pi/n$

$\sigma_h =$ reflection through a plane of symmetry, perpendicular to the principal axis of symmetry (axis with largest n)

$\sigma_v =$ reflection through a plane that contains the principal axis

$\sigma_d =$ reflection through a plane that contains the principal axis and bisects the angle between two 2-fold axes perpendicular to the principal axis

$S_n =$ clockwise rotation about an axis by $2\pi/n$ followed by a reflection through a plane perpendicular to the same axis

$I = S_2 =$ inversion through the center of symmetry

Ordinary rotation operations are referred to as proper rotations, reflections and inversions as improper rotations.

The following 32 point groups can be formed:

I. Rotation Groups

These groups have one symmetry axis of a higher degree than any other symmetry axis. This preferred direction is called *the principal*

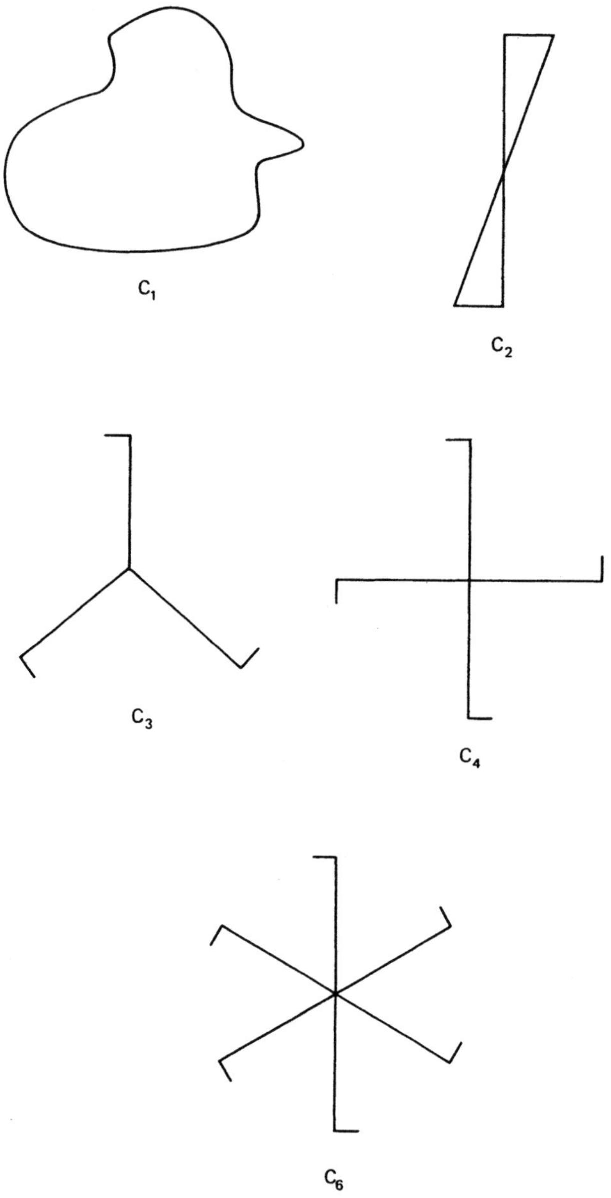

Fig. 4.2. Planar figures illustrating the point groups of proper rotations in two dimensions.

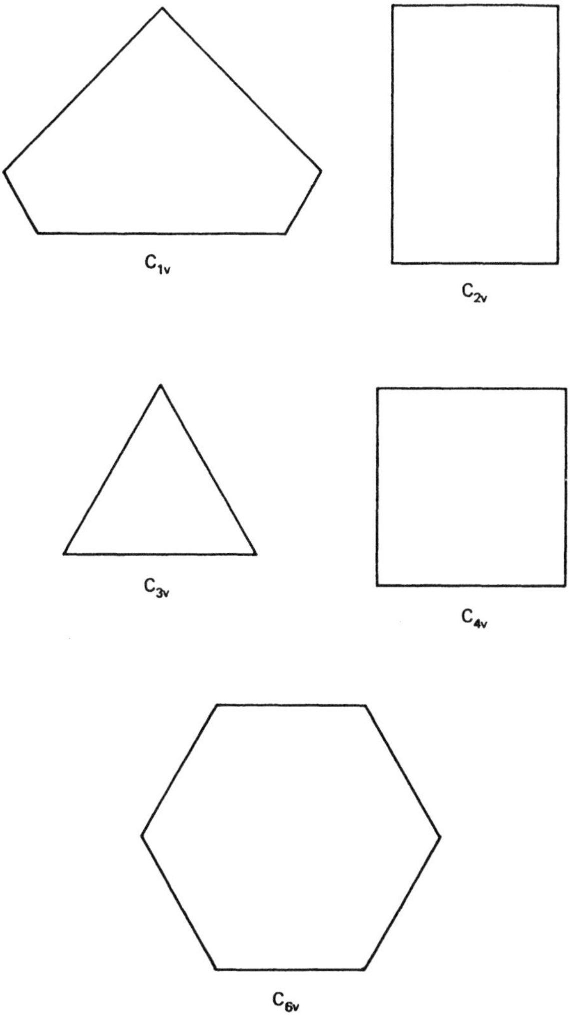

Fig. 4.3. Planar figures illustrating the point groups of proper and improper rotations in two dimensions.

axis and is generally considered to be the Z coordinate axis. The rotation groups include:

a. Groups of complexes with only one symmetry axis,

$$C_1, C_2, C_3, C_4, C_6$$

These groups are *cyclic* since $(C_n)^n = E$ and Abelian. The group C_6 contains the operations:

$$E, C_6, C_3, C_2, C_3^2, C_6^5$$

b. Groups of complexes with symmetry operations C_n and σ_v,

$$C_{2v}, C_{3v}, C_{4v}, C_{6v}$$

c. Groups of complexes with symmetry operations C_n and σ_h,

$$C_{1h}, C_{2h}, C_{3h}, C_{4h}, C_{6h}$$

d. Groups of complexes with symmetry operations S_n,

$$S_2, S_4, S_6$$

e. Groups of complexes with n 2-fold axes perpendicular to the principal n-fold axis,

$$D_2, D_3, D_4, D_6$$

f. Groups of complexes with the symmetry operations of the group D_n plus the symmetry operation σ_d,

$$D_{2d}, D_{3d}$$

g. Groups with the symmetry operations of the groups D_n plus the symmetry operation σ_d and σ_h,

$$D_{2h}, D_{3h}, D_{4h}, D_{6h}$$

II. Cubic Groups

These groups have no unique axis of higher symmetry but have more than one n-fold axis with $n > 2$.

a. $T =$ group of operations that leave a tetrahedron invariant
b. $T_h = T \times S_2$
c. $O =$ group of operations that leave a regular octahedron invariant
d. $O_h = O \times S_2$
e. $T_d =$ group of operations having T as a subgroup of proper rotations. If I indicates the ensemble of improper rotations, it is

$$T + (-I) = O.$$

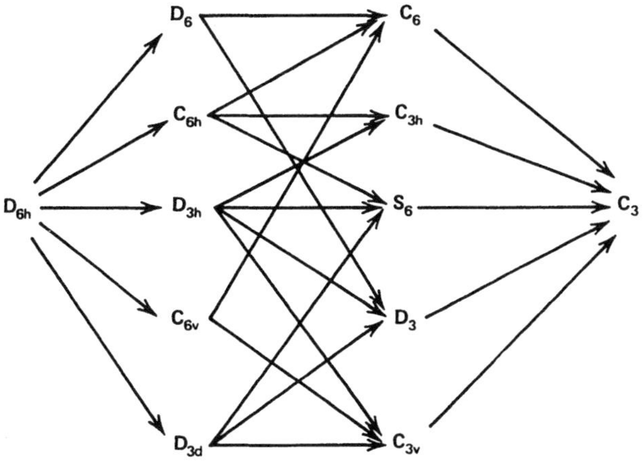

Fig. 4.4. The chain of hexagonal and trigonal groups.

Character tables for the groups described above can be found in Refs. 1–3.

The 32 point groups are related among themselves in the sense that several groups are subgroups of other groups. If a group H is a subgroup of a group G, we indicate this by writing $G \to H$. A sequence of relations $G \to H_1, H_1 \to H_2, H_2 \to H_3$, and so on forms a *chain* and is indicated by writing $G \to H_1 \to H_2 \to H_3 \cdots$. Figures 4.4 and 4.5 show some typical chains.

4.2.3. *Site Groups*

The ensemble of all operations that describe the point (rotational) symmetry at a certain point in the crystal forms the so called *site group*, which is in general a subgroup of the crystallographic point group. Only at special points in the crystal will the site group contain operations other than the identity E.

Example 1. Strontium titanate has the perovskite structure shown in Fig. 4.6. The space group is the symmorphic 0_h^1. The crystallographic point group is 0_h. The unit cell contains one molecule: one strontium, one titanium, and three oxygen atoms.

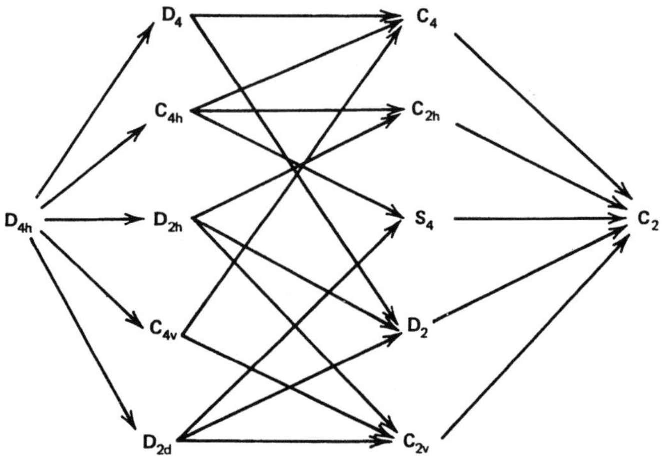

Fig. 4.5. The chain of tetragonal and orthorhombic groups.

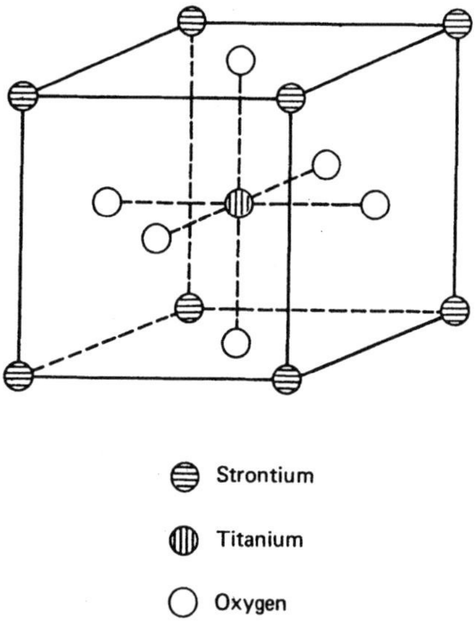

Fig. 4.6. The unit cell of $SrTiO_3$. The coordinates for the inequivalent sites are: strontium: 0, 0, 0; titanium: $\frac{1}{2}, \frac{1}{2}, \frac{1}{2}$; oxygen: $\frac{1}{2}, \frac{1}{2}, 0$; $\frac{1}{2}, 0, \frac{1}{2}$; $0, \frac{1}{2}, \frac{1}{2}$.

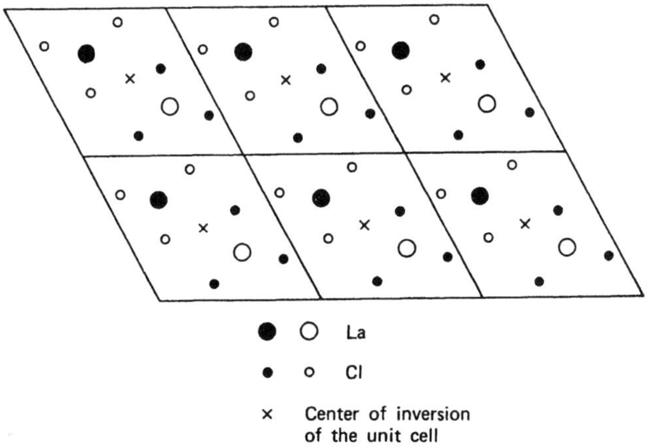

Fig. 4.7. The unit cell of $LaCl_3$. Open circles represent atoms in the plane of the paper, shaded circles represent atoms above or below the plane of the paper.

The site groups at different points in the unit cell are given by

Points	Site group
$0\ 0\ 0$	0_h
$\frac{1}{2}\ \frac{1}{2}\ \frac{1}{2}$	0_h
$\frac{1}{2}\ \frac{1}{2}\ 0$	D_{4h}
$\frac{1}{2}\ 0\ \frac{1}{2}$	D_{4h}
$0\ \frac{1}{2}\ \frac{1}{2}$	D_{4h}

Example 2. Lanthanum chloride has the structure represented in Fig. 4.7. The space group is the asymmorphic C_{6h}^2. The crystallographic point group is C_{6h} and contains the following operations:

a. The operations of the site group C_{3h} at the corner of the unit cell and at the La site.
b. A screw operation combining a C_6 rotation (about an axis perpendicular to the plane of the paper and located at the corner of the

unit cell) and a translation by one half the primitive translation (along the above axis).
c. Operations that combine (a) with (b).
d. Operation inversion.

4.3. The Invariant Subgroup of Primitive Translations: Bravais Lattices

A Bravais lattice is an array of points that establishes the skeleton of a space group in that it represents all the possible purely translational operations of the group.

If $\underset{\sim}{T}_n$ is a primitive translation and $\{R\,|\,\underset{\sim}{t}\}$ is an operation of a space group, we may write

$$\{E\,|\,R\underset{\sim}{T}_n\} = \{R\,|\underset{\sim}{t}\}\{E\,|\,\underset{\sim}{T}_n\}\{R^{-1}\,|\,-R^{-1}\underset{\sim}{t}\}. \tag{4.3.1}$$

Since the primitive translations form an invariant subgroup of the space group, $\{E\,|\,R\underset{\sim}{T}_n\}$ is also a primitive translation.

Therefore, the lattice produced by the primitive translations must be invariant with respect to the operations of the crystallographic point group. This fact imposes some restrictions on the lengths and orientations of the basic translations $\underset{\sim}{a}, \underset{\sim}{b}$, and $\underset{\sim}{c}$ of the unit cell and limits the number of distinct Bravais lattices to 5 in two dimensions and 14 in three dimensions.

A crystal system may, therefore, be described either by the relative lengths of the three basic vectors $\underset{\sim}{a}, \underset{\sim}{b}$, and $\underset{\sim}{c}$ and by the angles among them or by its point symmetry or *syngony*. In Tables 4.1 and 4.2 we report the point symmetries of two- and three-dimensional Bravais lattices, respectively.

Table 4.1. The point symmetries of the Bravais lattices in two dimensions.

Bravais lattice	Point symmetry
Oblique, P	C_2
Rectangular, P, I	C_{2v}
Square, P	C_{4v}
Hexagonal, P	C_{6v}

Table 4.2. The point symmetries of the Bravais lattices in three dimensions.

Bravais lattice	Point symmetry
Triclinic, P	S_2
Monoclinic, P, C	C_{2h}
Orthorhombic, P, C, F, I	D_{2h}
Tetragonal, P, I	D_{4h}
Cubic, P, F, I	O_h
Trigonal, P	D_{3d}
Hexagonal, P	D_{6h}

4.4. The Compatibility of Rotational and Translational Symmetries and Its Relevance to Space Groups

The arguments for the compatibility of rotational and translational symmetries can be summarized as follows:

1. Space groups with the generic operations $\{R \,|\, \underset{\sim}{t}\}$ have as an invariant subgroup the group of all primitive translations $\{E \,|\, \underset{\sim}{T}_n\}$.
2. The rotational parts R of the operations $\{R \,|\, \underset{\sim}{T}\}$ also form a group that is called the crystallographic point group. Only a limited number of point groups leave a crystal lattice (subgroup of primitive translations) invariant. The translational operations restrict the number of crystallographic point groups to 10 in two dimensions and 32 in three dimensions. A space group may be classified according to its crystallographic point group; therefore, the two-dimensional space groups can be divided in 10 *classes* and the three-dimensional space groups in 32 *classes.*
3. Restrictions are placed on the crystal lattices by the crystallographic point groups, as only a limited number of (Bravais) lattices are left invariant by the space groups of a certain class. There are 5 such lattices in two dimensions and 14 in three dimensions. Several classes may allow the same Bravais lattices as illustrated in Tables 4.3 and 4.4. For example. Table 4.4 shows that the classes C_2, C_{1h}, and C_{2h} are compatible with the Bravais lattices: monoclinic P and monoclinic C. The space groups in these classes are said to belong to the same *crystal system.*

Table 4.3. The distribution of the 17 two-dimensional space groups.

System	Bravais lattices	Point group (class)	Number of space groups	Total number of space groups in the system
Oblique	Oblique, P	C_1	1	
		C_2	1	2
Rectangular	Rectangular, P, I	C_{1v}	3	
		C_{2v}	4	7
Square	Square, P	C_4	1	
		C_{4v}	2	3
Hexagonal	Hexagonal, P	C_3	1	
		C_{3v}	2	
		C_6	1	
		C_{6v}	1	5
Total	5	10	17	17

4. Finally, it is possible to consider the point groups and find what space groups can be associated with each point group. When all the space groups are enumerated, totals of 17 space groups in two dimensions and of 230 in three dimension are found. The distributions of these groups over the different classes are illustrated in Tables 4.3 and 4.4.

The distribution of the space groups over the different classes is made according to the following considerations. We know that a crystal system may be defined in terms of its maximal point group or syngony; we can call, for example, the trigonal system, the system of syngony D_{3d}.

Let us consider two crystal systems A and B with syngonies a and b, respectively. The system A is said to be *subordinate* to system $B(A < B$ or $B > A)$ if

1. a is a subgroup of b,
2. Every Bravais lattice of B can be transformed into one of the Bravais lattices of A by an arbitrarily small deformation of the basic lattice vectors.

Table 4.4. The distribution of the 230 three-dimensional space groups.

System	Bravais lattices	Point group (class)	Number of space groups	Total number of space groups in the system
Triclinic	Triclinic	C_1	1	
		S_2	1	2
Monoclinic	Monoclinic, P, C	C_2	3	
		C_{1h}	4	
		C_{2h}	6	13
Orthorhombic	Orthorhombic, P, C, F, I	C_{2v}	22	
		D_2	9	
		D_{2h}	28	59
Tetragonal	Tetragonal, P, I	C_4	6	
		S_4	2	
		C_{4h}	6	
		C_{4v}	12	
		D_{2d}	12	
		D_4	10	
		D_{4h}	20	68
Cubic	Cubic, P, F, I	T	5	
		T_h	7	
		T_d	6	
		O	8	
		O_h	10	36
Trigonal	Trigonal, R	C_3	1	
		S_6	1	
		C_{3v}	2	
		D_3	1	
		D_{3d}	2	7
Hexagonal	Hexagonal, P	C_3	3	
		S_6	1	
		C_{3v}	4	
		D_3	6	
		D_{3d}	4	
		C_6	6	
		C_{3h}	1	
		C_{6h}	2	
		C_{6v}	4	
		D_{3h}	4	
		D_6	6	
		D_{6h}	4	45
Total	14	32	230	230

According to this definition we can write down the following relations:

cubic > tetragonal > orthorhombic > monoclinic > triclinic
hexagonal > orthorhombic
trigonal > monoclinic

Let us consider now a Bravais lattice, say the cubic P (with syngony O_h). Any point group that is a subgroup of O_h is compatible with the Bravais lattice cubic P; for example, the point group D_4 is compatible with the cubic P lattice. Why, then, do we assign the class D_4 to the tetragonal systems? The fact is that D_4 is also a subgroup of the syngony D_{4h} of the tetragonal system, which is subordinate to the cubic system, and we rather assign the D_4 class to the subordinate system. We can then write down the following rules for the assignment of a class to a system:

a. Given a certain point group (class), consider first all the syngonies with which this group is compatible. These syngonies will identify the crystal systems with which the class is compatible.
b. Choose among these crystal systems the one that has no subordinates and assign to it the class.

It is worth noting the case of the trigonal and hexagonal systems. The point groups C_3, S_6, C_{3v}, D_3, and D_{3d} are compatible with the syngonies D_{3d} and D_{6h} of the trigonal and hexagonal system, respectively. However, the trigonal system is not subordinate to the hexagonal system and therefore the classes C_3, S_6, C_{3v}, D_3, and D_{3d} have space groups in both the trigonal and the hexagonal systems.

4.5. The Irreducible Representations of a Group of Primitive Translations Brillouin Zones

Consider a crystal defined by a primitive unit cell with basis vectors $\underset{\sim}{a}_1, \underset{\sim}{a}_2$, and $\underset{\sim}{a}_3$ and the group of primitive translations

$$\{E \,|\, \underset{\sim}{T}_n\} = \{E \,|\, n_1\underset{\sim}{a}_1 + n_2\underset{\sim}{a}_2 + n_3\underset{\sim}{a}_3\}, \tag{4.5.1}$$

with n_1, n_2, and n_3 positive or negative integers or zero. This group is Abelian, because its operations commute; therefore, the irreducible representations are one dimensional. We indicate this group by T.

In order to make the group finite we impose the periodic boundary conditions

$$\{E\,|\,\underset{\sim}{a}_1\}^N = \{E\,|\,\underset{\sim}{a}_2\}^N = \{E\,|\,\underset{\sim}{a}_3\}^N = \{E\,|\,\underset{\sim}{0}\}. \tag{4.5.2}$$

We can define three groups of translations: the group formed by $\{E\,|\,\underset{\sim}{a}_1\}$ and its powers, the group formed by $\{E\,|\,\underset{\sim}{a}_2\}$ and its powers, and the group formed by $\{E\,|\,\underset{\sim}{a}_3\}$ and its powers. Since the operations of the three groups all commute with each other, the group T of primitive translations $\{E\,|\,\underset{\sim}{T}_n\}$ can be thought of as the product group of the three groups described above. (For a definition of product group, see Section 2.10.) The irreducible representations of the group T are then given by the direct product of the representations of the three constituent groups.

The irreducible representations of the group of $\{E\,|\,\underset{\sim}{a}_1\}$ are easily found taking into account the fact that

$$\{E\,|\,n\underset{\sim}{a}_1\}\{E\,|\,m\underset{\sim}{a}_1\} = \{E\,|\,(n+m)\underset{\sim}{a}_1\}, \tag{4.5.3}$$

and

$$\{E\,|\,\underset{\sim}{a}_1\}^N = \{E\,|\,N\underset{\sim}{a}_1\} = \{E\,|\,\underset{\sim}{0}\}. \tag{4.5.4}$$

Since $\{E\,|\,\underset{\sim}{0}\}$ must be represented by 1, $\{E\,|\,\underset{\sim}{a}_1\}$ must be represented by

$$\exp\left[i\left(\frac{2\pi}{a_1}\frac{s_1}{N}\right)a_1\right], \tag{4.5.5}$$

where $s_1 = 0, 1, 2, \ldots, N-1$.

The generic operation $\{E\,|\,n\underset{\sim}{a}_1\}$ is then represented by

$$\exp\left[i\left(\frac{2\pi}{a_1}\frac{s_1}{N}\right)na_1\right] = \exp\left[i\left(2\pi\frac{s_1}{N}\right)n\right] = e^{ikna_1}, \tag{4.5.6}$$

where

$$k = \frac{s_1}{N}\frac{2\pi}{a_1} \quad (s_1 = 0, 1, 2, \ldots, N-1). \tag{4.5.7}$$

Table 4.5. Irreducible representations of a group of one-dimensional translations.

s_1	k	$\{E\,|\,0\}$	$\{E\,|\,a_1\}$	$\{E\,|\,2a_1\}$	$\{E\,|\,3a_1\}$	...	$\{E\,|\,na_1\}$	...
0	0	1	i	1	1	...	1	...
1	$\frac{2\pi}{a_1 N}$	1	$e^{i(2\pi/N)}$	$e^{i(4\pi/N)}$	$e^{i(6\pi/N)}$	...	$e^{i(2\pi n/N)}$	...
2	$\frac{4\pi}{a_1 N}$	1	$e^{i(4\pi/N)}$	$e^{i(8\pi/N)}$	$e^{i(12\pi/N)}$	...	$e^{i(4\pi n/N)}$	...
3	$\frac{6\pi}{a_1 N}$	1	$e^{i(6\pi/N)}$	$e^{i(12\pi/N)}$	$e^{i(18\pi/N)}$	...	$e^{i(6\pi n/N)}$	...
s_1	$\frac{2\pi s_1}{N a_1}$	1	$e^{i(2\pi s_1/N)}$	$e^{i(4\pi s_1/N)}$	$e^{i(6\pi s_1/N)}$	...	$e^{i(2\pi s_1/N)}$	...

It then follows that

$$0 \leq k < \frac{2\pi}{a_1}. \tag{4.5.8}$$

The irreducible representations of the one-dimensional translational group of $\{E\,|\,\underset{\sim}{a}_1\}$ are given in Table 4.5.

The group of three-dimensional translations $\{E\,|\,\underset{\sim}{T}_n\}$ is the product of the three translational groups in the directions $\underset{\sim}{a}_1, \underset{\sim}{a}_2$, and $\underset{\sim}{a}_3$. It has N^3 irreducible representations of dimension one; the representation of the operation $\{E\,|\,\underset{\sim}{T}_n\}$ is given by

$$\exp\left\{i\left[\left(\frac{2\pi}{a_1}\frac{s_1}{N}\right)na_1 + \left(\frac{2\pi}{a_2}\frac{s_2}{N}\right)na_2 + \left(\frac{2\pi}{a_3}\frac{s_3}{N}\right)na_3\right]\right\} = e^{i\underset{\sim}{k}\cdot\underset{\sim}{T}_n}, \tag{4.5.9}$$

where $s_i = 0, 1, 2, \ldots, N-1$.

We may define at this point a *reciprocal lattice* by means of three basic primitive vectors $\underset{\sim}{b}_1, \underset{\sim}{b}_2$, and $\underset{\sim}{b}_3$ defined by

$$\underset{\sim}{a}_i \cdot \underset{\sim}{b}_j = 2\pi\delta_{ij}. \tag{4.5.10}$$

The vector $\underset{\sim}{k}$ may be expressed in the reciprocal space as

$$\underset{\sim}{k} = k_1\underset{\sim}{b}_1 + k_2\underset{\sim}{b}_2 + k_3\underset{\sim}{b}_3. \tag{4.5.11}$$

From (4.5.10) and (4.5.11) we derive

$$\underset{\sim}{k} \cdot \underset{\sim}{a}_1 = (k_1\underset{\sim}{b}_1 + k_2\underset{\sim}{b}_2 + k_3\underset{\sim}{b}_3) \cdot \underset{\sim}{a}_1 = 2\pi k_1. \tag{4.5.12}$$

On the other hand, from (4.5.7)

$$\underset{\sim}{k} \cdot \underset{\sim}{a}_1 = \frac{s_1}{N} 2\pi. \tag{4.5.13}$$

Therefore

$$k_i = \frac{s_i}{N} \quad (s_i = 0, 1, 2, \ldots, N-1). \tag{4.5.14}$$

The representation of the generic operation $\{E \,|\, \underset{\sim}{T}_n\}$ is then given by $e^{i\underset{\sim}{k} \cdot \underset{\sim}{T}_n}$ where $\underset{\sim}{k}$ is given by (4.5.11) with k_i restricted to the values (4.5.14).

If $\underset{\sim}{K}_q$ is a primitive translation in the reciprocal space, then because of (4.5.10),

$$e^{i\underset{\sim}{K}_q \cdot \underset{\sim}{T}_n} = 1. \tag{4.5.15}$$

This means that the distinct representations correspond only to the N^3 values of $\underset{\sim}{k}$ given by (4.5.14) and that two $\underset{\sim}{k}$ vectors that differ by a primitive vector of the reciprocal space produce the same irreducible representation.

The values of k_i in (4.5.14), $0 \leq k_1 < 1, 0 \leq k_2 < 1, 0 \leq k_3 < 1$, define a fundamental parallelepiped; each vector $\underset{\sim}{k}$ within this solid in the reciprocal space produces a distinct representation.

Other fundamental solids could be defined in the reciprocal space. In particular, we shall define the so-called *first Brillouin zone* in the following way. Starting from a point in the reciprocal lattice we draw lines connecting this point with all the other points of the reciprocal lattice; then we intersect each line with a plane perpendicular to it at midpoint between the starting point and the lattice point reached by the line. The smallest volume enclosed by all these planes is the first Brillouin zone.

Example. Let us consider the hexagonal Bravais lattice in two dimensions represented in Fig. 4.8. The basic primitive translations for this lattice are

$$\underset{\sim}{a}_1 = \begin{pmatrix} \frac{1}{2}a \\ \\ \frac{\sqrt{3}}{2}a \end{pmatrix}, \quad \underset{\sim}{a}_2 = \begin{pmatrix} a \\ 0 \end{pmatrix}.$$

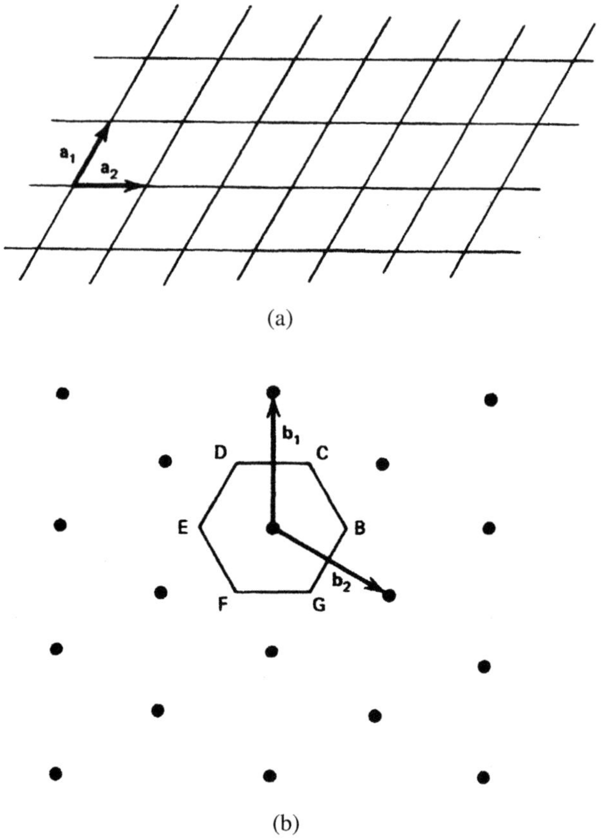

Fig. 4.8. (a) Two-dimensional hexagonal Bravais lattice; (b) reciprocal space and Brillouin zone.

The basic primitive vectors of the reciprocal lattice are given by

$$\underset{\sim}{b}_1 = \frac{4\pi}{\sqrt{3}a}\begin{pmatrix}0\\1\end{pmatrix}, \quad \underset{\sim}{b}_2 = \frac{4\pi}{\sqrt{3}a}\begin{pmatrix}\frac{\sqrt{3}}{2}\\ -\frac{1}{2}\end{pmatrix}.$$

We can check that

$$\underset{\sim}{b}_i \cdot \underset{\sim}{a}_j = 2\pi\delta_{ij}.$$

Using the $\underset{\sim}{b}_i$ vectors it is possible to construct the reciprocal lattice and the Brillouin zone that is represented by the hexagon in Fig. 4.8(b). Every allowed point in the interior of the hexagon represents a distinct representation of the group of primitive translations $\{E\,|\,\underset{\sim}{T}_n\}$.

The following observations can be made at this point for three-dimensional crystals:

1. The basic vectors of the reciprocal lattice may be expressed from (4.5.10) as follows:

$$\begin{aligned} \underset{\sim}{b}_1 &= 2\pi\frac{\underset{\sim}{a}_2 \times \underset{\sim}{a}_3}{\underset{\sim}{a}_1 \cdot \underset{\sim}{a}_2 \times \underset{\sim}{a}_3}, \\ \underset{\sim}{b}_2 &= 2\pi\frac{\underset{\sim}{a}_3 \times \underset{\sim}{a}_1}{\underset{\sim}{a}_1 \cdot \underset{\sim}{a}_2 \times \underset{\sim}{a}_3}, \\ \underset{\sim}{b}_3 &= 2\pi\frac{\underset{\sim}{a}_1 \times \underset{\sim}{a}_2}{\underset{\sim}{a}_1 \cdot \underset{\sim}{a}_2 \times \underset{\sim}{a}_3}. \end{aligned} \tag{4.5.16}$$

2. The volume of the unit cell of the reciprocal lattice is given by

$$\underset{\sim}{b}_1 \cdot \underset{\sim}{b}_2 \times \underset{\sim}{b}_3 = \frac{(2\pi)^3}{\underset{\sim}{a}_1 \cdot \underset{\sim}{a}_2 \times \underset{\sim}{a}_3}, \tag{4.5.17}$$

where $\underset{\sim}{a}_1 \cdot \underset{\sim}{a}_2 \times \underset{\sim}{a}_3$ is the volume of the unit cell of direct lattice.
3. The volume of the unit cell of the reciprocal lattice is equal to the volume of the first Brillouin zone.
4. The reciprocal of the reciprocal lattice is the direct lattice.
5. The reciprocal lattice must be one of the Bravais lattices. In particular, the reciprocal lattice has the same point symmetry as the direct lattice and, therefore, the two lattices belong to the same crystal system.
6. The Brillouin zone fills all the reciprocal space if displaced by all the primitive translations of the reciprocal lattice. Also, the Brillouin zone has the same point symmetry as the direct and reciprocal lattices.

4.6. The Irreducible Representations of Space Groups

Let us consider a space group G and let us call T its invariant subgroup of primitive translations. Let us also call $\{\alpha\,|\,\underset{\sim}{t}\}$ the generic operation of the space group and G_o the crystallographic point group formed by the operations α and $\underset{\sim}{D}\{\alpha \mid \underset{\sim}{t}\}$ the matrix of the element $\{\alpha \mid \underset{\sim}{t}\}$ of a unitary irreducible representation.

4.6.1. *Effects of Translational Symmetry*

Let us first examine the consequences of the translational symmetry of G on the representations and on the basis functions.

Any irreducible representation of G is also a reducible representation of the subgroup T of G and would reduce into some (one-dimensional) irreducible representations of T. These representations are of the form $e^{i\underset{\sim}{k}\cdot\underset{\sim}{T}_n}$ and several representations with the same $\underset{\sim}{k}$ could appear in the reduction of a space group representation.

We may assume that the space group representation under consideration has been brought by proper similarity transformations to such form that the matrices $\underset{\sim}{D}\{E \mid \underset{\sim}{T}_n\}$ representing pure translations are in diagonal form. Then the basis functions of this representation transform under translation as follows:

$$\{E \mid \underset{\sim}{T}_n\}\psi_i(\underset{\sim}{x}) = \psi_i(\underset{\sim}{x} + \underset{\sim}{T}_n) = e^{i\underset{\sim}{k}\cdot\underset{\sim}{T}_n}\psi_i(\underset{\sim}{x}), \quad (i = 1, 2, \ldots, n). \tag{4.6.1}$$

Any function $\psi_i(\underset{\sim}{x})$ that satisfies (4.6.1) can be written as

$$\psi_i(\underset{\sim}{x}) = u(\underset{\sim}{x})e^{i\underset{\sim}{k}\cdot\underset{\sim}{x}}, \tag{4.6.2}$$

where $u(\underset{\sim}{x})$ is a function periodic in the lattice, namely

$$u(\underset{\sim}{x} + \underset{\sim}{T}_n) = u(\underset{\sim}{x}), \tag{4.6.3}$$

for all $\underset{\sim}{T}_n$ in the operations $\{E \mid \underset{\sim}{T}_n\}$ of G. A function that has the property defined by (4.6.1) is called a *Bloch function*. We notice here that the basis functions for the space group G are Bloch functions (rather than linear combinations of Bloch functions) because we are considering the situation in which the matrices representing the pure translations $\{E \mid \underset{\sim}{T}_n\}$ are diagonal. Actually, the translational symmetry of the space group requires only that the basis functions be linear combinations of Bloch functions. We can, however, without loss of generality, assume that the basis functions for the space group are indeed Bloch functions.

4.6.2. *Effects of Rotational Symmetry*

We can now consider the effects of the symmetry, other than purely translational symmetry, on the representations and basis functions. Let $\psi_i(\underset{\sim}{k}, \underset{\sim}{x})$ be a Bloch basis function transforming as

$$\{E \,|\, \underset{\sim}{T}_n\}\psi_i(\underset{\sim}{k}, \underset{\sim}{x}) = e^{i\underset{\sim}{k}\cdot\underset{\sim}{T}_n}\psi_i(\underset{\sim}{k}, \underset{\sim}{x}). \tag{4.6.4}$$

If we now consider the function $\{\alpha \,|\, \underset{\sim}{t}\}\psi_i(\underset{\sim}{k}, \underset{\sim}{x})$ and apply to it the operation $\{E \,|\, \underset{\sim}{T}_n\}$ we find

$$\begin{aligned}\{E \,|\, \underset{\sim}{T}_n\}\{\alpha \,|\, \underset{\sim}{t}\}\psi_i(\underset{\sim}{k}, \underset{\sim}{x}) &= \{\alpha \,|\, \underset{\sim}{t}\}\{E \,|\, \alpha^{-1}\underset{\sim}{T}_n\}\psi_i(\underset{\sim}{k}, \underset{\sim}{x}) \\ &= e^{i\underset{\sim}{k}\cdot\alpha^{-1}\underset{\sim}{T}_n}\{\alpha \,|\, \underset{\sim}{t}\}\psi_i(\underset{\sim}{k}, \underset{\sim}{x}) \\ &= e^{i\alpha\underset{\sim}{k}\cdot\underset{\sim}{T}_n}\{\alpha \,|\, \underset{\sim}{t}\}\psi_i(\underset{\sim}{k}, \underset{\sim}{x}).\end{aligned} \tag{4.6.5}$$

Therefore we can state the following:

An operation $\{\alpha \,|\, \underset{\sim}{t}\}$ applied to a Bloch basis function $\psi_i(\underset{\sim}{k}, \underset{\sim}{x})$ transforms it into a Bloch function $\psi_i'(\alpha\underset{\sim}{k}, \underset{\sim}{x})$.

Let us consider the elements $\{\beta \,|\, \underset{\sim}{b}\}$ of G that have the following property:

$$e^{i\beta\underset{\sim}{k}\cdot\underset{\sim}{T}_n} = e^{i\underset{\sim}{k}\cdot\underset{\sim}{T}_n}, \tag{4.6.6}$$

for all $\underset{\sim}{T}_n$, which implies that

$$\beta\underset{\sim}{k} = \underset{\sim}{k} + \underset{\sim}{K}_j, \tag{4.6.7}$$

where $\underset{\sim}{K}_j$ is a primitive vector of the reciprocal lattice. The ensemble of all the $\{\beta \,|\, \underset{\sim}{b}\}$ elements forms a group that is a subgroup of G; this group is called the *group of the* $\underset{\sim}{k}$ *vector*, and we shall indicate it by the symbol K.

The group K contains all the $\{E \,|\, \underset{\sim}{T}_n\}$ elements of G. The rotational operations appearing in the elements $\{\beta \,|\, \underset{\sim}{b}\}$ of K form also a point group, which we shall call $G_O(k)$.

All the operations of the group K transform a Bloch basis function $\psi_i(\underset{\sim}{k}, \underset{\sim}{x})$ into a Bloch function $\psi_i'(\underset{\sim}{k}, \underset{\sim}{x})$ characterized by the same $\underset{\sim}{k}$ vector which can be in general expressed as a linear combination of Bloch basis functions with the same $\underset{\sim}{k}$ vector. Therefore we can state the following:

Bloch basis functions with the same $\underset{\sim}{k}$ vector transform irreducibly under the operations of the subgroup K of G.

The space group G can be decomposed into its left cosets with respect to its subgroup K:

$$G = K + \{\alpha_2|\underset{\sim}{t}_2\}K + \{\alpha_3\,|\,\underset{\sim}{t}_3\}K + \cdots + \{\alpha_q|\underset{\sim}{t}_q\}K, \qquad (4.6.8)$$

where the $\alpha_2, \alpha_3, \ldots, \alpha_q$ operations have the property

$$\alpha_i\underset{\sim}{k} = \underset{\sim}{k}_i \quad (i = 1, 2, \ldots, q). \qquad (4.6.9)$$

If $\alpha_i = E$, then $\underset{\sim}{k}_i = \underset{\sim}{k}$.

If the number of rotations α in the operations $\{\alpha\,|\,\underset{\sim}{t}\}$ of G is g_o and if the number of rotations β in the operations $\{\beta\,|\,\underset{\sim}{b}\}$ of K is g'_o, then the number of operations α_i in (4.6.9) and of cosets in the decomposition of G is

$$q = \frac{g_o}{g'_o} \qquad (4.6.10)$$

The elements of the ith coset in (4.6.8) transform a Bloch basis function characterized by a wave vector $\underset{\sim}{k}$ into a Bloch function characterized by $\alpha_i\underset{\sim}{k} = \underset{\sim}{k}_i$. The nonequivalent $\underset{\sim}{k}$ vectors produced by the different q rotations α_i are said to form a *star* $\{\underset{\sim}{k}\}$. A star is said to be nondegenerate if the number of $\underset{\sim}{k}$ vectors in $\{\underset{\sim}{k}\}$ is equal to the number of operations in G_O. The different $\underset{\sim}{k}$ vectors in $\{\underset{\sim}{k}\}$ are also called *arms of the star* $\{\underset{\sim}{k}\}$.

The rotations β of the crystallographic point group G_O that leave a wave vector $\underset{\sim}{k}$ invariant except for a primitive vector of the reciprocal lattice according to (4.6.6) correspond to the same arm of the star $\{\underset{\sim}{k}\}$.

4.6.3. *General Properties of the Irreducible Representations*

If the dimension of an irreducible representation of the space group is n, each distinct arm of the star, that is, each distinct $\underset{\sim}{k}$ vector in $\{\underset{\sim}{k}\}$, is associated with $d = n/q$ Bloch basis functions. A consequence of this fact is that the elements of the matrix $D\{E\,|\,\underset{\sim}{T}_n\}$ of an irreducible representation of G can be arranged in the following way:

$$\mathbf{D}\{E\,|\,\mathbf{T}_n\} = \begin{pmatrix} \boxed{e^{i\mathbf{k}\cdot\mathbf{T}_n}\cdot\mathbf{1}} & & & \bigcirc \\ & \boxed{e^{i\alpha_2\mathbf{k}\cdot\mathbf{T}_n}\cdot\mathbf{1}} & & \\ \bigcirc & & \boxed{e^{i\alpha_3\mathbf{k}\cdot\mathbf{T}_n}\cdot\mathbf{1}} & \\ & & & \ddots \end{pmatrix}. \tag{4.6.11}$$

Here the matrix has been divided in diagonal blocks of dimensions $d \times d$, with $d = n/q$, and $\mathbf{1}$ is the $d \times d$ unit matrix; that is, the elements in the matrix $\mathbf{D}\{E\,|\,\mathbf{T}_n\}$ are such that every element $e^{i\mathbf{k}_i\cdot\mathbf{T}_n}$ is repeated the same number of times along the diagonal.

The matrix for the generic operation $\{\alpha\,|\,\mathbf{t}\}$ can also be subdivided in blocks of dimensions $d \times d$:

$$\mathbf{D}\{\alpha\,|\,\mathbf{t}\} = \begin{pmatrix} \mathbf{D}_{11}\{\alpha\,|\,\mathbf{t}\} & \mathbf{D}_{12}\{\alpha\,|\,\mathbf{t}\} & \cdots & \mathbf{D}_{1q}\{\alpha\,|\,\mathbf{t}\} \\ \mathbf{D}_{21}\{\alpha\,|\,\mathbf{t}\} & \mathbf{D}_{22}\{\alpha\,|\,\mathbf{t}\} & \cdots & \mathbf{D}_{2q}\{\alpha\,|\,\mathbf{t}\} \\ \vdots & \vdots & \vdots & \vdots \\ \mathbf{D}_{q1}\{\alpha\,|\,\mathbf{t}\} & \mathbf{D}_{q2}\{\alpha\,|\,\mathbf{t}\} & \cdots & \mathbf{D}_{qq}\{\alpha\,|\,\mathbf{t}\} \end{pmatrix}. \tag{4.6.12}$$

With this notation the ij-th block of the $\mathbf{D}\{E\,|\,\mathbf{T}_n\}$ matrix is given by

$$\mathbf{D}_{ij}\{E\,|\,\mathbf{T}_n\} = e^{i\alpha_i\mathbf{k}\cdot\mathbf{T}_n}\mathbf{1}\delta_{ij}. \tag{4.6.13}$$

Consider now the matrices of the operations $\{\beta\,|\,\mathbf{b}\}$ of K. The set of Bloch basis functions with the same wave vector $\mathbf{k}$ forms an invariant subspace under the operations of K; therefore the matrices $\mathbf{D}\{\beta\,|\,\mathbf{b}\}$ form a reducible representation of the group K.

The matrix $\mathbf{D}\{\beta\,|\,\mathbf{b}\}$, when blocked off as in (4.6.12), has the form

$$\mathbf{D}\{\beta\,|\,\mathbf{b}\} = \left(\begin{array}{c|c} \mathbf{D}_{11}\{\beta\,|\,\mathbf{b}\} & \circ \\ \hline \circ & \equiv \end{array}\right). \tag{4.6.14}$$

It turns out that the $d \times d$ matrices $\mathbf{D}_{11}\{\beta\,|\,\mathbf{b}\}$ form an irreducible representation of the group K.[4,5] The irreducible representations of the group K are also called the *small representations* of G.

For any element $\{\alpha \mid \underset{\sim}{t}\}$ of G, and for any $\{\alpha_\ell \mid \underset{\sim}{t}_\ell\}$ we can find an element $\{\alpha_m \mid \underset{\sim}{t}_m\}$ such that

$$e^{i\alpha\alpha_\ell \underset{\sim}{k}\cdot \underset{\sim}{T}_n} = e^{i\alpha_m \underset{\sim}{k}\cdot \underset{\sim}{T}_n}. \tag{4.6.15}$$

This implies that

$$\alpha\alpha_\ell \underset{\sim}{k} = \alpha_m \underset{\sim}{k} + \underset{\sim}{K}_n, \tag{4.6.16}$$

where $\underset{\sim}{K}_n$ is a primitive vector of the reciprocal lattice, or

$$\alpha_m^{-1}\alpha\alpha_\ell \underset{\sim}{k} = \underset{\sim}{k} + \alpha_m^{-1}\underset{\sim}{K}_n. \tag{4.6.17}$$

That is, $\alpha_m^{-1}\alpha\alpha_\ell$ must be the rotational part of some element $\{\beta \mid \underset{\sim}{b}\}$ in K:

$$\{\alpha \mid \underset{\sim}{t}\}\{\alpha_\ell \mid \underset{\sim}{t}_\ell\} = \{\alpha_m \mid \underset{\sim}{t}_m\}\{\beta \mid \underset{\sim}{b}\}. \tag{4.6.18}$$

Two important results can now be reported:

1. The $m\ell$th block of the matrix $\underset{\sim}{D}\{\alpha \mid \underset{\sim}{t}\}$ is given by

$$\underset{\sim}{D}_{m\ell}\{\alpha \mid \underset{\sim}{t}\} = \underset{\sim}{D}_{11}\{\beta \mid \underset{\sim}{b}\}, \tag{4.6.19}$$

   where $\{\beta \mid \underset{\sim}{b}\}$ is the element of K that respects the relation (4.6.18).
2. The relevant blocks $\underset{\sim}{D}_{11}\{\alpha \mid \underset{\sim}{b}\}$ are given, for points within the Brillouin zone, by

$$\underset{\sim}{D}_{11}\{\beta \mid \underset{\sim}{b}\} = e^{i\underset{\sim}{k}\cdot\underset{\sim}{b}}\underset{\sim}{\Gamma}_j(\beta), \tag{4.6.20}$$

   where $\{\beta \mid \underset{\sim}{b}\}$ is an element of the group K and where $\underset{\sim}{\Gamma}_j(\beta)$ is an irreducible representation of the group $G_O(\underset{\sim}{k})$.

The above relation is also valid for points on the surface of the Brillouin zone provided the group K in correspondence to these points contains operations $\{\beta \mid \underset{\sim}{b}\}$ with vectors $\underset{\sim}{b}$ representing primitive translations, that is, provided the group K is symmorphic.

The dimension d of the small representation matrix D_{11} is equal to the dimension of $\Gamma_j(\beta)$. The character of D_{11} is given by

$$\chi_{\underset{\sim}{k},j}\{\beta \mid \underset{\sim}{b}\} = e^{i\underset{\sim}{k}\cdot\underset{\sim}{b}}\chi_j(\beta). \tag{4.6.21}$$

As we scan through the Brillouin zone, we get all the irreducible representations of the space group G. However, in order t get *distinct* representations, we have to limit ourselves to consider only distinct stars, that is, only that part of the Brillouin zone where no two vectors $\underset{\sim}{k}$ and $\underset{\sim}{k}'$ can be found such that

$$\underset{\sim}{k}' = \alpha \underset{\sim}{k} + \underset{\sim}{K}_j, \tag{4.6.22}$$

$\underset{\sim}{K}_j$ being a primitive vector of the reciprocal lattice and α any element of the crystallographic point group.

We can make the following observations:

a. At the point $\underset{\sim}{k} = 0$ the star $\{\underset{\sim}{k}\}$ has the highest possible degeneracy: $q = 1$. The space group G coincides with the group of $\underset{\sim}{k}$ vector K, also $G_O \equiv G_O(\underset{\sim}{k})$. The dimensions of the irreducible representations of G are $n = d$, that is, they are equal to the dimensions of the representations of $G_O \equiv G_O(\underset{\sim}{k})$. The characters are given in this case by

$$\begin{aligned} \chi_{\{\underset{\sim}{0}\},j}\{E \mid \underset{\sim}{T}_n\} &= d, \\ \chi_{\{\underset{\sim}{0}\},j}\{\gamma \mid \underset{\sim}{b}\} &= \chi_j(\gamma). \end{aligned} \tag{4.6.23}$$

b. At a generic point $\underset{\sim}{k}$ within the Brillouin zone, each operation of G_O changes $\underset{\sim}{k}$ to a nonequivalent vector. The star $\{\underset{\sim}{k}\}$ is nondegenerate: the number of $\underset{\sim}{k}$ vectors $\{\underset{\sim}{k}\}$ is equal to the number of elements in G_O. Also, the point group $G_O(\underset{\sim}{k})$ contains only the element E (identity). Therefore, since $d = 1$ and $q = g_o$ is the order of the point group G_o,

$$n = dq = g_o. \tag{4.6.24}$$

4.6.4. *Small Representations for Different Points of the Brillouin Zone*

We have stated that the matrices $D_{11}\{\beta \mid \underset{\sim}{b}\}$ which form an irreducible representation for the group K are given for points inside the Brillouin zone by

$$\underset{\sim}{D}_{11}\{\beta \mid \underset{\sim}{b}\} = e^{i\underset{\sim}{k}\cdot\underset{\sim}{b}}\underset{\sim}{\Gamma}_j(\beta). \tag{4.6.25}$$

The above is also true for points on the surface of the zone, provided the group K is symmorphic, that is, provided all the vectors

$\underset{\sim}{b}$ in $\{\beta\,|\,\underset{\sim}{b}\}$ are primitive translations. The reason for this provision derives from the fact that if the matrices $\underset{\sim}{D}_{11}\{\beta\,|\,\underset{\sim}{b}\}$ represent a valid representation of K they must satisfy the relation

$$\underset{\sim}{D}_{11}\{\beta\,|\,\underset{\sim}{b}\}\underset{\sim}{D}_{11}\{\beta'\,|\,\underset{\sim}{b}'\} = \underset{\sim}{D}_{11}\{\beta\beta'\,|\,\beta\underset{\sim}{b}' + \underset{\sim}{b}\}, \tag{4.6.26}$$

where $\{\beta\,|\,\underset{\sim}{b}\}$ and $\{\beta'\,|\,\underset{\sim}{b}'\}$ are two operations of K.

If $\underset{\sim}{D}_{11}\{\beta\,|\,\underset{\sim}{b}\}$ is given by $e^{i\underset{\sim}{k}\cdot\underset{\sim}{b}}\underset{\sim}{\Gamma}_j(\beta)$, then

$$\underset{\sim}{D}_{11}\{\beta\,|\,\underset{\sim}{b}\}\underset{\sim}{D}_{11}\{\beta'\,|\,\underset{\sim}{b}'\} = e^{i\underset{\sim}{k}\cdot(\underset{\sim}{b}+\underset{\sim}{b}')}\underset{\sim}{\Gamma}_j(\beta\beta'). \tag{4.6.27}$$

Also, because of (4.6.26) we must have

$$\begin{aligned}\underset{\sim}{D}_{11}\{\beta\,|\,\underset{\sim}{b}\}\underset{\sim}{D}_{11}\{\beta'\,|\,\underset{\sim}{b}'\} &= e^{i\underset{\sim}{k}\cdot(\beta\underset{\sim}{b}'+\underset{\sim}{b})}\underset{\sim}{\Gamma}_j(\beta\beta')\\ &= e^{i\underset{\sim}{k}\cdot\underset{\sim}{b}}e^{i\beta^{-1}\underset{\sim}{k}\cdot\underset{\sim}{b}'}\underset{\sim}{\Gamma}_j(\beta\beta')\\ &= e^{i\underset{\sim}{k}\cdot(\underset{\sim}{b}+\underset{\sim}{b}')}e^{i\underset{\sim}{K}\cdot\underset{\sim}{b}'}\underset{\sim}{\Gamma}_j(\beta\beta'),\end{aligned} \tag{4.6.28}$$

since $\beta^{-1}\underset{\sim}{k} = \underset{\sim}{k}+\underset{\sim}{K}$ with $\underset{\sim}{K}$ a primitive vector of the reciprocal lattice. If $\underset{\sim}{b}'$ is not a primitive vector of the real space

$$e^{i\underset{\sim}{K}\cdot\underset{\sim}{b}'} \neq 1, \tag{4.6.29}$$

and the results of (4.6.27) and (4.6.28) do not coincide. The relation (4.6.25), however, still holds for points inside the Brillouin zone, even if K is not symmorphic, because in this case $\underset{\sim}{K} = \underset{\sim}{O}$ in (4.6.28). Therefore the relation (4.6.25) can be used for constructing the irreducible representations of K, except for the case in which $\underset{\sim}{k}$ is on the boundary surfaces of the Brillouin zone and K is at the same time nonsymmorphic.

We show now that the representations of K so obtained are irreducible. The character of such a representation is, from (4.6.25),

$$\chi_{11}\{\beta\,|\,\underset{\sim}{b}\} = e^{i\underset{\sim}{k}\cdot\underset{\sim}{b}}\chi_i(\beta), \tag{4.6.30}$$

where $\chi_i(\beta)$ represents the character of $\underset{\sim}{\Gamma}_i(\beta)$. Then

$$\sum_{\{\beta\,|\,\underset{\sim}{b}\}} |\chi_{11}\{\beta\,|\,\underset{\sim}{b}\}|^2 = N\sum_{\beta} |\chi_i(\beta)|^2 = Ng'_o, \tag{4.6.31}$$

where $\{\beta \mid \underset{\sim}{b}\}$ is the operation in K, N is the number of operations in T, and g'_O is the number of operations in $G_O(\underset{\sim}{k})$. But Ng'_O is the order of the group K and (4.6.31) ensures that the representations of K are irreducible [see (2.5.8)].

If $\underset{\sim}{k}$ is on the surface of the Brillouin zone and the group K is nonsymmorphic, the small representations are not given by (4.6.25) and the so-called *factor group method* can be used to obtain the irreducible representations of K. This method consists essentially of the following:

Consider first the subgroup of K that consists of all the translations $\{E \mid \underset{\sim}{T}_k\}$ of K such that $e^{i\underset{\sim}{k}\cdot\underset{\sim}{T}_k} = 1$, that is, such that

$$\underset{\sim}{k} \cdot \underset{\sim}{T}_k = 2\pi \times \text{integer}. \tag{4.6.32}$$

The group consisting of all these translations is an invariant subgroup of K. We shall call it T_k.

The factor group of K with respect to T_k is given by

$$\frac{K}{T_k} : T_k, \{\beta \mid \underset{\sim}{b}\}T_k, \{\beta' \mid \underset{\sim}{b}'\}T_k, \ldots \tag{4.6.33}$$

Assume that we know the irreducible representations of the factor group; for each representation the matrix that corresponds to a certain coset is used to represent all the operations of K in that coset:

$$\underset{\sim}{D}_{11}\{\beta \mid \underset{\sim}{b}\} = \underset{\sim}{\Gamma}(\{\beta \mid \underset{\sim}{b}\}T_k). \tag{4.6.34}$$

The representations so obtained are valid representations of K since the elements multiply correctly:

$$\begin{aligned}
\underset{\sim}{D}_{11}\{\beta \mid \underset{\sim}{b}\}\underset{\sim}{D}_{11}\{\beta' \mid \underset{\sim}{b}'\} &= \underset{\sim}{\Gamma}(\{\beta \mid \underset{\sim}{b}\}T_k)\Gamma(\{\beta' \mid \underset{\sim}{b}'\}T_k)\\
&= \underset{\sim}{\Gamma}(\{\beta \mid \underset{\sim}{b}\}T_k \cdot \{\beta' \mid \underset{\sim}{b}'\}T_k)\\
&= \underset{\sim}{\Gamma}(\{\beta \mid \underset{\sim}{b}\}\{\beta' \mid \underset{\sim}{b}'\}T_k)\\
&= \underset{\sim}{\Gamma}(\{\beta\beta' \mid \beta\underset{\sim}{b}' + \underset{\sim}{b}\}T_k)\\
&= \underset{\sim}{D}_{11}\{\beta\beta' \mid \beta\underset{\sim}{b}' + \underset{\sim}{b}\}.
\end{aligned} \tag{4.6.35}$$

The representations so induced are also irreducible representations of K. In fact, let t be the order of the group T_k; the order of K/T_k

is then Ng'_O/t. Then

$$\sum_{\{\beta\,|\,\underset{\sim}{b}\}} |\chi_{11}\{\beta\,|\,\underset{\sim}{b}\}|^2 = t \sum_{\{\gamma\,|\,\underset{\sim}{c}\}} |\chi \text{ of } \Gamma(\{\gamma\,|\,\underset{\sim}{c}\}\,T_k)|^2$$

$$= t\frac{Ng'_O}{t} = Ng'_O, \tag{4.6.36}$$

where $\{\beta\,|\,\underset{\sim}{b}\}$ is the operation in K and $\{\gamma\,|\,\underset{\sim}{c}\}$ is the operation in K/T_k.

All the irreducible representations of K can be found by the use of the representations so induced. This can be shown considering the fact that in an irreducible representation of K, all the elements in T_k are represented by a unit matrix and therefore all the elements in a coset $\{\beta\,|\,\underset{\sim}{b}\}\,T_k$ are represented by the same matrix.

Finally, it should be noted that not all the representations so obtained are *allowable* representations of K. The fact that $e^{i\underset{\sim}{k}\cdot\underset{\sim}{T}_k} = 1$ implies that also $e^{in\underset{\sim}{k}\cdot\underset{\sim}{T}_k} = 1$ with n an integer. Some of the induced representations will therefore correspond to $\underset{\sim}{k}$ vectors that are integer multiples of the $\underset{\sim}{k}$ vector under consideration. The unwanted representations of K/T_k are eliminated by imposing the condition that the only allowable factor group representations are those for which every element $\{E\,|\,\underset{\sim}{T}_n\}\,T_k$ in K/T_k is represented by a matrix of the form $e^{i\underset{\sim}{k}\cdot\underset{\sim}{T}_n}\underset{\sim}{1}$ with $\underset{\sim}{1}$ being the unit matrix and excluding the representations for which $\{E\,|\,\underset{\sim}{T}_n\}\,T_k$ is represented by $e^{in\underset{\sim}{k}\cdot\underset{\sim}{T}_n}\underset{\sim}{1}$.

4.7. Example I. Symmorphic Group C^1_{4v}

Let us now consider the symmorphic space group represented in Fig. 4.9, which we shall designate C^1_{4v}. The unit cell is defined by the two basic vectors

$$\underset{\sim}{a}_1 = a\hat{x},$$
$$\underset{\sim}{a}_2 = a\hat{y}.$$

The Bravais lattice for such group is square P. The Brillouin zone is also square and is represented in Fig. 4.10.

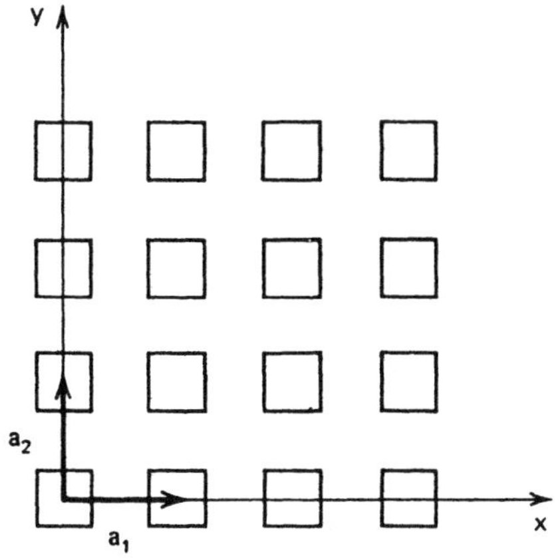

Fig. 4.9. Symmorphic two-dimensional space group with square Bravais lattice and crystallographic point group C_{4v}.

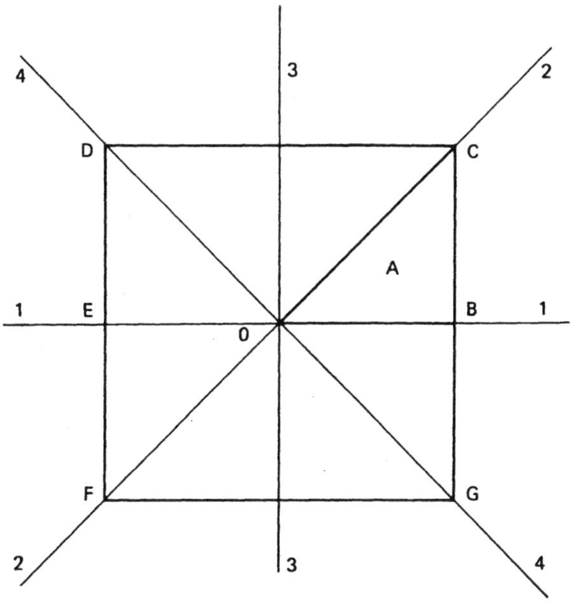

Fig. 4.10. Brillouin zone for a square Bravais lattice.

The crystallographic point group is C_{4v} and consists of the following operations (see Fig. 4.9):

$\alpha_1 = E$:	identity
$\alpha_2 = C_4, \alpha_4 = C_4^3$:	clockwise rotations through 90° and 270°, respectively
$\alpha_3 = C_2$:	clockwise rotation through 180°
$\alpha_5 = \sigma_1, \alpha_7 = \sigma_3$:	reflections through the 1–1 and 3–3 axes, respectively
$\alpha_6 = \sigma_2, \alpha_8 = \sigma_4$:	reflections through the 2–2 and 4–4 axes, respectively

The multiplication table for these operations follows:

	α_1	α_2	α_3	α_4	α_5	α_6	α_7	α_8
α_1	α_1	α_2	α_3	α_4	α_5	α_6	α_7	α_8
α_2	α_2	α_3	α_4	α_1	α_8	α_5	α_6	α_7
α_3	α_3	α_4	α_1	α_2	α_7	α_8	α_5	α_6
α_4	α_4	α_1	α_2	α_3	α_6	α_7	α_8	α_5
α_5	α_5	α_6	α_7	α_8	α_1	α_2	α_3	α_4
α_6	α_6	α_7	α_8	α_5	α_4	α_1	α_2	α_3
α_7	α_7	α_8	α_5	α_6	α_3	α_4	α_1	α_2
α_8	α_8	α_5	α_6	α_7	α_2	α_3	α_4	α_1

The character table of the group C_{4v} is given by

C_{4v}	E	C_4, C_4^3	C_2	σ_1, σ_3	σ_2, σ_4
$A_1 \equiv \Gamma_1$	1	1	1	1	1
$A_2 \equiv \Gamma_2$	1	1	1	−1	−1
$B_1 \equiv \Gamma_3$	1	−1	1	1	−1
$B_2 \equiv \Gamma_4$	1	−1	1	−1	1
$E \equiv \Gamma_5$	2	0	−2	0	0

In order to find the distinct irreducible representations of the space group we need to consider only the section OBC of the Brillouin zone in Fig. 4.10. Let us consider different sections in this zone separately.

ORIGIN O. $\underset{\sim}{k} = \underset{\sim}{0}$

At this point the group K coincides with the entire space group

$$G = \{E \,|\, \underset{\sim}{0}\} K,$$

because

$$e^{i\beta \underset{\sim}{k} \cdot \underset{\sim}{T}_n} = e^{i \underset{\sim}{k} \cdot \underset{\sim}{T}_n} = 1,$$

for all $\underset{\sim}{T}_n$. The point group $G_O(\underset{\sim}{k})$ coincides with G_O which is C_{4v}. Also

$$\underset{\sim}{D}\{\beta \,|\, \underset{\sim}{T}_n\} = \underset{\sim}{D}_{11}\{\beta \,|\, \underset{\sim}{T}_n\} = \underset{\sim}{\Gamma}_i(\beta),$$

where $\underset{\sim}{\Gamma}_i(\beta)$ is an irreducible representation of C_{4v}. The star $\{\underset{\sim}{k}\}$ has its maximum degeneracy ($q = 1$).

At this point we therefore have four one-dimensional and one two-dimensional irreducible representations. The matrices for the operations $\{\alpha_i \,|\, 0\}$ are given in Table 4.6.

POINTS A. $\underset{\sim}{k} \equiv (k_x, k_y)$

For a point A inside the section OBC of the Brillouin zone the group K coincides with the group T, since the only β operation is E:

$$\underset{\sim}{D}_{11}\{\beta \,|\, \underset{\sim}{T}_n\} = \underset{\sim}{D}_{11}\{E \,|\, \underset{\sim}{T}_n\} = e^{i \underset{\sim}{k} \cdot \underset{\sim}{T}_n}.$$

The star $\{\underset{\sim}{k}\}$ has $q = 8$ arms and is nondegenerate, as shown in Fig. 4.11. The group G can be divided in left cosets with respect to $K = T$:

$$G = \sum_{i=1}^{8} \{\alpha_i | \underset{\sim}{0}\} T,$$

where

$$\alpha_1 = E, \quad \alpha_2 = C_4, \quad \alpha_3 = C_4^2 = C_2, \quad \alpha_4 = C_4^3, \quad \alpha_5 = \sigma_1,$$
$$\alpha_6 = \sigma_2, \quad \alpha_7 = \sigma_3, \quad \alpha_8 = \sigma_4$$

Table 4.6. Irreducible Representations of the Space Group C^1_{4v} for the Point 0 of the Brillouin Zone [$\underset{\sim}{k} \equiv (0,0)$].

	$\{E \mid \underset{\sim}{T}_n\}$	$\{C_4 \mid \underset{\sim}{0}\}$	$\{C_2 \mid \underset{\sim}{0}\}$	$\{C_4^3 \mid \underset{\sim}{0}\}$	$\{\sigma_1 \mid \underset{\sim}{0}\}$	$\{\sigma_2 \mid \underset{\sim}{0}\}$	$\{\sigma_3 \mid \underset{\sim}{0}\}$	$\{\sigma_4 \mid \underset{\sim}{0}\}$
Γ_1	1	1	1	1	1	1	1	1
Γ_2	1	1	1	1	−1	−1	−1	−1
Γ_3	1	−1	1	−1	1	−1	1	−1
Γ_4	1	−1	1	−1	−1	1	−1	1
Γ_5	$\begin{pmatrix} 1 & 0 \\ 0 & 1 \end{pmatrix}$	$\begin{pmatrix} 0 & 1 \\ -1 & 0 \end{pmatrix}$	$\begin{pmatrix} -1 & 0 \\ 0 & -1 \end{pmatrix}$	$\begin{pmatrix} 0 & -1 \\ 1 & 0 \end{pmatrix}$	$\begin{pmatrix} -1 & 0 \\ 0 & 1 \end{pmatrix}$	$\begin{pmatrix} 0 & -1 \\ -1 & 0 \end{pmatrix}$	$\begin{pmatrix} 1 & 0 \\ 0 & -1 \end{pmatrix}$	$\begin{pmatrix} 0 & 1 \\ 1 & 0 \end{pmatrix}$

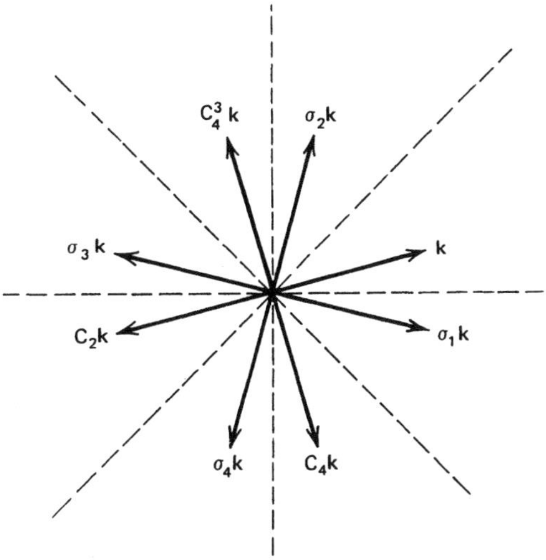

Fig. 4.11. The star $\{\underset{\sim}{k}\}$ at points A in the Brillouin zone of Fig. 4.10.

The matrix for a primitive translation is given, according to (4.6.11), by

$$\underset{\sim}{D}\{E\,|\,\underset{\sim}{T}_n\} = \begin{pmatrix} e^{i\underset{\sim}{k}\cdot\underset{\sim}{T}_n} & & & & & & & \bigcirc \\ & e^{i\alpha_2\underset{\sim}{k}\cdot\underset{\sim}{T}_n} & & & & & & \\ & & e^{i\alpha_3\underset{\sim}{k}\cdot\underset{\sim}{T}_n} & & & & & \\ & & & e^{i\alpha_4\underset{\sim}{k}\cdot\underset{\sim}{T}_n} & & & & \\ & & & & e^{i\alpha_5\underset{\sim}{k}\cdot\underset{\sim}{T}_n} & & & \\ & & & & & e^{i\alpha_6\underset{\sim}{k}\cdot\underset{\sim}{T}_n} & & \\ \bigcirc & & & & & & e^{i\alpha_7\underset{\sim}{k}\cdot\underset{\sim}{T}_n} & \\ & & & & & & & e^{i\alpha_8\underset{\sim}{k}\cdot\underset{\sim}{T}_n} \end{pmatrix}$$

Let us consider now the matrix $D\{\alpha_i\,|\,\underset{\sim}{0}\}$. The kjth block of this matrix is given by

$$D_{kj}\{\alpha_i\,|\,\underset{\sim}{0}\} = D_{11}\{\beta\,|\,\underset{\sim}{0}\} = D_{11}\{E\,|\,\underset{\sim}{0}\},$$

provided the relation (4.6.13) is respected,

$$\{\alpha_i\,|\,\underset{\sim}{0}\}\{\alpha_j\,|\,\underset{\sim}{0}\} = \{\alpha_k\,|\,\underset{\sim}{0}\}\{E\,|\,\underset{\sim}{0}\}.$$

This means that the only element in the j-th column that is different from zero is the k-th where k is given by

$$\alpha_i\alpha_j = \alpha_k.$$

The products $\alpha_i\alpha_j$ are now listed:

$$\begin{array}{llll}
\alpha_1\alpha_1 = \alpha_1 & \alpha_2\alpha_1 = \alpha_2 & \alpha_3\alpha_1 = \alpha_3 & \alpha_4\alpha_1 = \alpha_4 \\
\alpha_1\alpha_2 = \alpha_2 & \alpha_2\alpha_2 = \alpha_3 & \alpha_3\alpha_2 = \alpha_4 & \alpha_4\alpha_2 = \alpha_1 \\
\alpha_1\alpha_3 = \alpha_3 & \alpha_2\alpha_3 = \alpha_4 & \alpha_3\alpha_3 = \alpha_1 & \alpha_4\alpha_3 = \alpha_2 \\
\alpha_1\alpha_4 = \alpha_4 & \alpha_2\alpha_4 = \alpha_1 & \alpha_3\alpha_4 = \alpha_2 & \alpha_4\alpha_4 = \alpha_3 \\
\alpha_1\alpha_5 = \alpha_5 & \alpha_2\alpha_5 = \alpha_8 & \alpha_3\alpha_5 = \alpha_7 & \alpha_4\alpha_5 = \alpha_6 \\
\alpha_1\alpha_6 = \alpha_6 & \alpha_2\alpha_6 = \alpha_5 & \alpha_3\alpha_6 = \alpha_8 & \alpha_4\alpha_6 = \alpha_7 \\
\alpha_1\alpha_7 = \alpha_7 & \alpha_2\alpha_7 = \alpha_6 & \alpha_3\alpha_7 = \alpha_5 & \alpha_4\alpha_7 = \alpha_8 \\
\alpha_1\alpha_8 = \alpha_8 & \alpha_2\alpha_8 = \alpha_7 & \alpha_3\alpha_8 = \alpha_6 & \alpha_4\alpha_8 = \alpha_5 \\
\alpha_5\alpha_1 = \alpha_5 & \alpha_6\alpha_1 = \alpha_6 & \alpha_7\alpha_1 = \alpha_7 & \alpha_8\alpha_1 = \alpha_8 \\
\alpha_5\alpha_2 = \alpha_6 & \alpha_6\alpha_2 = \alpha_7 & \alpha_7\alpha_2 = \alpha_8 & \alpha_8\alpha_2 = \alpha_5 \\
\alpha_5\alpha_3 = \alpha_7 & \alpha_6\alpha_3 = \alpha_8 & \alpha_7\alpha_3 = \alpha_5 & \alpha_8\alpha_3 = \alpha_6 \\
\alpha_5\alpha_4 = \alpha_8 & \alpha_6\alpha_4 = \alpha_5 & \alpha_7\alpha_4 = \alpha_6 & \alpha_8\alpha_4 = \alpha_7 \\
\alpha_5\alpha_5 = \alpha_1 & \alpha_6\alpha_5 = \alpha_4 & \alpha_7\alpha_5 = \alpha_3 & \alpha_8\alpha_5 = \alpha_2 \\
\alpha_5\alpha_6 = \alpha_2 & \alpha_6\alpha_6 = \alpha_1 & \alpha_7\alpha_6 = \alpha_4 & \alpha_8\alpha_6 = \alpha_3 \\
\alpha_5\alpha_7 = \alpha_3 & \alpha_6\alpha_7 = \alpha_2 & \alpha_7\alpha_7 = \alpha_1 & \alpha_8\alpha_7 = \alpha_4 \\
\alpha_5\alpha_8 = \alpha_4 & \alpha_6\alpha_8 = \alpha_3 & \alpha_7\alpha_8 = \alpha_2 & \alpha_8\alpha_8 = \alpha_1
\end{array}$$

The matrices for all the operations $\{\alpha_i \,|\, \underset{\sim}{0}\}$ are given in Table 4.7. This representation actually corresponds to the so-called *regular representation* of the point group C_{4v}.

The matrix for the generic operation $\{\alpha \,|\, \underset{\sim}{T}_n\}$ can be obtained by the product

$$\underset{\sim}{D}\{\alpha|\underset{\sim}{T}_n\} = \underset{\sim}{D}\{E|\underset{\sim}{T}_n\}\underset{\sim}{D}\{\alpha|\underset{\sim}{0}\}.$$

Points on OB. $\underset{\sim}{k} \equiv (k_x, 0)$

For the points on OB the group *K* consists of the operations $\{E \mid \underset{\sim}{T}_n\}$ and $\{\sigma_1 \mid \underset{\sim}{T}_n\}$. The point group $G_o(\underset{\sim}{k})$ consists of the operations E and σ_1 and can be identified as C_{1h}; its representations

Table 4.7. Irreducible Representations of the Space Group c^1_{4v} for the Points A of the Brillouin Zone $[\underset{\sim}{k} \equiv (k_x, k_y)]$.

$\{E\vert\underset{\sim}{0}\}$	$\{C_4\vert\underset{\sim}{0}\}$	$\{C_2\vert\underset{\sim}{0}\}$	$\{C_4{}^3\vert\underset{\sim}{0}\}$
$\begin{pmatrix} 10000000 \\ 01000000 \\ 00100000 \\ 00010000 \\ 00001000 \\ 00000100 \\ 00000010 \\ 00000001 \end{pmatrix}$	$\begin{pmatrix} 0001 & \\ 1000 & \bigcirc \\ 0100 & \\ 0010 & \\ & 0100 \\ \bigcirc & 0010 \\ & 0001 \\ & 1000 \end{pmatrix}$	$\begin{pmatrix} 0010 & \\ 0001 & \bigcirc \\ 1000 & \\ 0100 & \\ & 0010 \\ \bigcirc & 0001 \\ & 1000 \\ & 0100 \end{pmatrix}$	$\begin{pmatrix} 0100 & \\ 0010 & \bigcirc \\ 0001 & \\ 1000 & \\ & 0001 \\ \bigcirc & 1000 \\ & 0100 \\ & 0010 \end{pmatrix}$
$\{\sigma_1 \mid \underset{\sim}{0}\}$	$\{\sigma_2 \mid \underset{\sim}{0}\}$	$\{\sigma_3 \mid \underset{\sim}{0}\}$	$\{\sigma_4 \mid \underset{\sim}{0}\}$
$\begin{pmatrix} & 1000 \\ \bigcirc & 0100 \\ & 0010 \\ & 0001 \\ 1000 & \\ 0100 & \bigcirc \\ 0010 & \\ 0001 & \end{pmatrix}$	$\begin{pmatrix} & 0100 \\ \bigcirc & 0010 \\ & 0001 \\ & 1000 \\ 0001 & \\ 1000 & \bigcirc \\ 0100 & \\ 0010 & \end{pmatrix}$	$\begin{pmatrix} & 0010 \\ \bigcirc & 0001 \\ & 1000 \\ & 0100 \\ 0010 & \\ 0001 & \bigcirc \\ 1000 & \\ 0100 & \end{pmatrix}$	$\begin{pmatrix} & 0001 \\ \bigcirc & 1000 \\ & 0100 \\ & 0010 \\ 0100 & \\ 0010 & \bigcirc \\ 0001 & \\ 1000 & \end{pmatrix}$

are given by

C_{1h}	E	σ_1
Γ_1	1	1
Γ_2	1	-1

We have also

$$\underset{\sim}{D}_{11}\{\beta \mid \underset{\sim}{b}\} = e^{i\underset{\sim}{k}\cdot\underset{\sim}{b}}\underset{\sim}{\Gamma}_m(\beta) = e^{in_1 ak_x}\underset{\sim}{\Gamma}_m(\beta),$$

where $m = 1, 2$ and $\beta = E, \sigma_1$ and $\underset{\sim}{b} = n_1\underset{\sim}{\alpha}_1 + n_2\underset{\sim}{\alpha}_2$. The star $\{\underset{\sim}{k}\}$ has $q = 4$ arms, as shown in Fig. 4.12 and therefore is degenerate.

The group G can be divided into left cosets with respect to K:

$$\begin{aligned} G &= \{\alpha_1|\underset{\sim}{0}\}K + \{\alpha_2|\underset{\sim}{0}\}K + \{\alpha_3 \mid \underset{\sim}{0}\}K + \{\alpha_4|\underset{\sim}{0}\}K \\ &= \{E|\underset{\sim}{0}\}K + \{C_4|\underset{\sim}{0}\}K + \{C_2|\underset{\sim}{0}\}K + \{C_4{}^3|\underset{\sim}{0}\}K. \end{aligned}$$

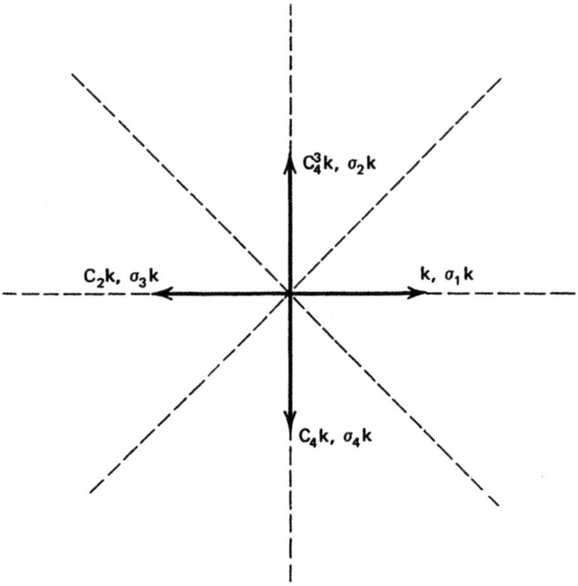

Fig. 4.12. The star $\{\underset{\sim}{k}\}$ at points on OB in the Brillouin zone of Fig. 4.10.

The matrix for a primitive translation is given according to (4.6.11) by:

$$\underset{\sim}{D}\{E|\underset{\sim}{T}_n\} = \begin{pmatrix} e^{i\underset{\sim}{k}\cdot\underset{\sim}{T}_n} & & & \bigcirc \\ & e^{i\alpha_2\underset{\sim}{k}\cdot\underset{\sim}{T}_n} & & \\ & & e^{i\alpha_3\underset{\sim}{k}\cdot\underset{\sim}{T}_n} & \\ \bigcirc & & & e^{i\alpha_4\underset{\sim}{k}\cdot\underset{\sim}{T}_n} \end{pmatrix}$$

Considering the matrix $\underset{\sim}{D}\{\alpha_i|\underset{\sim}{0}\}$, its kjth block is given by

$$\underset{\sim}{D}_{kj}\{\alpha_i|\underset{\sim}{0}\} = \underset{\sim}{D}_{11}\{\beta|\underset{\sim}{0}\} = \underset{\sim}{\Gamma}_m(\beta),$$

where β is given by

$$\{\alpha_i|\underset{\sim}{0}\}\{\alpha_j|\underset{\sim}{0}\} = \{\alpha_k|\underset{\sim}{0}\}\{\beta|\underset{\sim}{0}\},$$

which is, by

$$\alpha_i\alpha_j = \alpha_k\beta,$$

where

$$j, k = 1, 2, 3, 4,$$
$$\beta = \alpha_1, \alpha_5.$$

Let us consider $\alpha_2 = C_4$, by letting $i = 2$:

$$\begin{aligned} \alpha_2\alpha_1 &= \alpha_2 = \alpha_2\alpha_1, \\ \alpha_2\alpha_2 &= \alpha_3 = \alpha_3\alpha_1, \\ &\qquad\qquad\qquad \beta = \alpha_1 = E \\ \alpha_2\alpha_3 &= \alpha_4 = \alpha_4\alpha_1, \\ \alpha_2\alpha_4 &= \alpha_1 = \alpha_1\alpha_1. \end{aligned}$$

Therefore the only element different from zero and equal to $\Gamma_m(E) = 1$ is in the first column, the second; in the second column, the third; in the third column, the fourth; and in the fourth column, the first:

$$\underset{\sim}{D}\{C_4|\underset{\sim}{0}\} = \begin{pmatrix} 0 & 0 & 0 & 1 \\ 1 & 0 & 0 & 0 \\ 0 & 1 & 0 & 0 \\ 0 & 0 & 1 & 0 \end{pmatrix}.$$

Let us consider $\alpha_3 = C_2$ by letting $i = 3$. We obtain

$$\begin{aligned} \alpha_3\alpha_1 &= \alpha_3 = \alpha_3\alpha_1, \\ \alpha_3\alpha_2 &= \alpha_4 = \alpha_4\alpha_1, \\ &\qquad\qquad\qquad \beta = \alpha_1 = E \\ \alpha_3\alpha_3 &= \alpha_1 = \alpha_1\alpha_1, \\ \alpha_3\alpha_4 &= \alpha_2 = \alpha_2\alpha_1, \end{aligned}$$

and

$$\underset{\sim}{D}\{C_2|\underset{\sim}{0}\} = \begin{pmatrix} 0 & 0 & 1 & 0 \\ 0 & 0 & 0 & 1 \\ 1 & 0 & 0 & 0 \\ 0 & 1 & 0 & 0 \end{pmatrix}.$$

Let us consider $\alpha_4 = C_4{}^3$ by letting $i = 4$. We obtain

$$\alpha_4\alpha_1 = \alpha_4 = \alpha_4\alpha_1,$$
$$\alpha_4\alpha_2 = \alpha_1 = \alpha_1\alpha_1,$$
$$\beta = \alpha_1 = E$$
$$\alpha_4\alpha_3 = \alpha_2 = \alpha_2\alpha_1,$$
$$\alpha_4\alpha_4 = \alpha_3 = \alpha_3\alpha_1,$$

and

$$\underset{\sim}{D}\{C_4{}^3|\underset{\sim}{0}\} = \begin{pmatrix} 0 & 1 & 0 & 0 \\ 0 & 0 & 1 & 0 \\ 0 & 0 & 0 & 1 \\ 1 & 0 & 0 & 0 \end{pmatrix}.$$

Let us consider $\alpha_5 = \sigma_1$ by letting $i = 5$. We obtain

$$\alpha_5\alpha_1 = \alpha_5 = \alpha_1\alpha_5,$$
$$\alpha_5\alpha_2 = \alpha_6 = \alpha_4\alpha_5,$$
$$\beta = \alpha_5 = \sigma_1$$
$$\alpha_5\alpha_3 = \alpha_7 = \alpha_3\alpha_5,$$
$$\alpha_5\alpha_4 = \alpha_8 = \alpha_2\alpha_5,$$

and

$$\underset{\sim}{D}\{\sigma_1|\underset{\sim}{0}\} = \begin{pmatrix} \pm 1 & 0 & 0 & 0 \\ 0 & 0 & 0 & \pm 1 \\ 0 & 0 & \pm 1 & 0 \\ 0 & \pm 1 & 0 & 0 \end{pmatrix}.$$

Let us consider $\alpha_6 = \sigma_2$ by letting $i = 6$. We obtain

$$\alpha_6\alpha_1 = \alpha_6 = \alpha_4\alpha_5,$$
$$\alpha_6\alpha_2 = \alpha_7 = \alpha_3\alpha_5,$$
$$\beta = \alpha_5 = \sigma_1$$
$$\alpha_6\alpha_3 = \alpha_8 = \alpha_2\alpha_5,$$
$$\alpha_6\alpha_4 = \alpha_5 = \alpha_1\alpha_5,$$

and

$$\underset{\sim}{D}\{\sigma_2|\underset{\sim}{0}\} = \begin{pmatrix} 0 & 0 & 0 & \pm 1 \\ 0 & 0 & \pm 1 & 0 \\ 0 & \pm 1 & 0 & 0 \\ \pm 1 & 0 & 0 & 0 \end{pmatrix}.$$

Let us consider $\alpha_7 = \sigma_3$ by letting $i = 7$. We obtain

$$\begin{aligned} \alpha_7\alpha_1 &= \alpha_7 = \alpha_3\alpha_5, \\ \alpha_7\alpha_2 &= \alpha_8 = \alpha_2\alpha_5, \\ &\qquad\qquad \beta = \alpha_5 = \sigma_1 \\ \alpha_7\alpha_3 &= \alpha_5 = \alpha_1\alpha_5, \\ \alpha_7\alpha_4 &= \alpha_6 = \alpha_4\alpha_5, \end{aligned}$$

and

$$\underset{\sim}{D}\{\sigma_3|\underset{\sim}{0}\} = \begin{pmatrix} 0 & 0 & \pm 1 & 0 \\ 0 & \pm 1 & 0 & 0 \\ \pm 1 & 0 & 0 & 0 \\ 0 & 0 & 0 & \pm 1 \end{pmatrix}.$$

Let us consider $\alpha_8 = \sigma_4$ by letting $i = 8$. We obtain

$$\begin{aligned} \alpha_8\alpha_1 &= \alpha_8 = \alpha_2\alpha_5, \\ \alpha_8\alpha_2 &= \alpha_5 = \alpha_1\alpha_5, \\ &\qquad\qquad \beta = \alpha_5 = \sigma_1 \\ \alpha_8\alpha_3 &= \alpha_6 = \alpha_4\alpha_5, \\ \alpha_8\alpha_4 &= \alpha_7 = \alpha_3\alpha_5, \end{aligned}$$

and

$$\underset{\sim}{D}\{\sigma_4|\underset{\sim}{0}\} = \begin{pmatrix} 0 & \pm 1 & 0 & 0 \\ \pm 1 & 0 & 0 & 0 \\ 0 & 0 & 0 & \pm 1 \\ 0 & 0 & \pm 1 & 0 \end{pmatrix}.$$

The irreducible representations of the space group in this region of the Brillouin zone are given in Table 4.8.

Table 4.8. Irreducible Representations of the Space Group of C_{4v}^1 in Correspondence to the Points OB of the Brillouin Zone $[\underset{\sim}{k} \equiv (k_x, 0)]$.

	$\{E\|\underset{\sim}{T}_n\}$	$\{C_4\|\underset{\sim}{0}\}$	$\{C_2\|\underset{\sim}{0}\}$	$\{C_4{}^3\|\underset{\sim}{0}\}$	$\{\sigma_1\|\underset{\sim}{0}\}$	$\{\sigma_2\|\underset{\sim}{0}\}$	$\{\sigma_3\|\underset{\sim}{0}\}$	$\{\sigma_4\|\underset{\sim}{0}\}$
Γ_1	$\begin{pmatrix} e^{i\underset{\sim}{k}\cdot\underset{\sim}{T}_n} & & & \bigcirc \\ & e^{iC_4\underset{\sim}{k}\cdot\underset{\sim}{T}_n} & & \\ & & e^{iC_2\underset{\sim}{k}\cdot\underset{\sim}{T}_n} & \\ \bigcirc & & & e^{iC_4{}^3\underset{\sim}{k}\cdot\underset{\sim}{T}_n} \end{pmatrix}$	$\begin{pmatrix} 0001 \\ 1000 \\ 0100 \\ 0010 \end{pmatrix}$	$\begin{pmatrix} 0010 \\ 0001 \\ 1000 \\ 0100 \end{pmatrix}$	$\begin{pmatrix} 0100 \\ 0010 \\ 0001 \\ 1000 \end{pmatrix}$	$\begin{pmatrix} 1&0&0&0 \\ 0&0&0&1 \\ 0&0&1&0 \\ 0&1&0&0 \end{pmatrix}$	$\begin{pmatrix} 0&0&0&1 \\ 0&0&1&0 \\ 0&1&0&0 \\ 1&0&0&0 \end{pmatrix}$	$\begin{pmatrix} 0&0&1&0 \\ 0&1&0&0 \\ 1&0&0&0 \\ 0&0&0&1 \end{pmatrix}$	$\begin{pmatrix} 0&1&0&0 \\ 1&0&0&0 \\ 0&0&0&1 \\ 0&0&1&0 \end{pmatrix}$
Γ_2	$\begin{pmatrix} e^{i\underset{\sim}{k}\cdot\underset{\sim}{T}_n} & & & \bigcirc \\ & e^{iC_4\underset{\sim}{k}\cdot\underset{\sim}{T}_n} & & \\ & & e^{iC_2\underset{\sim}{k}\cdot\underset{\sim}{T}_n} & \\ \bigcirc & & & e^{iC_4{}^3\underset{\sim}{k}\cdot\underset{\sim}{T}_n} \end{pmatrix}$	$\begin{pmatrix} 0001 \\ 1000 \\ 0100 \\ 0010 \end{pmatrix}$	$\begin{pmatrix} 0010 \\ 0001 \\ 1000 \\ 0100 \end{pmatrix}$	$\begin{pmatrix} 0100 \\ 0010 \\ 0001 \\ 1000 \end{pmatrix}$	$\begin{pmatrix} -1&0&0&0 \\ 0&0&0&-1 \\ 0&0&-1&0 \\ 0&-1&0&0 \end{pmatrix}$	$\begin{pmatrix} 0&0&0&-1 \\ 0&0&-1&0 \\ 0&-1&0&0 \\ -1&0&0&0 \end{pmatrix}$	$\begin{pmatrix} 0&0&-1&0 \\ 0&-1&0&0 \\ -1&0&0&0 \\ 0&0&0&-1 \end{pmatrix}$	$\begin{pmatrix} 0&-1&0&0 \\ -1&0&0&0 \\ 0&0&0&-1 \\ 0&0&-1&0 \end{pmatrix}$

POINTS ON OC. $\underset{\sim}{k} \equiv (k_x, k_y = k_x)$

For the points on OC the group K consists of the operations $\{E|\underset{\sim}{T}_n\}$ and $\{\sigma_2|\underset{\sim}{T}_n\}$. The point group $G_o(\underset{\sim}{k})$ consists of the operations E and σ_2:

$$\beta = E,\ \sigma_2.$$

We can work out the irreducible representations for the points in this region in a way similar to that followed for the points on OB, taking into account the fact that now σ_2 replaces σ_1.

POINTS ON BC. $\underset{\sim}{k} \equiv (k_x, \pi/a)$

The operation σ_3 sends a point in BC into a point in DE; the two points differ by a vector of the reciprocal lattice and therefore correspond to the same representation. Therefore, the group K consists of $\{E|\underset{\sim}{T}_n\}$ and $\{\sigma_3|\underset{\sim}{T}_n\}$ and $G_o(\underset{\sim}{k})$ of E and σ_3. The representations can then be obtained following the procedure outlined for the points on OB.

POINT B. $\underset{\sim}{k} \equiv (\pi/a, 0)$

This point is sent into itself by the operations E and σ_1 and into another point that differs from B by a primitive vector of the reciprocal space by the operations C_2 and σ_3. Therefore the group $G_o(\underset{\sim}{k})$ consists of the operations E, C_2, σ_1, and σ_3 and can be identified as C_{2v}; its representations are given by

C_{2v}	E	C_2	σ_1	σ_3
Γ_1	1	1	1	1
Γ_2	1	1	−1	−1
Γ_3	1	−1	1	−1
Γ_4	1	−1	−1	1

We have also

$$\underset{\sim}{D}_{11}\{\beta|\underset{\sim}{b}\} = e^{i\underset{\sim}{k}\cdot\underset{\sim}{b}}\underset{\sim}{\Gamma}_\gamma(\beta) = \pm\underset{\sim}{\Gamma}_\gamma(\beta),$$

where $\gamma = 1, 2, 3, 4$ and $\beta = E, C_2, \sigma_1$ and σ_3. Also $\underset{\sim}{b} = n_1\underset{\sim}{a}_1 + n_2\underset{\sim}{a}_2$ and the plus sign corresponds to n_1 even and the minus sign to n_1 odd. The star $\{\underset{\sim}{k}\}$ is degenerate and has two arms as shown in Fig. 4.13.

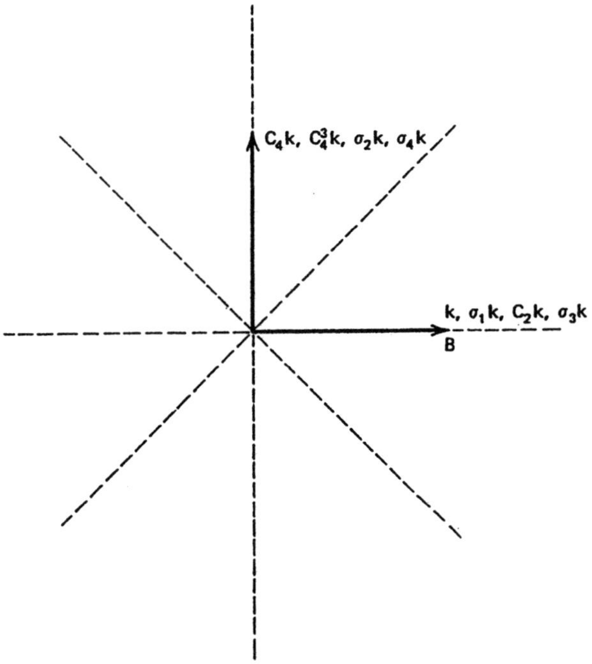

Fig. 4.13. The star $\{\underset{\sim}{k}\}$ at point B in the Brillouin zone of Fig. 4.10.

The group G can be decomposed into left cosets with respect to K:

$$\begin{aligned} G &= \{\alpha_1|\underset{\sim}{0}\}K + \{\alpha_2|\underset{\sim}{0}\}K \\ &= \{E|\underset{\sim}{0}\}K + \{C_4|\underset{\sim}{0}\}K. \end{aligned}$$

The matrix for a primitive translation is given, according to (4.6.11) by

$$\underset{\sim}{D}\{E|\underset{\sim}{T}_n\} = \begin{pmatrix} e^{i\underset{\sim}{k}\cdot\underset{\sim}{T}_n} & 0 \\ 0 & e^{i\alpha_2\underset{\sim}{k}\cdot\underset{\sim}{T}_n} \end{pmatrix}.$$

Considering the matrix $\underset{\sim}{D}\{\alpha_i|\underset{\sim}{0}\}$, its k-jth block is given by

$$D_{kj}\{\alpha_i|\underset{\sim}{0}\} = D_{11}\{\beta|\underset{\sim}{0}\} = \underset{\sim}{\Gamma}_\gamma(\beta),$$

where β is given

$$\{\alpha_i|\underset{\sim}{0}\}\{\alpha_j|\underset{\sim}{0}\} = \{\alpha_k|\underset{\sim}{0}\}\{\beta|\underset{\sim}{0}\},$$

which is, by

$$\alpha_i\alpha_j = \alpha_k\beta,$$

where

$$j, k = 1, 2,$$
$$\beta = \alpha_1, \alpha_3, \alpha_5, \alpha_7.$$

Let us consider $\alpha_2 = C_4$, by setting $i = 2$. We obtain

$$\begin{aligned} \alpha_2\alpha_1 &= \alpha_2 = \alpha_2\alpha_1, \quad \beta = \alpha_1 = E, \\ \alpha_2\alpha_2 &= \alpha_3 = \alpha_1\alpha_3, \quad \beta = \alpha_3 = C_2, \end{aligned}$$

and

$$\underset{\sim}{D}\{C_4|\underset{\sim}{0}\} = \begin{pmatrix} 0 & \Gamma_\gamma(C_2) \\ \Gamma_\gamma(E) & 0 \end{pmatrix} = \begin{pmatrix} 0 & \pm 1 \\ 1 & 0 \end{pmatrix},$$

where the plus sign has to be used for $\gamma = 1, 2$ and the minus sign for $\gamma = 3, 4$.

Let us consider $\alpha_3 = C_2$ by setting $i = 3$. We obtain

$$\left.\begin{aligned} \alpha_3\alpha_1 &= \alpha_3 = \alpha_1\alpha_3, \\ \alpha_3\alpha_2 &= \alpha_4 = \alpha_2\alpha_3, \end{aligned}\right. \quad \beta = \alpha_3 = C_2$$

and

$$\underset{\sim}{D}\{C_2|\underset{\sim}{0}\} = \begin{pmatrix} \Gamma_\gamma(C_2) & 0 \\ 0 & \Gamma_\gamma(C_2) \end{pmatrix} = \begin{pmatrix} \pm 1 & 0 \\ 0 & \pm 1 \end{pmatrix},$$

where the plus sign has to be used for $\gamma = 1, 2$ and the minus sign for $\gamma = 3, 4$.

Let us consider $\alpha_4 = {C_4}^3$ by setting $i = 4$. We obtain

$$\begin{aligned} \alpha_4\alpha_1 &= \alpha_4 = \alpha_2\alpha_3 \quad \beta = \alpha_3 = C_2, \\ \alpha_4\alpha_2 &= \alpha_1 = \alpha_1\alpha_1 \quad \beta = \alpha_1 = E, \end{aligned}$$

and

$$\underset{\sim}{D}\{C_4{}^3|\underset{\sim}{0}\} = \begin{pmatrix} 0 & \Gamma_\gamma(E) \\ \Gamma_\gamma(C_2) & 0 \end{pmatrix} = \begin{pmatrix} 0 & 1 \\ \pm 1 & 0 \end{pmatrix},$$

where the plus sign has to be used for $\gamma = 1, 2$ and the minus sign for $\gamma = 3, 4$.

Let us consider $\alpha_5 = \sigma_1$ by setting $i = 5$. We obtain

$$\alpha_5\alpha_2 = \alpha_5 = \alpha_1\alpha_5, \quad \beta = \alpha_5 = \sigma_1,$$
$$\alpha_5\alpha_2 = \alpha_6 = \alpha_2\alpha_7, \quad \beta = \alpha_7 = \sigma_3,$$

and

$$\underset{\sim}{D}\{\sigma_1|\underset{\sim}{0}\} = \begin{pmatrix} \Gamma_\gamma(\sigma_1) & 0 \\ 0 & \Gamma_\gamma(\sigma_3) \end{pmatrix},$$

where $\Gamma_\gamma(\sigma_1) = 1$ for $\gamma = 1, 3$ and -1 for $\gamma = 2, 4$, and $\Gamma_\gamma(\sigma_3) = 1$ for $\gamma = 1, 4$, and -1 for $\gamma = 2, 3$

Let us consider $\alpha_6 = \sigma_2$ by setting $i = 6$. We obtain

$$\left.\begin{array}{l} \alpha_6\alpha_1 = \alpha_6 = \alpha_2\alpha_7, \\ \alpha_6\alpha_2 = \alpha_7 = \alpha_1\alpha_7, \end{array}\right. \quad \beta = \alpha_7 = \sigma_3$$

and

$$\underset{\sim}{D}\{\sigma_2|\underset{\sim}{0}\} = \begin{pmatrix} 0 & \Gamma_\gamma(\sigma_3) \\ \Gamma_\gamma(\sigma_3) & 0 \end{pmatrix} = \begin{pmatrix} 0 & \pm 1 \\ \pm 1 & 0 \end{pmatrix},$$

where the plus sign has to be used for $\gamma = 1, 4$ and the minus sign for $\gamma = 2, 3$.

Let us consider $\alpha_7 = \sigma_3$ by setting $i = 7$. We obtain

$$\alpha_7\alpha_1 = \alpha_7 = \alpha_1\alpha_7, \quad \beta = \alpha_7 = \sigma_3,$$
$$\alpha_7\alpha_2 = \alpha_8 = \alpha_2\alpha_5, \quad \beta = \alpha_5 = \sigma_1,$$

and

$$\underset{\sim}{D}\{\sigma_3|\underset{\sim}{0}\} = \begin{pmatrix} \Gamma_\gamma(\sigma_3) & 0 \\ 0 & \Gamma_\gamma(\sigma_1) \end{pmatrix},$$

where $\Gamma_\gamma(\sigma_3) = 1$ for $\gamma = 1, 4$ and -1 for $\gamma = 2, 3$ and $\gamma_\gamma(\sigma_1) = 1$ for $\gamma = 1, 3$ and -1 for $\gamma = 2, 4$.

Let us consider $\alpha_8 = \sigma_4$ by setting $i = 8$. We obtain

$$\left.\begin{array}{l}\alpha_8\alpha_1 = \alpha_8 = \alpha_2\alpha_5, \\ \alpha_8\alpha_2 = \alpha_5 = \alpha_1\alpha_5,\end{array}\right. \qquad \beta = \alpha_5 = \sigma_1$$

and

$$\underset{\sim}{D}\{\sigma_4|\underset{\sim}{0}\} = \begin{pmatrix} 0 & \Gamma_\gamma(\sigma_1) \\ \Gamma_\gamma(\sigma_1) & 0 \end{pmatrix} = \begin{pmatrix} 0 & \pm 1 \\ \pm 1 & 0 \end{pmatrix},$$

where the plus sign has to be used for $\gamma = 1, 3$ and the minus sign for $\gamma = 2, 4$.

The matrix for the generic operation $\{\alpha_i|\underset{\sim}{T}_n\}$ can be obtained by the product

$$\underset{\sim}{D}\{E|\underset{\sim}{T}_n\}\underset{\sim}{D}\{\alpha_i|\underset{\sim}{0}\} = \underset{\sim}{D}\{\alpha_i|\underset{\sim}{T}_n\}.$$

The irreducible representations for the space group in correspondence to the point B of the Brillouin zone are given in Table 4.9.

POINT C. $\underset{\sim}{k} \equiv (\pi/a, \pi/a)$

The operations of the point group C_{4v} send this point into itself or into another point that differs from C by a primitive vector of the reciprocal lattice. Therefore $K = G$ and $G_o(\underset{\sim}{k}) = G_o = C_{4v}$:

$$G = \{E|\underset{\sim}{0}\}K,$$

and the irreducible representations are given by

$$\underset{\sim}{D}\{\beta|\underset{\sim}{T}_n\} = D_{11}\{\beta|\underset{\sim}{T}_n\} = e^{i\underset{\sim}{k}\cdot\underset{\sim}{T}_n}\underset{\sim}{\Gamma}_\gamma(\beta) = \pm\underset{\sim}{\Gamma}_\gamma(\beta),$$

where $\underset{\sim}{\Gamma}_\gamma(\beta)$ are the irreducible representations of C_{4v}. Also $\underset{\sim}{T}_n = n_1\underset{\sim}{a}_1 + n_2\underset{\sim}{a}_2$ and the plus sign corresponds to $(n_1 + n_2)$ even and the minus sign to $(n_1 + n_2)$ odd.

The representations for the present example are summarized in Table 4.10.

Table 4.9. Irreducible Representations of the Space Group C^1_{4v} in Correspondence to the Point B of the Brillouin Zone $[\underset{\sim}{k} \equiv (\pi/a, 0)]$.

	$\{E\|\underset{\sim}{T}_n\}$	$\{C_4\|\underset{\sim}{0}\}$	$\{C_2\|\underset{\sim}{0}\}$	$\{C_4{}^3\|\underset{\sim}{0}\}$	$\{\sigma_1\|\underset{\sim}{0}\}$	$\{\sigma_2\|\underset{\sim}{0}\}$	$\{\sigma_3\|\underset{\sim}{0}\}$	$\{\sigma_4\|\underset{\sim}{0}\}$
Γ_1	$\begin{pmatrix} e^{i\underset{\sim}{k}\cdot\underset{\sim}{T}_n} & 0 \\ 0 & e^{iC_4\underset{\sim}{k}\cdot\underset{\sim}{T}_n} \end{pmatrix}$	$\begin{pmatrix} 0 & 1 \\ 1 & 0 \end{pmatrix}$	$\begin{pmatrix} 1 & 0 \\ 0 & 1 \end{pmatrix}$	$\begin{pmatrix} 0 & 1 \\ 1 & 0 \end{pmatrix}$	$\begin{pmatrix} 1 & 0 \\ 0 & 1 \end{pmatrix}$	$\begin{pmatrix} 0 & 1 \\ 1 & 0 \end{pmatrix}$	$\begin{pmatrix} 1 & 0 \\ 0 & 1 \end{pmatrix}$	$\begin{pmatrix} 0 & 1 \\ 1 & 0 \end{pmatrix}$
Γ_2	$\begin{pmatrix} e^{i\underset{\sim}{k}\cdot\underset{\sim}{T}_n} & 0 \\ 0 & e^{iC_4\underset{\sim}{k}\cdot\underset{\sim}{T}_n} \end{pmatrix}$	$\begin{pmatrix} 0 & 1 \\ 1 & 0 \end{pmatrix}$	$\begin{pmatrix} 1 & 0 \\ 0 & 1 \end{pmatrix}$	$\begin{pmatrix} 0 & 1 \\ 1 & 0 \end{pmatrix}$	$\begin{pmatrix} -1 & 0 \\ 0 & -1 \end{pmatrix}$	$\begin{pmatrix} 0 & -1 \\ -1 & 0 \end{pmatrix}$	$\begin{pmatrix} -1 & 0 \\ 0 & -1 \end{pmatrix}$	$\begin{pmatrix} 0 & -1 \\ -1 & 0 \end{pmatrix}$
Γ_3	$\begin{pmatrix} e^{i\underset{\sim}{k}\cdot\underset{\sim}{T}_n} & 0 \\ 0 & e^{iC_4\underset{\sim}{k}\cdot\underset{\sim}{T}_n} \end{pmatrix}$	$\begin{pmatrix} 0 & -1 \\ 1 & 0 \end{pmatrix}$	$\begin{pmatrix} -1 & 0 \\ 0 & -1 \end{pmatrix}$	$\begin{pmatrix} 0 & 1 \\ -1 & 0 \end{pmatrix}$	$\begin{pmatrix} 1 & 0 \\ 0 & -1 \end{pmatrix}$	$\begin{pmatrix} 0 & -1 \\ -1 & 0 \end{pmatrix}$	$\begin{pmatrix} -1 & 0 \\ 0 & 1 \end{pmatrix}$	$\begin{pmatrix} 0 & 1 \\ 1 & 0 \end{pmatrix}$
Γ_4	$\begin{pmatrix} e^{i\underset{\sim}{k}\cdot\underset{\sim}{T}_n} & 0 \\ 0 & e^{iC_4\underset{\sim}{k}\cdot\underset{\sim}{T}_n} \end{pmatrix}$	$\begin{pmatrix} 0 & -1 \\ 1 & 0 \end{pmatrix}$	$\begin{pmatrix} -1 & 0 \\ 0 & -1 \end{pmatrix}$	$\begin{pmatrix} 0 & 1 \\ -1 & 0 \end{pmatrix}$	$\begin{pmatrix} -1 & 0 \\ 0 & 1 \end{pmatrix}$	$\begin{pmatrix} 0 & 1 \\ 1 & 0 \end{pmatrix}$	$\begin{pmatrix} 1 & 0 \\ 0 & -1 \end{pmatrix}$	$\begin{pmatrix} 0 & -1 \\ -1 & 0 \end{pmatrix}$

Table 4.10. Summary of the Irreducible Representations of the Space Group C^1_{4v}.

Region	K	$G_o(\underset{\sim}{k})$	Number of Representations	Dimension of Representations	Observations
O	$K = G$	$G_o(\underset{\sim}{k}) = G_o = C_{4v}$	4	$1 \begin{cases} d = 1 \\ q = 1 \end{cases}$	$\underset{\sim}{D}\{\beta\|\underset{\sim}{T}_n\} = e^{i\underset{\sim}{k}\cdot\underset{\sim}{T}_n}\underset{\sim}{\Gamma}_\gamma(\beta)$
C	$K = G$	$G_o(\underset{\sim}{k}) = G_o = C_{4v}$	1	$2 \begin{cases} d = 2 \\ q = 1 \end{cases}$	$\underset{\sim}{\Gamma}_\gamma(\beta) =$ representations of C_{4v}
A	$K = T$	$G_o(\underset{\sim}{k}) = C_1{:}E$	1	$8 \begin{cases} d = 1 \\ q = 8 \end{cases}$	Regular representation for $\{\alpha\|\underset{\sim}{0}\}$ operations
OB	$K : \{E\|\underset{\sim}{T}_n\}, \{\sigma_1\|\underset{\sim}{T}_n\}$	$G_o(\underset{\sim}{k}) = C_{1h} : E, \sigma_1$	2	$4 \begin{cases} d = 1 \\ q = 4 \end{cases}$	
OC	$K : \{E\|\underset{\sim}{T}_n\}, \{\sigma_2\|\underset{\sim}{T}_n\}$	$G_o(\underset{\sim}{k}) = C_{1h} : E, \sigma_2$	2	$4 \begin{cases} d = 1 \\ q = 4 \end{cases}$	
BC	$K : \{E\|\underset{\sim}{T}_n\}, \{\sigma_3\|\underset{\sim}{T}_n\}$	$G_o(\underset{\sim}{k}) = C_{1h} : E, \sigma_3$	2	$4 \begin{cases} d = 1 \\ q = 4 \end{cases}$	
B	$K : \{E\|\underset{\sim}{T}_n\}, \{C_2\|\underset{\sim}{T}_n\},$ $\{\sigma_1\|\underset{\sim}{T}_n\}, \{\sigma_3\|\underset{\sim}{T}_n\}$	$G_o(\underset{\sim}{k}) = C_{2V}$: E, C_2, σ_1, σ_3	4	$2 \begin{cases} d = 1 \\ q = 2 \end{cases}$	

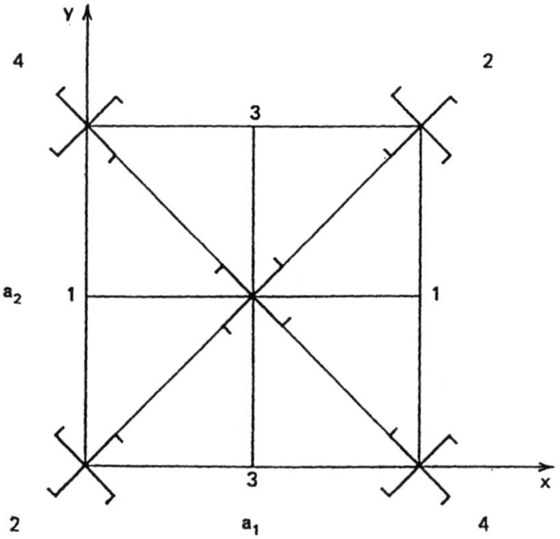

Fig. 4.14. Nonsymmorphic two-dimensional space group with square Bravais lattice and crystallographic point group C_{4v}.

4.8. Example II.* Nonsymmorphic Group C_{4v}^2

Let us now consider the nonsymmorphic space group represented in Fig. 4.14, which we designate C_{4v}^2. The unit cell is defined by the two basic vectors

$$\underset{\sim}{a}_1 = a\hat{x},$$
$$\underset{\sim}{a}_2 = a\hat{y}.$$

The crystallographic point group is C_{4v}. The Bravais lattice is square P and the Brillouin zone is also square (with C_{4v} symmetry) and is represented in Fig. 4.10.

The space group consists of the following operations:

$\{E|\underset{\sim}{0}\}$: identity
$\{C_4|\underset{\sim}{0}\}$, $\{C_2|\underset{\sim}{0}\}$, $\{C_4{}^3|\underset{\sim}{0}\}$: clockwise rotations through $\pi/2$, π, and $3\pi/2$, respectively
$\{\sigma_i|\underset{\sim}{A}\}$, with $i = 1, 2, 3, 4$: reflection through axis $i - i$
$\{E|\underset{\sim}{T}_n\}$: primitive translations

---

*This example was worked out by Dr. W. A. Wall.

products of primitive translations with operations $\{C_4|\underset{\sim}{0}\}$, $\{C_2|\underset{\sim}{0}\}$, $\{C_4{}^3|\underset{\sim}{0}\}$, $\{\sigma_i|\underset{\sim}{A}\}$, where

$$\underset{\sim}{A} = \frac{1}{2}(\underset{\sim}{a}_1 + \underset{\sim}{a}_2).$$

The multiplication table for these operations is given in Table 4.11.

In order to find the distinct irreducible representations of the space group we need to consider only the section OBC of the Brillouin zone in Fig. 4.10. Let us consider the different points of this section separately.

The method used for finding the irreducible representations within the OBC part of the Brillouin zone is the same as that used in the first example and consists of using the formula

$$\underset{\sim}{D}_{11}\{\beta|\underset{\sim}{b}\} = e^{i\underset{\sim}{k}\cdot\underset{\sim}{b}}\underset{\sim}{\Gamma}_j(\beta),$$

where $\{\beta|\underset{\sim}{b}\}$ is an element of the group K and $\underset{\sim}{\Gamma}_j(\beta)$ is an irreducible representation of the Group $G_o(\underset{\sim}{k})$.

POINT O. $\underset{\sim}{k} \equiv (0, 0)$

At this point the group K coincides with the entire space group because

$$e^{i\beta\underset{\sim}{k}\cdot\underset{\sim}{T}_n} = e^{i\underset{\sim}{k}\cdot\underset{\sim}{T}_n} = 1,$$

for all $\underset{\sim}{T}_n$. Also

$$G_o(\underset{\sim}{k}) = G_o = C_{4v} : E,\ C_4,\ C_2,\ C_4^3,\ \sigma_1,\ \sigma_2,\ \sigma_3,\ \sigma_4.$$

The generic matrix is given by

$$\underset{\sim}{D}\{\beta|\underset{\sim}{b}\} = \underset{\sim}{D}_{11}^{\{\beta|\underset{\sim}{b}\}} = \underset{\sim}{\Gamma}_j(\beta),$$

where $\underset{\sim}{\Gamma}_j(\beta)$ is an irreducible representation of C_{4v}. The matrices for the operations of the space group are given in Table 4.12.

POINTS A. $\underset{\sim}{k} \equiv (k_x, k_y)$

For a point A inside the Brillouin zone the group K coincides with the group T and

$$\underset{\sim}{D}_{11}{}^{\{\beta|\underset{\sim}{b}\}} = \underset{\sim}{D}_{11}{}^{\{E|\underset{\sim}{T}_n\}} = e^{i\underset{\sim}{k}\cdot\underset{\sim}{T}_n}.$$

Table 4.11. Multiplication Table of the Elements of the Space Group C_{4v}^2.

	$\{E\|\mathbf{o}\}$	$\{C_4\|\mathbf{o}\}$	$\{C_4\|\mathbf{o}\}$	$\{C_4{}^3\|\mathbf{o}\}$	$\{\sigma_1\|\mathbf{A}\}$	$\{\sigma_2\|\mathbf{A}\}$	$\{\sigma_3\|\mathbf{A}\}$	$\{\sigma_4\|\mathbf{A}\}$
$\{E\|\mathbf{o}\}$	$\{E\|\mathbf{o}\}$	$\{C_4\|\mathbf{o}\}$	$\{C_2\|\mathbf{o}\}$	$\{C_4{}^3\|\mathbf{o}\}$	$\{\sigma_1\|\mathbf{A}\}$	$\{\sigma_2\|\mathbf{A}\}$	$\{\sigma_3\|\mathbf{A}\}$	$\{\sigma_4\|\mathbf{A}\}$
$\{C_4\|\mathbf{o}\}$	$\{C_4\|\mathbf{o}\}$	$\{C_2\|\mathbf{o}\}$	$\{C_4{}^3\|\mathbf{o}\}$	$\{E\|\mathbf{o}\}$	$\{\sigma_4\|\mathbf{A}-\mathbf{a}_2\}$	$\{\sigma_1\|\mathbf{A}-\mathbf{a}_2\}$	$\{\sigma_2\|\mathbf{A}-\mathbf{a}_2\}$	$\{\sigma_3\|\mathbf{A}-\mathbf{a}_2\}$
$\{C_2\|\mathbf{o}\}$	$\{C_2\|\mathbf{o}\}$	$\{C_4{}^3\|\mathbf{o}\}$	$\{E\|\mathbf{o}\}$	$\{C_4\|\mathbf{o}\}$	$\{\sigma_3\|-\mathbf{A}\}$	$\{\sigma_4\|-\mathbf{A}\}$	$\{\sigma_1\|-\mathbf{A}\}$	$\{\sigma_2\|-\mathbf{A}\}$
$\{C_4{}^3\|\mathbf{o}\}$	$\{C_4{}^3\|\mathbf{o}\}$	$\{E\|\mathbf{o}\}$	$\{C_4\|\mathbf{o}\}$	$\{C_2\|\mathbf{o}\}$	$\{\sigma_2\|\mathbf{A}-\mathbf{a}_1\}$	$\{\sigma_3\|\mathbf{A}-\mathbf{a}_1\}$	$\{\sigma_4\|\mathbf{A}-\mathbf{a}_1\}$	$\{\sigma_1\|\mathbf{A}-\mathbf{a}_1\}$
$\{\sigma_1\|\mathbf{A}\}$	$\{\sigma_1\|\mathbf{A}\}$	$\{\sigma_2\|\mathbf{A}\}$	$\{\sigma_3\|\mathbf{A}\}$	$\{\sigma_4\|\mathbf{A}\}$	$\{E\|\mathbf{a}_1\}$	$\{C_4\|\mathbf{a}_1\}$	$\{C_2\|\mathbf{a}_1\}$	$\{C_4{}^3\|\mathbf{a}_1\}$
$\{\sigma_2\|\mathbf{A}\}$	$\{\sigma_2\|\mathbf{A}\}$	$\{\sigma_3\|\mathbf{A}\}$	$\{\sigma_4\|\mathbf{A}\}$	$\{\sigma_1\|\mathbf{A}\}$	$\{C_4{}^3\|2\mathbf{A}\}$	$\{E\|2\mathbf{A}\}$	$\{C_4\|2\mathbf{A}\}$	$\{C_2\|2\mathbf{A}\}$
$\{\sigma_3\|\mathbf{A}\}$	$\{\sigma_3\|\mathbf{A}\}$	$\{\sigma_4\|\mathbf{A}\}$	$\{\sigma_1\|\mathbf{A}\}$	$\{\sigma_2\|\mathbf{A}\}$	$\{C_2\|\mathbf{a}_2\}$	$\{C_4{}^3\|\mathbf{a}_2\}$	$\{E\|\mathbf{a}_2\}$	$\{C_4\|\mathbf{a}_2\}$
$\{\sigma_4\|\mathbf{A}\}$	$\{\sigma_4\|\mathbf{A}\}$	$\{\sigma_1\|\mathbf{A}\}$	$\{\sigma_2\|\mathbf{A}\}$	$\{\sigma_3\|\mathbf{A}\}$	$\{C_4\|\mathbf{o}\}$	$\{C_2\|\mathbf{o}\}$	$\{C_4{}^3\|\mathbf{o}\}$	$\{E\|\mathbf{o}\}$

$\mathbf{A} = \frac{1}{2}(\mathbf{a}_1 + \mathbf{a}_2)$

Table 4.12. Irreducible Representations of the Space Group C^2_{4v} for the Point O of the Brillouin Zone $[\underset{\sim}{k} = (o, o)]$

	$\{E\|\underset{\sim}{T}_n\}$	$\{C_4\|\underset{\sim}{o}\}$	$\{C_2\|\underset{\sim}{o}\}$	$\{C_4^3\|\underset{\sim}{o}\}$	$\{\sigma_1\|\underset{\sim}{A}\}$	$\{\sigma_2\|\underset{\sim}{A}\}$	$\{\sigma_3\|\underset{\sim}{A}\}$	$\{\sigma_4\|\underset{\sim}{A}\}$
Γ_1	1	1	1	1	1	1	1	1
Γ_2	1	1	1	1	−1	−1	−1	−1
Γ_3	1	−1	1	−1	1	−1	1	−1
Γ_4	1	−1	1	−1	−1	1	−1	1
Γ_5	$\begin{pmatrix} 1 & 0 \\ 0 & 1 \end{pmatrix}$	$\begin{pmatrix} 0 & 1 \\ -1 & 0 \end{pmatrix}$	$\begin{pmatrix} -1 & 0 \\ 0 & -1 \end{pmatrix}$	$\begin{pmatrix} 0 & -1 \\ 1 & 0 \end{pmatrix}$	$\begin{pmatrix} -1 & 0 \\ 0 & 1 \end{pmatrix}$	$\begin{pmatrix} 0 & -1 \\ -1 & 0 \end{pmatrix}$	$\begin{pmatrix} 1 & 0 \\ 0 & -1 \end{pmatrix}$	$\begin{pmatrix} 0 & 1 \\ 1 & 0 \end{pmatrix}$

$\underset{\sim}{A} = \frac{1}{2}(\underset{\sim}{a}_1 + \underset{\sim}{a}_2)$

The group G can be divided in left cosets with respect to $K = T$:

$$G = \sum_{i=1}^{8} \{\alpha_i | \underset{\sim}{t}_i\} T,$$

where

$$\alpha_1 = E, \quad \alpha_2 = C_4, \quad \alpha_3 = C_2, \quad \alpha_4 = C_4^3,$$

$$\alpha_5 = \sigma_1, \quad \alpha_6 = \sigma_2, \quad \alpha_7 = \sigma_3, \quad \alpha_8 = \sigma_4,$$

and

$$\underset{\sim}{t}_1 = \underset{\sim}{t}_2 = \underset{\sim}{t}_3 = \underset{\sim}{t}_4 = \underset{\sim}{0},$$
$$\underset{\sim}{t}_5 = \underset{\sim}{t}_6 = \underset{\sim}{t}_7 = \underset{\sim}{t}_8 = \underset{\sim}{A}.$$

The ij-th block of the generic matrix $\underset{\sim}{D}\{\alpha|\underset{\sim}{t}\}$ is given by

$$\underset{\sim}{D}_{ij}\{\alpha|\underset{\sim}{t}\} = \underset{\sim}{D}_{11}\{E|\underset{\sim}{T}_n\},$$

where $\{E|\underset{\sim}{T}_n\}$ is given by

$$\{\alpha|\underset{\sim}{t}\}\{\alpha_j|\underset{\sim}{t}_j\} = \{\alpha_i|\underset{\sim}{t}_i\}\{E|\underset{\sim}{T}_n\}.$$

The calculations follow very closely those of the preceding example and the matrices for the operations of the space group are given in Table 4.13.

Points on OB. $\underset{\sim}{k} \equiv (k_x, 0)$

For the points on OB the group K consists of the operations $\{E|\underset{\sim}{T}_n\}$ and $\{\sigma_1|\underset{\sim}{A} + \underset{\sim}{T}_n\}$. The group $G_o(\underset{\sim}{k})$ contains the operations E and σ_1 and can be identified as C_{1h}; its representations are given by

C_{1h}	E	σ_1
Γ_1	1	1
Γ_2	1	-1

Also

$$\underset{\sim}{D}_{11}\{\beta|\underset{\sim}{b}\} = e^{i\underset{\sim}{k}\cdot\underset{\sim}{b}}\underset{\sim}{\Gamma}_m(\beta).$$

Table 4.13. Irreducible Representation of the Space Group C_{4v}^2 for the Points A of the Brillouin Zone $[\underset{\sim}{k} \equiv (k_x, k_y)]$.

$\{E \vert \underset{\sim}{T}_n\}$	$\{C_4 \vert \underset{\sim}{o}\}$
$\begin{pmatrix} e^{i\underset{\sim}{k}\cdot\underset{\sim}{T}_n} & & & & & & & \\ & e^{iC_4\underset{\sim}{k}\cdot\underset{\sim}{T}_n} & & & & & \bigcirc & \\ & & e^{iC_2\underset{\sim}{k}\cdot\underset{\sim}{T}_n} & & & & & \\ & & & e^{iC_4{}^3\underset{\sim}{k}\cdot\underset{\sim}{T}_n} & & & & \\ & & & & e^{i\underset{\sim}{\sigma}_1\underset{\sim}{k}\cdot\underset{\sim}{T}_n} & & & \\ & \bigcirc & & & & e^{i\underset{\sim}{\sigma}_2\underset{\sim}{k}\cdot\underset{\sim}{T}_n} & & \\ & & & & & & e^{i\underset{\sim}{\sigma}_3\underset{\sim}{k}\cdot\underset{\sim}{T}_n} & \\ & & & & & & & e^{i\underset{\sim}{\sigma}_4\underset{\sim}{k}\cdot\underset{\sim}{T}_n} \end{pmatrix}$	$\begin{pmatrix} 0\,0\,0\,1 & & & & \\ 1\,0\,0\,0 & & \bigcirc & & \\ 0\,1\,0\,0 & & & & \\ 0\,0\,1\,0 & & & & \\ & 0 & e^{i\underset{\sim}{k}\cdot\underset{\sim}{a}_2} & 0 & 0 \\ \bigcirc & 0 & 0 & e^{-i\underset{\sim}{k}\cdot\underset{\sim}{a}_1} & 0 \\ & 0 & 0 & 0 & e^{-i\underset{\sim}{k}\cdot\underset{\sim}{a}_2} \\ & e^{i\underset{\sim}{k}\cdot\underset{\sim}{a}_1} & 0 & 0 & 0 \end{pmatrix}$
$\{\underset{\sim}{\sigma}_1 \vert \underset{\sim}{A}\}$	$\{\underset{\sim}{\sigma}_2 \vert \underset{\sim}{A}\}$
$\begin{pmatrix} & e^{i\underset{\sim}{k}\cdot\underset{\sim}{a}_1} & & & \\ & & e^{i\underset{\sim}{k}\cdot\underset{\sim}{a}_2} & & \\ \bigcirc & & & e^{-i\underset{\sim}{k}\cdot\underset{\sim}{a}_1} & \\ & & & & e^{-i\underset{\sim}{k}\cdot\underset{\sim}{a}_2} \\ 1\,0\,0\,0 & & & & \\ 0\,1\,0\,0 & & \bigcirc & & \\ 0\,0\,1\,0 & & & & \\ 0\,0\,0\,1 & & & & \end{pmatrix}$	$\begin{pmatrix} & 0 & e^{i\underset{\sim}{k}\cdot(\underset{\sim}{a}_1+\underset{\sim}{a}_2)} & 0 & 0 \\ \bigcirc & 0 & 0 & e^{-i\underset{\sim}{k}\cdot(\underset{\sim}{a}_1-\underset{\sim}{a}_2)} & 0 \\ & 0 & 0 & 0 & e^{-i\underset{\sim}{k}\cdot(\underset{\sim}{a}_1+\underset{\sim}{a}_2)} \\ & e^{i\underset{\sim}{k}\cdot(\underset{\sim}{a}_1-\underset{\sim}{a}_2)} & 0 & 0 & 0 \\ 0\,0\,0\,1 & & & & \\ 1\,0\,0\,0 & & & \bigcirc & \\ 0\,1\,0\,0 & & & & \\ 0\,0\,1\,0 & & & & \end{pmatrix}$

(*Continued*)

Table 4.13. (*Continued*)

$\{C_2\|\mathbf{0}\}$	$\{C_4{}^3\|\mathbf{0}\}$
$\begin{pmatrix} \begin{matrix} 0&0&1&0\\ 0&0&0&1\\ 1&0&0&0\\ 0&1&0&0 \end{matrix} & \bigcirc \\ \bigcirc & \begin{matrix} 0 & 0 & e^{-i\mathbf{k}\cdot(\mathbf{a}_1-\mathbf{a}_2)} & 0\\ 0 & 0 & 0 & e^{-i\mathbf{k}\cdot(\mathbf{a}_1+\mathbf{a}_2)}\\ e^{i\mathbf{k}\cdot(\mathbf{a}_1-\mathbf{a}_2)} & 0 & 0 & 0\\ 0 & e^{i\mathbf{k}\cdot(\mathbf{a}_1+\mathbf{a}_2)} & 0 & 0 \end{matrix} \end{pmatrix}$	$\begin{pmatrix} \begin{matrix} 0&1&0&0\\ 0&0&1&0\\ 0&0&0&1\\ 1&0&0&0 \end{matrix} & \bigcirc \\ \bigcirc & \begin{matrix} 0 & 0 & 0 & e^{-i\mathbf{k}\cdot\mathbf{a}_1}\\ e^{-i\mathbf{k}\cdot\mathbf{a}_2} & 0 & 0 & 0\\ 0 & e^{i\mathbf{k}\cdot\mathbf{a}_1} & 0 & 0\\ 0 & 0 & e^{i\mathbf{k}\cdot\mathbf{a}_2} & 0 \end{matrix} \end{pmatrix}$
$\{\sigma_3\|\mathbf{A}\}$	$\{\sigma_4{}^3\|\mathbf{A}\}$
$\begin{pmatrix} \bigcirc & \begin{matrix} 0 & 0 & e^{i\mathbf{k}\cdot\mathbf{a}_2} & 0\\ 0 & 0 & 0 & e^{-i\mathbf{k}\cdot\mathbf{a}_1}\\ e^{-i\mathbf{k}\cdot\mathbf{a}_2} & 0 & 0 & 0\\ 0 & e^{i\mathbf{k}\cdot\mathbf{a}_1} & 0 & 0 \end{matrix} \\ \begin{matrix} 0&0&1&0\\ 0&0&0&1\\ 1&0&0&0\\ 0&1&0&0 \end{matrix} & \bigcirc \end{pmatrix}$	$\begin{pmatrix} \bigcirc & \begin{matrix} 0&0&0&1\\ 1&0&0&0\\ 0&1&0&0\\ 0&0&1&0 \end{matrix} \\ \begin{matrix} 0&1&0&0\\ 0&0&1&0\\ 0&0&0&1\\ 1&0&0&0 \end{matrix} & \bigcirc \end{pmatrix}$

$\mathbf{A} = \frac{1}{2}(\mathbf{a}_1 + \mathbf{a}_2)$

where m = 1, 2 and $\beta = E, \sigma_1$. The group G can be divided into left cosets with respect to K:

$$G = \{E|\underset{\sim}{0}\}K + \{C_4|\underset{\sim}{0}\}K + \{C_2|\underset{\sim}{0}\}K + \{C_4{}^3|\underset{\sim}{0}\}K.$$

Considering the generic matrix $\{\underset{\sim}{D}\{\alpha|\underset{\sim}{t}\}$, its ijth block is given by

$$\{\underset{\sim}{D}_{ij}\{\alpha|\underset{\sim}{t}\} = \{\underset{\sim}{D}_{11}\{\beta|\underset{\sim}{b}\} = e^{i\underset{\sim}{k}\cdot\underset{\sim}{b}}\underset{\sim}{\Gamma}_m(\beta),$$

where $\{\beta|\underset{\sim}{b}\}$ is given by

$$\{\alpha|\underset{\sim}{t}\}\{\alpha_j|\underset{\sim}{0}\} = \{\alpha_i|\underset{\sim}{0}\}\{\beta|\underset{\sim}{b}\},$$

or

$$\alpha\alpha_j = \alpha_i\beta,$$
$$\underset{\sim}{t} = \alpha_i\underset{\sim}{b},$$

with i, j = 1, 2, 3, 4. The matrices for the operations of the space group are given in Table 4.14.

POINTS ON OC. $\underset{\sim}{k} \equiv (k_x, k_y = k_x)$

For the points on OC the group K consists of the operations $\{E|\underset{\sim}{T}_n\}$ and $\{\sigma_2|\underset{\sim}{A} + \underset{\sim}{T}_n\}$. The point group $G_o(\underset{\sim}{k})$ consists of the operations E and σ_2.

The irreducible representations can be worked out in a way similar to that used for the points on OB.

POINT C. $\underset{\sim}{k} \equiv (\pi/a, \pi/a)$

In correspondence to this point

$$G_o(\underset{\sim}{k}) = C_{4v} = E,\ C_4,\ C_2,\ C_4{}^3,\ \sigma_1,\ \sigma_2,\ \sigma_3,\ \sigma_4,$$

and the group K contains the operations

$$\{\beta|\underset{\sim}{b}\} : \{E|\underset{\sim}{T}_n\},\ \{C_4|\underset{\sim}{T}_n\},\ \{C_2|\underset{\sim}{T}_n\},\ \{C_4{}^3|\underset{\sim}{T}_n\},$$
$$\{\sigma_1|\underset{\sim}{A} + \underset{\sim}{T}_n\},\ \{\sigma_2|\underset{\sim}{A} + \underset{\sim}{T}_n\},\ \{\sigma_3|\underset{\sim}{A} + \underset{\sim}{T}_n\},\ \{\sigma_4|\underset{\sim}{A} + \underset{\sim}{T}_n\}.$$

The group G coincides with the group K:

$$G = \{E|\underset{\sim}{0}\}K.$$

Table 4.14. Irreducible Representations of the Space Group C_{4v}^2 for the Points OB of the Brillouin Zone $[\underset{\sim}{k} \equiv (k_x, 0)]$.

	$\{E \mid m_1\underset{\sim}{a}_1 + m_2\underset{\sim}{a}_2\}$	$\{C_4 \mid \underset{\sim}{0}\}$	$\{C_2 \mid \underset{\sim}{0}\}$	$\{C_4{}^3 \mid \underset{\sim}{0}\}$
Γ_1	$\begin{pmatrix} e^{m_1 a k_x} & & & \bigcirc \\ & e^{-im_2 a k_x} & & \\ & & e^{-im_1 a k_x} & \\ \bigcirc & & & e^{im_2 a k_x} \end{pmatrix}$	$\begin{pmatrix} 0 & 0 & 0 & 1 \\ 1 & 0 & 0 & 0 \\ 0 & 1 & 0 & 0 \\ 0 & 0 & 1 & 0 \end{pmatrix}$	$\begin{pmatrix} 0 & 0 & 1 & 0 \\ 0 & 0 & 0 & 1 \\ 1 & 0 & 0 & 0 \\ 0 & 1 & 0 & 0 \end{pmatrix}$	$\begin{pmatrix} 0 & 1 & 0 & 0 \\ 0 & 0 & 1 & 0 \\ 0 & 0 & 0 & 1 \\ 1 & 0 & 0 & 0 \end{pmatrix}$
Γ_2	$\begin{pmatrix} e^{im_1 a k_x} & & & \bigcirc \\ & e^{-im_2 a k_x} & & \\ & & e^{-im_1 a k_x} & \\ \bigcirc & & & e^{im_2 a k_x} \end{pmatrix}$	$\begin{pmatrix} 0 & 0 & 0 & 1 \\ 1 & 0 & 0 & 0 \\ 0 & 1 & 0 & 0 \\ 0 & 0 & 1 & 0 \end{pmatrix}$	$\begin{pmatrix} 0 & 0 & 1 & 0 \\ 0 & 0 & 0 & 1 \\ 1 & 0 & 0 & 0 \\ 0 & 1 & 0 & 0 \end{pmatrix}$	$\begin{pmatrix} 0 & 1 & 0 & 0 \\ 0 & 0 & 1 & 0 \\ 0 & 0 & 0 & 1 \\ 1 & 0 & 0 & 0 \end{pmatrix}$

	$\{\sigma_1 \mid \underset{\sim}{A}\}$	$\{\sigma_2 \mid \underset{\sim}{A}\}$
Γ_1	$\begin{pmatrix} e^{iak_x/2} & 0 & 0 & 0 \\ 0 & 0 & 0 & e^{-iak_x/2} \\ 0 & 0 & e^{-iak_x/2} & 0 \\ 0 & e^{iak_x/2} & 0 & 0 \end{pmatrix}$	$\begin{pmatrix} 0 & 0 & 0 & e^{iak_x/2} \\ 0 & 0 & e^{-iak_x/2} & 0 \\ 0 & e^{-iak_x/2} & 0 & 0 \\ e^{iak_x/2} & 0 & 0 & 0 \end{pmatrix}$
Γ_2	$\begin{pmatrix} -e^{iak_x/2} & 0 & 0 & 0 \\ 0 & 0 & 0 & -e^{-iak_x/2} \\ 0 & 0 & -e^{-iak_x/2} & 0 \\ 0 & -e^{iak_x/2} & 0 & 0 \end{pmatrix}$	$\begin{pmatrix} 0 & 0 & 0 & -e^{iak_x/2} \\ 0 & 0 & -e^{-iak_x/2} & 0 \\ 0 & -e^{-iak_x/2} & 0 & 0 \\ -e^{iak_x/2} & 0 & 0 & 0 \end{pmatrix}$

	$\{\sigma_3 \mid \underset{\sim}{A}\}$	$\{\sigma_4 \mid \underset{\sim}{A}\}$
Γ_1	$\begin{pmatrix} 0 & 0 & e^{iak_x/2} & 0 \\ 0 & e^{-iak_x/2} & 0 & 0 \\ e^{-iak_x/2} & 0 & 0 & 0 \\ 0 & 0 & 0 & e^{iak_x/2} \end{pmatrix}$	$\begin{pmatrix} 0 & e^{iak_x/2} & 0 & 0 \\ e^{-iak_x/2} & 0 & 0 & 0 \\ 0 & 0 & 0 & e^{-iak_x/2} \\ 0 & 0 & e^{iak_x/2} & 0 \end{pmatrix}$
Γ_2	$\begin{pmatrix} 0 & 0 & -e^{iak_x/2} & 0 \\ 0 & -e^{-iak_x/2} & 0 & 0 \\ -e^{-iak_x/2} & 0 & 0 & 0 \\ 0 & 0 & 0 & -e^{iak_x/2} \end{pmatrix}$	$\begin{pmatrix} 0 & -e^{iak_x/2} & 0 & 0 \\ -e^{-iak_x/2} & 0 & 0 & 0 \\ 0 & 0 & 0 & -e^{-iak_x/2} \\ 0 & 0 & -e^{iak_x/2} & 0 \end{pmatrix}$

Let us consider the invariant subgroup T_K of $K = \mathrm{G}$ of all the translations $\{E|\underset{\sim}{T}_k\}$ such that $e^{i\underset{\sim}{k}\cdot\underset{\sim}{T}_k} = 1$. It is, in general,

$$\underset{\sim}{k} \cdot \underset{\sim}{T}_m = \underset{\sim}{k} \cdot (m_1\underset{\sim}{a}_1 + m_2\underset{\sim}{a}_2) = m_1\pi + m_2\pi$$
$$= (m_1 + m_2)\pi,$$

and

$$e^{i\underset{\sim}{k}\cdot\underset{\sim}{T}_m} = \begin{matrix} +1 \text{ for } (m_1 + m_2) \text{ even} \\ -1 \text{ for } (m_1 + m_2) \text{ odd.} \end{matrix}$$

The primitive vectors $\underset{\sim}{T}_k$ are such that $(m_1 + m_2)$ is even; that is, they must be of either of the two possible forms:

$$\underset{\sim}{T}_k = 2n_1\underset{\sim}{a}_1 + 2n_2\underset{\sim}{a}_2,$$
$$\underset{\sim}{T}_k = (2n_1 + 1)\underset{\sim}{a}_1 + (2n_2 + 1)\underset{\sim}{a}_2.$$

The factor group K/T_k consists of the following elements:

$$\begin{array}{ll} F_1 = \{E|\underset{\sim}{a}_e\} & F_2 = \{E|\underset{\sim}{a}_0\} \\ F_3 = \{C_4|\underset{\sim}{a}_e\} & F_4 = \{C_4|\underset{\sim}{a}_0\} \\ F_5 = \{C_2|\underset{\sim}{a}_e\} & F_6 = \{C_2|\underset{\sim}{a}_0\} \\ F_7 = \{C_4{}^3|\underset{\sim}{a}_e\} & F_8 = \{C_4{}^3|\underset{\sim}{a}_0 + \underset{\sim}{A}\} \\ F_9 = \{\sigma_1|\underset{\sim}{a}_e + \underset{\sim}{A}\} & F_{10} = \{\sigma_1|\underset{\sim}{a}_0 + \underset{\sim}{A}\} \\ F_{11} = \{\sigma_2|\underset{\sim}{a}_e + \underset{\sim}{A}\} & F_{12} = \{\sigma_2|\underset{\sim}{a}_0 + \underset{\sim}{A}\} \\ F_{13} = \{\sigma_3|\underset{\sim}{a}_e + \underset{\sim}{A}\} & F_{14} = \{\sigma_3|\underset{\sim}{a}_0 + \underset{\sim}{A}\} \\ F_{15} = \{\sigma_4|\underset{\sim}{a}_e + \underset{\sim}{A}\} & F_{16} = \{\sigma_4|\underset{\sim}{a}_0 + \underset{\sim}{A}\} \end{array}$$

where

$$\underset{\sim}{a}_e = 2n_1\underset{\sim}{a}_1 + 2n_2\underset{\sim}{a}_2 \quad \text{or} \quad (2n_1+1)\underset{\sim}{a}_1 + (2n_2+1)\underset{\sim}{a}_2,$$
$$\underset{\sim}{a}_0 = 2n_1\underset{\sim}{a}_1 + (2n_2+1)\underset{\sim}{a}_2 \quad \text{or} \quad (2n_1+1)\underset{\sim}{a}_1 + 2n_2\underset{\sim}{a}_2.$$

The multiplication table for these elements is given in Table 4.15. From the multiplication table we can derive the existence of the following ten classes:

$$F_1;\ F_2;\ F_5;\ F_6;\ F_3;\ F_8;\ F_4;\ F_7;$$
$$F_9;\ F_{14};\ F_{10};\ F_{13};\ F_{11};\ F_{16};\ F_{12};\ F_{15}.$$

The factor group K/T_k has therefore ten irreducible representations: eight one-dimensional and two two-dimensional. However, four one-dimensional and one two-dimensional representations are eliminated by the condition that $\underset{\sim}{D}\{E|\underset{\sim}{a}_e\} = -\underset{\sim}{D}\{E|\underset{\sim}{a}_0\}$. The allowed representations of the factor group can be easily calculated and are given in Table 4.16.

The irreducible representations of K are obtained by assigning to each element of K the appropriate matrix. Since $K = G$ the representations so obtained are also the irreducible representations of the space group. Table 4.17 gives these representations.

POINT B. $\underset{\sim}{k} \equiv (\pi/a, 0)$

In correspondence to this point

$$G_o(\underset{\sim}{k}) = C_{2V} : E,\ C_2,\ \sigma_1,\ \sigma_3.$$

Table 4.15. Multiplication Table of the Elements of the Factor Group K/T_k for the Point C of the Brillouin Zone of the Space Group C_{4v}^2. [$\underset{\sim}{k} \equiv (\pi/a, \pi/a)$].

	F_1	F_2	F_3	F_4	F_5	F_6	F_7	F_8	F_9	F_{10}	F_{11}	F_{12}	F_{13}	F_{14}	F_{15}	F_{16}
F_1	F_1	F_2	F_3	F_4	F_5	F_6	F_7	F_8	F_9	F_{10}	F_{11}	F_{12}	F_{13}	F_{14}	F_{15}	F_{16}
F_2	F_2	F_1	F_4	F_3	F_6	F_5	F_8	F_7	F_{10}	F_9	F_{12}	F_{11}	F_{14}	F_{13}	F_{16}	F_{15}
F_3	F_3	F_4	F_5	F_6	F_7	F_8	F_1	F_2	F_{16}	F_{15}	F_{10}	F_9	F_{12}	F_{11}	F_{14}	F_{13}
F_4	F_4	F_3	F_5	F_5	F_8	F_7	F_2	F_1	F_{15}	F_{16}	F_9	F_{10}	F_{11}	F_{12}	F_{13}	F_{14}
F_5	F_5	F_6	F_7	F_8	F_1	F_2	F_3	F_4	F_{13}	F_{14}	F_{15}	F_{16}	F_9	F_{10}	F_{11}	F_{12}
F_6	F_6	F_5	F_3	F_7	F_2	F_1	F_4	F_3	F_{14}	F_{13}	F_{16}	F_{15}	F_{10}	F_9	F_{12}	F_{11}
F_7	F_7	F_8	F_1	F_2	F_3	F_4	F_5	F_6	F_{12}	F_{11}	F_{14}	F_{13}	F_{16}	F_{15}	F_{10}	F_9
F_8	F_8	F_7	F_2	F_1	F_4	F_3	F_6	F_5	F_{11}	F_{12}	F_{13}	F_{14}	F_{15}	F_{16}	F_9	F_{10}
F_9	F_9	F_{10}	F_{11}	F_{12}	F_{13}	F_{14}	F_{15}	F_{16}	F_2	F_1	F_4	F_3	F_6	F_5	F_8	F_7
F_{10}	F_{10}	F_9	F_{12}	F_{11}	F_{14}	F_{13}	F_{16}	F_{15}	F_1	F_2	F_3	F_4	F_5	F_6	F_7	F_8
F_{11}	F_{11}	F_{12}	F_{13}	F_{14}	F_{15}	F_{16}	F_9	F_{10}	F_7	F_8	F_1	F_2	F_3	F_4	F_5	F_6
F_{12}	F_{12}	F_{11}	F_{14}	F_{13}	F_{16}	F_{15}	F_{10}	F_9	F_8	F_7	F_2	F_1	F_4	F_3	F_6	F_5
F_{13}	F_{13}	F_{14}	F_{15}	F_{16}	F_9	F_{10}	F_{11}	F_{12}	F_6	F_5	F_8	F_7	F_2	F_1	F_4	F_3
F_{14}	F_{14}	F_{13}	F_{16}	F_{15}	F_{10}	F_9	F_{12}	F_{11}	F_5	F_6	F_7	F_8	F_1	F_2	F_3	F_4
F_{15}	F_{15}	F_{16}	F_9	F_{10}	F_{11}	F_{12}	F_{13}	F_{14}	F_3	F_4	F_5	F_6	F_7	F_8	F_1	F_2
F_{16}	F_{16}	F_{15}	F_{10}	F_9	F_{12}	F_{11}	F_{14}	F_{13}	F_4	F_3	F_6	F_5	F_8	F_7	F_2	F_1

Table 4.16. Allowed Irreducible Representations of the Factor Group K/T_k for the Point C of the Brillouin Zone of the Space Group C_{4v}^2 $[\underset{\sim}{k} = (\pi/a, \pi/a)]$.

	$\{E\|\underset{\sim}{a}_e\}$	$\{E\|\underset{\sim}{a}_0\}$	$\{C_4\|\underset{\sim}{a}_e\}$	$\{C_4\|\underset{\sim}{a}_0\}$	$\{C_2\|\underset{\sim}{a}_e\}$	$\{C_2\|\underset{\sim}{a}_0\}$	$\{C_4{}^3\|\underset{\sim}{a}_e\}$	$\{C_4{}^3\|\underset{\sim}{a}_e\}$
	F_1	F_2	F_3	F_4	F_5	F_6	F_7	F_8
Γ_1	1	−1	i	i	1	1	i	i
Γ_2	1	−1	i	−i	−1	1	−i	i
Γ_3	1	−1	−i	i	−1	1	i	−i
Γ_4	1	−1	−i	i	−1	1	i	−i
Γ_5	$\begin{pmatrix}1 & 0\\ 0 & 1\end{pmatrix}$	$\begin{pmatrix}-1 & 0\\ 0 & -1\end{pmatrix}$	$\begin{pmatrix}1 & 0\\ 0 & -1\end{pmatrix}$	$\begin{pmatrix}-1 & 0\\ 0 & 1\end{pmatrix}$	$\begin{pmatrix}1 & 0\\ 0 & 1\end{pmatrix}$	$\begin{pmatrix}-1 & 0\\ 0 & -1\end{pmatrix}$	$\begin{pmatrix}1 & 0\\ 0 & -1\end{pmatrix}$	$\begin{pmatrix}-1 & 0\\ 0 & 1\end{pmatrix}$
	$\{\sigma_1\|\underset{\sim}{A}+\underset{\sim}{a}_e\}$	$\{\sigma_1\|\underset{\sim}{A}+\underset{\sim}{a}_0\}$	$\{\sigma_2\|\underset{\sim}{A}+\underset{\sim}{a}_e\}$	$\{\sigma_2\|\underset{\sim}{A}+\underset{\sim}{a}_0\}$	$\{\sigma_3\|\underset{\sim}{A}+\underset{\sim}{a}_e\}$	$\{\sigma_3\|\underset{\sim}{A}+\underset{\sim}{a}_0\}$	$\{\sigma_4\|\underset{\sim}{A}+\underset{\sim}{a}_e\}$	$\{\sigma_4\|\underset{\sim}{A}+\underset{\sim}{a}_0\}$
	F_9	F_{10}	F_{11}	F_{12}	F_{13}	F_{14}	F_{15}	F_{16}
	i	i	1	1	i	−i	−1	1
	i	−i	−1	1	−i	i	1	−1
	i	−i	1	−1	−i	i	−1	1
	−i	i	−1	1	i	−i	1	−1
	$\begin{pmatrix}0 & 1\\ -1 & 0\end{pmatrix}$	$\begin{pmatrix}0 & -1\\ 1 & 0\end{pmatrix}$	$\begin{pmatrix}0 & -1\\ -1 & 0\end{pmatrix}$	$\begin{pmatrix}0 & 1\\ 1 & 0\end{pmatrix}$	$\begin{pmatrix}0 & 1\\ -1 & 0\end{pmatrix}$	$\begin{pmatrix}0 & -1\\ 1 & 0\end{pmatrix}$	$\begin{pmatrix}0 & -1\\ -1 & 0\end{pmatrix}$	$\begin{pmatrix}0 & 1\\ 1 & 0\end{pmatrix}$

$\underset{\sim}{a}_e = 2n_1\underset{\sim}{a}_1 + 2n_2\underset{\sim}{a}_2$ or $(2n_1+1)\underset{\sim}{a}_1 + (2n_2+1)\underset{\sim}{a}_2$
$\underset{\sim}{a}_0 = 2n_1\underset{\sim}{a}_1 + (2n_2+1)\underset{\sim}{a}_2$ or $(2n_1+1)\underset{\sim}{a}_1 + 2n_2\underset{\sim}{a}_2$
$\underset{\sim}{A} = \frac{1}{2}(\underset{\sim}{a}_1 + \underset{\sim}{a}_2)$

Table 4.17. Irreducible Representations of the Space Group C_{4v}^2 for the Point C of the Brillouin Zone, $[k \equiv (\pi/a, \pi/a)]$.

	$\{E\|m_1 \underset{\sim}{a}_1 + m_2 \underset{\sim}{a}_2\}$		$\{C_4\|\underset{\sim}{o}\}$	$\{C_2\|\underset{\sim}{o}\}$	$\{C_4{}^3\|\underset{\sim}{o}\}$	$\{\sigma_1\|\underset{\sim}{A}\}$	$\{\sigma_2\|\underset{\sim}{A}\}$	$\{\sigma_3\|\underset{\sim}{A}\}$	$\{\sigma_4\|\underset{\sim}{A}\}$
	$m_1 + m_2$ even	$m_1 + m_2$ odd							
Γ_1	1	−1	i	−1	−i	−i	1	i	−1
Γ_2	1	−1	i	−1	−i	i	−1	−i	1
Γ_3	1	−1	−i	−1	i	i	1	−i	−1
Γ_4	1	−1	−i	−1	i	−i	−1	i	1
Γ_5	$\begin{pmatrix} 1 & 0 \\ 0 & 1 \end{pmatrix}$	$\begin{pmatrix} -1 & 0 \\ 0 & -1 \end{pmatrix}$	$\begin{pmatrix} 1 & 0 \\ 0 & -1 \end{pmatrix}$	$\begin{pmatrix} 1 & 0 \\ 0 & 1 \end{pmatrix}$	$\begin{pmatrix} 1 & 0 \\ 0 & -1 \end{pmatrix}$	$\begin{pmatrix} 0 & 1 \\ -1 & 0 \end{pmatrix}$	$\begin{pmatrix} 0 & -1 \\ -1 & 0 \end{pmatrix}$	$\begin{pmatrix} 0 & 1 \\ -1 & 0 \end{pmatrix}$	$\begin{pmatrix} 0 & -1 \\ -1 & 0 \end{pmatrix}$

$\underset{\sim}{A} = \frac{1}{2}(\underset{\sim}{a}_1 + \underset{\sim}{a}_2)$

and the group K contains the operations

$$\{\beta|\underset{\sim}{b}\} : \{E|\underset{\sim}{T}_n\},\ \{C_2|\underset{\sim}{T}_n\},\ \{\sigma_1|\underset{\sim}{A}+\underset{\sim}{T}_n\},\ \{\sigma_3|\underset{\sim}{A}+\underset{\sim}{T}_n\}.$$

The group G can be decomposed into left cosets with respect to K:

$$G = \{E|\underset{\sim}{0}\}K + \{C_4|\underset{\sim}{0}\}K.$$

Considering the generic matrix $\underset{\sim}{D}\{\alpha|\underset{\sim}{t}\}$, its kjth block $(k, j = 1, 2)$ is given by

$$\underset{\sim}{D}_{kj}\{\alpha|\underset{\sim}{t}\} = \underset{\sim}{D}_{11}\{\beta|\underset{\sim}{b}\},$$

where $\{\beta|\underset{\sim}{b}\}$ is given by

$$\{\alpha|\underset{\sim}{t}\}\{\alpha_j|\underset{\sim}{0}\} = \{\alpha_k|\underset{\sim}{0}\}\{\beta|\underset{\sim}{b}\},$$

or

$$\alpha\alpha_j = \alpha_k\beta,$$
$$\underset{\sim}{t} = a_k\underset{\sim}{b}.$$

In order to find the irreducible representations $\underset{\sim}{D}_{11}$ of K let us consider the invariant subgroup T_k of K that contains all the translations $\{E|\underset{\sim}{T}_k\}$ such that $e^{i\underset{\sim}{k}\cdot\underset{\sim}{T}_k} = 1$. It is

$$\underset{\sim}{k}\cdot\underset{\sim}{T}_k = \frac{\pi}{a}\hat{x}\cdot(m_1\underset{\sim}{a}_1 + m_2\underset{\sim}{a}_2) = m_1\pi,$$

and

$$e^{i\underset{\sim}{k}\cdot\underset{\sim}{T}_k} = e^{im_1\pi} = \begin{cases} +1 & m_1 \text{ even} \\ -1 & m_1 \text{ odd} \end{cases}.$$

The condition has to be imposed that every element $\{E|\underset{\sim}{T}_k\}$ is represented by a matrix of the form $e^{i\underset{\sim}{k}\cdot\underset{\sim}{T}_k}\underset{\sim}{1}$:

$$\underset{\sim}{D}_{11}\{E|\underset{\sim}{T}_k\} = e^{i\underset{\sim}{k}\cdot\underset{\sim}{T}_k}\underset{\sim}{1}.$$

Therefore only those representations of the factor group K/T_k for which translations not in T_k are represented by matrices that are the negative of the matrices representing translations in T_k will generate allowable representations of K.

The factor group K/T_k contains the following eight elements:

$$
\begin{array}{ll}
D_1 = \{E|\underset{\sim}{a}_e\} & D_2 = \{E|\underset{\sim}{a}_0\} \\
D_3 = \{C_2|\underset{\sim}{a}_e\} & D_4 = \{C_2|\underset{\sim}{a}_0\} \\
D_5 = \{\sigma_1|\underset{\sim}{A} + \underset{\sim}{a}_e\} & D_6 = \{\sigma_1|\underset{\sim}{A} + \underset{\sim}{a}_0\} \\
D_7 = \{\sigma_3|\underset{\sim}{A} + \underset{\sim}{a}_e\} & D_8 = \{\sigma_3|\underset{\sim}{A} + \underset{\sim}{a}_0\}
\end{array}
$$

where

$$\underset{\sim}{A} = \frac{1}{2}(\underset{\sim}{a}_1 + \underset{\sim}{a}_2),$$

$$\underset{\sim}{a}_e = 2n_1\underset{\sim}{a}_1 + \underset{\sim}{a}_2,$$

$$\underset{\sim}{a}_0 = (2n_1 + 1)\underset{\sim}{a}_1 + \underset{\sim}{a}_2.$$

The multiplication table for these elements follows:

	D_1	D_2	D_3	D_4	D_5	D_6	D_7	D_8
D_1	D_1	D_2	D_3	D_4	D_5	D_6	D_7	D_8
D_2	D_2	D_1	D_4	D_3	D_6	D_5	D_8	D_7
D_3	D_3	D_4	D_1	D_2	D_8	D_7	D_6	D_5
D_4	D_4	D_3	D_2	D_1	D_7	D_8	D_5	D_6
D_5	D_5	D_6	D_7	D_8	D_2	D_1	D_4	D_3
D_6	D_6	D_5	D_8	D_7	D_1	D_2	D_3	D_4
D_7	D_7	D_8	D_5	D_6	D_3	D_4	D_1	D_2
D_8	D_8	D_7	D_6	D_5	D_4	D_3	D_3	D_1

This group is isomorphic with the group C_{4v} with the following correspondence:

$$E \leftrightarrow D_1,\ C_2 \leftrightarrow D_2,\ \sigma_1 \leftrightarrow D_3,\ \sigma_3 \leftrightarrow D_4$$
$$C_4 \leftrightarrow D_5,\ {C_4}^3 \leftrightarrow D_6,\ \sigma_4 \leftrightarrow D_7,\ \sigma_2 \leftrightarrow D_8$$

The factor group K/T_k has, therefore, the same representations as C_{4v}. The representations of these groups are given in Table 4.6.

Looking at this table we see that only the two-dimensional representation meets the condition

$$D\{E|\underset{\sim}{a}_e\} = -D\{E|\underset{\sim}{a}_0\}.$$

The irreducible representation of *K*, obtained by assigning the proper matrix to each element of *K*, is given in Table 4.18.

Having found the representation of *K*, the irreducible representation of the space group in correspondence to the point B of the

Table 4.18. Irreducible Representation of the Group K for the Point B of the Brillouin Zone of the Space Group C_{4v}^2. $[\underset{\sim}{k} \equiv (\pi/a, 0)]$.

$\{E\|\underset{\sim}{a}_e\}$	$\{E\|\underset{\sim}{a}_0\}$	$\{C_2\|\underset{\sim}{a}_e\}$	$\{C_2\|\underset{\sim}{a}_0\}$
$\begin{pmatrix} 1 & 0 \\ 0 & 1 \end{pmatrix}$	$\begin{pmatrix} -1 & 0 \\ 0 & -1 \end{pmatrix}$	$\begin{pmatrix} -1 & 0 \\ 0 & 1 \end{pmatrix}$	$\begin{pmatrix} 1 & 0 \\ 0 & -1 \end{pmatrix}$
$\{\sigma_1\|\underset{\sim}{A} + \underset{\sim}{a}_e\}$	$\{\sigma_1\|\underset{\sim}{A} + \underset{\sim}{a}_0\}$	$\{\sigma_3\|\underset{\sim}{A} + \underset{\sim}{a}_e\}$	$\{\sigma_3\|\underset{\sim}{A} + \underset{\sim}{a}_0\}$
$\begin{pmatrix} 0 & 1 \\ -1 & 0 \end{pmatrix}$	$\begin{pmatrix} 0 & -1 \\ 1 & 0 \end{pmatrix}$	$\begin{pmatrix} 0 & 1 \\ 1 & 0 \end{pmatrix}$	$\begin{pmatrix} 0 & -1 \\ -1 & 0 \end{pmatrix}$

$\underset{\sim}{a}_e = 2n_1\underset{\sim}{a}_1 + n_2\underset{\sim}{a}_2$
$\underset{\sim}{a}_0 = (2n_1 + 1)\underset{\sim}{a}_1 + n_2\underset{\sim}{a}_2$
$\underset{\sim}{A} = \frac{1}{2}(\underset{\sim}{a}_1 + \underset{\sim}{a}_2)$

Brillouin zone can be also be found. This representation is given in Table 4.19.

POINTS ON BC. $\underset{\sim}{k} \equiv (\pi/a, k_y)$

For these points

$$G_o(\underset{\sim}{k}) = C_{1h} : E, \sigma_3,$$

and the group K contains the operations

$$\{\beta|\underset{\sim}{b}\} : \{E|\underset{\sim}{T}_n\}, \{\sigma_3|\underset{\sim}{A} + \underset{\sim}{T}_n\}.$$

The space group G can be decomposed into left cosets with respect to K as follows:

$$G = \{E|\underset{\sim}{0}\}K + \{C_4|\underset{\sim}{0}\}K + \{C_2|\underset{\sim}{0}\}K + \{C_4{}^3|\underset{\sim}{0}\}K.$$

The i, j-th $(i, j = 1, 2, 3, 4)$ block of the generic matrix $\underset{\sim}{D}\{\alpha|\underset{\sim}{t}\}$ is given by

$$\underset{\sim}{D}_{ij}\{\alpha|\underset{\sim}{t}\} = \underset{\sim}{D}_{11}\{\beta|\underset{\sim}{b}\},$$

where $\{\beta|\underset{\sim}{b}\}$ is given by

$$\{\alpha|\underset{\sim}{t}\}\{\alpha_j|\underset{\sim}{0}\} = \{\alpha_i|\underset{\sim}{0}\}\{\beta|\underset{\sim}{b}\}.$$

Table 4.19. Irreducible Representations of the Space Group C_{4v}^2 for the Point B of the Brillouin Zone $[\underset{\sim}{k} \equiv (\pi/a, 0)]$.

$\{E\|m_1\underset{\sim}{a}_1 + m_2\underset{\sim}{a}_2\}$	$\{E\|m_1\underset{\sim}{a}_1 + m_2\underset{\sim}{a}_2\}$	$\{E\|m_1\underset{\sim}{a}_1 + m_2\underset{\sim}{a}_2\}$	$\{E\|m_1\underset{\sim}{a}_1 + m_2\underset{\sim}{a}_2\}$
m_1 even, m_2 even	m_1 even, m_2 odd	m_1 odd, m_2 even	m_1 odd, m_2 odd
$\begin{pmatrix} 1 & 0 & & \\ 0 & 1 & \bigcirc & \\ & & 1 & 0 \\ \bigcirc & & 0 & 1 \end{pmatrix}$	$\begin{pmatrix} 1 & 0 & & \\ 0 & 1 & \bigcirc & \\ & & -1 & 0 \\ \bigcirc & & 0 & -1 \end{pmatrix}$	$\begin{pmatrix} -1 & 0 & & \\ 0 & -1 & \bigcirc & \\ & & 1 & 0 \\ \bigcirc & & 0 & 1 \end{pmatrix}$	$\begin{pmatrix} -1 & 0 & & \\ 0 & -1 & \bigcirc & \\ & & -1 & 0 \\ \bigcirc & & 0 & -1 \end{pmatrix}$

$\{C_4\|\underset{\sim}{0}\}$	$\{C_2\|\underset{\sim}{0}\}$	$\{C_4{}^3\|\underset{\sim}{0}\}$	$\{\sigma_1\|\underset{\sim}{A}\}$	$\{\sigma_2\|\underset{\sim}{A}\}$	$\{\sigma_3\|\underset{\sim}{A}\}$	$\{\sigma_4\|\underset{\sim}{A}\}$
$\begin{pmatrix} & & -1 & 0 \\ \bigcirc & & 0 & 1 \\ 1 & 0 & & \\ 0 & 1 & \bigcirc & \end{pmatrix}$	$\begin{pmatrix} -1 & 0 & & \\ 0 & 1 & \bigcirc & \\ & & -1 & 0 \\ \bigcirc & & 0 & 1 \end{pmatrix}$	$\begin{pmatrix} & & 1 & 0 \\ \bigcirc & & 0 & 1 \\ -1 & 0 & & \\ 0 & 1 & \bigcirc & \end{pmatrix}$	$\begin{pmatrix} 0 & 1 & & \\ -1 & 0 & \bigcirc & \\ & & 0 & -1 \\ \bigcirc & & -1 & 0 \end{pmatrix}$	$\begin{pmatrix} & & 0 & 1 \\ \bigcirc & & 1 & 0 \\ 0 & -1 & & \\ -1 & 0 & \bigcirc & \end{pmatrix}$	$\begin{pmatrix} 0 & 1 & & \\ 1 & 0 & \bigcirc & \\ & & 0 & -1 \\ \bigcirc & & 1 & 0 \end{pmatrix}$	$\begin{pmatrix} & & 0 & 1 \\ \bigcirc & & -1 & 0 \\ 0 & -1 & & \\ 1 & 0 & \bigcirc & \end{pmatrix}$

$\underset{\sim}{A} = \frac{1}{2}(\underset{\sim}{a}_1 + \underset{\sim}{a}_1)$

or

$$\alpha\alpha_j = \alpha\ \beta,$$
$$\underset{\sim}{t} = \alpha_i \underset{\sim}{b}.$$

The irreducible representations of K may be found in the following way. Consider the invariant subgroup T_k of K that consists of all the primitive translations $\{E|\underset{\sim}{T}_k\}$ for which $e^{i\underset{\sim}{k}\cdot\underset{\sim}{T}_k} = 1$. Since the group K consists of operations $\{E|\underset{\sim}{T}_n\}$ and $\{\sigma_3|\underset{\sim}{A}+\underset{\sim}{T}_n\}$, the number of cosets in the factor group K/T_k is 2N, if N is the number of cosets in the factor group T/T_k.

For an irreducible representation Γ of K/T_k the sum of the squares of the characters is given by

$$\Sigma|X_\Gamma(\{\beta|\underset{\sim}{b}\}T_k)|^2 = 2N,$$

where the sum extends over all the cosets in K/T_k. On the other hand if we sum only over the cosets in T/T_k for the same representation of K/T_k we get

$$\Sigma|X_\Gamma(\{\beta|\underset{\sim}{b}\}T_k)|^2 = Nl^2,$$

if l is the dimension of the representation Γ. Therefore $l = 1$ and the dimension of all the irreducible representations of K/T_k is one.

Taking into account the fact that in an irreducible representation of K

$$\underset{\sim}{D}\{E|\underset{\sim}{T}_n\} = e^{i\underset{\sim}{k}\cdot\underset{\sim}{T}_n} = (-1)^{n_1}e^{in_2k_ya},$$

if $\underset{\sim}{T}_n = n_1\underset{\sim}{a}_1 + n_2\underset{\sim}{a}_2$, and that

$$[\{\sigma_3|\underset{\sim}{A}\}]^2 = \{E|\underset{\sim}{a}_2\},$$

we may find the irreducible representations of K that are given in Table 4.20.

Given the representations of K it is easy to find the representations of the space group corresponding to the points considered. These representations are given in Table 4.21.

A summary of all the calculations performed in the present example is given in Table 4.22.

Table 4.20. Irreducible Representations of the Group K for the Points BC of the Brillouin Zone of the Space Group C_{4v}^2. $[\underset{\sim}{k} \equiv (\pi/a, k_y)]$.

	$\{E\|m_1\underset{\sim}{a}_1 + m_2\underset{\sim}{a}_2\}$	$\{\sigma_3\|\underset{\sim}{A}\}$
Γ_1	$(-1)^{m_1} e^{im_2k_ya}$	$e^{ik_ya/2}$
Γ_2	$(-1)^{m_1} e^{im_2k_ya}$	$-e^{ik_ya/2}$

$\underset{\sim}{A} = \frac{1}{2}(\underset{\sim}{a}_1 + \underset{\sim}{a}_2)$

Table 4.21. Irreducible Representations of the Group C_{4v}^2 for the Points BC of the Brillouin Zone $[\underset{\sim}{k} \equiv (\pi/a, k_y)]$.

	$\{E\|m_1\underset{\sim}{a}_1 + m_2\underset{\sim}{a}_2\}$	$\{C_4\|\underset{\sim}{0}\}$	$\{C_2\|\underset{\sim}{0}\}$	$\{C_4{}^3\|\underset{\sim}{0}\}$
Γ_1	$\begin{pmatrix} (-1)^{m_1}e^{im_2k_ya} & & & \bigcirc \\ & (-1)^{m_2}e^{-im_1k_ya} & & \\ & & (-1)^{m_1}e^{-im_2k_ya} & \\ \bigcirc & & & (-1)^{m_2}e^{im_1k_ya} \end{pmatrix}$	$\begin{pmatrix} 0 & 0 & 0 & 1 \\ 1 & 0 & 0 & 0 \\ 0 & 1 & 0 & 0 \\ 0 & 0 & 1 & 0 \end{pmatrix}$	$\begin{pmatrix} 0 & 0 & 1 & 0 \\ 0 & 0 & 0 & 1 \\ 1 & 0 & 0 & 0 \\ 0 & 1 & 0 & 0 \end{pmatrix}$	$\begin{pmatrix} 0 & 1 & 0 & 0 \\ 0 & 0 & 1 & 0 \\ 0 & 0 & 0 & 1 \\ 1 & 0 & 0 & 0 \end{pmatrix}$
Γ_2	$\begin{pmatrix} (-1)^{m_1}e^{im_2k_ya} & & & \bigcirc \\ & (-1)^{m_2}e^{im_1k_ya} & & \\ & & (-1)^{m_1}e^{-im_2k_ya} & \\ \bigcirc & & & (-1)^{m_2}e^{-im_1k_ya} \end{pmatrix}$	$\begin{pmatrix} 0 & 0 & 0 & 1 \\ 1 & 0 & 0 & 0 \\ 0 & 1 & 0 & 0 \\ 0 & 0 & 1 & 0 \end{pmatrix}$	$\begin{pmatrix} 0 & 0 & 1 & 0 \\ 0 & 0 & 0 & 1 \\ 1 & 0 & 0 & 0 \\ 0 & 1 & 0 & 0 \end{pmatrix}$	$\begin{pmatrix} 0 & 1 & 0 & 0 \\ 0 & 0 & 1 & 0 \\ 0 & 0 & 0 & 1 \\ 1 & 0 & 0 & 0 \end{pmatrix}$

	$\{\sigma_1\|\underset{\sim}{A}\}$	$\{\sigma_2\|\underset{\sim}{A}\}$
Γ_1	$\begin{pmatrix} 0 & 0 & e^{ik_ya/2} & 0 \\ 0 & -e^{ik_ya/2} & 0 & 0 \\ -e^{-ik_ya/2} & 0 & 0 & 0 \\ 0 & 0 & 0 & e^{-ik_ya/2} \end{pmatrix}$	$\begin{pmatrix} 0 & e^{ik_ya/2} & 0 & 0 \\ -e^{ik_ya/2} & 0 & 0 & 0 \\ 0 & 0 & 0 & -e^{-ik_ya/2} \\ 0 & 0 & e^{-ik_ya/2} & 0 \end{pmatrix}$
Γ_2	$\begin{pmatrix} 0 & 0 & -e^{ik_ya/2} & 0 \\ 0 & e^{ik_ya/2} & 0 & 0 \\ e^{-ik_ya/2} & 0 & 0 & 0 \\ 0 & 0 & 0 & -e^{-ik_ya/2} \end{pmatrix}$	$\begin{pmatrix} 0 & -e^{ik_ya/2} & 0 & 0 \\ e^{-ik_ya/2} & 0 & 0 & 0 \\ 0 & 0 & 0 & e^{-ik_ya/2} \\ 0 & 0 & -e^{-ik_ya/2} & 0 \end{pmatrix}$

	$\{\sigma_4\|\underset{\sim}{A}\}$	$\{\sigma_4\|\underset{\sim}{A}\}$
Γ_1	$\begin{pmatrix} e^{ik_ya/2} & 0 & 0 & 0 \\ 0 & 0 & 0 & -e^{ik_ya/2} \\ 0 & 0 & -e^{-ik_ya/2} & 0 \\ 0 & e^{-ik_ya/2} & 0 & 0 \end{pmatrix}$	$\begin{pmatrix} 0 & 0 & 0 & e^{ik_ya/2} \\ 0 & 0 & -e^{ik_ya/2} & 0 \\ 0 & -e^{-ik_ya/2} & 0 & 0 \\ e^{-ik_ya/2} & 0 & 0 & 0 \end{pmatrix}$
Γ_2	$\begin{pmatrix} -e^{ik_ya/2} & 0 & 0 & 0 \\ 0 & 0 & 0 & e^{ik_ya/2} \\ 0 & 0 & e^{-ik_ya/2} & 0 \\ 0 & -e^{-ik_ya/2} & 0 & 0 \end{pmatrix}$	$\begin{pmatrix} 0 & 0 & 0 & -e^{ik_ya/2} \\ 0 & 0 & e^{ik_ya/2} & 0 \\ 0 & e^{-ik_ya/2} & 0 & 0 \\ -e^{-ik_ya/2} & 0 & 0 & 0 \end{pmatrix}$

$\underset{\sim}{A} = \frac{1}{2}(\underset{\sim}{a}_1 + \underset{\sim}{a}_2)$

Table 4.22. Summary of the Irreducible Representations of the Space Group C_{4v}^2.

Region	K	$G_o(\underset{\sim}{k})$	Number of Representations	Dimension of Representations	Observations
0	$K = G$	$G_o(\underset{\sim}{k}) = G_o = C_{4v}$	4	$1\begin{cases} d = 1 \\ q = 1 \end{cases}$	$\underset{\sim}{D}\{\alpha\|\underset{\sim}{t}\} = \underset{\sim}{\Gamma}_j(\alpha)$
			1	$2\begin{cases} d = 2 \\ q = 1 \end{cases}$	$\underset{\sim}{\Gamma}_j(\alpha) =$ representation of C_{4v}
A	$K = T$	$G_o(\underset{\sim}{k}) = C_1 : E$	1	$8\begin{cases} d = 1 \\ q = 8 \end{cases}$	
OB	$K : \{E\|\underset{\sim}{T}_n\}$, $\{\sigma_1\|\underset{\sim}{A} + \underset{\sim}{T}_n\}$	$G_o(\underset{\sim}{k}) = C_{1h}, E, \sigma_1$	2	$4\begin{cases} d = 1 \\ q = 4 \end{cases}$	
OC	$K : \{E\|\underset{\sim}{T}_n\}$, $\{\sigma_2\|\underset{\sim}{A} + \underset{\sim}{T}_n\}$	$G_o(\underset{\sim}{k}) = C_{1h}, E, \sigma_2$	2	$4\begin{cases} d = 1 \\ q = 4 \end{cases}$	

(*Continued*)

Table 4.22. (*Continued*)

Region	K	$G_o(\underset{\sim}{k})$	Number of Representations	Dimension of Representations	Observations
C	$K = G$,	$G_o(\underset{\sim}{k}) = G_o = C_{4v}$	4	$1\begin{cases} d = 1 \\ q = 1 \end{cases}$	One-dimensional Representations not real
			1	$2\begin{cases} d = 2 \\ q = 1 \end{cases}$	
B	$K : \{E\vert\underset{\sim}{T}_n\}$, $\{C_2\vert\underset{\sim}{T}_n\}, \{\sigma_1\vert\underset{\sim}{A} + \underset{\sim}{T}_n\}$, $\{\sigma_3\vert\underset{\sim}{A} + \underset{\sim}{T}_n\}$,	$G_o(\underset{\sim}{k}) = C_{2v} : E, C_2, \sigma_1\sigma_3$	1	$4\begin{cases} d = 2 \\ q = 2 \end{cases}$	
BC	$K : \{E\vert\underset{\sim}{T}_n\}$, $\{\sigma_3\vert\underset{\sim}{A} + \underset{\sim}{T}_n\}$,	$G_o(\underset{\sim}{k}) = C_{1h} : E, \sigma_3$	2	$4\begin{cases} d = 1 \\ q = 4 \end{cases}$	

References

1. J.L. Prather, *Atomic Energy Levels in Crystals*, National Bureau of Standards Monograph 19, February 1961.
2. H. Eyring, J. Walter and G. B. Kimball, *Quantum Chemistry*, Wiley, New York, 1957.
3. F.A. Cotton, *Chemical Applications of Group Theory*, Inter-science, New York, 1964.
4. G.F. Koster, *Notes on Group Theory, Techn. Rep. No. B*, March 1956, Solid State and Molecular Theory Group, Massachusetts Institute of Technology, Cambridge, Massachusetts.
5. G.F. Koster, "Space Groups and Their Representations," in Solid state Physics, Vol. 5, edited by F. Seitz and D. Turnbull, Academic, New York, 1957, p. 173.

Chapter 5

Scattering of X-Rays by Crystals

This chapter describes the experimental methods based on X-ray diffraction that lead to the study of crystal symmetries.

5.1. Introduction

The diffraction of X-rays is based on the interaction of electromagnetic waves with the electrons of the atoms of a physical system. The scattered wave is observed at a point far from the "interaction point" and is due to the superposition of the electromagnetic waves produced by the induced electron motion.

In 1920 Laue, Fredrich, and Knipping demonstrated with one experiment the periodic nature of a crystal and the wave nature of X-rays. A great deal of information about the structure of atoms and solids was derived by the use of X-ray techniques.

An impinging electromagnetic wave striking a crystal causes the electrons of the atoms to vibrate and emit radiation. The scattered wave is a superposition of waves coming from different electrons and furnishes information about the structure of the crystal. In order to get this information the wavelength of the incident radiation has to have the same magnitude as the lattice spacing.

We shall make the following assumptions:

1. Frequency of incident wave $\gg \nu_{\text{atom}}$ where ν_{atom} is the characteristic frequency of the motion of electrons in the atom.
2. No lattice vibrations: atoms clamped in equilibrium position (for the moment).

3. The waves radiated from the electrons are not scattered within crystal. This can be achieved by using a very thin crystal.
4. The field acting on each electron is due only to the incident wave. If we do not make this assumption we have the problem of evaluating the field inside the crystal; this problem is the subject of the "dynamical theory" due to Ewald.
5. No absorption and no dispersion in the crystal

$$|\underset{\sim}{k}_{\text{input}}| = |\underset{\sim}{k}_{\text{output}}|. \tag{5.1.1}$$

6. No relativistic effect, that is, no Compton scattering.

5.2. Scattering from a Single Electron

Let us assume that an electromagnetic wave given by

$$\underset{\sim}{E} = \underset{\sim}{E}_O e^{i(-\omega t + \underset{\sim}{k}\cdot\underset{\sim}{r})}, \tag{5.2.1}$$

is striking an electron (see Fig. 5.1). The electron will experience a force

$$m\ddot{\underset{\sim}{r}}' = -e\underset{\sim}{E}_O e^{i(-\omega t + \underset{\sim}{k}\cdot\underset{\sim}{r}')}. \tag{5.2.2}$$

The motion of the electron is the solution of the above equation. If we make the assumption

$$|\underset{\sim}{r}'(t) - \underset{\sim}{r}'(\underset{\sim}{O})| \ll \lambda, \tag{5.2.3}$$

where λ is the wavelength of incident radiation, then we can consider $e^{i\underset{\sim}{k}\cdot\underset{\sim}{r}'}$ as a constant in the integration and we can write

$$\underset{\sim}{r}'(t) - \underset{\sim}{r}'(O) = \frac{e\underset{\sim}{E}_O}{m\omega^2} e^{i(\underset{\sim}{k}\cdot\underset{\sim}{r}' - \omega t)}. \tag{5.2.4}$$

The assumption (5.2.3) implies

$$\frac{e\underset{\sim}{E}_O}{m\omega^2} \ll \lambda, \tag{5.2.5}$$

or

$$E_O \ll \frac{m\omega^2\lambda}{e} = \frac{m\lambda}{e}\frac{4\pi^2c^2}{\lambda^2} = \frac{4\pi^2c^2m}{\lambda e}. \tag{5.2.6}$$

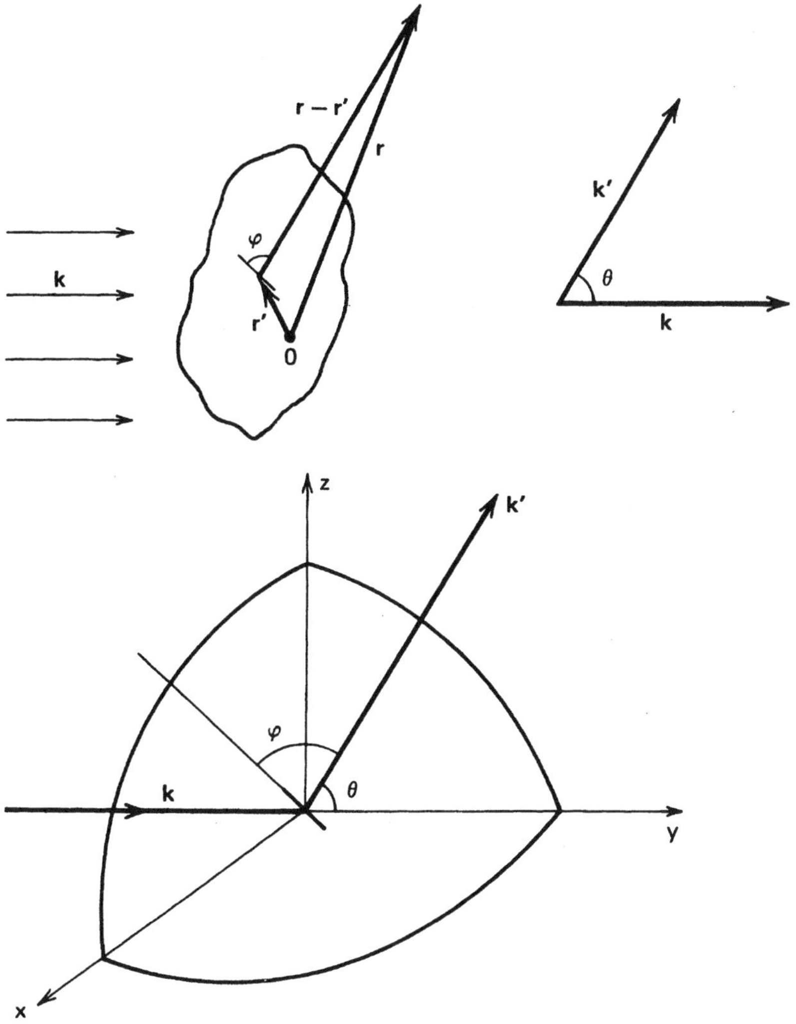

Fig. 5.1. Scattering of X-rays from an electron.

Since

$$\begin{aligned}
\lambda &= 1.54 \times 10^{-8} \text{ (wavelength of Cu K}\alpha_1\text{ line)}\\
e &= 4.8 \times 10^{-10} \text{ esu}\\
m &= 9.1 \times 10^{-28} \simeq 10^{-27} \text{ g}\\
c &= 3 \times 10^{10} \text{ cm/sec,}
\end{aligned}$$

therefore

$$E_O \ll \frac{4\pi^2 c^2 m}{\lambda e} \simeq 14.5 \times 10^{14}\,\text{volts/cm},$$

and condition (5.2.6) is easily fulfilled.

The induced dipole moment is

$$\begin{aligned} \underset{\sim}{P}(t) &= e[\underset{\sim}{r}'(t) - \underset{\sim}{r}'(0)] \\ &= \frac{e^2 \underset{\sim}{E}_O}{m\omega^2} e^{i(\underset{\sim}{k}\cdot\underset{\sim}{r}' - \omega t)}. \end{aligned} \tag{5.2.7}$$

It is also, with the assumption (5.2.3),

$$\begin{aligned} \ddot{\underset{\sim}{P}}(t) &= \frac{e^2 \underset{\sim}{E}_O}{m\omega^2}\, \omega^2 e^{i[\underset{\sim}{k}\cdot\underset{\sim}{r}' - \omega t]} \\ &= \frac{e^2 \underset{\sim}{E}_O}{m}\, e^{i[\underset{\sim}{k}\cdot\underset{\sim}{r}' - \omega t]}. \end{aligned} \tag{5.2.8}$$

The radiation field at the point $\underset{\sim}{r}$ produced by the dipole $\underset{\sim}{P}$ is given by (see Fig. 5.2)

$$\begin{aligned} \underset{\sim}{E}(\underset{\sim}{r}, t) &= -\hat{1}_\phi \frac{\ddot{\underset{\sim}{P}} \sin\phi}{c^2|\underset{\sim}{r} - \underset{\sim}{r}'|}\, e^{i\omega|\underset{\sim}{r} - \underset{\sim}{r}'|/c} \\ &= \hat{1}_\phi \frac{e^2 E_O \sin\phi}{mc^2|\underset{\sim}{r} - \underset{\sim}{r}'|}\, e^{i[\underset{\sim}{k}\cdot\underset{\sim}{r}' - \omega(t - |\underset{\sim}{r} - \underset{\sim}{r}'|/c)]}. \end{aligned} \tag{5.2.9}$$

But

$$\frac{\omega}{c}|\underset{\sim}{r} - \underset{\sim}{r}'| = \underset{\sim}{k}' \cdot (\underset{\sim}{r} - \underset{\sim}{r}'). \tag{5.2.10}$$

Then

$$e^{i[\underset{\sim}{k}\cdot\underset{\sim}{r}' - \omega(t - |\underset{\sim}{r} - \underset{\sim}{r}'|/c)]} = e^{i(\underset{\sim}{k}'\cdot\underset{\sim}{r} - \omega t)} e^{i(\underset{\sim}{k} - \underset{\sim}{k}')\cdot\underset{\sim}{r}'}. \tag{5.2.11}$$

Therefore

$$\underset{\sim}{E}(\underset{\sim}{r}, t) = \hat{1}_\phi \frac{e^2 E_O \sin\phi}{mc^2|\underset{\sim}{r} - \underset{\sim}{r}'|} e^{i(\underset{\sim}{k}'\cdot\underset{\sim}{r} - \omega t)} e^{i(\underset{\sim}{k} - \underset{\sim}{k}')\cdot\underset{\sim}{r}'}$$

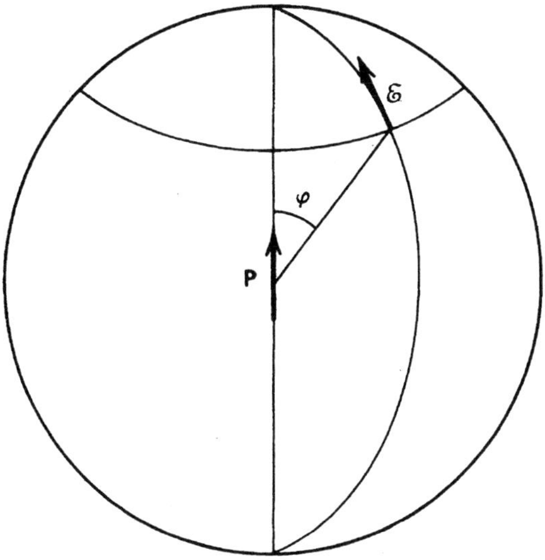

Fig. 5.2. Electric field due to an electric dipole.

$$= \hat{1}_\phi \frac{e^2 E_O \sin\phi}{mc^2} e^{i(\underset{\sim}{k}'\cdot\underset{\sim}{r}-\omega t)} \int \frac{e^{i(\underset{\sim}{k}'-\underset{\sim}{k}')\cdot r''}}{|\underset{\sim}{r}-\underset{\sim}{r}''|}\delta(\underset{\sim}{r}'-\underset{\sim}{r}'')d\underset{\sim}{r}''$$

$$\simeq \hat{1}_\phi \frac{e^2 E_O \sin\phi}{mc^2}\frac{e^{i(\underset{\sim}{k}'\cdot\underset{\sim}{r}-\omega t)}}{r} \int e^{i(\underset{\sim}{k}-\underset{\sim}{k}')r''}\delta(\underset{\sim}{r}'-\underset{\sim}{r}'')d\underset{\sim}{r}'', \tag{5.2.12}$$

having taken

$$|\underset{\sim}{r} - r''| \approx \underset{\sim}{r}. \tag{5.2.13}$$

5.3. Scattering from a Single Atom

We can treat this case by simply making the substitution

$$e\delta(\underset{\sim}{r}'-\underset{\sim}{r}'') \to e|\psi(\underset{\sim}{r}'')|^2, \tag{5.3.1}$$

where

$$\int |\psi(\underset{\sim}{r}'')|^2 d\underset{\sim}{r}'' = Z = \text{number of electrons.}$$

We can then write

$$\underset{\sim}{E}(\vec{\underset{\sim}{r}},t) = \hat{1}_\phi \frac{e^2 E_O \sin\phi}{mc^2}\frac{e^{i(\underset{\sim}{k}'\cdot\underset{\sim}{r}-\omega t)}}{r}\int |\psi|^2 e^{i(\underset{\sim}{k}-\underset{\sim}{k}')\cdot\underset{\sim}{r}''}d\underset{\sim}{r}''. \tag{5.3.2}$$

We set

$$\underset{\sim}{K} = \underset{\sim}{k} - \underset{\sim}{k}', \tag{5.3.3}$$

with

$$|\underset{\sim}{K}| = 2k \sin \frac{\theta}{2} = \frac{4\pi}{\lambda} \sin \frac{\theta}{2}, \tag{5.3.4}$$

and define the *atomic scattering factor* as

$$\begin{aligned} f_a &= \int |\psi(\underset{\sim}{r}'')|^2 \, e^{i(\underset{\sim}{k} - \underset{\sim}{k}') \cdot \underset{\sim}{r}''} d\underset{\sim}{r}'' \\ &= \int |\psi(\underset{\sim}{r}'')|^2 \, e^{i\underset{\sim}{K} \cdot \underset{\sim}{r}''} d\underset{\sim}{r}''. \end{aligned} \tag{5.3.5}$$

We also assume that $|\psi|^2$ is spherically symmetric:

$$|\psi(\underset{\sim}{r}'')|^2 = |\psi(r'')|^2. \tag{5.3.6}$$

Then

$$\begin{aligned} f_a &= \int_0^\infty dr'' r''^2 \int_0^\pi d\alpha \sin\alpha \int_0^{2\pi} d\beta |\psi(r'')|^2 \, e^{i\underset{\sim}{K} \cdot \underset{\sim}{r}''} \\ &= 2\pi \int_0^\infty dr'' |\psi(r'')|^2 r''^2 \int_0^\pi d\alpha \sin\alpha \, e^{iKr''} \cos\alpha, \end{aligned} \tag{5.3.7}$$

having taken the z direction in the direction of $\underset{\sim}{K}$. Let us solve the integral

$$\int_0^\pi \sin\alpha \, e^{i(\cos\alpha)Kr''} d\alpha.$$

Set

$$\cos\alpha = s$$
$$-\sin\alpha \, d\alpha = ds$$

$$\begin{aligned} \int_0^\pi \sin\alpha \, e^{iK(\cos\alpha)r''} d\alpha &= \int_{-1}^1 e^{iKr''s} ds \\ &= \left[\frac{e^{iKr''s}}{iKr''} \right]_{-1}^1 \\ &= \frac{e^{iKr''} - e^{-iKr''}}{iKr''} = \frac{2 \sin Kr''}{Kr''}. \end{aligned} \tag{5.3.8}$$

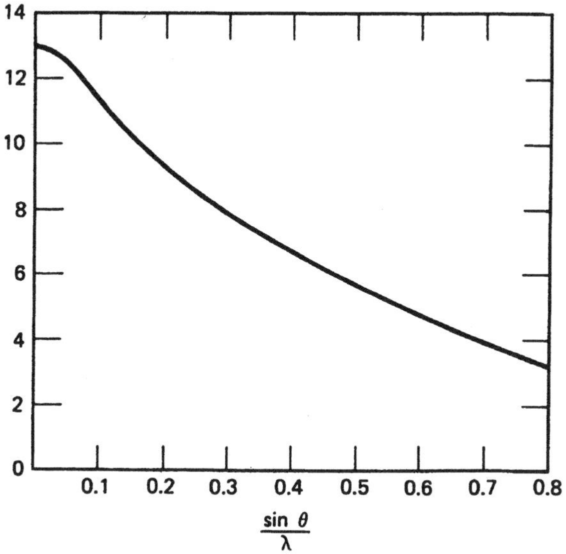

Fig. 5.3. Atomic scattering factor for metallic aluminum.[1]

Then

$$f_a(\theta) = 4\pi \int r''^2 |\psi(r'')|^2 dr'' \frac{\sin Kr''}{Kr''}, \tag{5.3.9}$$

where K= $4\pi/\lambda \sin\theta/2$. In the forward direction, corresponding to $\theta = 0$,

$$f_a(0) = 4\pi \int |\psi(r'')|^2 r''^2 dr'' = Z. \tag{5.3.10}$$

The integration to find $f_a(\theta)$ can be performed if we know $\psi(r)$ as a function of r (as given for example by the Hartree-Fock solution or by the Thomas-Fermi model). f_a has been calculated for all atoms; see Fig. 5.3 for an example.[1–3]

By X-ray scattering from gases we can get some information about the distribution of electrons in an atom. We can write the expression of the scattered field as follows:

$$\underset{\sim}{E}(\underset{\sim}{r}, t) = \hat{1}_\phi \frac{e^2 E_O \sin\phi}{mc^2} \frac{e^{i(\underset{\sim}{k}' \cdot \underset{\sim}{r} - \omega t)}}{r} f_a(\theta). \tag{5.3.11}$$

5.4. Scattering from the Atoms in the Unit Cell of a Crystal

The wave scattered from an atom is given by

$$\underset{\sim}{E}(\underset{\sim}{r}, t) = \hat{1}_\phi \frac{e^2 E_O}{mc^2} \sin\phi \frac{e^{i(\underset{\sim}{k}' \cdot \underset{\sim}{r} - \omega t)}}{r} f_a(\theta). \tag{5.4.1}$$

Consider now the unit cell of a crystal, and assume that it contains J atoms. If an atom is located at the distance $\underset{\sim}{r}_s$ from the origin, then [see (7.2.11)]

1. the amplitude of the incident wave compared to that at the origin contains the factor $e^{i\underset{\sim}{k} \cdot \underset{\sim}{r}_s}$ and
2. the amplitude of the scattered wave compared to that of an atom at the origin contains the factor $e^{-i\underset{\sim}{k}' \cdot \underset{\sim}{r}_s}$.

This means that we must multiply $\underset{\sim}{E}(\underset{\sim}{r}, t)$ in (5.4.1) by the factor

$$e^{i(\underset{\sim}{k} - \underset{\sim}{k}') \cdot \underset{\sim}{r}_s} = e^{-i\underset{\sim}{K} \cdot \underset{\sim}{r}_s}. \tag{5.4.2}$$

Summing over the atoms in the unit cell we obtain

$$\underset{\sim}{E}(\underset{\sim}{r}, t) = \hat{1}_\phi \frac{e^2 E_O}{mc^2} \sin\phi \frac{e^{i(\underset{\sim}{k}' \cdot \underset{\sim}{r} - \omega t)}}{r} \sum_{s=1}^{J} e^{i\underset{\sim}{K} \cdot \underset{\sim}{r}_s} f_{as}(\theta), \tag{5.4.3}$$

where f_{as} is the atomic scattering factor of the atom located at $\underset{\sim}{r}_s$. We call the quantity

$$S = \sum_{s=1}^{J} e^{i\underset{\sim}{K} \cdot \underset{\sim}{r}_s} f_{as}(\theta) = \sum_{s=1}^{J} e^{i\underset{\sim}{K} \cdot \underset{\sim}{r}_s} \int |\psi_s(r'')|^2 \, e^{i\underset{\sim}{K} \cdot \underset{\sim}{r}''} dr'', \tag{5.4.4}$$

the *structure amplitude* and the quantity S^2 the *structure factor*. We notice that S is the Fourier transform of the electron density due to all the atoms in the unit cell.

5.5. Scattering from a Crystal

Consider a crystal with the primitive basic vectors $\underset{\sim}{a}_1, \underset{\sim}{a}_2, \underset{\sim}{a}_3$, and N_i unit cells in the $\underset{\sim}{a}_i$ direction. The generic unit cell will be at a position

$$\underset{\sim}{R}_n = n_1\underset{\sim}{a}_1 + n_2\underset{\sim}{a}_2 + n_3\underset{\sim}{a}_3, \tag{5.5.1}$$

where $n_i = 1, 2, \ldots, N_i$. Assume that the unit cell contains J atoms. The scattered X-ray wave is given by

$$\underset{\sim}{E}(\underset{\sim}{r}, t) = \hat{1}_\phi \frac{e^2 E_O}{mc^2} \sin\phi \frac{e^{i(\underset{\sim}{k}'\cdot\underset{\sim}{r}-\omega t)}}{r} \sum_{s=1}^{J} e^{i\underset{\sim}{K}\cdot\underset{\sim}{r}_s} f_{as}(\theta) \sum_{n_1, n_2, n_3} e^{i\underset{\sim}{K}\cdot\underset{\sim}{R}_n}, \tag{5.5.2}$$

where

$$K = \underset{\sim}{k} - \underset{\sim}{k}',$$
$$n_i = 1, 2, \ldots, N_i.$$

The structure amplitude is a slowly varying function of θ. The quantity

$$I = \sum_{n_1, n_2, n_3} e^{i\underset{\sim}{K}\cdot\underset{\sim}{R}_n}, \tag{5.5.3}$$

is defined as *intensity function*, and, considering the very large number of atoms in a crystal ($N \simeq 10^{23}$), is a very rapidly varying function. We can write

$$\begin{aligned} I &= \sum_{n_1=1}^{N_1} e^{i\underset{\sim}{K}\cdot n_1\underset{\sim}{a}_1} \sum_{n_2=1}^{N_2} e^{i\underset{\sim}{K}\cdot n_2\underset{\sim}{a}_2} \sum_{n_3=1}^{N_3} e^{i\underset{\sim}{K}\cdot n_3\underset{\sim}{a}_3} \\ &= e^{i\underset{\sim}{K}\cdot(\underset{\sim}{a}_1+\underset{\sim}{a}_2+\underset{\sim}{a}_3)} \frac{1-e^{iN_1\underset{\sim}{K}\cdot\underset{\sim}{a}_1}}{1-e^{i\underset{\sim}{K}\cdot\underset{\sim}{a}_1}} \frac{1-e^{iN_2\underset{\sim}{K}\cdot\underset{\sim}{a}_2}}{1-e^{i\underset{\sim}{K}\cdot\underset{\sim}{a}_2}} \frac{1-e^{iN_3\underset{\sim}{K}\cdot\underset{\sim}{a}_3}}{1-e^{i\underset{\sim}{K}\cdot\underset{\sim}{a}_3}}, \end{aligned} \tag{5.5.4}$$

and

$$\begin{aligned} I_o = |I|^2 &= \frac{1-\cos N_1\underset{\sim}{K}\cdot\underset{\sim}{a}_1}{1-\cos\underset{\sim}{K}\cdot\underset{\sim}{a}_1} \frac{1-\cos N_2\underset{\sim}{K}\cdot\underset{\sim}{a}_2}{1-\cos\underset{\sim}{K}\cdot\underset{\sim}{a}_2} \frac{1-\cos N_3\underset{\sim}{K}\cdot\underset{\sim}{a}_3}{1-\cos\underset{\sim}{K}\cdot\underset{\sim}{a}_3} \\ &= \frac{\sin^2\frac{1}{2}N_1\underset{\sim}{K}\cdot a_1}{\sin^2\frac{1}{2}\underset{\sim}{K}\cdot\underset{\sim}{a}_1} \frac{\sin^2\frac{1}{2}N_2\underset{\sim}{K}\cdot\underset{\sim}{a}_2}{\sin^2\frac{1}{2}\underset{\sim}{K}\cdot\underset{\sim}{a}_2} \frac{\sin^2\frac{1}{2}N_3\underset{\sim}{K}\cdot\underset{\sim}{a}_3}{\sin^2\frac{1}{2}\underset{\sim}{K}\cdot\underset{\sim}{a}_3}. \end{aligned} \tag{5.5.5}$$

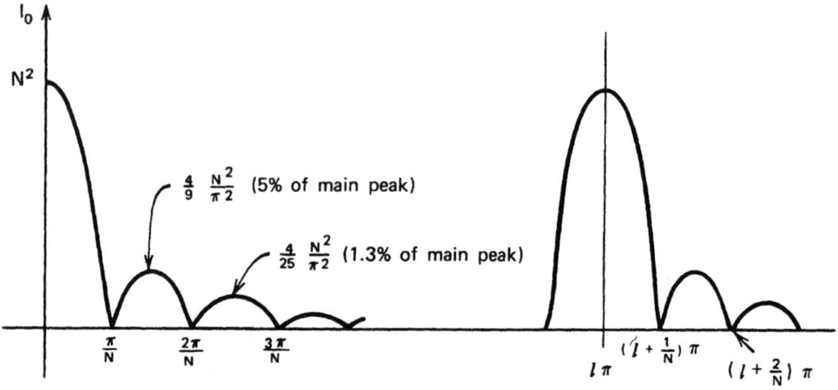

Fig. 5.4. The function $|I_O|^2$.

The function $\sin^2 Nx/\sin^2 x$ has maxima at $x = 0, \pi, 2\pi, \ldots, \ell\pi, \ldots$ and for these values of x it takes the value N^2. This function is graphically represented in Fig. 5.4.

The directions in which there will be a constructive superposition are given by

$$\begin{aligned} \frac{1}{2}\underset{\sim}{K}\cdot\underset{\sim}{a}_1 &= \pi s', \\ \frac{1}{2}\underset{\sim}{K}\cdot\underset{\sim}{a}_2 &= \pi t', \\ \frac{1}{2}\underset{\sim}{K}\cdot\underset{\sim}{a}_3 &= \pi u', \end{aligned} \tag{5.5.6}$$

where s', t', u' are integer numbers. The above equations can be written in the following way:

$$\begin{aligned} \underset{\sim}{K}\cdot\underset{\sim}{a}_1 &= 2\pi ns, \\ \underset{\sim}{K}\cdot\underset{\sim}{a}_2 &= 2\pi nt, \\ \underset{\sim}{K}\cdot\underset{\sim}{a}_3 &= 2\pi nu, \end{aligned} \tag{5.5.7}$$

where n is the maximum common factor of s', t' and u':

$$\begin{aligned} ns &= s', \\ nt &= t', \\ nu &= u', \end{aligned} \tag{5.5.8}$$

The equations (5.5.7) above are called *Laue equations*. We know that

$$|\underset{\sim}{K}| = \frac{4\pi}{\lambda} \sin \frac{1}{2}\theta. \tag{5.5.9}$$

Then we can rewrite the Laue equations in the following form:

$$\begin{aligned} \hat{1}_{\underset{\sim}{K}} \cdot \frac{\underset{\sim}{a}_1}{s} &= \frac{n\lambda}{2\sin\frac{1}{2}\theta}, \\ \hat{1}_{\underset{\sim}{K}} \cdot \frac{\underset{\sim}{a}_2}{t} &= \frac{n\lambda}{2\sin\frac{1}{2}\theta}, \\ \hat{1}_{\underset{\sim}{K}} \cdot \frac{\underset{\sim}{a}_3}{u} &= \frac{n\lambda}{2\sin\frac{1}{2}\theta}, \end{aligned} \tag{5.5.10}$$

$\underset{\sim}{a}_1/s, \underset{\sim}{a}_2/t$, and $\underset{\sim}{a}_3/u$ may be thought of as the intercepts of a plane with the three axes $\underset{\sim}{a}_1, \underset{\sim}{a}_2$ and $\underset{\sim}{a}_3$, respectively. This plane is perpendicular to the vector $\underset{\sim}{K}$ and is therefore the reflecting plane of $\underset{\sim}{k}$ and $\underset{\sim}{k}'$. From (5.5.9) the distance of this plane from the origin is given by

$$\hat{1}_{\underset{\sim}{K}} \cdot \frac{\underset{\sim}{a}_1}{s} = \hat{1}_{\underset{\sim}{K}} \cdot \frac{\underset{\sim}{a}_2}{t} = \hat{1}_{\underset{\sim}{K}} \cdot \frac{\underset{\sim}{a}_3}{u} = \frac{n\lambda}{2\sin\frac{1}{2}\theta}. \tag{5.5.11}$$

The reinforcement of the scattered wave occurs when the incident wave is reflected off a plane whose Miller indices are s, t, and u (see Fig. 5.5).

It is interesting to calculate the distance between two adjacent reflecting planes, which is simply given by

$$\begin{aligned} \hat{1}_{\underset{\sim}{K}} \cdot \left[\frac{2\underset{\sim}{a}_1}{s} - \frac{\underset{\sim}{a}_1}{s}\right] &= \hat{1}_{\underset{\sim}{K}} \cdot \left[\frac{2\underset{\sim}{a}_2}{t} - \frac{\underset{\sim}{a}_2}{t}\right] \\ &= \hat{1}_{\underset{\sim}{K}} \cdot \left[\frac{2\underset{\sim}{a}_3}{u} - \frac{\underset{\sim}{a}_3}{u}\right] \\ &= \frac{n\lambda}{2\sin\frac{1}{2}\theta} = d(stu). \end{aligned} \tag{5.5.12}$$

We can then write

$$n\lambda = 2d(stu)\ \sin \frac{1}{2}\theta. \tag{5.5.13}$$

The above formula is the expression of the so-called *Bragg's Law* (see Fig. 5.6). The number n gives the order of the reflection.

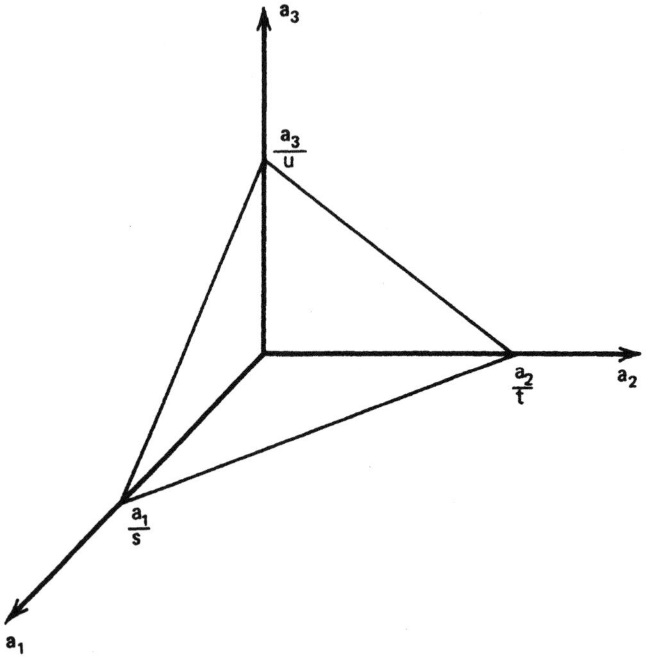

Fig. 5.5. A reflecting plane.

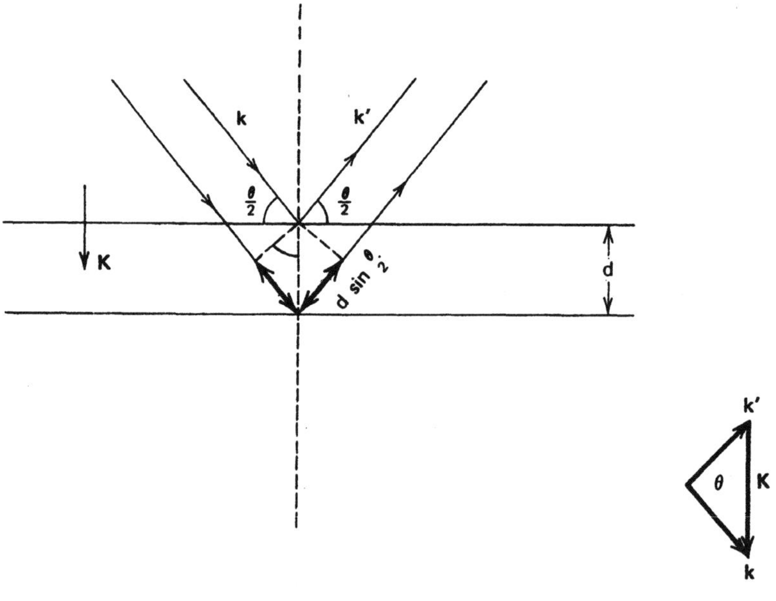

Fig. 5.6. Illustration of Bragg's law.

5.6. Interpretation of Laue Equations in Reciprocal Space

The basic primitive vectors of the reciprocal lattice are

$$\begin{aligned} \underset{\sim}{b}_1 &= 2\pi \frac{\underset{\sim}{a}_2 \times \underset{\sim}{a}_3}{\underset{\sim}{a}_1 \cdot \underset{\sim}{a}_2 \times \underset{\sim}{a}_3}, \\ \underset{\sim}{b}_2 &= 2\pi \frac{\underset{\sim}{a}_3 \times \underset{\sim}{a}_1}{\underset{\sim}{a}_2 \cdot \underset{\sim}{a}_3 \times \underset{\sim}{a}_1}, \\ \underset{\sim}{b}_3 &= 2\pi \frac{\underset{\sim}{a}_1 \times \underset{\sim}{a}_2}{\underset{\sim}{a}_3 \cdot \underset{\sim}{a}_1 \times \underset{\sim}{a}_2}, \end{aligned} \tag{5.6.1}$$

and have the property

$$\underset{\sim}{b}_i \cdot \underset{\sim}{a}_j = 2\pi\delta_{ij}. \tag{5.6.2}$$

We can express $\underset{\sim}{K}$ as a vector of the reciprocal space as follows:

$$\underset{\sim}{K} = \nu_1 \underset{\sim}{b}_1 + \nu_2 \underset{\sim}{b}_2 + \nu_3 \underset{\sim}{b}_3 \tag{5.6.3}$$

The Laue equations can now be written

$$\begin{aligned} \underset{\sim}{K} \cdot \underset{\sim}{a}_1 &= \nu_1 2\pi \frac{\underset{\sim}{a}_1 \cdot \underset{\sim}{a}_2 \times \underset{\sim}{a}_3}{\underset{\sim}{a}_1 \cdot \underset{\sim}{a}_2 \times \underset{\sim}{a}_3} = 2\pi\nu_1 = 2\pi ns, \\ \underset{\sim}{K} \cdot \underset{\sim}{a}_2 &= \nu_2 2\pi \frac{\underset{\sim}{a}_2 \cdot \underset{\sim}{a}_3 \times \underset{\sim}{a}_1}{\underset{\sim}{a}_2 \cdot \underset{\sim}{a}_3 \times \underset{\sim}{a}_1} = 2\pi\nu_2 = 2\pi nt, \\ \underset{\sim}{K} \cdot \underset{\sim}{a}_3 &= \nu_3 2\pi \frac{\underset{\sim}{a}_3 \cdot \underset{\sim}{a}_1 \times \underset{\sim}{a}_2}{\underset{\sim}{a}_3 \cdot \underset{\sim}{a}_1 \times \underset{\sim}{a}_2} = 2\pi\nu_3 = 2\pi nu. \end{aligned} \tag{5.6.4}$$

or

$$\begin{aligned} \nu_1 &= ns, \\ \nu_2 &= nt, \\ \nu_3 &= nu. \end{aligned} \tag{5.6.5}$$

Since n, s, t, and u are integer numbers, the vector $\underset{\sim}{K}$ in the Laue equations is a primitive vector of the reciprocal lattice. This finding says that $\underset{\sim}{K} = \underset{\sim}{k} - \underset{\sim}{k}'$ *must be a primitive vector of the reciprocal lattice in order for* $\underset{\sim}{k}'$ *to be in the direction of a scattering maximum.*

A simple geometrical construction due to Ewald and shown in Fig. 5.7 tells us which are the possible directions of maximum scattering. These directions are simply given by the intersection of the

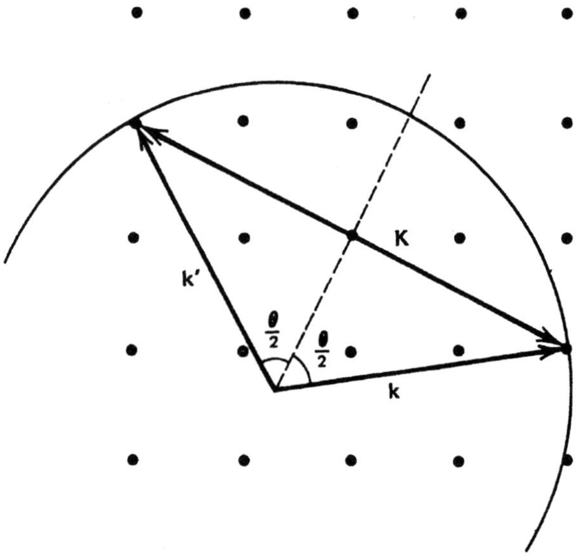

Fig. 5.7. The Ewald sphere in reciprocal space.

sphere with points of the reciprocal lattice. If the sphere corresponding to a certain $\underset{\sim}{k}$ does not pass over any point of the reciprocal lattice, that particular $\underset{\sim}{k}$ does not produce any scattering maximum. We notice also that the sphere, because of the interference pattern shown in Fig. 5.4, has some *thickness*.

At this point we make some consideration in regard to the vector $\underset{\sim}{K}$ that fulfills the Laue equations:

1. $\underset{\sim}{K}$ is perpendicular to all the planes in real spaces whose Miller indices are s, t, and u.
2. The length of $\underset{\sim}{K}$ is

$$\underset{\sim}{K} \cdot \frac{\underset{\sim}{a}_1}{s} = 2\pi n = |\underset{\sim}{K}| \left[\hat{1}_{\underset{\sim}{K}} \cdot \frac{\underset{\sim}{a}_1}{s}\right]. \tag{5.6.6}$$

But

$$\hat{1}_{\underset{\sim}{K}} \cdot \frac{\underset{\sim}{a}_1}{s} = d(stu). \tag{5.6.7}$$

Then

$$|\underset{\sim}{K}| = \frac{2\pi n}{d(stu)}. \tag{5.6.8}$$

where n represents the order of the reflection. A reflection of order 2 can be considered as a reflection from planes spaced by d/2.

3. We cannot have *any* order of reflection. Since

$$|\underset{\sim}{K}| = \frac{4\pi}{\lambda} \sin \frac{1}{2}\theta,$$

the value of $|\underset{\sim}{K}|$ is limited to a maximum value $4\pi/\lambda$, which corresponds to $\theta = \pi$ or backward scattering.

4. For n = 1 it is

$$\nu_1 = s$$
$$\nu_2 = t$$
$$\nu_3 = u$$

and

$$\underset{\sim}{K} = s\underset{\sim}{b}_1 + t\underset{\sim}{b}_2 + u\underset{\sim}{b}_3$$

5. The smallest $\underset{\sim}{K}$ is of the order of $2\pi/a$, where a is the lattice constant of the order of 10^{-8} cm; also, $|\underset{\sim}{k}| = |\underset{\sim}{k}'|$ must be at least one-half the length of the smallest $\underset{\sim}{K}$. For this reason we can say that the lattice does not scatter light unless the wavelength is in the X-ray region or shorter.

5.7. Methods of X-Ray Diffraction

It may be interesting at this point to illustrate briefly some experimental methods used to perform X-ray diffraction measurements.

5.7.1. *The Laue Method (see Fig. 5.8a)*

The beam is not monochromatic. There is a range of values of $\underset{\sim}{k}$ available. The Ewald sphere has a continuous range of radii and every time this sphere passes through a point in the reciprocal lattice we get a scattering maximum.

The advantage of this method is that it may be used very efficiently for the determination of the crystal symmetry, scattering point results in the scattering in a particular direction; a photographic plate records a network of points belonging to one point of the reciprocal lattice.

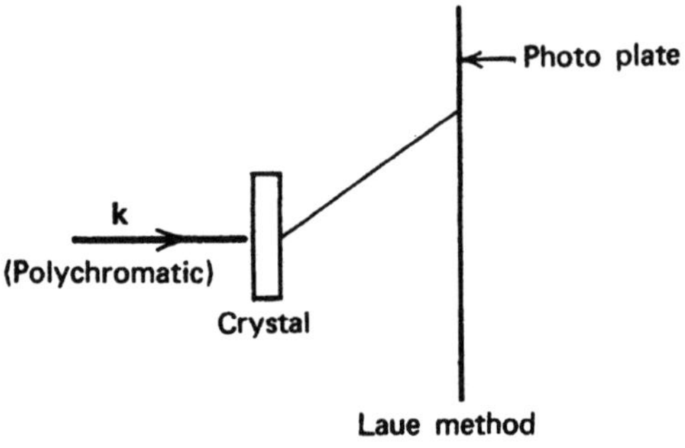

(a)

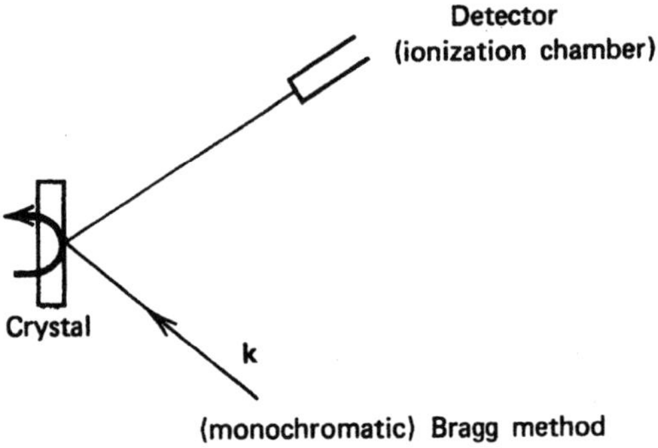

(b)

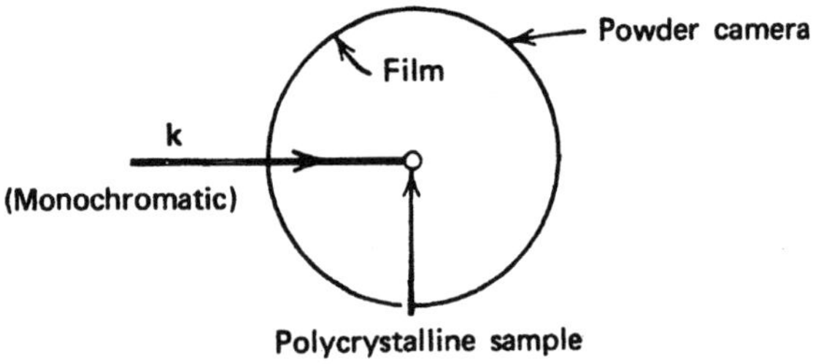

(c)

Fig. 5.8. Methods of X-ray diffraction.

5.7.2. *The Bragg Method (see Fig. 5.8b)*

This method uses monochromatic radiation, but varies the direction of the crystal. In actuality the crystal is rotated about a suitable axis. The effect of this is to keep the Ewald sphere fixed but to rotate the reciprocal lattice. During the rotation some lattice points will cross the sphere, thus making the scattering possible.

5.7.3. *The Debye-Scherrer Method (see Fig. 5.8c)*

The sample used in this case is not a single crystal, but consists of either a polycrystalline material or a powder. In either case we have here many small crystals with random orientations; this is equivalent to having the same reciprocal lattice in all possible different orientations. In this case each reciprocal lattice vector $\underset{\sim}{K}$ yields scattered radiation covering a cone with the direction of the incident radiation as axis. The radiation used is in this case monochromatic.

References

1. B. W. Batterman, D. R. Chipman and J. J. DeMarco, *Phys. Rev.* 122, 68 (1961).
2. L. D. Jennings, D. R. Chipman and J. J. DeMarco, *Phys. Rev.* 135, 1612 (1964).
3. von S. Göttlicher and E. Wolfel, *Z. Electrochem.* 63, 891 (1959).

Part II

Lattice Vibrations of Crystals

In the second part of this book we first introduce models of lattice vibrations in crystals of different dimensions. We then apply group theory to the treatment of lattice vibrations and present important examples of this application. We then deal with the theories of specific heat of solids and with the Lindemann law of melting. We follow with a treatment of the effects of vibrations on the X-ray scattering signal. Finally we present a theory that is at the basis of the neutron scattering technique and examine the application of this technique to the study of lattice vibrations.

Chapter 6

Lattice Vibrations of Crystals

In this chapter we introduce models of lattice vibrations in linear crystals and in three-dimensional crystals. We then apply group theory to the treatment of lattice vibrations and present two important examples of the group-theoretical treatment of two two-dimensional array of atoms.

6.1. The Infinite Linear Crystal

Let us consider a unidimensional infinite crystal consisting of an array of atoms of equal masses.

Let us call d_n the position coordinate of the nth particle given by

$$d_n = R_n + u_n, \tag{6.1.1}$$

where $R_n = na$ is the equilibrium position and a is the equilibrium distance between two successive particles as indicated in Fig. 6.1. The reciprocal lattice and the Brillouin zone for such a crystal are given in Fig. 6.2.

Let us assume that the potential energy between two particles is given by a function

$$v_{ij} = v(d_i - d_j) = v(|d_i - d_j|) = v(d_j - d_i). \tag{6.1.2}$$

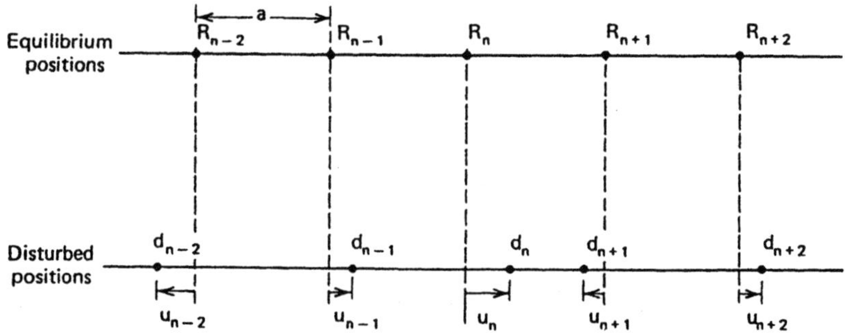

Fig. 6.1. A linear array of identical atoms.

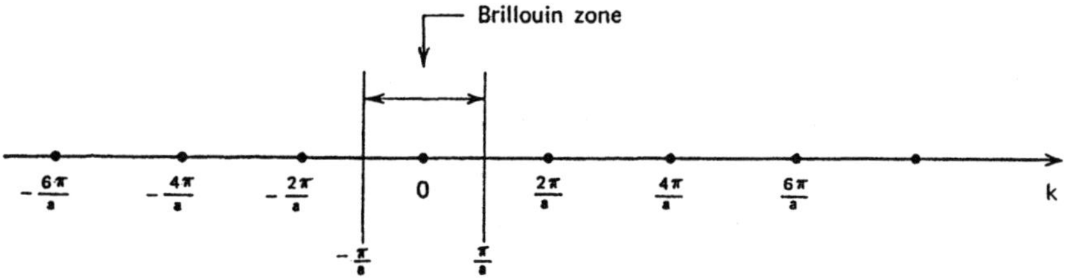

Fig. 6.2. Reciprocal lattice and Brillouin zone of a unidimensional crystal.

The total potential energy is given by

$$V = V(d_1, d_2, \ldots, d_i, \ldots) = \sum_{\substack{i,j \\ i \neq j}} \frac{1}{2} v(d_i - d_j). \tag{6.1.3}$$

We may expand V_{ij} in a power series as follows:

$$\begin{aligned} v_{ij} = v(d_i - d_j) = v(R_i - R_j) + u_i \left(\frac{\partial v_{ij}}{\partial d_i}\right)_o - u_j \left(\frac{\partial v_{ij}}{\partial d_j}\right)_o \\ + \frac{1}{2} u_i^2 \left(\frac{\partial^2 v_{ij}}{\partial d_i^2}\right)_o + \frac{1}{2} u_j^2 \left(\frac{\partial^2 v_{ij}}{\partial d_j^2}\right)_o \\ + u_i u_j \left(\frac{\partial^2 v_{ij}}{\partial d_i \partial d_j}\right)_o + \cdots . \end{aligned} \tag{6.1.4}$$

Neglecting terms higher than quadratic in u (harmonic approximation),[1]

$$
\begin{aligned}
V &= \frac{1}{2}\sum_{\substack{i,j\\ i\neq j}} v(d_i - d_j)\\
&= \frac{1}{2}\sum_{\substack{i,j\\ i\neq j}} v(R_i - R_j) + \frac{1}{2}\sum_{\substack{i,j\\ i\neq j}} (u_i - u_j)\left(\frac{\partial v_{ij}}{\partial d_i}\right)_o\\
&\quad + \frac{1}{2}\sum_{\substack{i,j\\ i\neq j}} u_i^2\left(\frac{\partial^2 v_{ij}}{\partial d_i^2}\right)_o + \frac{1}{2}\sum_{\substack{i,j\\ i\neq j}} u_i u_j\left(\frac{\partial^2 v_{ij}}{\partial d_i \partial d_j}\right)_o .
\end{aligned} \tag{6.1.5}
$$

Since the forces are zero at equilibrium, we may drop the second term in the sum and write

$$
V = V_o + \frac{1}{2}\sum_{ij} A_{ij} u_i u_j, \tag{6.1.6}
$$

where

$$
V_o = \frac{1}{2}\sum_{\substack{i,j\\ i\neq j}} v(R_i - R_j), \tag{6.1.7}
$$

and

$$
\begin{aligned}
A_{ij} &= \left(\frac{\partial^2 v_{ij}}{\partial d_i\, \partial d_j}\right)_o = A_{ji}\\
A_{ii} &= \sum_{\substack{j\\ j\neq i}} \left(\frac{\partial^2 v_{ij}}{\partial d_i^2}\right)_o .
\end{aligned} \tag{6.1.8}
$$

The elements A_{ij} define a matrix $\underset{\sim}{A}$, which has the following properties:

1. It is real and symmetrical:

$$
A_{ij} = A_{ij}^* = A_{ji}. \tag{6.1.9}
$$

2. The force acting on the sth atom is given by

$$F_s = -\frac{\partial V}{\partial u_s} = -\sum_{s'} A_{ss'} u_{s'}. \tag{6.1.10}$$

Therefore $A_{ss'}$ represents the absolute value of the force acting on the atom at the R_s site, because of a unit displacement of the atom at the $R_{s'}$ site. Since the force between two atoms depends only on their relative position, we can write

$$A_{ss'} = A^{s-s'} = A^n, \tag{6.1.11}$$

where

$$R_n = R_s - R_{s'}. \tag{6.1.12}$$

3. If we add a constant, arbitrary displacement c to all the $u_{s'}$ in (6.1.10) we get

$$F_s = -\sum_{s'} A_{ss'}(u_{s'} + c). \tag{6.1.13}$$

Subtracting (6.1.10) from (6.1.13) we find

$$\sum_{s'} A_{ss'} = 0. \tag{6.1.14}$$

The equation of motion of the generic atom is given by

$$m\ddot{u}_s = -\sum_{s'} A_{ss'} u_{s'}. \tag{6.1.15}$$

In order to decouple the above equations we may use wave solutions of the type

$$u_s(t) = u_o(k) e^{-i\omega t + ikR_s}. \tag{6.1.16}$$

More general solutions can be built up by making linear combinations of the expressions (6.1.16).

Putting (6.1.16) in (6.1.15) we obtain the dispersion relation

$$m\omega^2 = \sum_{s'} A_{ss'} e^{-ik(R_s - R_{s'})}, \tag{6.1.17}$$

or

$$\omega^2 = G(k), \tag{6.1.18}$$

where

$$\begin{aligned} G(k) &= \frac{1}{m}\sum_{s'} A_{ss'}e^{-ik(R_s-R_{s'})} \\ &= \frac{1}{m}\sum_{n} A^n e^{-ikR_n}. \end{aligned} \tag{6.1.19}$$

G(k) has the following properties:

1. It is real:

$$\begin{aligned} G^*(k) &= \frac{1}{m}\sum_{n} A^n e^{ikR_n} \\ &= \frac{1}{m}\sum_{n} A^{-n} e^{-ikR_n} = \frac{1}{m}\sum_{n} A^n e^{-ikR_n} = G(k). \end{aligned} \tag{6.1.20}$$

   Therefore $\omega^2(k) = \omega^2(-k)$.

2. It is periodical in the reciprocal lattice. If

$$K_r = r\frac{2\pi}{a}, \tag{6.1.21}$$

   represents a primitive translation in the reciprocal space,

$$\begin{aligned} G(k+K_r) &= \frac{1}{m}\sum_{n} A_n e^{-i(k+K_r)R_n} \\ &= \frac{1}{m}\sum_{n} A_n e^{-ikR_n} e^{-ir(2\pi/a)na} \\ &= \frac{1}{m}\sum_{n} A_n e^{-ikR_n} e^{-i(rn)2\pi} \\ &= \frac{1}{m}\sum_{n} A_n e^{-ikR_n} = G(k), \end{aligned} \tag{6.1.22}$$

   where we have taken into account the fact that $R_n = na$.

From the first property it follows that there is one value of ω^2 corresponding to every value of k. We can choose $\omega(k)$ to be positive. It is also clear that $\omega(k)$ is periodical in the reciprocal lattice.

We can write the expression for the generic displacement as a superposition of wave solutions,

$$u_s(t) = \sum_{BZ} C(k) e^{ikR_s - i\omega(k)t}, \tag{6.1.23}$$

where C(k) are, in general, complex constants and the sum is extended over the Brillouin zone, because of the periodicity of C(k) and $\omega(k)$.

Finally, because of (6.1.14),

$$G(0) = \frac{1}{m} \sum_n A^n = 0 = \omega(0). \tag{6.1.24}$$

From (6.1.16) it can be seen that all the displacements are equal in correspondence to $\omega = \omega(0) = 0$.

6.2. The Finite Linear Crystal

We consider now a finite crystal with N equal atoms.

1. We identify the equilibrium configuration of the finite crystal with the equilibrium configuration of N atoms of the infinite linear lattice.
2. We impose the so-called *Born-Von Karman boundary conditions:*

$$u_n(t) = u_{n+mN}(t), \tag{6.2.1}$$

   where m is an integer number.

Because of the conditions above, we have from (6.1.16)

$$e^{ik(R_n + mNa)} = e^{ikR_n}, \tag{6.2.2}$$

or

$$e^{ikmNa} = 1, \tag{6.2.3}$$

that is,

$$\mathrm{Nka} = 2\pi \times \text{integer}. \tag{6.2.4}$$

Let us call the integer in (6.2.4) r; then

$$\mathrm{k} = \frac{2\pi \mathrm{r}}{\mathrm{Na}}, \tag{6.2.5}$$

with $\mathrm{r} = 0, \pm 1, \pm 2, \ldots, \pm \mathrm{N}/2$ and

$$-\frac{\pi}{\mathrm{a}} \leq \mathrm{k} \leq \frac{\pi}{\mathrm{a}}. \tag{6.2.6}$$

Every k corresponds to a distinct point in the Brillouin zone; the density of these points is greater the greater is the length of the crystal.

Let us consider now the dispersion relation (6.1.17). If we put $\mathrm{s}' - \mathrm{s} = \ell$ we obtain

$$\begin{aligned} \mathrm{m}\omega^2 &= \sum_{\mathrm{s}'} \mathrm{A}_{\mathrm{ss}'} \mathrm{e}^{\mathrm{i}(\mathrm{s}'-\mathrm{s})\mathrm{ka}} \\ &= \sum_{\ell=1-\mathrm{s}}^{\mathrm{N}-\mathrm{s}} \mathrm{A}_\ell \mathrm{e}^{\mathrm{i}\ell\mathrm{ka}}. \end{aligned} \tag{6.2.7}$$

The value of ω^2 is independent of s; therefore taking $\mathrm{s} = 1$ we get

$$\mathrm{m}\omega^2 = \sum_{\ell=0}^{\mathrm{N}-1} \mathrm{A}_\ell \mathrm{e}^{\mathrm{i}\ell\mathrm{ka}} = \sum_{\ell=0}^{\mathrm{N}-1} \mathrm{A}_\ell \cos \ell\mathrm{ka}. \tag{6.2.8}$$

Therefore $\omega^2(-\mathrm{k}) = \omega^2(\mathrm{k})$. On the other hand we can write

$$\mathrm{A}_\mathrm{o} = \sum_{\ell=1}^{\mathrm{N}-1} \left(\frac{\partial^2 \mathrm{v}(\mathrm{d})}{\partial \mathrm{d}^2}\right)_{\mathrm{d}=\ell\mathrm{a}} \tag{6.2.9}$$

$$\mathrm{A}_\ell = \left(\frac{\partial^2 \mathrm{v}(\mathrm{d})}{\partial \mathrm{d}_\mathrm{s} \partial \mathrm{d}_{\mathrm{s}'}}\right)_\mathrm{o} = -\left(\frac{\partial^2 \mathrm{v}(\mathrm{d})}{\partial \mathrm{d}^2}\right)_{\mathrm{d}=\ell\mathrm{a}}, \tag{6.2.10}$$

where d is the relative distance between atoms s and s′. Using (6.2.9) and (6.2.10) in (6.2.8) we find

$$\begin{aligned} m\omega^2 &= A_o + \sum_{\ell=1}^{N-1} A_\ell e^{i\ell ka} \\ &= \sum_{\ell=1}^{N-1} \left[\left(\frac{\partial^2 v}{\partial d^2} \right)_{d=\ell a} (1 - e^{i\ell ka}) \right], \end{aligned} \tag{6.2.11}$$

or

$$m\omega^2 = \sum_{\ell=0}^{N-1} \left[\left(\frac{\partial^2 v}{\partial d^2} \right)_{d=\ell a} (1 - e^{i\ell ka}) \right]. \tag{6.2.12}$$

Example. Let us consider now the case in which only the forces between nearest neighbors are non-negligible. In this case

$$\begin{aligned} m\omega^2 &= 2 \left(\frac{\partial^2 v}{\partial d^2} \right)_{d=a} (1 - \cos ka) \\ &= 2\beta(1 - \cos ka), \end{aligned} \tag{6.2.13}$$

where

$$\beta = \left(\frac{\partial^2 v}{\partial d^2} \right)_{d=a} = \text{force constant}. \tag{6.2.14}$$

We may express (6.2.13) as

$$m\omega^2 = 2\beta(1 - \cos ka) = 4\beta \sin^2 \frac{1}{2}ka, \tag{6.2.15}$$

and

$$\omega = 2\sqrt{\frac{\beta}{m}} \sin \frac{ka}{2}. \tag{6.2.16}$$

The sinusoidal dependence of ω on k is represented in Fig. 6.3.

The group velocity of the running waves is given by

$$c = \frac{d\omega}{dk} = a\sqrt{\frac{\beta}{m}} \cos \frac{1}{2}ka, \tag{6.2.17}$$

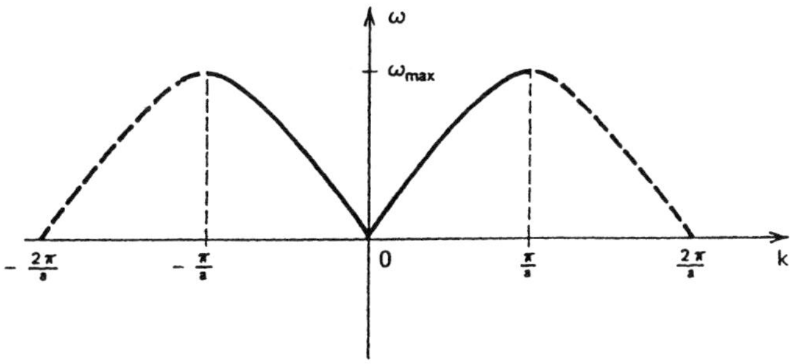

Fig. 6.3. Dispersion relation for a linear crystal.

and

$$\left(\frac{d\omega}{dk}\right)_{k=0} = a\sqrt{\frac{\beta}{m}} = c_{max}. \tag{6.2.18}$$

Then

$$\omega = \frac{2c_{max}}{a}\sin\frac{ka}{2}, \tag{6.2.19}$$

and

$$\omega_{max} = \frac{2c_{max}}{a}. \tag{6.2.20}$$

The smallest wavelength is given by

$$\lambda_{min} = \frac{2\pi}{k_{max}} = \frac{2\pi}{\pi/a} = 2a. \tag{6.2.21}$$

6.3. Normal Modes of Vibration of a Linear Crystal

The skeleton solution for the displacement of the atom at the position s is given, according to (6.1.16), by

$$u_s(t) = u_o(k)e^{-i\omega t}e^{iska}. \tag{6.3.1}$$

We may express $u_s(t)$ as

$$u_s(t) = q_k(t)e^{iska}. \tag{6.3.2}$$

The most general displacement is given by

$$u_s(t) = \sum_{k=-\pi/a}^{k=\pi/a} q_k(t)e^{iska}. \tag{6.3.3}$$

Since u_s must be real,

$$q_{-k}(t) = q_k^*(t).$$

Therefore only one-half of the coordinates q_k are independent but each q_k contains two real variables, its absolute value and its phase and the total number of independent variables is N.

Let us express now the kinetic and potential energies in terms of these new coordinates q_k. We have for the kinetic energy

$$T = \frac{1}{2}m\sum_{s=0}^{N-1}\dot{u}_s^2 = \frac{1}{2}m\sum_k\sum_{k'}\dot{q}_k\dot{q}_{k'}\sum_{s=0}^{N-1}e^{is(k+k')a}. \tag{6.3.4}$$

Let us consider the sum

$$S = \sum_{s=0}^{N-1}e^{is(k+k')a}. \tag{6.3.5}$$

This sum is extended to all the atoms of the crystal. The displacement of each atomic site by the interatomic distance results in the relabeling of the terms of the sum, but does not change the sum. Because of this displacement, each term in (6.3.5) is multiplied by $e^{i(k+k')a}$ and

$$S = Se^{i(k+k')a}. \tag{6.3.6}$$

Therefore $S = 0$ unless

$$k + k' = 0 \quad \text{or} \quad k = -k'. \tag{6.3.7}$$

The condition $k = k' + K$ with K a primitive translation of the reciprocal lattice has to be excluded because both k and k' are taken within the Brillouin zone. The sum S, on the other hand, is equal to

$$S = \sum_{s=0}^{N-1}e^{is(k+k')a} = \frac{1-e^{iN(k+k')a}}{1-e^{i(k+k')a}}. \tag{6.3.8}$$

As $k \to -k'$, this sum tends to N. Therefore we can write

$$S = \sum_{s=0}^{N-1} e^{is(k+k')a} = N\delta_{k,-k'}. \tag{6.3.9}$$

Using the above equality in (6.3.4) we get

$$\begin{aligned} T &= \frac{1}{2} m \sum_{k}\sum_{k'} \dot{q}_k \dot{q}_{k'} N\delta_{k,-k'} \\ &= \frac{1}{2} Nm \sum_{k=-\pi/a}^{\pi/a} \dot{q}_k \dot{q}_{-k} = \frac{1}{2} M \sum_{k=-\pi/a}^{\pi/a} |\dot{q}_k|^2, \end{aligned} \tag{6.3.10}$$

where $M = Nm$ is the total mass of the crystal.

The potential energy is given by

$$\begin{aligned} V &= V_o + \frac{1}{2} \sum_{ss'} A_{ss'} u_s u_{s'} \\ &= V_o + \frac{1}{2} \sum_{ss'} A_{ss'} \sum_{kk'} q_k q_{k'} e^{i(ska+s'k'a)} \\ &= V_o + \frac{1}{2} \sum_{kk'} q_k q_{k'} \sum_{s'=0}^{N-1} e^{is'(k+k')a} \sum_{s=0}^{N-1} A_{ss'} e^{i(s-s')ka} \\ &= V_o + \frac{1}{2} \sum_{kk'} q_k q_{k'} \sum_{s'=0}^{N-1} e^{is'(k+k')a} \sum_{\ell=-s'}^{N-s'-1} A_\ell e^{i\ell ka} \\ &= V_o + \frac{1}{2} \sum_{kk'} q_k q_{k'} N\delta_{k,-k'} m\omega_k^2 \\ &= V_o + \frac{1}{2} Nm \sum_{k=-\pi/a}^{\pi/a} \omega_k^2 |q_k|^2 \\ &= V_o + \frac{1}{2} M \sum_{k=-\pi/a}^{\pi/a} \omega_k^2 |q_k|^2, \end{aligned} \tag{6.3.11}$$

where we have taken into account (6.3.9) and (6.2.7).

The Lagrangian of the system is

$$L = T - V = \frac{1}{2}M \sum_{k=-\pi/a}^{\pi/a} |\dot{q}_k|^2 - V_o - \frac{1}{2}M \sum_{k=-\pi/a}^{\pi/a} \omega_k^2 |q_k|^2. \tag{6.3.12}$$

The equations of motion arc given by

$$\frac{d}{dt}\frac{\partial L}{\partial \dot{q}_{-k}} - \frac{\partial L}{\partial q_{-k}} = 0. \tag{6.3.13}$$

Then

$$\frac{\partial L}{\partial \dot{q}_k} = M\dot{q}_{-k} = p_k = \text{generalized momentum}, \tag{6.3.14}$$

$$\frac{\partial L}{\partial q_k} = -M\omega_k^2 q_{-k}, \tag{6.3.15}$$

and the (6.3.13) becomes

$$M\ddot{q}_k + M\omega_k^2 q_k = 0. \tag{6.3.16}$$

Therefore the q_k's are *normal coordinates.*

We may introduce the Hamiltonian of the system as follows:

$$H = \sum_k (p_k \dot{q}_k - L) = H(p_k, q_k). \tag{6.3.17}$$

Differentiating (6.3.17) we get

$$\begin{aligned} dH &= \sum_k \left(\frac{\partial H}{\partial q_k} dq_k + \frac{\partial H}{\partial p_k} dp_k \right) \\ &= \sum_k \left(p_k d\dot{q}_k + \dot{q}_k dp_k - \frac{\partial L}{\partial q_k} dq_k - \frac{\partial L}{\partial \dot{q}_k} d\dot{q}_k \right) \\ &= \sum_k \left(\dot{q}_k dp_k - \frac{\partial L}{\partial q_k} dq_k \right) = \sum_k (\dot{q}_k dp_k - \dot{p}_k dq_k). \end{aligned} \tag{6.3.18}$$

The canonical or Hamilton's equations are given by

$$\dot{q}_k = \frac{\partial H}{\partial p_k},$$
$$\dot{p}_k = -\frac{\partial H}{\partial q_k}. \tag{6.3.19}$$

On the other hand,

$$p_k = \frac{\partial L}{\partial \dot{q}_k} = \frac{\partial T}{\partial \dot{q}_k}, \tag{6.3.20}$$

and

$$\sum_k p_k \dot{q}_k = \sum_k \left(\frac{\partial T}{\partial \dot{q}_k} \dot{q}_k \right) = M \sum_k \dot{q}_k \dot{q}_{-k} = 2T. \tag{6.3.21}$$

Using this equality in (6.3.17) we get

$$H = \sum_k (p_k \dot{q}_k - L) = T + V$$
$$= \sum_k \left(\frac{|p_k|^2}{2M} + \frac{M}{2} \omega_k^2 |q_k|^2 \right) + V_o. \tag{6.3.22}$$

The Hamilton's equations give

$$\dot{p}_k = -\frac{\partial H}{\partial q_k} = -M\omega_k^2 q_{-k}, \tag{6.3.23}$$

$$\dot{q}_k = \frac{\partial H}{\partial p_k} = \frac{1}{M} p_{-k}. \tag{6.3.24}$$

We can then write

$$u_s = \sum_k e^{iska} q_k,$$
$$P_s = m\dot{u}_s = m \sum_k \dot{q}_k e^{iska} = \frac{1}{N} \sum_k e^{-iska} p_k, \tag{6.3.25}$$

and the inverse relations

$$
\begin{aligned}
q_k &= \frac{1}{N}\sum_s u_s e^{-iska}, \\
p_k &= \sum_s p_s e^{iska}.
\end{aligned}
\tag{6.3.26}
$$

We can now make a transition to quantum mechanics by considering u_s, p_s, q_k, and p_k as operators. We start from the commutation relation

$$
[u_s, p_{s'}] = i\hbar\delta_{ss'}, \tag{6.3.27}
$$

and using also (6.3.26) we find

$$
\begin{aligned}
[q_k, p_k] &= \frac{1}{N}\sum_{ss'}(u_s\, e^{-iska} p_{s'}\, e^{is'ka} - p_{s'}\, e^{is'ka} u_s\, e^{-iska}) \\
&= \frac{1}{N}\sum_{ss'} e^{i(s'-s)ka}[u_s, p_{s'}] \\
&= \frac{1}{N}\sum_{ss'} e^{i(s'-s)ka}\delta_{ss'}\, i\hbar \\
&= i\hbar, \qquad (6.3.28)\\
[q_k, p_{k'}] &= i\hbar\delta_{kk'}. \qquad (6.3.29)
\end{aligned}
$$

At this point, in order to solve the Schrödinger equation and find the energy levels of the lattice vibrations, we introduce the dimensionless operators

$$
\begin{aligned}
a_k &= \left(\frac{M\omega_k}{2\hbar}\right)^{1/2}\left(q_k + \frac{i}{M\omega_k}p_{-k}\right), \\
a_k^+ &= \left(\frac{M\omega_k}{2\hbar}\right)^{1/2}\left(q_{-k} - \frac{i}{M\omega_k}p_k\right).
\end{aligned}
\tag{6.3.30}
$$

These operators are called *annihilation* and *creation* operators, respectively. q_k and p_k can be expressed in terms of these operators as

$$
q_k = \left(\frac{\hbar}{2M\omega_k}\right)^{1/2}(a_k + a_{-k}^+), \tag{6.3.31}
$$

$$
p_k = \left(\frac{\hbar M\omega_k}{2}\right)^{1/2}\frac{1}{i}(a_{-k} - a_k^+).
$$

The commutation relations of the operators a_k and a_k^+ can be derived from (6.3.28). We find

$$[a_k, a_{k'}^+] = \delta_{kk'}, \tag{6.3.32}$$

and

$$[a_k, a_{k'}] = [a_k^+, a_{k'}^+] = 0. \tag{6.3.33}$$

The Hamiltonian can now be expressed as

$$\begin{aligned} H &= \sum_k \hbar\omega_k \left(a_k^+ a_k + \frac{1}{2} \right) \\ &= \sum_k \frac{\hbar\omega_k}{2} (a_k a_k^+ + a_k^+ a_k), \end{aligned} \tag{6.3.34}$$

where we have neglected the term V_o, which adds only a constant to the energy. Therefore the system is equivalent to a collection of N independent harmonic oscillators. Let us examine a single oscillator represented by the Hamiltonian

$$H = \frac{\hbar\omega}{2}(aa^+ + a^+a) = \hbar\omega \left(a^+a + \frac{1}{2} \right). \tag{6.3.35}$$

The commutators of a and a^+ with H are given by

$$\begin{aligned} [H, a] &= \hbar\omega[a^+a, a] \\ &= \hbar\omega\{[a^+, a]a + a^+[a, a]\} = -\hbar\omega a, \end{aligned} \tag{6.3.36}$$

$$\begin{aligned} [H, a^+] &= \hbar\omega[a^+a, a^+] \\ &= \hbar\omega\{[a^+, a^+]a + a^+[a, a^+]\} = \hbar\omega a^+. \end{aligned} \tag{6.3.37}$$

We have also the following relations:

$$\begin{aligned} \langle m|[H, a]|n\rangle &= \langle m|Ha - aH|n\rangle \\ &= \langle m|Ha|n\rangle - \langle m|aH|n\rangle \\ &= (E_m - E_n)\langle m|a|n\rangle = -\hbar\omega\langle m|a|n\rangle, \end{aligned}$$

or

$$(E_m - E_n + \hbar\omega)\langle m|a|n\rangle = 0. \tag{6.3.38}$$

Using the same procedure for $[\mathrm{H}, \mathrm{a}^+]$ we get

$$(\mathrm{E_m} - \mathrm{E_n} - \hbar\omega)\langle \mathrm{m}|\mathrm{a}^+|\mathrm{n}\rangle = 0. \tag{6.3.39}$$

We have also

$$\mathrm{a}^+\mathrm{a} = \frac{\mathrm{H}}{\hbar\omega} - \frac{1}{2} = \frac{1}{\hbar\omega}\left(\mathrm{H} - \frac{\hbar\omega}{2}\right), \tag{6.3.40}$$

and

$$\begin{aligned}\langle \mathrm{n}|\mathrm{a}^+\mathrm{a}|\mathrm{n}\rangle &= \frac{1}{\hbar\omega}\left(\mathrm{E_n} - \frac{\hbar\omega}{2}\right)\\ &= \sum_{\mathrm{t}} \langle \mathrm{n}|\mathrm{a}^+|\mathrm{t}\rangle\langle \mathrm{t}|\mathrm{a}|\mathrm{n}\rangle\\ &= \sum_{\mathrm{t}} |\langle \mathrm{n}|\mathrm{a}^+|\mathrm{t}\rangle|^2 \geq 0.\end{aligned} \tag{6.3.41}$$

Let us now show that the energy $\mathrm{E_o}$ of the lowest level is $\hbar\omega/2$. First, we can say that, because of (6.3.41), $\mathrm{E_o} \geq \hbar\omega/2$. If we assume that $\mathrm{E_o} > \hbar\omega/2$,

$$\sum_{\mathrm{t}} |\mathrm{a}^+_{\mathrm{ot}}|^2 = \frac{1}{\hbar\omega}\left(\mathrm{E_o} - \frac{\hbar\omega}{2}\right) > 0, \tag{6.3.42}$$

which implies that there is some t for which $\mathrm{a_{ot}}$ is $\neq 0$.

On the other hand, because of (6.3.39)

$$(\mathrm{E_o} - \mathrm{E_t} - \hbar\omega)\mathrm{a}^+_{\mathrm{ot}} = 0, \tag{6.3.43}$$

and

$$\mathrm{E_o} = \mathrm{E_t} + \hbar\omega, \tag{6.3.44}$$

which is contrary to our assumption that $\mathrm{E_o}$ was the lowest state. With this we have then proved that the energy of the lowest state is $\hbar\omega/2$.

Let us call now $\mathrm{E_1}$ the energy of the first excited state. From (6.3.41)

$$\sum_{\mathrm{t}} |\mathrm{a}^+_{1\mathrm{t}}|^2 = \frac{1}{\hbar\omega}\left(\mathrm{E_1} - \frac{\hbar\omega}{2}\right) > 0, \tag{6.3.45}$$

which implies that there is at least one t for which $a^+_{1t} \neq 0$. But from (6.3.39)

$$(E_1 - E_t - \hbar\omega)a^+_{1t} = 0. \tag{6.3.46}$$

Then

$$E_1 - E_t - \hbar\omega = 0, \tag{6.3.47}$$

and

$$E_t = E_1 - \hbar\omega. \tag{6.3.48}$$

E_t can only be $\hbar\omega/2$. Then

$$E_1 = \left(1 + \frac{1}{2}\right)\hbar\omega. \tag{6.3.49}$$

We can, in the same way, show that

$$E_n = \left(n + \frac{1}{2}\right)\hbar\omega. \tag{6.3.50}$$

If the energy of the oscillator is E_n, we say that the oscillator is excited to the nth level. From (6.3.38) we can now derive the fact that $\langle m|a|n\rangle \neq 0$ only when $m = n - 1$, and from (6.3.39) the fact that $\langle m|a^+|n\rangle \neq 0$ only when $m = n + 1$. Thus,

$$\begin{aligned} E_n = \langle n|H|n\rangle &= \left(n + \frac{1}{2}\right)\hbar\omega \\ &= \hbar\omega\langle n|a^+a|n\rangle + \frac{1}{2}\hbar\omega \\ &= \hbar\omega\sum_m(\langle n|a^+|m\rangle\langle m|a|n\rangle) + \frac{1}{2}\hbar\omega \\ &= \hbar\omega(\langle n|a^+|n-1\rangle\langle n-1|a|n\rangle) + \frac{1}{2}\hbar\omega \\ &= \hbar\omega|\langle n-1|a|n\rangle|^2 + \frac{1}{2}\hbar\omega, \end{aligned} \tag{6.3.51}$$

and

$$\begin{aligned} E_{n+1} = \langle n+1|H|n+1\rangle &= \left(n + 1 + \frac{1}{2}\right)\hbar\omega \\ &= \hbar\omega\langle n+1|a^+a|n+1\rangle + \frac{1}{2}\hbar\omega \end{aligned}$$

$$= \hbar\omega \sum_m (\langle n+1|a^+|m\rangle\langle m|a|n+1\rangle) + \frac{1}{2}\hbar\omega$$
$$= \hbar\omega(\langle n+1|a^+|n\rangle\langle n|a|n+1\rangle) + \frac{1}{2}\hbar\omega$$
$$= \hbar\omega|\langle n+1|a^+|n\rangle|^2 + \frac{1}{2}\hbar\omega. \qquad (6.3.52)$$

Then

$$\langle n-1|a|n\rangle = \sqrt{n},$$
$$\langle n+1|a^+|n\rangle = \sqrt{n+1}, \qquad (6.3.53)$$

and

$$a|n\rangle = \sqrt{n}|n-1\rangle,$$
$$a^+|n\rangle = \sqrt{n+1}|n+1\rangle. \qquad (6.3.54)$$

Also

$$\langle n|a^+a|n\rangle = n. \qquad (6.3.55)$$

The enerqy levels of the whole system represented by the Hamiltonian (6.3.34) are given by

$$E_{\{n_1,n_2,\ldots\}} = \sum_k \left(n_k + \frac{1}{2}\right)\hbar\omega_k, \qquad (6.3.56)$$

and the eigenfunction by

$$|n_1, n_2 \cdots n_{k_i} \cdots n_N\rangle = |n_1\rangle|n_2\rangle \cdots |n_{k_i}\rangle \cdots |n_N\rangle. \qquad (6.3.57)$$

The displacement and the momentum of the generic atom in terms of the a_k and a_k^+ operators are given by

$$u_s = \sum_k q_k e^{iska} = \sum_k \left(\frac{\hbar}{2M\omega_k}\right)^{1/2} e^{iska}(a_k + a_{-k}^+) \qquad (6.3.58)$$

$$m\dot{u}_s = m\sum_k \dot{q}_k e^{iska} = \frac{1}{N}\sum_k p_k e^{-iska}$$
$$= \sum_k \left(\frac{\hbar m\omega_k}{2N}\right)^{1/2} \frac{1}{i} e^{-iska}(a_{-k} - a_k^+). \qquad (6.3.59)$$

6.4. Linear Crystal with a Basis

Let us consider now a linear crystal with N unit cells and J atoms in the unit cell. The J atoms in a cell form the *basis* of the crystal. Let m_ν be the mass of the νth atom ($\nu = 1, 2, \ldots, J$) and $u_{s\nu}$ the displacement of the νth atom in the unit cell at $R_s = sa$.

The kinetic and potential energies are now given by

$$T = \frac{1}{2}\sum_{s=1}^{N}\sum_{\nu=1}^{J} m_\nu \dot{u}_{s\nu}^2, \tag{6.4.1}$$

$$V = \frac{1}{2}\sum_{s=1}^{N}\sum_{s'=1}^{N}\sum_{\nu=1}^{J}\sum_{\nu'=1}^{J} A_{\nu\nu'}^{ss'} u_{s\nu} u_{s'\nu'} + V_o, \tag{6.4.2}$$

where

$$A_{\nu\nu'}^{ss'} = \left[\frac{\partial^2 V}{\partial u_{s\nu}\partial u_{s'\nu'}}\right]_o. \tag{6.4.3}$$

As in the previous case of one atom per unit cell, the $NJ \times NJ$ matrix $\underset{\sim}{A}$ is symmetrical and real:

$$A_{\nu\nu'}^{ss'} = A_{\nu'\nu}^{s's} = A_{\nu\nu'}^{n}, \tag{6.4.4}$$

where n is the subscript of $R_n = R_s - R_{s'}$.

The NJ equations of motion are given by

$$m_\nu \ddot{u}_{s\nu} = -\sum_{s'\nu'} A_{\nu\nu'}^{ss'} u_{s'\nu'}. \tag{6.4.5}$$

Similarly to (6.1.14)

$$\sum_{s'\nu'} A_{\nu\nu'}^{ss'} = 0. \tag{6.4.6}$$

In order to decouple these equations (6.4.5) we look for solutions of the type

$$u_{s\nu} = V_\nu e^{-i\omega t + ikR_s}. \tag{6.4.7}$$

By using this expression in (6.4.5) we obtain

$$m_\nu \omega^2 V_\nu = \sum_{s'\nu'} V_{\nu'} e^{ik(R_{s'} - R_s)} A_{\nu\nu'}^{ss'}, \tag{6.4.8}$$

or

$$m_\nu \omega^2 V_\nu = \sum_{\nu'} G_{\nu\nu'}(k)\, V_{\nu'}, \tag{6.4.9}$$

where

$$G_{\nu\nu'}(k) = \sum_{s'} A^{ss'}_{\nu\nu'} e^{ik(R_{s'}-R_s)}. \tag{6.4.10}$$

The matrix $\underset{\sim}{G}$ has dimensions $J \times J$ and has the following properties:

1. It is Hermitian:

$$\begin{aligned} G_{\nu\nu'}(k)^* &= \sum_n A^n_{\nu\nu'} e^{ikR_n} \\ &= \sum_n A^{-n}_{\nu\nu'} e^{-ikR_n} \\ &= \sum_n A^n_{\nu'\nu} e^{-ikR_n} = G_{\nu'\nu}(k). \end{aligned} \tag{6.4.11}$$

2. It is periodic in the reciprocal lattice. If K is a primitive translation in the reciprocal space.

$$G_{\nu\nu'}(k+K) = G_{\nu\nu'}(k)e^{-iKR_n} = G_{\nu\nu'}(k). \tag{6.4.12}$$

The equations (6.4.9) give J positive values for ω^2 in correspondence to each value of k. This means that the vibrational spectrum consists of J *branches*.

Let us examine what happens to the J branches of the vibrational spectrum at $k = 0$. Equation (6.4.9) becomes

$$m_\nu \omega^2_{o\nu} V_\nu = \sum_{s'\nu'} A^{ss'}_{\nu\nu'} V_{\nu'}. \tag{6.4.13}$$

One solution of this equation is simply given when all V_ν are equal to some constant number c. In this case

$$m_\nu \omega^2_{o\nu} = \sum_{s'\nu'} A^{ss'}_{\nu\nu'} = 0. \tag{6.4.14}$$

Therefore, for zero frequency, (6.4.13) has one solution corresponding to a uniform displacement of all atoms.

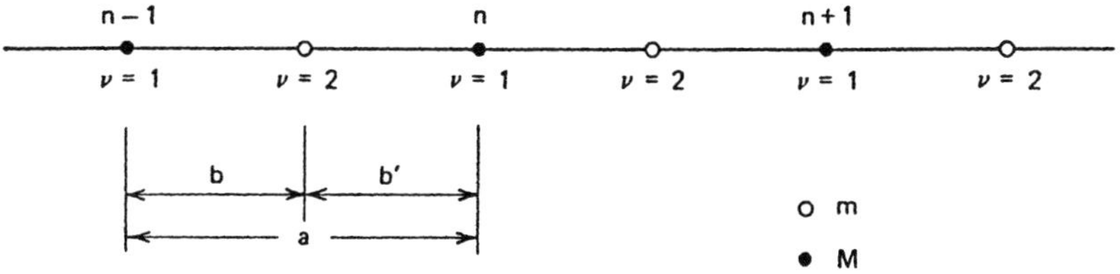

Fig. 6.4. Linear crystal with a two-atom basis.

Example. Let us consider now the crystal in Fig. 6.4 with two different types of atoms and b = b′. Let us assume that only the nearest neighbor interactions are non-negligible. The equations of motion for the two types of atoms are given by

$$\begin{aligned} M\ddot{u}_{n1} &= \beta u_{n2} + \beta u_{n-1,2} - 2\beta u_{n1}, \\ M\ddot{u}_{n2} &= \beta u_{n+1,1} + \beta u_{n1} - 2\beta u_{n2}. \end{aligned} \tag{6.4.15}$$

It is

$$\begin{aligned} G_{11} &= \sum_{s'} A_{11}^{ss'} e^{ik(R_{s'}-R_s)} = +2\beta, \\ G_{22} &= \sum_{s'} A_{22}^{ss'} e^{ik(R_{s'}-R_s)} = +2\beta, \\ G_{12} &= \sum_{s'} A_{12}^{ss'} e^{ik(R_{s'}-R_s)} = -\beta(1+e^{-ika}), \\ G_{21} &= \sum_{s'} A_{21}^{ss'} e^{ik(R_{s'}-R_s)} = -\beta(1+e^{ika}). \end{aligned} \tag{6.4.16}$$

The dispersion relation can be derived from (6.4.9), which gives the condition

$$\begin{pmatrix} G_{11} - M\omega^2 & G_{12} \\ G_{21} & G_{22} - m\omega^2 \end{pmatrix} = \begin{pmatrix} 2\beta - M\omega^2 & -\beta(1+e^{-ika}) \\ -\beta(1+e^{-ika}) & 2\beta - m\omega^2 \end{pmatrix} = 0. \tag{6.4.17}$$

We find

$$\omega^2 = \beta\left(\frac{1}{M} + \frac{1}{m}\right) \pm \beta\left[\left(\frac{1}{M} + \frac{1}{m}\right)^2 - \frac{2}{mM}(1 - \cos ka)\right]^{\frac{1}{2}}. \tag{6.4.18}$$

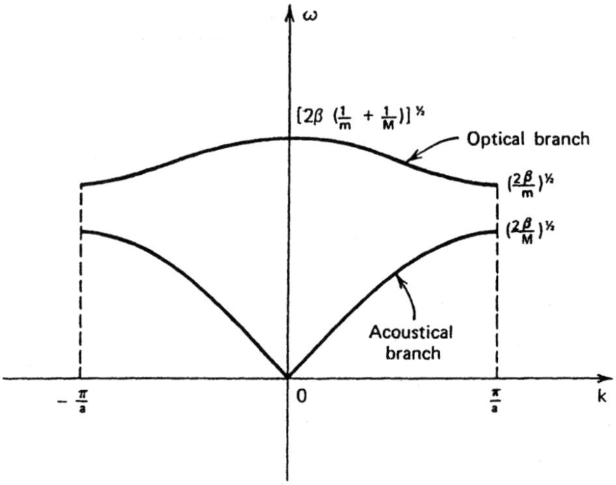

Fig. 6.5. Dispersion curves for a linear crystal with a two-atom basis.

The function ω^2 is periodic in k with period $2\pi/a$. The Brillouin zone is given by the interval $-\pi/a \leq k \leq \pi/a$. The dispersion relation $\omega = \omega(k)$ is represented in Fig. 6.5. The two solutions in (6.4.18) in correspondence to the + and − signs are called *optical* and *acoustical* branches, respectively.

For long wavelengths (small k),

$$\omega_+^2 = 2\beta\left(\frac{1}{M} + \frac{1}{m}\right) \quad \text{optical,} \tag{6.4.19}$$

$$\omega_-^2 = \frac{2\beta k^2 b^2}{m + M} \quad \text{acoustical.} \tag{6.4.20}$$

For $k = \pm\frac{\pi}{a}$,

$$\omega_+^2 = \frac{2\beta}{m} \quad \text{optical,} \tag{6.4.21}$$

$$\omega_-^2 = \frac{2\beta}{M} \quad \text{acoustical.} \tag{6.4.22}$$

We may say a few words on the shape of the dispersion curves.

1. The slope of the $\omega = \omega(k)$ branches is zero for values of k corresponding to the edges of the Brillouin zone. This follows from the following facts: (a) The curves $\omega = \omega(k)$ axe symmetric in k,

(b) ω is periodical in the reciprocal lattice, (c) $d\omega/dk$ is a continuous function of k.

2. The slope of the $\omega = \omega(k)$ branch other than the acoustical is zero in correspondence to $k = 0$.

It is interesting to investigate the two modes of vibration in correspondence to some special points of the Brillouin zone, such as $k = 0$ and $k = \pi/a$.

The equations that give the eigenvalues of the frequency and the polarization vectors V_1 and V_2 are given by

$$\begin{aligned} M\omega^2 V_1 &= G_{11}V_1 + G_{12}V_2, \\ m\omega^2 V_2 &= G_{21}V_1 + G_{22}V_2, \end{aligned} \tag{6.4.23}$$

where

$$\begin{aligned} G_{11} &= G_{22} = 2\beta, \\ G_{12} &= -\beta(1 + e^{-ika}), \quad G_{21} = -\beta(1 + e^{ika}). \end{aligned}$$

For $k = 0, G_{12} = G_{21} = -2\beta$ and the equations above become

$$\begin{aligned} (2\beta - M\omega^2)V_1 - 2\beta V_2 &= 0, \\ -2\beta V_1 + (2\beta - m\omega^2)V_2 &= 0. \end{aligned} \tag{6.4.24}$$

The determinant of these equations must be zero:

$$(2\beta - M\omega^2)(2\beta - m\omega^2) - 4\beta^2 = 0, \tag{6.4.25}$$

or

$$\omega^4 - 2\beta\frac{m + M}{mM}\omega^2 = 0, \tag{6.4.26}$$

with the two solutions

$$\begin{aligned} \omega^2 &= 0, \\ \omega^2 &= 2\beta\frac{m + M}{mM} = 2\beta\left(\frac{1}{M} + \frac{1}{m}\right). \end{aligned} \tag{6.4.27}$$

In correspondence to $\omega^2 = 0$:

$$\begin{aligned} 2\beta V_1 - 2\beta V_2 &= 0, \\ -2\beta V_1 + 2\beta V_2 &= 0, \end{aligned} \tag{6.4.28}$$

or

$$V_1 = V_2.$$

In correspondence to $\omega^2 = 2\beta(1/M + 1/m)$,

$$\begin{aligned} \left[2\beta - 2\beta M\left(\frac{1}{M} + \frac{1}{m}\right)\right] V_1 - 2\beta V_2 = 0, \\ -2\beta V_1 + \left[2\beta - 2\beta m\left(\frac{1}{M} + \frac{1}{m}\right)\right] V_2 = 0, \end{aligned} \tag{6.4.29}$$

or

$$V_1 = -\frac{m}{M} V_2. \tag{6.4.30}$$

Using the normalization condition [see (6.5.46)]

$$M|V_1(k,\lambda)|^2 + m|V_2(k,\lambda)|^2 = M + m, \tag{6.4.31}$$

we can put the polarization vectors in the matrix form:

$$k = 0, \quad \omega = 0: \begin{pmatrix} V_1 \\ V_2 \end{pmatrix} = \begin{pmatrix} 1 \\ 1 \end{pmatrix},$$

$$k = 0, \quad \omega = \left[2\beta\left(\frac{1}{m} + \frac{1}{M}\right)\right]^{\frac{1}{2}}: \begin{pmatrix} V_1 \\ V_2 \end{pmatrix} = \begin{pmatrix} \sqrt{m/M} \\ -\sqrt{M/m} \end{pmatrix}.$$

The two modes of vibration in correspondence to $k = 0$ are represented in Fig. 6.6.

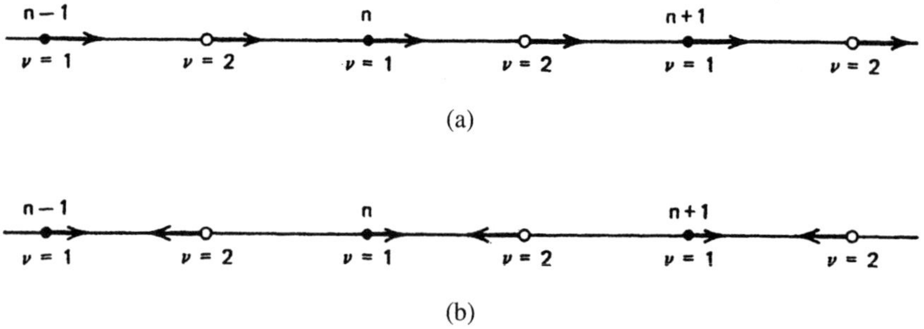

Fig. 6.6. Normal modes of a diatomic linear crystal for $k = 0$. (a) $\omega = 0$; $V_1 = V_2$; Γ_1. (b) $\omega = [2\beta(1/m + 1/M)]^{\frac{1}{2}}$; $V_1 = -(m/M)V_2$; Γ_2.

For $k = \pi/a$, $G_{12} = G_{21} = 0$ and the equations (6.4.23) become

$$(2\beta - M\omega^2)V_1 = 0,$$
$$(2\beta - m\omega^2)V_2 = 0. \qquad (6.4.32)$$

The two eigenvalues of the frequency are given by

$$\omega = \sqrt{2\beta/M},$$
$$\omega = \sqrt{2\beta/m}. \qquad (6.4.33)$$

In correspondence to $\omega = \sqrt{2\beta/M}$ or $\omega^2 = 2\beta/M$,

$$V_1 \text{ arbitrary},$$
$$V_2 = 0. \qquad (6.4.34)$$

In correspondence to $\omega = \sqrt{2\beta/m}$ or $\omega^2 = 2\beta/m$,

$$V_1 = 0,$$
$$V_2 \text{ arbitrary}. \qquad (6.4.35)$$

This means that for the latter mode the atoms labeled 1 in the unit cells will not move and for the former mode the atoms labeled 2 in the unit cells will not move. Using the normalization condition above we can put the polarization vectors in the matrix form

$$k = \frac{\pi}{a}, \quad \omega = \sqrt{2\beta/m} : \begin{pmatrix} V_1 \\ V_2 \end{pmatrix} = \begin{pmatrix} \left(1 + \frac{m}{M}\right)^{\frac{1}{2}} \\ 0 \end{pmatrix},$$

$$k = \frac{\pi}{a}, \quad \omega = \sqrt{2\beta/M} : \begin{pmatrix} V_1 \\ V_2 \end{pmatrix} = \begin{pmatrix} 0 \\ \left(1 + \frac{M}{m}\right)^{\frac{1}{2}} \end{pmatrix}.$$

The two modes of vibration in correspondence to $k = \pi/a$ are represented in Fig. 6.7.

6.5. Lattice Vibrations in Three Dimensions

6.5.1. *The Equations of Motion*

Let us consider now a three-dimensional crystal with a primitive unit cell defined by the three vectors $\underset{\sim}{a}_1, \underset{\sim}{a}_2$, and $\underset{\sim}{a}_3$ and J atoms in the unit cell. Let us also assume that the crystal has N_1 cells in the

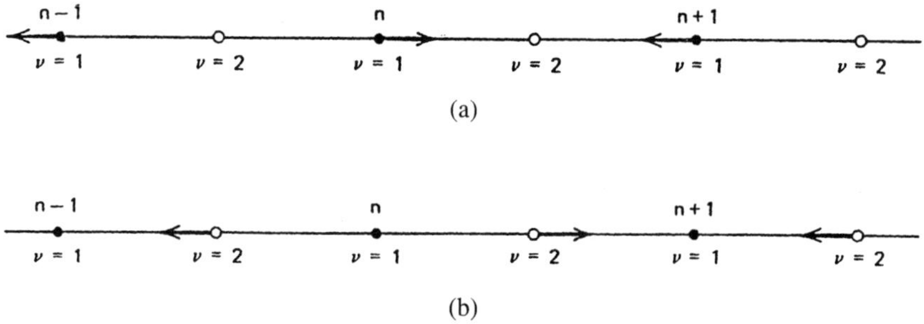

Fig. 6.7. Normal modes of a diatomic linear crystal for $k = \pi/a$. (a) $\omega = \sqrt{2\beta/M}$; V_1 arbitrary, $V_2 = 0$; Γ_2. (b) $\omega = \sqrt{2\beta/m}$; $V_1 = 0$, V_2 arbitrary; Γ_1.

direction $\underset{\sim}{a}_1, N_2$ in the direction $\underset{\sim}{a}_2$, and N_3 in the direction $\underset{\sim}{a}_3$, with $N = N_1N_2N_3$. Let m_ν be the mass of the ν-th atom $(\nu = 1, 2, \ldots, J)$ and $\underset{\sim}{u}_{s\nu}$ the displacement of the νth atom of the unit cell at

$$\underset{\sim}{R}_s = s_1\underset{\sim}{a}_1 + s_2\underset{\sim}{a}_2 + s_3\underset{\sim}{a}_3, \tag{6.5.1}$$

where s_1, s_2, and s_3 are integer numbers.

The kinetic and potential energies are

$$T = \frac{1}{2}\sum_{s=1}^{N}\sum_{\nu=1}^{J}\sum_{\alpha=1}^{3} m_\nu \dot{u}_{s\nu\alpha}^2, \tag{6.5.2}$$

$$V = \frac{1}{2}\sum_{s=1}^{N}\sum_{s'=1}^{N}\sum_{\nu=1}^{J}\sum_{\nu'=1}^{J}\sum_{\alpha=1}^{3}\sum_{\alpha'=1}^{3} \overset{ss'}{\underset{\alpha\alpha'}{A\nu\nu'}}\, u_{s\nu\alpha}u_{s'\nu'\alpha'} + V_o, \tag{6.5.3}$$

where

$$\overset{ss'}{\underset{\alpha\alpha'}{A\nu\nu'}} = \left[\frac{\partial^2 V}{\partial u_{s\nu\alpha}\partial u_{s'\nu'\alpha'}}\right]_o. \tag{6.5.4}$$

The $3NJ \times 3NJ$ matrix $\underset{\sim}{A}$ is symmetrical and real:

$$\overset{ss'}{\underset{\alpha\alpha'}{A\nu\nu'}} = \overset{s's}{\underset{\alpha'\alpha}{A\nu'\nu}} = \overset{ss'*}{\underset{\alpha\alpha'}{A\nu\nu'}} = \overset{n}{\underset{\alpha\alpha'}{A\nu\nu'}}, \tag{6.5.5}$$

where n is the subscript of $\underset{\sim}{R}_n = \underset{\sim}{R}_s - \underset{\sim}{R}_{s'}$. The 3NJ equations of motion are given by

$$m_\nu \ddot{u}_{s\nu\alpha} = -\sum_{s'\nu'\alpha'} \overset{ss'}{\underset{\alpha\alpha'}{A\nu\nu'}}\, u_{s'\nu'\alpha'}. \tag{6.5.6}$$

$\overset{ss'}{\underset{\alpha\alpha'}{A\nu\nu'}}$ is the absolute value of the α component of the force acting on the νth atom of the cell at $\underset{\sim}{R}_s$, because of a unit displacement of the ν'th atom of the cell at $\underset{\sim}{R}_{s'}$ in the α' direction.

Similarly to (6.4.6)

$$\sum_{s'\nu'} \overset{ss'}{\underset{\alpha\alpha'}{A\nu\nu'}} = 0. \tag{6.5.7}$$

In order to decouple the equations of motion we look for solutions of the type

$$\underset{\sim}{u}_{s\nu} = \underset{\sim}{V}_{\nu} e^{-i\omega t + i\underset{\sim}{k}\cdot\underset{\sim}{R}_s}. \tag{6.5.8}$$

By using this expression in (6.5.6) we obtain

$$m_\nu \omega^2 V_{\nu\alpha} = \sum_{\nu'\alpha'} G^{\nu\nu'}_{\alpha\alpha'}(\underset{\sim}{k})\, V_{\nu'\alpha'}, \tag{6.5.9}$$

where

$$G^{\alpha\alpha'}_{\nu\nu'}(\underset{\sim}{k}) = \sum_{s'} \overset{ss'}{\underset{\alpha\alpha'}{A\nu\nu'}}\, e^{i\underset{\sim}{k}\cdot(\underset{\sim}{R}_{s'}-\underset{\sim}{R}_s)} = \sum_{n} \overset{n}{\underset{\alpha\alpha'}{A\nu\nu'}}\, e^{-i\underset{\sim}{k}\cdot\underset{\sim}{R}_n}. \tag{6.5.10}$$

The $3J \times 3J$ matrix $\underset{\sim}{G}$ has the following properties:

1. It is Hermitian:

$$\begin{aligned} G^{\alpha\alpha'}_{\nu\nu'}(\underset{\sim}{k})^* &= \sum_{n} \overset{n}{\underset{\alpha\alpha'}{A\nu\nu'}}\, e^{i\underset{\sim}{k}\cdot\underset{\sim}{R}_n} = \sum_{n} \overset{-n}{\underset{\alpha\alpha'}{A\nu\nu'}}\, e^{-i\underset{\sim}{k}\cdot\underset{\sim}{R}_n} \\ &= \sum_{n} \overset{n}{\underset{\alpha'\ \alpha}{A\nu'\ \nu}}\, e^{-\underset{\sim}{k}\cdot\underset{\sim}{R}_n} = G^{\alpha'\alpha}_{\nu'\nu}(\underset{\sim}{k}). \end{aligned} \tag{6.5.11}$$

2. It is periodical in the reciprocal lattice. If $\underset{\sim}{K}$ is a primitive vector of the reciprocal lattice,

$$G^{\alpha\alpha'}_{\nu\nu'}(\underset{\sim}{k}+\underset{\sim}{K}) = G^{\alpha\alpha'}_{\nu\nu'}(\underset{\sim}{k}) e^{-i\underset{\sim}{K}\cdot\underset{\sim}{R}_n} = G^{\alpha\alpha'}_{\nu\nu'}(\underset{\sim}{k}). \tag{6.5.12}$$

3. It is real if the crystal has a center of symmetry

$$\begin{aligned} G^{\alpha\alpha'}(\underset{\sim}{k})^* &= \sum_{n} A^{n}_{\alpha\alpha'} e^{i\underset{\sim}{k}\cdot\underset{\sim}{R}_n} = G^{\alpha\alpha'}(-\underset{\sim}{k}) = \sum_{n} A^{-n}_{\alpha\alpha'} e^{-i\underset{\sim}{k}\cdot\underset{\sim}{R}_n} \\ &= \sum_{n} A^{n}_{\alpha\alpha'} e^{-i\underset{\sim}{k}\cdot\underset{\sim}{R}_n} = G^{\alpha\alpha'}(\underset{\sim}{k}). \end{aligned} \tag{6.5.13}$$

Equations (6.5.9) give 3J positive values for ω^2 in correspondence to each value of $\underset{\sim}{k}$; that is, the vibrational spectrum consists of 3J branches. For each frequency $\omega = \omega_{\underset{\sim}{k}\lambda}(\lambda = 1, 2, \ldots, 3J)$ we obtain J vectors $\underset{\sim}{V}_\nu(\underset{\sim}{k}, \lambda)$, which may be considered to be the *polarization vectors* of the displacements of the J atoms in the unit cell in correspondence to a vibrational mode of frequency to $\omega_{\underset{\sim}{k}\lambda}$ and propagation vector $\underset{\sim}{k}$.

6.5.2. *Allowed Values of* $\underset{\sim}{k}$. *Density of Phonon Modes*

As in the unidimensional case, we identify the equilibrium configuration of N atoms of the infinite crystal. We also impose the periodic boundary conditions:

$$\underset{\sim}{u}_{s\nu} = \underset{\sim}{u}_\nu(\underset{\sim}{R}_s) = \underset{\sim}{u}_\nu(\underset{\sim}{R}_s + \underset{\sim}{L}_1 + \underset{\sim}{L}_2 + \underset{\sim}{L}_3), \tag{6.5.14}$$

where $\underset{\sim}{L}_i = N_i\underset{\sim}{a}_i$ are the dimensions of the crystal. By using (6.5.8) in (6.5.14) we get

$$\exp\left[e^{i\underset{\sim}{k}\cdot(\underset{\sim}{R}_s + \sum_i \underset{\sim}{L}_i)}\right] = e^{i\underset{\sim}{k}\cdot\underset{\sim}{R}_s}, \tag{6.5.15}$$

or

$$\exp\left(e^{i\underset{\sim}{k}\cdot\sum_i \underset{\sim}{L}_i}\right) = 1, \tag{6.5.16}$$

that is,

$$\underset{\sim}{k} \cdot \sum_i \underset{\sim}{L}_i = 2\pi \times \text{integer}. \tag{6.5.17}$$

The above relation defines the allowed values of $\underset{\sim}{k}$ in the *reciprocal space*. We introduce the notion of *reciprocal lattice* as the lattice of the reciprocal space with the basic primitive vectors

$$\begin{aligned}
\underset{\sim}{b}_1 &= 2\pi\frac{\underset{\sim}{a}_2 \times \underset{\sim}{a}_3}{\underset{\sim}{a}_1 \cdot \underset{\sim}{a}_2 \times \underset{\sim}{a}_3}, \\
\underset{\sim}{b}_2 &= 2\pi\frac{\underset{\sim}{a}_3 \times \underset{\sim}{a}_1}{\underset{\sim}{a}_1 \cdot \underset{\sim}{a}_2 \times \underset{\sim}{a}_3}, \\
\underset{\sim}{b}_3 &= 2\pi\frac{\underset{\sim}{a}_1 \times \underset{\sim}{a}_2}{\underset{\sim}{a}_1 \cdot \underset{\sim}{a}_2 \times \underset{\sim}{a}_3},
\end{aligned} \tag{6.5.18}$$

or

$$\underset{\sim}{b}_j \cdot \underset{\sim}{a}_i = 2\pi\delta_{ij}. \tag{6.5.19}$$

These vectors define a primitive cell in the reciprocal space whose volume is given by

$$\Omega_k = \underset{\sim}{b}_1 \cdot \underset{\sim}{b}_2 \times \underset{\sim}{b}_3 = \frac{(2\pi)^3}{\Omega_a}, \tag{6.5.20}$$

where $\Omega_a = \underset{\sim}{a}_1 \cdot \underset{\sim}{a}_2 \times \underset{\sim}{a}_3$ is the volume of the primitive cell of the direct lattice.

A primitive vector of the reciprocal lattice is defined as follows:

$$\underset{\sim}{K}_\sigma = \sigma_1\underset{\sim}{b}_1 + \sigma_2\underset{\sim}{b}_2 + \sigma_3\underset{\sim}{b}_3, \tag{6.5.21}$$

where σ_1, σ_2, and σ_3 are integer numbers. It is easy to verify that

$$\begin{aligned} \underset{\sim}{K}_\sigma \cdot \underset{\sim}{R}_s &= (\sigma_1\underset{\sim}{b}_1 + \sigma_2\underset{\sim}{b}_2 + \sigma_3\underset{\sim}{b}_3) \cdot (s_1\underset{\sim}{a}_1 + s_2\underset{\sim}{a}_2 + s_3\underset{\sim}{a}_3) \\ &= 2\pi\sum_i \sigma_i s_i = 2\pi \times \text{integer}, \end{aligned} \tag{6.5.22}$$

and therefore

$$e^{i\underset{\sim}{K}_\sigma \cdot \underset{\sim}{R}_s} = 1. \tag{6.5.23}$$

A position vector $\underset{\sim}{k}$ of the reciprocal space may be expressed as

$$\underset{\sim}{k} = k_1 \underset{\sim}{1}_{\underset{\sim}{b}_1} + k_2 \underset{\sim}{1}_{\underset{\sim}{b}_2} + k_3 \underset{\sim}{1}_{\underset{\sim}{b}_3}, \tag{6.5.24}$$

where $\underset{\sim}{1}_{\underset{\sim}{b}_i}$ is the unit vector in the direction $\underset{\sim}{b}_i$. The condition (6.5.17), when the relation (6.5.24) is used, gives

$$\underset{\sim}{k} \cdot \sum_i \underset{\sim}{L}_i = (k_1 \underset{\sim}{1}_{\underset{\sim}{b}_1} + k_2 \underset{\sim}{1}_{\underset{\sim}{b}_2} + k_3 \underset{\sim}{1}_{\underset{\sim}{b}_3}) \cdot (N_1\underset{\sim}{a}_1 + N_2\underset{\sim}{a}_2 + N_2\underset{\sim}{a}_3)$$

$$= k_1 N_1 1_{\underset{\sim}{b}_1} \cdot \underset{\sim}{a}_1 + k_2 N_2 1_{\underset{\sim}{b}_2} \cdot \underset{\sim}{a}_2 + k_3 N_3 1_{\underset{\sim}{b}_3} \cdot \underset{\sim}{a}_3$$
$$= 2\pi \times \text{integer}. \tag{6.5.25}$$

Then the allowed values for the coefficients k_i are given by

$$k_1 = \frac{2\pi r_1}{N_1 1_{\underset{\sim}{b}_1} \cdot \underset{\sim}{a}_1} \quad (r_1 = 1, 2, \ldots, N_1),$$
$$k_2 = \frac{2\pi r_2}{N_2 1_{\underset{\sim}{b}_2} \cdot \underset{\sim}{a}_2} \quad (r_2 = 1, 2, \ldots, N_2),$$
$$k_3 = \frac{2\pi r_3}{N_3 1_{\underset{\sim}{b}_3} \cdot \underset{\sim}{a}_3} \quad (r_3 = 1, 2, \ldots, N_3), \tag{6.5.26}$$

and are $N_1 \cdot N_2 \cdot N_3 = N$ in number.

Given a certain lattice we may define the so-called *Wigner–Seitz unit cell* in the following way. We consider a generic lattice point at the center of the cell and draw lines connecting this point with all the other points of the lattice; then we intersect each line with a plane perpendicular to it at midpoint between the center of the cell and the lattice point reached by the line. The volume enclosed by all these planes is the Wigner–Seitz unit cell. The Brillouin zone can be defined as the Wigner–Seitz unit cell of the reciprocal lattice.

Since the Brillouin zone and the primitive unit cell of the reciprocal lattice have the same volume, they both contain the same number of allowed k points. If we consider the Brillouin zone as unit cell of the reciprocal lattice the coefficients k_i are given by

$$k_1 = \frac{2\pi r_1}{N_1 1_{\underset{\sim}{b}_1} \cdot \underset{\sim}{a}_1} \quad \left(r_1 = \pm 1, \pm 2, \ldots, \pm \frac{N_1}{2}\right),$$
$$k_2 = \frac{2\pi r_2}{N_2 1_{\underset{\sim}{b}_2} \cdot \underset{\sim}{a}_2} \quad \left(r_2 = \pm 1, \pm 2, \ldots, \pm \frac{N_2}{2}\right), \tag{6.5.27}$$
$$k_3 = \frac{2\pi r_3}{N_3 1_{\underset{\sim}{b}_3} \cdot \underset{\sim}{a}_3} \quad \left(r_3 = \pm 1, \pm 2, \ldots, \pm \frac{N_3}{2}\right).$$

The total number of allowed k values is N.

It is interesting to find out the density of the allowed values of k in the reciprocal space. The number of k values with k_1 in $(k_1, k_1 + dk_1)$,

k_2 in $(k_2, k_2 + dk_2)$ and k_3 in $(k_3, k_3 + dk_3)$ is given by

$$dr_1 dr_2 dr_3 = \frac{N_1 N_2 N_3}{(2\pi)^3} dk_1 dk_2 dk_3 (\underset{\sim}{a}_1 \cdot 1_{\underset{\sim}{b}_1})(\underset{\sim}{a}_2 \cdot 1_{\underset{\sim}{b}_2})(\underset{\sim}{a}_3 \cdot 1_{\underset{\sim}{b}_3}). \tag{6.5.28}$$

The volume element in k space is given by

$$|d\underset{\sim}{k}| = dk_1 dk_2 dk_3 \, \underset{\sim}{1}_{\underset{\sim}{b}_1} \cdot (\underset{\sim}{1}_{\underset{\sim}{b}_2} \times \underset{\sim}{1}_{\underset{\sim}{b}_3}).$$

Then

$$\begin{aligned} dr_1 dr_2 dr_3 &= \frac{N_1 N_2 N_3}{(2\pi)^3} |d\underset{\sim}{k}| \frac{(\underset{\sim}{a}_1 \cdot \underset{\sim}{1}_{\underset{\sim}{b}_1})(\underset{\sim}{a}_2 \cdot \underset{\sim}{1}_{\underset{\sim}{b}_2})(\underset{\sim}{a}_3 \cdot \underset{\sim}{1}_{\underset{\sim}{b}_3})}{\underset{\sim}{1}_{\underset{\sim}{b}_1} \cdot (\underset{\sim}{1}_{\underset{\sim}{b}_2} \times \underset{\sim}{1}_{\underset{\sim}{b}_3})} \\ &= \frac{N}{(2\pi)^3} \underset{\sim}{a}_k \cdot (\underset{\sim}{a}_2 \times \underset{\sim}{a}_3) |d\underset{\sim}{k}| = \frac{V}{(2\pi)^3} |d\underset{\sim}{k}|, \end{aligned} \tag{6.5.29}$$

where $\underset{\sim}{a}_1 \cdot (\underset{\sim}{a}_2 \times \underset{\sim}{a}_3)$ is the volume of the unit cell and V is the volume of the crystal. Therefore the density of $\underset{\sim}{k}$ points is given by

$$\frac{dr_1 dr_2 dr_3}{|d\underset{\sim}{k}|} = \frac{V}{(2\pi)^3}. \tag{6.5.30}$$

6.5.3. *Normal Modes of Vibration*

The most general motion of the atoms is given by

$$\underset{\sim}{u}_{s\nu} = \sum_{\underset{\sim}{k}}^{N} \sum_{\lambda}^{3J} q_{\underset{\sim}{k}}^{\lambda} \underset{\sim}{V}_{\nu}(\underset{\sim}{k}, \lambda) e^{i\underset{\sim}{k} \cdot \underset{\sim}{R}_s}, \tag{6.5.31}$$

where $\underset{\sim}{V}(\underset{\sim}{k}, \lambda)$ are the polarization vectors that are obtained from the eigenvalue equations (6.5.9). Since

$$G_{\nu\nu'}^{\alpha\alpha'}(-\underset{\sim}{k}) = G_{\nu\nu'}^{\alpha\alpha'}(\underset{\sim}{k})^*, \tag{6.5.32}$$

then

$$\underset{\sim}{V}_{\nu}(-\underset{\sim}{k}, \lambda) = \underset{\sim}{V}_{\nu}(\underset{\sim}{k}, \lambda)^*. \tag{6.5.33}$$

Also, the reality condition for $u_{s\nu}$ requires

$$q_{-\underset{\sim}{k}}^{\lambda} = q_{\underset{\sim}{k}}^{\lambda^*}. \tag{6.5.34}$$

Arguments similar to the one used to prove (6.3.9) lead us to the relation

$$\sum_{s} e^{i(\underset{\sim}{k}-\underset{\sim}{k}')\cdot \underset{\sim}{R}_s} = N\delta_{\underset{\sim}{k}\underset{\sim}{k}'}. \tag{6.5.35}$$

Let us consider now the expression

$$\sum_{\nu}\sum_{s} m_\nu [\underset{\sim}{V}_\nu(\underset{\sim}{k},\lambda) e^{i\underset{\sim}{k}\cdot\underset{\sim}{R}_s}]^* \cdot [\underset{\sim}{V}_\nu(\underset{\sim}{k}',\lambda') e^{i\underset{\sim}{k}'\cdot\underset{\sim}{R}_s}]. \tag{6.5.36}$$

Taking (6.5.35) into account this expression becomes

$$\begin{aligned}&\left[\sum_{s} e^{i(\underset{\sim}{k}'-\underset{\sim}{k})\cdot\underset{\sim}{R}_s}\right] \sum_{\nu} m_\nu [\underset{\sim}{V}_\nu(\underset{\sim}{k},\lambda)^* \cdot \underset{\sim}{V}_\nu(\underset{\sim}{k}',\lambda')]\\ &\qquad = N\delta_{\underset{\sim}{k}\underset{\sim}{k}'} \sum_{\nu} m_\nu \underset{\sim}{V}_\nu(\underset{\sim}{k},\lambda)^* \cdot \underset{\sim}{V}_\nu(\underset{\sim}{k}',\lambda').\end{aligned} \tag{6.5.37}$$

Let us consider now equation (6.5.9)

$$m_\nu \omega^2_{\underset{\sim}{k}\lambda} V_{\nu\alpha}(\underset{\sim}{k},\lambda) = \sum_{\nu'\alpha'} G^{\nu\nu'}_{\alpha\alpha'}(\underset{\sim}{k}) V_{\nu'\alpha'}(\underset{\sim}{k},\lambda), \tag{6.5.38}$$

and take the complex conjugate of both members:

$$m_\nu \omega^2_{\underset{\sim}{k}\lambda} V_{\nu\alpha}(\underset{\sim}{k},\lambda)^* = \sum_{\nu'\alpha'} G^{\nu\nu'}_{\alpha\alpha'}(\underset{\sim}{k})^* V_{\nu'\alpha'}(\underset{\sim}{k},\lambda)^*. \tag{6.5.39}$$

Let us multiply by $V_{\nu\alpha}(\underset{\sim}{k},\lambda')$ and sum over ν and α

$$\begin{aligned}&\sum_{\nu\alpha} m_\nu \omega^2_{\underset{\sim}{k}\lambda} V_{\nu\alpha}(\underset{\sim}{k},\lambda)^* V_{\nu\alpha}(\underset{\sim}{k},\lambda')\\ &\qquad = \sum_{\nu\alpha}\sum_{\nu'\alpha'} G^{\nu\nu'}_{\alpha\alpha'}(\underset{\sim}{k})^* V_{\nu'\alpha'}(\underset{\sim}{k},\lambda)^* V_{\nu\alpha}(\underset{\sim}{k},\lambda').\end{aligned} \tag{6.5.40}$$

Changing λ into λ' and λ' into λ, the above expression becomes

$$\begin{aligned}&\sum_{\nu\alpha} m_\nu \omega^2_{\underset{\sim}{k}\lambda'} V_{\nu\alpha}(\underset{\sim}{k},\lambda')^* V_{\nu\alpha}(\underset{\sim}{k},\lambda)\\ &\qquad = \sum_{\nu\alpha}\sum_{\nu'\alpha'} G^{\nu\nu'}_{\alpha\alpha'}(\underset{\sim}{k})^* V_{\nu'\alpha'}(\underset{\sim}{k},\lambda')^* V_{\nu\alpha}(\underset{\sim}{k},\lambda).\end{aligned} \tag{6.5.41}$$

The right member of (6.5.41) can be written

$$\sum_{\nu\alpha}\sum_{\nu'\alpha'} G^{\nu'\nu}_{\alpha'\alpha}(\underset{\sim}{k})\, V_{\nu'\alpha'}(\underset{\sim}{k},\lambda')^*\, V_{\nu\alpha}(\underset{\sim}{k},\lambda).$$

$$= \sum_{\nu\alpha}\sum_{\nu'\alpha'} G^{\nu\nu'}_{\alpha\alpha'}(\underset{\sim}{k})\, V_{\nu\alpha}(\underset{\sim}{k},\lambda')^*\, V_{\nu'\alpha'}(\underset{\sim}{k},\lambda). \qquad (6.5.42)$$

By taking the complex conjugate of (6.5.41) and using (6.5.42), we find

$$\sum_{\nu\alpha} m_\nu \omega^2_{\underset{\sim}{k}\lambda'}\, V_{\nu\alpha}(\underset{\sim}{k},\lambda)^*\, V_{\nu\alpha}(\underset{\sim}{k},\lambda')$$

$$= \sum_{\nu\alpha}\sum_{\nu'\alpha'} G^{\nu\nu'}_{\alpha\alpha'}(\underset{\sim}{k})^*\, V_{\nu'\alpha'}(\underset{\sim}{k},\lambda)^*\, V_{\nu\alpha}(\underset{\sim}{k},\lambda'). \qquad (6.5.43)$$

If we subtract now (6.5.43) from (6.5.40) we obtain

$$(\omega^2_{\underset{\sim}{k}\lambda} - \omega^2_{\underset{\sim}{k}\lambda'}) \sum_{\nu\alpha} m_\nu\, V_{\nu\alpha}(\underset{\sim}{k},\lambda)^*\, V_{\nu\alpha}(\underset{\sim}{k},\lambda') = 0, \qquad (6.5.44)$$

that is,

$$\sum_{\nu} m_\nu\, \underset{\sim}{V}_\nu(\underset{\sim}{k}\lambda)^* \cdot \underset{\sim}{V}_\nu(\underset{\sim}{k},\lambda') = 0, \qquad (6.5.45)$$

unless $\lambda = \lambda'$.

Since the equations (6.5.9), which define the vectors $\underset{\sim}{V}_\nu$, leave a constant factor free, we may choose to set

$$\sum_{\nu} m_\nu\, \underset{\sim}{V}_\nu(\underset{\sim}{k},\lambda)^* \cdot \underset{\sim}{V}_\nu(\underset{\sim}{k},\lambda) = M, \qquad (6.5.46)$$

where

$$N = \sum_{\nu} m_\nu, \qquad (6.5.47)$$

is the mass of the unit cell.

The expression (6.5.36), taking into account (6.5.45), (6.5.46), and (6.5.47), can now be written

$$\sum_{\nu}\sum_{s} m_\nu [\underset{\sim}{V}_\nu(\underset{\sim}{k},\lambda)e^{i\underset{\sim}{k}\cdot\underset{\sim}{R}_s}]^* \cdot [\underset{\sim}{V}_\nu(\underset{\sim}{k}',\lambda')e^{i\underset{\sim}{k}'\cdot\underset{\sim}{R}_s}] = M\delta_{\underset{\sim}{k}\underset{\sim}{k}'}\delta_{\lambda\lambda'}, \tag{6.5.48}$$

where $M = NN$ is the total mass of the crystal. Equation (6.5.48) expresses the orthogonality of the different nodes of vibration.

If we use a simpler notation for (6.5.48), indicating the subscripts ν, s, and a by μ, end the variables $\underset{\sim}{k}$ and λ by η, and if we set

$$V_{\nu\alpha}(\underset{\sim}{k},\lambda)e^{i\underset{\sim}{k}\cdot\underset{\sim}{R}_s} = V_\mu(\eta), \tag{6.5.49}$$

we obtain

$$\sum_{\mu} m_\mu V_\mu(\eta)^* V_\mu(\eta') = M\delta_{\eta\eta'}. \tag{6.5.50}$$

With this notation, the most general displacement is given by

$$u_\mu = \sum_{\eta} q(\eta) V_\mu(\eta), \tag{6.5.51}$$

with

$$q(\eta) = \frac{1}{M}\sum_{\mu} m_\mu V_\mu(\eta)^* u_\mu. \tag{6.5.52}$$

By using (6.5.51) in (6.5.52) we obtain

$$u_\mu = \sum_{\eta} V_\mu(\eta)\sum_{\mu'}\frac{m_{\mu'}}{M}V_{\mu'}(\eta)^* u_{\mu'}, \tag{6.5.53}$$

and therefore

$$\sum_{\eta} V_\mu(\eta) V_{\mu'}(\eta)^* = \frac{M}{m_\mu}\delta_{\mu\mu'}. \tag{6.5.54}$$

We may now express the kinetic energy in terms of the coordinates $q(\eta)$:

$$T = \frac{1}{2}\sum_{\mu} m_\mu \dot{u}_\mu^2$$

$$= \frac{1}{2}\sum_{\mu} m_{\mu} \sum_{\eta}\sum_{\eta'} \dot{q}(\eta)^* \dot{q}(\eta') V_{\mu}(\eta)^* V_{\mu}(\eta')$$

$$= \frac{1}{2}\sum_{\eta}\sum_{\eta'} M\delta_{\eta\eta'} \dot{q}(\eta)^* \dot{q}(\eta') - \frac{1}{2}\sum_{\eta} M|\dot{q}(\eta)|^2$$

$$= \frac{1}{2}\sum_{\underset{\sim}{k}\lambda} M|\dot{q}_{\underset{\sim}{k}}^{\lambda}|^2. \tag{6.5.55}$$

Similarly the potential energy is given by

$$V = V_o + \frac{1}{2}\sum_{\mu}\sum_{\mu'} A_{\mu\mu'} u_{\mu} u_{\mu'}$$

$$= V_o + \frac{1}{2}\sum_{\mu}\sum_{\mu'} A_{\mu\mu'} \sum_{\eta} q(\eta) V_{\mu}(\eta) \sum_{\mu'} q(\eta') V_{\mu'}(\eta')$$

$$= V_o + \frac{1}{2}\sum_{\eta}\sum_{\eta'}\sum_{\mu}\left[\sum_{\mu'} A_{\mu\mu'} V_{\mu'}(\eta')\right] V_{\mu}(\eta) q(\eta) q(\eta'). \tag{6.5.56}$$

The equation (6.5.9) can be written, in the shortened notation, as

$$m_{\mu}\omega^2(\eta) V_{\mu}(\eta) = \sum_{\mu'} A_{\mu\mu'} V_{\mu'}(\eta). \tag{6.5.57}$$

If we use (6.5.57) in (6.5.56) we obtain

$$V = V_o + \frac{1}{2}\sum_{\eta}\sum_{\eta'}\sum_{\mu} [m_{\mu}\omega^2(\eta') V_{\mu}(\eta')] V_{\mu}(\eta) q(\eta) q(\eta')$$

$$= V_o + \frac{1}{2}\sum_{\underset{\sim}{k}\lambda}\sum_{\underset{\sim}{k}'\lambda'} \omega^2_{\underset{\sim}{k}'\lambda'} q_{\underset{\sim}{k}}^{\lambda} q_{\underset{\sim}{k}'}^{\lambda'}$$

$$\times \left[\sum_{s}\sum_{\nu} m_{\nu} V_{\nu}(\underset{\sim}{k}', \lambda') e^{i\underset{\sim}{k}'\cdot \underset{\sim}{R}_s} V_{\nu}(\underset{\sim}{k}, \lambda) e^{i\underset{\sim}{k}\cdot \underset{\sim}{R}_s}\right]. \tag{6.5.58}$$

We can now use (6.5.48) to find

$$V = V_o + \frac{1}{2}\sum_{\underset{\sim}{k}\lambda} M\omega^2_{\underset{\sim}{k}\lambda} q_{\underset{\sim}{k}}^{\lambda} q_{-\underset{\sim}{k}}^{\lambda}. \tag{6.5.59}$$

6.5.4. *Energy Levels*

The kinetic and potential energies are given by

$$T = \frac{1}{2}\sum_{\mu} m_{\mu}\dot{u}_{\mu}^{2} = \frac{1}{2}\sum_{\underset{\sim}{k}\lambda} M\dot{q}_{\underset{\sim}{k}}^{\lambda}\dot{q}_{-\underset{\sim}{k}}^{\lambda}$$

$$V = V_o + \frac{1}{2}\sum_{\mu}\sum_{\mu'} A_{\mu\mu'}u_{\mu}u_{\mu'}$$

$$= V_o + \frac{1}{2}\sum_{\underset{\sim}{k}\lambda} M\omega_{\underset{\sim}{k}\lambda}^{2} q_{\underset{\sim}{k}}^{\lambda} q_{-\underset{\sim}{k}}^{\lambda}. \qquad (6.5.60)$$

The momentum conjugate to $q_{\underset{\sim}{k}}^{\lambda}$ is given by

$$p_{\underset{\sim}{k}}^{\lambda} = \frac{\partial L}{\partial \dot{q}_{\underset{\sim}{k}}^{\lambda}} = \frac{\partial (T - V)}{\partial \dot{q}_{\underset{\sim}{k}}^{\lambda}} = M\dot{q}_{-\underset{\sim}{k}}^{\lambda}. \qquad (6.5.61)$$

It is then

$$p_{-\underset{\sim}{k}}^{\lambda} = M\dot{q}_{\underset{\sim}{k}}^{\lambda} = p_{\underset{\sim}{k}}^{\lambda^*}. \qquad (6.5.62)$$

The Hamiltonian is then given by

$$H = \sum_{\underset{\sim}{k}\lambda} (p_{\underset{\sim}{k}}^{\lambda}\dot{q}_{\underset{\sim}{k}}^{\lambda} - L)$$

$$= \frac{1}{2M}\sum_{\underset{\sim}{k}\lambda} p_{\underset{\sim}{k}}^{\lambda} p_{-\underset{\sim}{k}}^{\lambda} + \frac{1}{2}M\sum_{\underset{\sim}{k}\lambda} \omega_{\underset{\sim}{k}\lambda}^{2} q_{\underset{\sim}{k}}^{\lambda} q_{-\underset{\sim}{k}}^{\lambda} + V_o. \qquad (6.5.63)$$

The Hamilton's equations are

$$\dot{p}_{\underset{\sim}{k}}^{\lambda} = -\frac{\partial H}{\partial q_{\underset{\sim}{k}}^{\lambda}} = -M\omega_{\underset{\sim}{k}\lambda}^{2} q_{-\underset{\sim}{k}}^{\lambda},$$

$$\dot{q}_{\underset{\sim}{k}}^{\lambda} = \frac{\partial H}{\partial p_{\underset{\sim}{k}}^{\lambda}} = \frac{1}{M} p_{-\underset{\sim}{k}}^{\lambda}. \qquad (6.5.64)$$

The two equations above give a solution

$$q_{\underset{\sim}{k}}^{\lambda}(t) = q_{\underset{\sim}{k}}^{\lambda}(0)e^{-i\omega_{\underset{\sim}{k}\lambda}t},$$

implying that each normal coordinate is associated with an eigenvalue of the dynamical matrix.

We can write

$$
\begin{aligned}
u_{s\nu\alpha} &= \sum_{\underset{\sim}{k}\lambda} q_{\underset{\sim}{k}}^{\lambda} V_{\nu\alpha}(\underset{\sim}{k}, \lambda) e^{i\underset{\sim}{k}\cdot\underset{\sim}{R}_s}, \\
P_{s\nu\alpha} &= m_\nu \dot{u}_{s\nu\alpha} = \sum_{\underset{\sim}{k}\lambda} m_\nu \dot{q}_{\underset{\sim}{k}}^{\lambda} V_{\nu\alpha}(\underset{\sim}{k}, \lambda) e^{i\underset{\sim}{k}\cdot\underset{\sim}{R}_s} \\
&= \sum_{\underset{\sim}{k}\lambda} \frac{m_\nu}{M} P_{\underset{\sim}{k}}^{\lambda} V_{\nu\alpha}(\underset{\sim}{k}, \lambda)^* e^{-i\underset{\sim}{k}\cdot\underset{\sim}{R}_s},
\end{aligned}
\tag{6.5.65}
$$

with the inverse relations

$$
\begin{aligned}
q_{\underset{\sim}{k}}^{\lambda} &= \frac{1}{M} \sum_{s\nu\alpha} m_\nu u_{s\nu\alpha} V_{\nu\alpha}(\underset{\sim}{k}, \lambda)^* e^{-i\underset{\sim}{k}\cdot\underset{\sim}{R}_s}, \\
P_{\underset{\sim}{k}}^{\lambda} &= M \dot{q}_{-\underset{\sim}{k}}^{\lambda} = \sum_{s\nu\alpha} (m_\nu \dot{u}_{s\nu\alpha}) V_{\nu\alpha}(\underset{\sim}{k}, \lambda) e^{i\underset{\sim}{k}\cdot\underset{\sim}{R}_s} \\
&= \sum_{s\nu\alpha} P_{s\nu\alpha} V_{\nu\alpha}(\underset{\sim}{k}, \lambda) e^{i\underset{\sim}{k}\cdot\underset{\sim}{R}_s}.
\end{aligned}
\tag{6.5.66}
$$

We can now make a transition to quantum mechanics by considering $\underset{\sim}{u}_{s\nu}$, $\underset{\sim}{P}_{s\nu}$, $q_{\underset{\sim}{k}}^{\lambda}$, and $p_{\underset{\sim}{k}}^{\lambda}$ as operators. We use here again the short notation, where μ stands for the subscripts s, ν, and α, and η for the variables $\underset{\sim}{k}$ and λ. We start from the commutation relations

$$
[u_\mu, P_{\mu'}] = i\hbar\delta_{\mu\mu'}. \tag{6.5.67}
$$

With this notation the relations (6.5.66) can be written

$$
\begin{aligned}
q(\eta) &= \frac{1}{M} \sum_{\mu} m_\mu u_\mu V_\mu(\eta)^*, \\
P(\eta) &= \sum_{\mu} P_\mu V_\mu(\eta),
\end{aligned}
\tag{6.5.68}
$$

where $V_\mu(\eta)$ is given by (6.5.49). We have now

$$
\begin{aligned}
[q(\eta), P(\eta')] &= q(\eta)P(\eta') - P(\eta')q(\eta)\\
&= \frac{1}{M}\sum_{\mu\mu'}[m_\mu u_\mu V_\mu(\eta)^* P_{\mu'} V_{\mu'}(\eta')\\
&\quad - P_{\mu'} V_{\mu'}(\eta') m_\mu u_\mu V_\mu(\eta)^*]\\
&= \frac{1}{M}\sum_{\mu\mu'} m_\mu (u_\mu P_{\mu'} - P_{\mu'} u_\mu) V_\mu(\eta)^* V_{\mu'}(\eta')\\
&= \frac{1}{M}\sum_{\mu\mu'} m_\mu i\hbar\delta_{\mu\mu'} V_\mu(\eta)^* V_{\mu'}(\eta')\\
&= \frac{i\hbar}{M}\sum_{\mu} m_\mu V_\mu(\eta)^* V_\mu(\eta') = i\hbar\delta_{\eta\eta'}.
\end{aligned}
\tag{6.5.69}
$$

Therefore

$$[q_{\underset{\sim}{k}}^{\lambda}, P_{\underset{\sim}{k}'}^{\lambda'}] = i\hbar\delta_{\underset{\sim}{k}\underset{\sim}{k}'}\delta_{\lambda\lambda'}. \tag{6.5.70}$$

At this point we introduce the dimensionless operators

$$
\begin{aligned}
a_{\underset{\sim}{k}}^{\lambda} &= \left(\frac{M\omega_{\underset{\sim}{k}\lambda}}{2\hbar}\right)^{\frac{1}{2}}\left(q_{\underset{\sim}{k}}^{\lambda} + \frac{i}{M\omega_{\underset{\sim}{k}\lambda}}P_{-\underset{\sim}{k}}^{\lambda}\right),\\
a_{\underset{\sim}{k}}^{\lambda+} &= \left(\frac{M\omega_{\underset{\sim}{k}\lambda}}{2\hbar}\right)^{\frac{1}{2}}\left(q_{-\underset{\sim}{k}}^{\lambda} - \frac{i}{M\omega_{\underset{\sim}{k}\lambda}}P_{\underset{\sim}{k}}^{\lambda}\right),
\end{aligned}
\tag{6.5.71}
$$

$q_{\underset{\sim}{k}}^{\lambda}$ and $P_{\underset{\sim}{k}}^{\lambda}$ can be expressed in terms of these operators as follows:

$$
\begin{aligned}
q_{\underset{\sim}{k}}^{\lambda} &= \left(\frac{\hbar}{2M\omega_{\underset{\sim}{k}\lambda}}\right)^{\frac{1}{2}}(a_{\underset{\sim}{k}}^{\lambda} + a_{-\underset{\sim}{k}}^{\lambda+}),\\
P_{\underset{\sim}{k}}^{\lambda} &= \left(\frac{\hbar M\omega_{\underset{\sim}{k}\lambda}}{2}\right)^{\frac{1}{2}}\frac{1}{i}(a_{-\underset{\sim}{k}}^{\lambda} - a_{\underset{\sim}{k}}^{\lambda+}).
\end{aligned}
\tag{6.5.72}
$$

The commutation relations of the operators $a_{\underset{\sim}{k}}^{\lambda}$ and $a_{\underset{\sim}{k}}^{\lambda+}$ can be derived from (6.5.71) and (6.5.70). We find

$$
\begin{aligned}
[a_{\underset{\sim}{k}}^{\lambda}, a_{\underset{\sim}{k}'}^{\lambda'+}] &= \delta_{\underset{\sim}{k}\underset{\sim}{k}'}\delta_{\lambda\lambda'},\\
[a_{\underset{\sim}{k}}^{\lambda}, a_{\underset{\sim}{k}'}^{\lambda'}] &= [a_{\underset{\sim}{k}}^{\lambda+}, a_{\underset{\sim}{k}'}^{\lambda'+}] = 0.
\end{aligned}
\tag{6.5.73}
$$

The Hamiltonian (6.5.63) can now be expressed as follows:

$$\begin{aligned} H &= \sum_{\underset{\sim}{k}\lambda}^{3NJ} \hbar\omega_{\underset{\sim}{k}\lambda}\left(a_{\underset{\sim}{k}}^{\lambda+}a_{\underset{\sim}{k}}^{\lambda} + \frac{1}{2}\right) \\ &= \sum_{\underset{\sim}{k}\lambda}^{3NJ} \frac{\hbar\omega_{\underset{\sim}{k}\lambda}}{2}(a_{\underset{\sim}{k}}^{\lambda}a_{\underset{\sim}{k}}^{\lambda+} + a_{\underset{\sim}{k}}^{\lambda+}a_{\underset{\sim}{k}}^{\lambda}), \end{aligned} \tag{6.5.74}$$

where we have neglected the term V_o that adds only a constant to the energy. Therefore the system is equivalent to a collection of 3NJ independent harmonic oscillators. The energy levels are then given by

$$E_{\{n_1,n_2,\ldots\}} = \sum_{\underset{\sim}{k}\lambda}\left(n_{\underset{\sim}{k}}^{\lambda} + \frac{1}{2}\right)\hbar\omega_{\underset{\sim}{k}\lambda}, \tag{6.5.75}$$

and the eigenfunctions by

$$|n_1, n_2 \ldots n_{3NJ}\rangle = |n_1\rangle|n_2\rangle\cdots|n_{3NJ}\rangle. \tag{6.5.76}$$

The displacement and the momentum of the generic atom in terms of the operators $a_{\underset{\sim}{k}}^{\lambda}$ and $a_{\underset{\sim}{k}}^{\lambda+}$ operators are given by

$$\begin{aligned} u_{s\nu\alpha} &= \sum_{\underset{\sim}{k}\lambda} q_{\underset{\sim}{k}}^{\lambda} V_{\nu\alpha}(\underset{\sim}{k},\lambda)e^{i\underset{\sim}{k}\cdot\underset{\sim}{R}_s} \\ &= \sum_{\underset{\sim}{k}\lambda} V_{\nu\alpha}(\underset{\sim}{k},\lambda)e^{i\underset{\sim}{k}\cdot\underset{\sim}{R}_s}\left(\frac{\hbar}{2M\omega_{\underset{\sim}{k}\lambda}}\right)^{1/2}(a_{\underset{\sim}{k}}^{\lambda} + a_{-\underset{\sim}{k}}^{\lambda+}) \\ &= \sum_{\underset{\sim}{k}\lambda}\left(\frac{\hbar}{2M\omega_{\underset{\sim}{k}\lambda}}\right)^{1/2}[V_{\nu\alpha}(\underset{\sim}{k},\lambda)e^{i\underset{\sim}{k}\cdot\underset{\sim}{R}_s}a_{\underset{\sim}{k}}^{\lambda} + V_{\nu\alpha}(\underset{\sim}{k},\lambda)^*e^{-i\underset{\sim}{k}\cdot\underset{\sim}{R}_s}a_{\underset{\sim}{k}}^{\lambda+}], \end{aligned} \tag{6.5.77}$$

and

$$\begin{aligned} P_{s\nu\alpha} &= m_\nu\dot{u}_{s\nu\alpha} \\ &= \sum_{\underset{\sim}{k}\lambda}\frac{m_\nu}{M}P_{\underset{\sim}{k}}^{\lambda} V_{\nu\alpha}(\underset{\sim}{k},\lambda)^*e^{-i\underset{\sim}{k}\cdot\underset{\sim}{R}_s} \end{aligned}$$

$$= \sum_{\underset{\sim}{k}\lambda} m_\nu \left(\frac{\hbar\omega_{\underset{\sim}{k},\lambda}}{2M}\right)^{1/2} V_{\nu\alpha}(\underset{\sim}{k},\lambda)^* e^{-i\underset{\sim}{k}\cdot\underset{\sim}{R}_s} \frac{1}{i}(a^\lambda_{-\underset{\sim}{k}} - a^{\lambda+}_{\underset{\sim}{k}})$$

$$= \sum_{\underset{\sim}{k}\lambda} m_\nu \left(\frac{\hbar\omega_{\underset{\sim}{k},\lambda}}{2M}\right)^{1/2} \frac{1}{i}[V_{\nu\alpha}(\underset{\sim}{k},\lambda) e^{i\underset{\sim}{k}\cdot\underset{\sim}{R}_s} a^\lambda_{\underset{\sim}{k}}$$

$$- V_{\nu\alpha}(\underset{\sim}{k},\lambda)^* e^{-i\underset{\sim}{k}\cdot\underset{\sim}{R}_s} a^{\lambda+}_{\underset{\sim}{k}}]. \tag{6.5.78}$$

6.5.5. *Particular Modes of Vibration*

The actual finding of the normal modes and frequencies is a very difficult task, even for systems with one atom per unit cell. Even if the force coefficients A^n_{ij} are known, the problem requires the solution of 3J equations, corresponding to J sets of three components for the polarization vectors. For the simpler case of $J = 1$ (one atom per unit cell) the following three equations have to be solved:

$$m\omega^2_{\underset{\sim}{k}\lambda} V_\alpha(\underset{\sim}{k},\lambda) = \sum_{\alpha'} G_{\alpha\alpha'}(\underset{\sim}{k}) V_{\alpha'}(\underset{\sim}{k},\lambda), \tag{6.5.79}$$

where

$$G_{\alpha\alpha'}(\underset{\sim}{k}) = \sum_{s'} A^{ss'}_{\alpha\alpha'} e^{i\underset{\sim}{k}\cdot(\underset{\sim}{R}_{s'} - \underset{\sim}{R}_s)}.$$

This gives a cubic equation for ω^2,

$$|G_{\alpha\alpha'}(\underset{\sim}{k}) - m\omega^2\delta_{\alpha\alpha'}| = 0. \tag{6.5.80}$$

For each solution ω_1, ω_2, ω_3, we obtain a ratio $V_\alpha : V_\beta : V_\gamma$ corresponding to three polarizations of the vibrations that propagate through the crystal. In general, these three directions of polarization are not orthogonal, except in special cases in which the direction of the propagation vector $\underset{\sim}{k}$ has a simple relation to the axes of symmetry of the crystal. In particular, for a cubic crystal with one atom per unit cell, the following can be said:

1. If $\underset{\sim}{k}$ is in the direction of a 4-fold axis, the solutions of (6.5.79) must have the polarization vector $\underset{\sim}{V}$ parallel either to the 4-fold axis, or to one of the two other 4-fold axes. In the latter occurrence, the

two solutions correspond to the same frequency (or, as it is said, are degenerate).

2. If $\underset{\sim}{k}$ is in a $\langle 111 \rangle$ direction (space diagonal), the equations (6.5.79) have one solution corresponding to longitudinal waves and two solutions corresponding to transverse degenerate waves.
3. If $\underset{\sim}{k}$ is in a $\langle 110 \rangle$ direction (face diagonal), equations (6.5.79) have one solution corresponding to longitudinal waves and two nondegenerate solutions corresponding to transverse waves.

Another simple case is that of infinite wavelength, $\underset{\sim}{k} = \underset{\sim}{0}$. In this case equations (6.5.79) become

$$m_\nu \omega^2_{\underset{\sim}{0}\lambda} V_{\nu\alpha}(\underset{\sim}{0}, \lambda) = \sum_{s'\nu'\alpha'} A^{ss'}_{\nu\nu'\,\alpha\alpha'} V_{\nu'\alpha'}(\underset{\sim}{0}, \lambda). \tag{6.5.81}$$

We can try the following solutions:

$$\underset{\sim}{V}_\nu(\underset{\sim}{0}, \lambda) = \underset{\sim}{V}, \tag{6.5.82}$$

with $\underset{\sim}{V}$ independent of ν, that is, the same for all the J atoms of the unit cell. Equations (6.5.81) become

$$m_\nu \omega^2_{\underset{\sim}{0}\lambda} V_\alpha = \sum_{\alpha'} \left[\sum_{s'\nu'} A^{ss'}_{\nu\nu'\,\alpha\alpha'} \right] V_{\alpha'} = 0, \tag{6.5.83}$$

because of (6.5.7). Therefore the solutions for zero frequency consist of a uniform displacement of all the atoms in the same direction. Since the direction of this uniform displacement $\underset{\sim}{V}$ is arbitrary, there are three independent solutions of this type, and therefore there are three branches of the vibrational spectrum which go to zero at $\underset{\sim}{k} = \underset{\sim}{0}$. We call these three branches the *acoustical branches* and the remaining $3J - 3$ the *optical branches.*

For $\underset{\sim}{k} = \underset{\sim}{0}$, the acoustical branches correspond to a uniform displacement of all the atoms; the optical branches correspond to vibrational modes in which atoms in different cells, but corresponding to the same position in the cell (same index ν), experience the same displacement, but atoms with different index ν within the same cell move relatively one to the other.

6.5.6. *Spectrum of Lattice Vibrations*

The relevant information about lattice dynamics is contained in the ω versus $\underset{\sim}{k}$ dispersion curves. In many cases, however, a detailed knowledge of these curves may not be necessary for the interpretation of the experimental data, since these data are an outcome of all the lattice modes. This is, for example, the case when one considers such thermodynamic properties as entropy or specific heat (see Chapter 7): the relevant information for the evaluation of these thermodynamics properties is contained in the *vibrational spectrum* of the crystal, which is the distribution in frequency of the normal modes of vibration. The knowledge of this spectrum is also important when one considers the interaction of electromagnetic radiation with phonons.

We call $g(\omega)$ the frequency distribution function, where $g(\omega)d\omega$ is the fraction of the total number of modes with frequency in $(\omega+d\omega)$. The distribution function $g(\omega)$ is actually the sum of the distribution functions $g_j(\omega)$ corresponding to the 3J branches:

$$\int_0^{\omega_{max}} g(\omega)d\omega = \sum_{j=1}^{3J} \int_0^{\omega_{max}} g_j(\omega)d\omega = 1, \tag{6.5.84}$$

where ω_{max} is the maximum frequency of the vibrational modes and J is the number of atoms in the unit cell, and where $g(\omega)$ is normalized to unity.

From (6.5.30) we know that the density of distinct $\underset{\sim}{k}$ vectors in the Brillouin zone is given by

$$\frac{V}{8\pi^3} = \frac{N\Omega_a}{8\pi^3}, \tag{6.5.85}$$

where N is the number of unit cells and Ω_a is the volume of the unit cell of the crystal. The frequency distribution of a single branch can then be expressed as

$$g_j(\omega)d\omega = \frac{\Omega_a}{(3J)(8\pi^3)} \int_{V'} d^3\underset{\sim}{k}, \tag{6.5.86}$$

where the summation is replaced by an integration, because of the high density of distinct $\underset{\sim}{k}$ vectors. The integration in (6.5.86) is through the volume V′ of a shell in the Brillouin zone defined by $\omega \leq \omega_{\underset{\sim}{k}j} \leq \omega + d\omega$. This equation can be put in a different form by changing the integral to one over frequency. If dS is an element of area on the surface of frequency ω and $dk_\perp$ is the distance between this surface and the surface corresponding to the frequency $(\omega + d\omega)$, then

$$d\omega = |\underset{\sim}{\nabla}_{\underset{\sim}{k}}\omega| dk_\perp, \tag{6.5.87}$$

and

$$d^3k = (ds)(dk_\perp) = \frac{dsd\omega}{|\underset{\sim}{\nabla}_{\underset{\sim}{k}}\omega|}. \tag{6.5.88}$$

The frequency distribution function $g_j(\omega)$ can now be written as

$$g_j(\omega) = \frac{\Omega_a}{(3J)(8\pi^3)} \int_{s'} \frac{ds_j}{|\underset{\sim}{\nabla}_{\underset{\sim}{k}}\omega_{\underset{\sim}{k}j}|}, \tag{6.5.89}$$

where the integral is now to be evaluated over the surface of constant frequency ω. $\underset{\sim}{\nabla}_{\underset{\sim}{k}}\omega_{\underset{\sim}{k}j}$ is the group velocity of the mode $(\underset{\sim}{k}, j)$ and represents the rate at which the energy is transmitted for this mode.

The above equation specifies points at which the function $g_j(\omega)$ presents singularities; these are the points of the Brillouin zone for which the group velocity $\underset{\sim}{\nabla}_{\underset{\sim}{k}}\omega_{\underset{\sim}{k}j}$ vanishes and are called *critical points.* The presence of critical points will result in a characteristic structure for $g_j(\omega)$ and $g(\omega)$.

The theory of critical points will not be treated here. The reader is referred to the simple treatment by Kittel,[2] the more complete treatment by Wall,[3] and the original works by Van Hove[4] and Phillips.[5]

6.6. Group Theory and Lattice Vibrations

6.6.1. *Properties of the Normal Coordinates*

A crystalline solid presents symmetries that can be described by a space group G. The kinetic energy T and the potential energy V of the atoms can be expressed as a sum of 3NJ (N is the number of unit cells; J is the number of atoms in the unit cell) terms squared in the normal coordinates. The potential and kinetic energies cannot be changed by performing any symmetry operation of G on the crystal.

In general some frequency eigenvalue ω_ℓ may be degenerate:

$$\omega_{\ell 1} = \omega_{\ell 2} = \cdots = \omega_{\ell r} = \omega_\ell. \tag{6.6.1}$$

Since T and V are independent of any symmetry operation, if r different coordinates q_ℓ correspond to the same frequency ω_ℓ, then

$$Rq_{\ell i} = \sum_{j=1}^{r} a_{ji} q_{\ell j}, \tag{6.6.2}$$

where the coefficients a_{ji} form an $r \times r$ unitary matrix. If we operate again with an operation S,

$$SRq_{\ell i} = \sum_{j=1}^{r} a_{ji} Sq_{\ell j} = \sum_{j=1}^{r} \sum_{m=1}^{r} a_{ji} b_{mj} q_{\ell m}. \tag{6.6.3}$$

Also, if $SR = T$

$$Tq_{\ell i} = \sum_{m=1}^{r} c_{mi} q_{\ell m}. \tag{6.6.4}$$

Then

$$c_{mi} = \sum_{j=1}^{r} b_{mj} a_{ji}. \tag{6.6.5}$$

The above relation expresses the following fact:

The normal coordinates corresponding to the same frequency eigenvalue must form a basis for an irreducible unitary representation of the space group.

This property of the normal coordinates is of great help in finding the normal modes of vibrations. We explore first a very impractical but perhaps educational method of performing this task. Assume that we attach to each of the 3NJ atoms a set of three vectors of the same length $\underset{\sim}{x}_i$, $\underset{\sim}{y}_i$ and $\underset{\sim}{z}_i$ representing the displacements of each atom from its equilibrium position, respectively, in the $\underset{\sim}{x}$, $\underset{\sim}{y}$ and $\underset{\sim}{z}$ directions. The set of these 3NJ vectors forms a basis for a representation of the space group G. This representation, which has dimensions $3NJ \times 3NJ$, is a reducible representation of G and can be reduced in terms of irreducible representations of G. Considering that a crystal with volume of $1\,cm^3$ contains, say, 10^{23} atoms, to call the task of reducing a $10^{23} \times 10^{23}$ matrix representation impractical is quite an understatement.

Things are made easier by a property of the basis functions of space groups that was reported in Section 4.6.2: Block functions that form a basis for a space group and have the same $\underset{\sim}{k}$ vector transform irreducibly under the operations of the group of the $\underset{\sim}{k}$ vector K. Therefore the normal coordinates that belong to the same frequency eigenvalue and correspond to the same $\underset{\sim}{k}$ vector transform according to an irreducible representation of K; then the problem of finding the transformation properties of the normal modes is reduced to that of finding the irreducible representations (of K) of the normal coordinates with a certain $\underset{\sim}{k}$.

The total number of normal modes in the crystal is 3NJ. Since there are N distinct values of $\underset{\sim}{k}$ in the Brillouin zone, the number of normal modes in correspondence to a certain $\underset{\sim}{k}$ is 3J. The sum of the dimensionalities of all the irreducible representations of the lattice vibrations in correspondence to a certain $\underset{\sim}{k}$ vector is then 3J.

6.6.2. *The Frequency Eigenvalues and the Polarization Vectors*

The frequency eigenvalues and the polarization vectors are obtained from the equation (6.5.9):

$$m_\nu \omega^2 V_{\nu\alpha} = \sum_{\nu'\alpha'} G^{\nu\nu'}_{\alpha\alpha'}(\underset{\sim}{k})\, V_{\nu'\alpha'}. \tag{6.6.6}$$

If we set

$$\begin{aligned} G^{\nu\nu'}_{\alpha\alpha'}(\underset{\sim}{k}) &= \sqrt{m_\nu m_{\nu'}}\, D^{\nu\nu'}_{\alpha\alpha'}(\underset{\sim}{k}), \\ V_{\nu\alpha}(\underset{\sim}{k},\lambda) &= \frac{e_{\nu\alpha}(\underset{\sim}{k},\lambda)}{\sqrt{m_\nu}}, \end{aligned} \tag{6.6.7}$$

equations (6.6.6) become

$$\sum_{\nu'\alpha'}^{3J} D_{\alpha\alpha'}^{\nu\nu'}(\underset{\sim}{k}) e_{\nu'\alpha'}(\underset{\sim}{k}, \lambda) = \omega_{\underset{\sim}{k}\lambda}^2 e_{\nu\alpha}(\underset{\sim}{k}, \lambda), \tag{6.6.8}$$

where the eigenvalues $\omega_{\underset{\sim}{k}\lambda}^2$ are given by the secular equation

$$|D_{\alpha\alpha'}^{\nu\nu'}(\underset{\sim}{k}) - \delta_{\alpha\alpha'}\delta_{\nu\nu'}\omega_{\underset{\sim}{k}\lambda}^2| = 0. \tag{6.6.9}$$

The 3J quantities $e_{\nu\alpha}$ can be considered to be the components of a 3J-dimensional vector $\underset{\sim}{e}$. The 3J vectors $\underset{\sim}{e}(\underset{\sim}{k}, \lambda)$, one for each λ, are orthogonal and normalized according to condition (6.5.46):

$$\underset{\sim}{e}(\underset{\sim}{k}, \lambda)^* \cdot \underset{\sim}{e}(\underset{\sim}{k}, \lambda) = M, \tag{6.6.10}$$

where $M = \sum_\nu m_\nu$.

A similarity exists between the eigenvalue equations above and the eigenvalue equation of a quantum-mechanical system. It is useful, then, to recount the effects of symmetry on a quantum-mechanical eigenvalue equation. Consider a system with a Hamiltonian H and assume that the eigenvalues and the eigenfunctions of H are known. Let us call $\{\psi_{n\ell}\}$ a set of m degenerate eigenfunctions with eigenvalue E_n:

$$H\psi_{n\ell} = E_n\psi_{n\ell} \quad (\ell = 1, 2, \ldots, m). \tag{6.6.11}$$

The space like symmetry operations that leave the Hamiltonian invariant form a group, called the *group of the Schrödinger equation*. By applying an operation R of this group on both sides of the eigenvalue equation (6.6.11) we find

$$RH\psi_{n\ell} = HR\psi_{n\ell} = E_n R\psi_{n\ell}. \tag{6.6.12}$$

Therefore $R\psi_{n\ell}$ is a solution of the eigenvalue equation and can be expressed as

$$R\Psi_{n\ell} = \sum_{i=1}^{m} \Gamma(R)_{i\ell}\psi_{ni}, \tag{6.6.13}$$

or

$$R(\Psi_{n1}\Psi_{n2}\cdots\Psi_{nm})$$

$$= (\psi_{n1}\psi_{n2}\cdots\psi_{nm})\begin{pmatrix}\Gamma_{11} & \Gamma_{12} & \cdots & \Gamma_{1m}\\ \Gamma_{21} & \Gamma_{22} & \cdots & \Gamma_{2m}\\ \vdots & \vdots & & \vdots\\ \Gamma_{m1} & \Gamma_{m2} & \cdots & \Gamma_{mm}\end{pmatrix}. \qquad (6.6.14)$$

The matrices obtained by operating on the set $\{\psi_{n\ell}\}$ with all the operations of G form an irreducible representation of G. Then we can conclude[6]

> "Eigenfunctions belonging to the same eigenvalue form a basis for an irreducible representation of the group of operations which leave the Hamiltonian invariant. The dimension of this representation is equal to the degree of the degeneracy."

Let us turn now our attention to the eigenvalue problem of equations (6.6.8). The equations can be rewritten in the following fashion:

$$\begin{pmatrix}D_{11} & D_{12} & \cdots & D_{1,3J}\\ D_{21} & D_{22} & \cdots & D_{2,3J}\\ \vdots & \vdots & & \vdots\\ D_{3J,1} & D_{3J,2} & \cdots & D_{3J,3J}\end{pmatrix}\begin{pmatrix}e_1\\ e_2\\ \vdots\\ e_{3J}\end{pmatrix} = \omega^2_{\underset{\sim}{k}\lambda}\begin{pmatrix}e_1\\ e_2\\ \vdots\\ e_{3J}\end{pmatrix}, \qquad (6.6.15)$$

or

$$\underset{\sim}{D}(\underset{\sim}{k})\underset{\sim}{e}(\underset{\sim}{k},\lambda) = \omega^2_{\underset{\sim}{k}\lambda}\underset{\sim}{e}(\underset{\sim}{k},\lambda). \qquad (6.6.16)$$

Here the column eigenvector $\underset{\sim}{e}$ with 3J components plays a role similar to that of the eigenfunction in (6.6.11). Let us assume that the eigenvalue $\omega^2_{\underset{\sim}{k}\lambda}$ is m-fold degenerate; in this case a set of m eigenvectors will correspond to the same eigenvalue. If $\underset{\sim}{R}$ is a $3J \times 3J$ matrix that commutes with $\underset{\sim}{D}$, then

$$\underset{\sim}{R}\underset{\sim}{e}_\ell(\underset{\sim}{k},\lambda) = \sum_{i=1}^{m}\Gamma(R)_{i\ell}\underset{\sim}{e}_i(\underset{\sim}{k},\lambda), \qquad (6.6.17)$$

where

$$\ell = 1, 2, \ldots, m.$$

It can be shown that (1) all the 3J × 3J matrices $\underset{\sim}{R}$ which commute with $\underset{\sim}{D}$ form a group that is isomorphic with the group of the $\underset{\sim}{k}$ vector K and (2) these matrices form a reducible representation of K that contains *all* the irreducible representations of the lattice vibrations in correspondence to a certain vector $\underset{\sim}{k}$.

A simple way of forming the matrices $\underset{\sim}{R}$ consists of using as basis the set of 3J vectors $\underset{\sim}{f}_1, \underset{\sim}{f}_2, \ldots, \underset{\sim}{f}_{3J}$ represented by

$$\begin{pmatrix} 1 \\ 0 \\ 0 \\ \cdots \\ \cdots \\ 0 \\ 0 \end{pmatrix}, \begin{pmatrix} 0 \\ 1 \\ 0 \\ \cdots \\ \cdots \\ 0 \\ 0 \end{pmatrix}, \begin{pmatrix} 0 \\ 0 \\ 1 \\ \cdots \\ \cdots \\ 0 \\ 0 \end{pmatrix}, \cdots \begin{pmatrix} 0 \\ 0 \\ 0 \\ \cdots \\ \cdots \\ 0 \\ 1 \end{pmatrix}, \tag{6.6.18}$$

and obtained by attaching to each atom in the unit cell three displacement vectors in the x, y, and z directions. The matrices $\underset{\sim}{R}$ are obtained by operating on these 3J vectors with the operations R of K:

$$R\underset{\sim}{f}_m = \sum_{j=1}^{3J} R_{jm}\underset{\sim}{f}_j. \tag{6.6.19}$$

It suffices, for the purpose of reducing the matrices $\underset{\sim}{R}$ and finding the irreducible representations of the lattice vibrations, to calculate the characters of these matrices, and this is easily accomplished by considering that a vector $\underset{\sim}{f}_m$ contributes to the character *only* if the operation R transforms it into the equivalent vector of a neighboring unit cell. This procedure will be illustrated by several examples in the following sections.

The reduction of the representation formed by the matrices $\underset{\sim}{R}$ gives the irreducible representations of the lattice vibrations for the particular $\underset{\sim}{k}$ vector under consideration. In turn, these irreducible representations give the degeneracies of the frequency eigenvalues and the transformation properties of the normal coordinates.

6.6.3. *Additional Degeneracies Not Due to Spacelike Symmetries*

As a rule the normal coordinates with the same frequency eigenvalue transform irreducibly under the operations of the space group; moreover, the normal coordinates with the same frequency eigenvalue and the same $\underset{\sim}{k}$ transform irreducibly under the operations of the group of the $\underset{\sim}{k}$ vector K.

There are essentially two exceptions to this rule:

I. *Accidental Degeneracy*

It may happen that the force constants and the masses of the atoms in the crystal have such values that two sets of normal coordinates with the same $\underset{\sim}{k}$ and belonging to two distinct representations of K have the same frequency eigenvalue. A slight change in the masses or force constants would remove this degeneracy without changing the spacelike symmetry of the crystal.

II. *Excess Degeneracy*

This type of degeneracy is due to the fact that the group $\mathcal{K}$ does not contain all the symmetry operations of the dynamical matrix.

Consider the eigenvalue equation (6.6.16) and replace $\underset{\sim}{k}$ with $-\underset{\sim}{k}$:

$$\underset{\sim}{D}(-\underset{\sim}{k})\underset{\sim}{e}(-\underset{\sim}{k},\lambda) = \omega^2_{-\underset{\sim}{k}\lambda}\underset{\sim}{e}(-\underset{\sim}{k},\lambda). \tag{6.6.20}$$

Taking the complex conjugate of both members we find

$$\underset{\sim}{D}(\underset{\sim}{k})\underset{\sim}{e}(-\underset{\sim}{k},\lambda)^* = \omega^2_{-\underset{\sim}{k}\lambda}\underset{\sim}{e}(-\underset{\sim}{k},\lambda)^*, \tag{6.6.21}$$

where the eigenvalues $\omega^2_{-\underset{\sim}{k}\lambda}$ are given by the secular equation

$$|D^{\nu\nu'}_{\alpha\alpha'}(\underset{\sim}{k}) - \delta_{\alpha\alpha'}\delta_{\nu\nu'}\omega^2_{-\underset{\sim}{k}\lambda}| = 0. \tag{6.6.22}$$

Comparing (6.6.22) with (6.6.9) we obtain

$$\omega^2_{-\underset{\sim}{k}\lambda} = \omega^2_{\underset{\sim}{k}\lambda}, \tag{6.6.23}$$

and also

$$\underset{\sim}{e}(-\underset{\sim}{k},\lambda)^* = \underset{\sim}{e}(\underset{\sim}{k},\lambda), \tag{6.6.24}$$

in agreement with (6.5.33). The degeneracy (6.6.23) is present in any case, whether the point group $G_o(\underset{\sim}{k})$ contains the operation I (inversion) or not. In the former case the degeneracy is determined by the spacelike symmetry of the crystal; in the latter case the degeneracy is not determined merely by the spacelike symmetry.

Given a certain $\underset{\sim}{k}$, the normal coordinates in correspondence to $\underset{\sim}{k}$ and to all the other arms of the star $\{\underset{\sim}{k}\}$ belong to the same irreducible representation of the space group and have therefore the same frequency eigenvalue. Considering this fact, together with the inversion symmetry, we have, in general

$$\omega_{\underset{\sim}{k}\lambda} = \omega_{\{\underset{\sim}{k}\}\lambda} = \omega_{\{-\underset{\sim}{k}\}\lambda}. \tag{6.6.25}$$

Also, referring back to (6.5.12), we have

$$\omega_{\underset{\sim}{k}\lambda} = \omega_{\underset{\sim}{k}+\underset{\sim}{K},\lambda}, \tag{6.6.26}$$

where $\underset{\sim}{K}$ is the primitive vector of the reciprocal lattice. The relations (6.6.25) and (6.6.26) above give the symmetries of the frequency eigenvalues.

A case of extra degeneracy is the one that may be present in correspondence to the point $\underset{\sim}{k}=0$ in the Brillouin zone. We have already found in Section 6.5.5 that the three acoustical branches go to zero *together* at $\underset{\sim}{k}=0$. On the other hand, the eigenvectors of these acoustical branches at $\underset{\sim}{k}=0$ transform as x, y, and z under the operations of the point group $G_o(\underset{\sim}{k})$, which at this point of the Brillouin zone coincides with the crystallo-graphic point group G_o. An additional degeneracy may be present here if the crystal has such low symmetry that G_o does not have any three-dimensional representation: in this case x, y, and z belong to different representations that are degenerate.

Another case of excess degeneracy is the one due to *time reversal*, discussed in the following section.

6.6.4. *Time-Reversal Degeneracy*

The concept of time reversal can be simply illustrated by considering the motion of a particle under the action of a force in the x direction. Let x_o and v_o be the position and the velocity of the particle at time $t = 0$ and allow the particle to proceed under the action of the force for a time t_1, at which the position and the velocity will be x_1 and v_1. At the time $t = t_1$ get an identical particle started with velocity $-v_1$ and let it proceed under the action of the force up to time $2t_1$. If at this time the position and the velocity of the second particle are x_o and $-v_o$, we say that the equation of motion is invariant under time reversal. In case of time reversal, the filming of the motion of the second particle in the time interval $(t_1, 2_{t_1})$ would be equal to the reverse filming of the motion of the particle in the time interval $(0, t_1)$.

The equation of motion of a system under the action of a force derivable from a potential presents time reversal. An example of a system that does not present time reversal is a charged particle in a magnetic "external" field (however, if the motion of the charged particles that produce the field is included, the time-reversal symmetry is restored).

Time reversal has relevance in lattice dynamics in that it can produce additional degeneracies besides those due to the space-like symmetry of the crystal.[7] Consider, as usual, a crystal with J atoms in the unit cell and invariant under the operations of a space group G. The frequency eigenvalues and the polarization vectors in correspondence to a particular r-degenerate mode $(\underset{\sim}{k}, \lambda)$ are given by

$$\underset{\sim}{D}(\underset{\sim}{k})\underset{\sim}{e}_\ell(\underset{\sim}{k}, \lambda) = \omega^2_{\underset{\sim}{k}\lambda}\underset{\sim}{e}_\ell(\underset{\sim}{k}, \lambda), \tag{6.6.27}$$

where $\ell = 1, 2, \ldots, r$. The matrix $\underset{\sim}{D}(\underset{\sim}{k})$ commutes with all the $(3J \times 3J)$ matrices $\underset{\sim}{R}$ that represent the operations R of *K*. Assume that the group G contains an operation $\{\psi_o|\underset{\sim}{t}\}$ such that

$$\psi_o\underset{\sim}{k} = -\underset{\sim}{k} + \underset{\sim}{K}, \tag{6.6.28}$$

where $\underset{\sim}{K}$ is the primitive vector of the reciprocal lattice. A time-reversal matrix operator [of dimensions $(3J \times 3J)$] can be constructed as follows:

$$\underset{\sim}{C} = C_o\underset{\sim}{\Gamma}(\underset{\sim}{k}; \{\psi_o|\underset{\sim}{t}\}), \tag{6.6.29}$$

where C_o is the complex conjugation operation and $\underset{\sim}{\Gamma}(\underset{\sim}{k}; \{\psi_o|\underset{\sim}{t}\})$ is the $(3J \times 3J)$ matrix that represents the operation $\{\psi_o|\underset{\sim}{t}\}$. We notice immediately that the matrices $\underset{\sim}{D}(\underset{\sim}{k})$ and $\underset{\sim}{C}$ commute:

$$\underset{\sim}{C}\underset{\sim}{D} = \underset{\sim}{D}\underset{\sim}{C}, \tag{6.6.30}$$

even if the operation $\{\psi_o|\underset{\sim}{t}\}$ is not an operation of the group K.

Let us apply now the matrix operator $\underset{\sim}{C}$ to both members of (6.6.27):

$$\begin{aligned}\underset{\sim}{C}\underset{\sim}{D}(\underset{\sim}{k})\underset{\sim}{e}_\ell(\underset{\sim}{k},\lambda) &= \underset{\sim}{D}(\underset{\sim}{k})[\underset{\sim}{C}\underset{\sim}{e}_\ell(\underset{\sim}{k},\lambda)]\\ &= \omega^2_{\underset{\sim}{k}\lambda}[\underset{\sim}{C}\underset{\sim}{e}_\ell(\underset{\sim}{k},\lambda)].\end{aligned} \tag{6.6.31}$$

Therefore the set of r vectors $\{\underset{\sim}{C}\underset{\sim}{e}_\ell(\underset{\sim}{k},\lambda)\}$ gives also a solution of the eigenvalue equation in correspondence to the eigenvalue $\omega^2_{\underset{\sim}{k}\lambda}$. If this set $\{\underset{\sim}{C}\underset{\sim}{e}_\ell(\underset{\sim}{k},\lambda)\}$ is linearly independent of the initial set $\{\underset{\sim}{e}_\ell(\underset{\sim}{k},\lambda)\}$, there must be another set of r degenerate eigenvectors of $\underset{\sim}{D}(\underset{\sim}{k})$ with the same eigenvalue $\omega^2_{\underset{\sim}{k}\lambda}$; in this case an additional degeneracy arises. The problem of finding any additional time-reversal degeneracy reduces to checking if the two sets $\{\underset{\sim}{e}_\ell(\underset{\sim}{k},\lambda)\}$ and $\{\underset{\sim}{C}\underset{\sim}{e}_\ell(\underset{\sim}{k},\lambda)\}$ are linearly independent.

We shall now state several facts without proof. The reader is referred to the proper sources as references:

1. If the set $\{\underset{\sim}{e}_\ell(\underset{\sim}{k},\lambda)\}$ generates an irreducible representation of K in which the operation $\{\beta|\underset{\sim}{b}\}$ is represented by an $(r \times r)$ matrix $\underset{\sim}{T}(\{\beta|\underset{\sim}{b}\})$, then the set $\{\underset{\sim}{C}\underset{\sim}{e}_\ell(\underset{\sim}{k},\lambda)\}$ generates a representation of K in which $\{\beta|\underset{\sim}{b}\}$ is represented by

$$\bar{\underset{\sim}{T}}(\{\beta|\underset{\sim}{b}\}) = \underset{\sim}{T}(\{\psi_o|\underset{\sim}{t}\}^{-1}\{\beta|\underset{\sim}{b}\}\{\psi_o|\underset{\sim}{t}\})^*. \tag{6.6.32}$$

   Notice that T and $\bar{T}$ have the same dimensionalities and correspond to the same $\underset{\sim}{k}$ and frequency eigenvalue.
2. If the two representations $\underset{\sim}{T}$ and $\bar{\underset{\sim}{T}}$ are inequivalent, the two sets of vector are linearly independent. In this case $\underset{\sim}{T}$ and $\bar{\underset{\sim}{T}}$ are said to be of *the third kind.* Since the existence of a representation $\underset{\sim}{T}$ postulates the existence of $\bar{\underset{\sim}{T}}$, the two representations must always occur in pairs and additional degeneracy occurs.
3. If the two representations $\underset{\sim}{T}$ and $\bar{\underset{\sim}{T}}$ are equivalent the two sets of vectors may or may not be linearly independent.[8] If they are, additional degeneracy occurs and the representation $\underset{\sim}{T}$ must occur

an even number of times: in this case $\underset{\sim}{T}$ and $\bar{\underset{\sim}{T}}$ are said to be of *the second kind.* If the two sets of eigenvectors are linearly dependent, no additional degeneracy occurs and the representation T may occur any number of times. In this last case, T and $\bar{\underset{\sim}{T}}$ are said to be of *the first kind.*

4. A simple criterion exists for deciding to which case T belongs[7]:

$$\sum_{\{\psi_i|\underset{\sim}{t}_i\}}' \chi(\{\psi_i|\underset{\sim}{t}_i\}^2) = \begin{cases} h & \text{first kind} \\ -h & \text{second kind} \\ 0 & \text{third kind,} \end{cases} \tag{6.6.33}$$

where χ is the character of an irreducible representation of K, and the sum is carried out over one arbitrarily chosen element $\{\psi_i|\underset{\sim}{t}_i\}$ from each of the h cosets of the factor group $[\{\psi_o|\underset{\sim}{t}\}K]/T$. It is evident that the choice of $\{\psi_o|\underset{\sim}{t}\}$ does not affect the results.

5. The criterion above can be specified for point $\underset{\sim}{k} = 0$ or other points on the surface of the Brillouin zone with $G_o(\underset{\sim}{k}) = G_o$. In this case any operation β of G_o changes, in general, $\underset{\sim}{k}$ into $-\underset{\sim}{k} + \underset{\sim}{K}$. The criterion above becomes then

$$\sum_{\{\beta|\underset{\sim}{b}\}}' \chi(\{\beta|\underset{\sim}{b}\}^2) = \begin{cases} h & \text{first kind} \\ -h & \text{second kind} \\ 0 & \text{third kind,} \end{cases} \tag{6.6.34}$$

where the sum is carried out over one arbitrarily chosen element from each of the h cosets of the factor group K/T.

The two criteria above can be simplified if the group of the $\underset{\sim}{k}$ vector K is symmorphic. In this case (6.6.33) reduces to

$$\sum_{\gamma} \chi(\gamma^2) = \begin{cases} g_o' & \text{first kind} \\ -g_o' & \text{second kind} \\ 0 & \text{third kind,} \end{cases} \tag{6.6.35}$$

where χ is the character of an irreducible representation of $G_o(\underset{\sim}{k})$ and g_o' is the order of the group $G_o(\underset{\sim}{k})$. The sum above is carried out over all rotational operations γ that bring $\underset{\sim}{k}$ into $-\underset{\sim}{k}+\underset{\sim}{K}$, where $\underset{\sim}{K}$ is the primitive vector of the reciprocal lattice.

For the special case in which K is symmorphic and $G_o(\underset{\sim}{k}) = G'_o$ the criterion (6.6.34) becomes

$$\sum_{\beta} \chi(\beta^2) = \begin{cases} g'_o & \text{first kind} \\ -g'_o & \text{second kind} \\ 0 & \text{third kind,} \end{cases} \tag{6.6.36}$$

where the sum is carried out over all the elements β of $G_o(\underset{\sim}{k})$.

6.7. Group-Theoretical Analysis of the Lattice Vibrations of a Linear Crystal

6.7.1. *Case of One Atom Per Unit Cell*

A linear crystal with one atom per unit cell [see Fig. 6.1] is invariant under all the operations of a one-dimensional space group G. The operations of this group are

$$\{E|na\}, \ \{I|na\},$$

where E represents identity, I represents inversion, to be performed with respect to an atomic site, and a is the equilibrium distance between two successive particles.

The Brillouin zone for such a crystal is given in Fig. 6.2. We need to consider only the section of the Brillouin zone from 0 to π/a, since the k values in the interval $(0, -\pi/a)$ can be obtained by operating on the k points in $(0, \pi/a)$ with the operation I. Without going into detailed calculations we give in Tables 6.1, 6.2 and 6.3 all the information regarding the irreducible representations of this space group.

We can now make the following points:

1. The normal coordinate corresponding to a point k of the Brillouin zone transforms according to an irreducible representation of the group K:

POINT $k = 0$

$$\{E|na\}q_o = q_o,$$
$$\{I|na\}q_o = \pm q_o.$$

Table 6.1. Irreducible representations of the group K for a linear monatomic crystal.

$k = 0$		$\{E\|na\}$	$\{I\|na\}$	G_o	E	I
$G_o(k) = G_o : E, I$	Γ_1	1	1	Γ_1	1	1
	Γ_2	1	-1	Γ_2	1	-1

$0 < k < \pi/a$		$\{E\|na\}$
$G_o(k) : E$	Γ	e^{ikna}

$k = \pi/a$		$\{E\|na\}$	$\{I\|na\}$
$G_o(k) = G_o : E, I$	Γ_1	$e^{i\pi n}$	$e^{i\pi n}$
	Γ_2	$e^{i\pi n}$	$-e^{i\pi n}$

Table 6.2. Irreducible representations of the space group for a linear monatomic crystal.

		$\{E\|na\}$	$\{I\|na\}$
$k = 0$	Γ_1	1	1
	Γ_2	1	-1

		$\{E\|na\}$	$\{I\|na\}$
$0 < k < \pi/a$	Γ	$\begin{pmatrix} e^{ikna} & \\ 0 & e^{-ikna} \end{pmatrix}$	$\begin{pmatrix} 0 & e^{ikna} \\ e^{-ikna} & 0 \end{pmatrix}$

		$\{E\|na\}$	$\{I\|na\}$
$k = \pi/a$	Γ_1	$e^{i\pi n}$	$e^{i\pi n}$
	Γ_2	$e^{i\pi n}$	$-e^{i\pi n}$

POINT $0 < k < \pi/a$

$$\{E|na\}q_k = e^{ikna}q_k.$$

POINT $k = \pi/a$

$$\{E|na\}q_{\pi/a} = e^{i\pi n}q_{\pi/a},$$
$$\{I|na\}q_{\pi/a} = \pm e^{i\pi n}q_{\pi/a}.$$

Table 6.3. Summary of the irreducible representations of the space group of a linear crystal.

Region	K	$G_o(k)$	Number of representations	Dimension of representations
$k = 0$	$K = G$	$G_o(k) = G_o : E, I$	2	1 $d = 1$, $q = 1$
$0 < k < \pi/a$	$K = T$	$G_o(k) : E$	1	2 $d = 1$, $q = 2$
$k = \pi/a$	$K = G$	$G_o(k) = G_o : E, I$	2	1 $d = 1$, $q = 1$

2. The normal coordinates representing the modes of vibration of a crystal with the same frequency form basis for an irreducible representation of the space group:

POINT $k = 0$

$$\{E|na\}q_o = q_o,$$
$$\{I|na\}q_o = \pm q_o.$$

POINTS $0 < k < \pi/a$

$$\{E|na\}(q_k\ q_{-k}) = (q_k\ q_{-k}) \begin{pmatrix} e^{ikna} & 0 \\ 0 & e^{-ikna} \end{pmatrix}$$
$$= (e^{ikna} q_k\quad e^{-ikna} q_{-k}),$$
$$\{I|na\}(q_k\ q_{-k}) = (q_k\ q_{-k}) \begin{pmatrix} 0 & e^{ikna} \\ e^{-ikna} & 0 \end{pmatrix}$$
$$= (e^{-ikna} q_{-k}\quad e^{ikna} q_k),$$

or

$$\{E|na\}q_k = e^{ikna} q_k,$$
$$\{E|na\}q_{-k} = e^{-ikna} q_{-k},$$
$$\{I|na\}q_k = e^{-ikna} q_{-k},$$
$$\{I|na\}q_{-k} = e^{ikna} q_k.$$

POINT $k = \pi/a$

$$\{E|na\}q_{\pi/a} = e^{i\pi n}q_{\pi/a},$$
$$\{I|na\}q_{\pi/a} = \pm e^{i\pi n}q_{\pi/a}.$$

3. In treating the lattice vibrations of a linear crystal with one atom per unit cell, the N equations of motion can be reduced to one equation (6.1.17). This means that in order to find the normal mode of vibration corresponding to a certain k, it is sufficient to consider the only component of the displacement in the unit cell. This corresponds also to the group-theoretical fact that this displacement forms a basis for a representation of the group *K*. The displacement of an atom occupying the position sa is given by

$$u_s = u_o e^{iska}. \tag{6.7.1}$$

Since the displacement is already in the Block form, it must transform according to some representation of the group $G_o(k)$. Let us consider now different points of the Brillouin zone.

POINT $k = 0$

For this value of k, $G_o(k) = G_o$. In order to find the representation set up by the displacement coordinate, we attach to the atom in the unit cell a vector along the coordinate axis and apply to it the operations of $G_o(k)$. The character table of the representation is found by using the following rules:

1. If the displacement transforms into itself it contributes +1 to the character.
2. If the displacement transforms into the opposite of itself it contributes −1 to the character.

By applying this procedure we find the following representation set up by the displacement coordinate:

E	I
1	−1

which corresponds to the representation Γ_2 of G_o.

POINTS $0 < k < \pi/a$

For these values of k the group $G_o(k)$ consists only of the operation E. The displacement transforms according to the only representation, which we shall call Γ, of $G_o(k)$.

POINT $k = \pi/a$

For this point, $G_o(k) = G_o$. The displacement coordinate transforms again according to the representation Γ_2 of G_o.

In summary, the normal coordinate will transform according to the following representations

$k = 0$:	$G_o(k) = G_o$,	Γ_2,
$0 < k < \pi/a$:	$G_o(k) : E$,	Γ,
$k = \pi/a$:	$G_o(k) = G_o$,	Γ_2.

6.7.2. *Case of Two Atoms Per Unit Cell*

Let us consider a linear crystal with two atoms per unit cell, represented in Fig. 6.4. Let us also consider the situation in which $b = b'$ $(a = 2b)$, that is, the case in which the spacing between neighboring atoms is the same. We will consider later the case of unequal spacing $(b \neq b')$.

The space group that leaves the crystal with $b = b'$ invariant is the same space group of Section 6.6.1 with the representations given in Tables 6.2 and 6.3. We may find the irreducible representations of the normal coordinates considering the displacements of the two atoms in the nth cell in correspondence to a particular k:

$$\begin{aligned} u_{n1} &= u_{01} e^{ikna}, \\ u_{n2} &= u_{02} e^{ikna}. \end{aligned} \tag{6.7.2}$$

We can now refer back to the property enunciated in Section 4.6.2 that Block functions which form a basis for a space group and have the same k vector transform irreducibly under the operations of the subgroup K of G. The two displacements above, being already in the Block form with the same k, must transform according to some *reducible* representation of K. Moreover, since in the present case K is symmorphic, they must transform, when operated upon by the operations of $G_o(k)$, according to some reducible representation of $G_o(k)$.

Since the crystallographic point group contains the operation I (inversion), we may limit ourselves to consider only the section of the Brillouin zone from 0 to π/a. We define I as the inversion with respect to the site n in Fig. 6.4

POINT $k = 0$

$$\begin{aligned} Eu_{n1} &= u_{n1}, \\ Eu_{n2} &= u_{n2}, \\ Iu_{n1} &= -u_{n1}, \\ Iu_{n2} &= -u_{n2}. \end{aligned}$$

The following (reducible) representation is set up by the displacement coordinates

E	I	Reduction
2	−2	$2\Gamma_2$

Therefore the normal coordinates transform according to two one-dimensional representations Γ_2.

POINTS $0 < k < \pi/a$

$$\begin{aligned} Eu_{n1} &= u_{n1}, \\ Eu_{n2} &= u_{n2}. \end{aligned}$$

The representation set up by the displacements reduces as 2Γ, where Γ is the (only) representation of $G_o(k)$.

POINT $k = \pi/a$

$$\begin{aligned} Eu_{n1} &= u_{n1}, \\ Eu_{n2} &= u_{n2}, \\ Iu_{n1} &= -u_{n1}, \\ Iu_{n2} &= u_{n2}. \end{aligned}$$

The following (reducible) representation is set up by the displacement coordinates

E	I	Reduction
2	0	$\Gamma_1 + \Gamma_2$

Therefore the normal coordinates transform according to the two one-dimensional representations Γ_1 and Γ_2.

In summary the normal coordinates will transform according to the following representations:

$$\begin{array}{lll} k = 0, & G_o(k) = G_o, & 2\Gamma_2 \text{ (2 nondegenerate modes)}, \\ 0 < k < \pi/a, & G_o(k) : E, & 2\Gamma \text{ (2 nondegenerate modes)}, \\ k = \pi/a, & G_o(k) = G_o, & \Gamma_1 + \Gamma_2 \text{ (2 nondegenerate modes)}. \end{array}$$

It is easy to show that the time-reversal symmetry does not introduce any additional degeneracy:

$$k = 0.$$

The point group $G_o(k)$ contains the operations E and I. Both these operations send k into $-k + K$, with K a primitive vector of the reciprocal lattice. Applying the criterion (6.6.36) we find

$$\sum_\beta \chi_j(\beta^2) = \chi_j(E^2) + \chi_j(I^2) = 2\chi_j(E) = 2 = g_o',$$

with $j = 1, 2$; g_o' is the order of $G_o(k)$. No time-reversal degeneracy is present.

$$0 < k < \pi/a.$$

The point group $G_o(k)$ contains only the operation E. It is

$$\sum_\gamma \chi(\gamma^2) = \chi(I^2) = \chi(E) = 1.$$

No time-reversal degeneracy is present.

$$k = \pi/a.$$

In this case also no time-reversal degeneracy is present.

Let us refer back to Fig. 6.4 and set $b \neq b'$. In this case the space group of the crystal consists only of the primitive translations $\{E|na\}$: I (inversion) is not an operation of the crystallographic point group G_o. Therefore we cannot limit our-selves to considering only the part of the Brillouin zone from 0 to π/a. Rather, we need to consider the zone from $-\pi/a$ to $+\pi/a$. The irreducible representation for the space group are simply given by e^{ikna} for the operation $\{E|na\}$, for any k. No degeneracy of ω_k and ω_{-k} should be expected on the basis of spacelike symmetry considerations. However, considering the symmetry of ω in (6.6.25) we know that actually $\omega_{\mathrm{k}} = \omega_{-\mathrm{k}}$: we have here an additional degeneracy which is *not* due to the spacelike symmetry of the crystal.

6.8. Group-Theoretical Analysis of the Lattice Vibrations of a Three-Dimensional Crystal

Let us consider now the case of a three-dimensional crystal with a primitive unit cell defined by the three vectors $\underset{\sim}{a}_1$, $\underset{\sim}{a}_2$, and $\underset{\sim}{a}_3$ and J atoms in the unit cell and invariant under the operations of a space group G. As usual we shall call G_o the point group of G, K the (space) group of the $\underset{\sim}{k}$ vector, and $G_o(\underset{\sim}{k})$ the point group of K.

We can make the following points:

1. The normal coordinates corresponding to the same frequency eigenvalue form a basis for an irreducible representation of the space group. Also

$$\omega_{\underset{\sim}{k}\lambda} = \omega_{\alpha\underset{\sim}{k},\lambda}, \tag{6.8.1}$$

   where λ ranges over the 3J branches of the dispersion relation and α is any operation of G_o. Notice that if α is an operation of $G_o(\underset{\sim}{k})$, then $\underset{\sim}{k}$ and $\alpha\underset{\sim}{k}$ are represented by the same arm in the star $\{\underset{\sim}{k}\}$. A consequence of this is that we need to consider only that part of the Brillouin zone where no two vectors $\underset{\sim}{k}$ and $\underset{\sim}{k}'$ can be found such that

$$\underset{\sim}{k}' = \alpha\underset{\sim}{k} + \underset{\sim}{K}_j, \tag{6.8.2}$$

   where $\underset{\sim}{K}_j$ is a vector of the reciprocal lattice. If the number of operations in G_o is g_o we need to consider only $1/g_o$ of the Brillouin zone.
2. The normal coordinates that belong to the same frequency eigenvalue and correspond to the same $\underset{\sim}{k}$ vector, transform irreducibly under the operations of the subgroup K of G.
3. Consider the expression for the displacement of an atom occupying the νth position in the sth unit cell:

$$u_{s\nu\alpha} = u_{o\nu\alpha}e^{i\underset{\sim}{k}\cdot\underset{\sim}{R}_s}, \tag{6.8.3}$$

where

$$\alpha = 1, 2, 3$$
$$\nu = 1, 2, \ldots, J$$
$$\underset{\sim}{R}_s = s_1\underset{\sim}{a}_1 + s_2\underset{\sim}{a}_2 + s_3\underset{\sim}{a}_3.$$

The 3J displacement components above, being already in the Block form with the same $\underset{\sim}{k}$ vector, must transform according to some (in general, reducible) representation of the group K of the vector $\underset{\sim}{k}$. The reducible representation of K set up by the displacement components can be reduced in terms of irreducible representations of K. This process of reduction gives the degeneracies of the frequency eigenvalues and the transformation properties of the normal coordinates.

Let us now consider the task of finding these representations by making use of the properties described above.

POINT $\underset{\sim}{k} = \underset{\sim}{0}$

The star $\{\underset{\sim}{k}\}$ has the highest degeneracy and the space group of the $\underset{\sim}{k}$ vector K coincides with the space group G; also $G_o(\underset{\sim}{k}) = G_o$. The irreducible representations of $G = K$ are formed by attributing to both the $\{\gamma|\underset{\sim}{0}\}$ and $\{\gamma|\underset{\sim}{b}\}$ operations the same matrix that represents the operation γ in an irreducible representation of $G_o(\underset{\sim}{k}) = G_o$:

$$D_{11}\{\gamma|\underset{\sim}{b}\} = \Gamma_i(\gamma). \tag{6.8.4}$$

Since the group K is symmorphic, in correspondence to $\underset{\sim}{k} = \underset{\sim}{0}$, the 3J displacement components $u_{s\nu\alpha}$ transform among themselves so as to form a *reducible* representation of $G_o(\underset{\sim}{k}) = G_o$. In order to find the characters of this representation we consider the effect of the operations in $G_o(\underset{\sim}{k})$ on the 3J vector components.

1. Considering an operation γ, if the rotated vector component $\gamma u_{s\nu\alpha}$ is equal to $u_{s\nu\alpha}$ itself or if it is equal to the *same* component belonging to an equivalent atom of a neighboring unit cell, the component $u_{s\nu\alpha}$ contributes $+1$ to the character.
2. Each vector component that transforms into the opposite of itself or of the same component of an equivalent atom of a neighboring unit cell, contributes -1 to the character.
3. Any other vector component does not contribute to the character.

POINTS $\underset{\sim}{k} \neq \underset{\sim}{0}$. [Within the Brillouin zone; $G_o(\underset{\sim}{k})$: E]

At a generic point within the relevant part of the Brillouin zone, the group of the $\underset{\sim}{k}$ vector *K* coincides with the subgroup *T* of G consisting of all the primitive translations. The only operations in $G_o(\underset{\sim}{k})$ is the identity E; the irreducible representations of *K* are given by

$$D_{11}\{\beta|\underset{\sim}{T}_n\} = D_{11}\{E|\underset{\sim}{T}_n\} = e^{i\underset{\sim}{k}\cdot\underset{\sim}{T}_n}. \qquad (6.8.5)$$

On the other hand,

$$\{E|\underset{\sim}{T}_n\}u_{s\nu\alpha} = e^{i\underset{\sim}{k}\cdot\underset{\sim}{T}_n}u_{s\nu\alpha}. \qquad (6.8.6)$$

Therefore each one of the 3J displacements $u_{s\nu\alpha}$ forms a basis for an irreducible (one-dimensional) representation of *K*. In this region, 3J eigenvalues of the frequency correspond to the same value of $\underset{\sim}{k}$.

POINTS $\underset{\sim}{k} \neq \underset{\sim}{0}$. [Points of special symmetry within the Brillouin zone; $G_o(\underset{\sim}{k})$: E, ...]

Some points inside the Brillouin zone may have a special symmetry (see, for example, points on OB in Fig. 4.10). In this case $G_o(\underset{\sim}{k})$ contains other operations besides the identity E, and the irreducible representations of *K* are given by

$$D_{11}\{\beta|\underset{\sim}{b}\} = e^{i\underset{\sim}{k}\cdot\underset{\sim}{b}}\underset{\sim}{\Gamma}_i(\beta), \qquad (6.8.7)$$

where $\underset{\sim}{\Gamma}_i(\beta)$ are the irreducible representations of $G_o(\underset{\sim}{k})$, and $\underset{\sim}{b}$ can be either a primitive or a nonprimitive vector of the real space. On the other hand, since

$$\{\beta|\underset{\sim}{b}\}u_{s\nu\alpha} = e^{i\underset{\sim}{k}\cdot\underset{\sim}{b}}(\beta u_{s\nu\alpha}), \qquad (6.8.8)$$

the 3J displacement components $u_{s\nu\alpha}$ transform, under the operations of $G_o(\underset{\sim}{k})$, so as to form a reducible representation of $G_o(\underset{\sim}{k})$.

The number of frequency eigenvalues in this region is determined by the degeneracies of the irreducible representations $\underset{\sim}{\Gamma}_i(\beta)$ present in the reduction of the representation of $G_o(\underset{\sim}{k})$ set up by the displacement components. In this case the number of frequency eigenvalues is equal or less than 3J.

POINTS $\underset{\sim}{k}$. [on the Surface of the Brillouin Zone]

For points on the surface of the Brillouin zone we have to distinguish the following two cases:

1. *K symmorphic.* As usual, the 3J displacement components $u_{s\nu\alpha}$ transform among themselves as to form a representation of $G_o(\underset{\sim}{k})$ that can be reduced in terms of irreducible representations of $G_o(\underset{\sim}{k})$ giving the degeneracies of the frequency eigenvalues.
2. *K nonsymmorphic.* In this case the irreducible representations of *K* are not given by (6.8.7) and the entire group *K* rather than its point group $G_o(\underset{\sim}{k})$ has to be considered. The displacement components set a reducible representation of *K* that can be reduced in terms of irreducible representations of *K* giving the degeneracies of the frequency eigenvalues.

6.9. Example I. Lattice Vibrations of a Two-Dimensional Crystal with Symmetry C^1_{4v}

Let us consider the two-dimensional crystal whose unit cell is shown in Fig. 6.8. The space group of this crystal is C^1_{4v} and was treated in Section 4.7. We shall refer to the Brillouin zone in Fig. 4.10 and in particular to the Section OBC. Group-theoretical properties of the dispersion curves are given in Table 6.4.

The displacement components are shown in Fig. 6.8. If R is the generic (rotational) operation of $G_o(\underset{\sim}{k})$, the effect of R on the four components can be described by a 4×4 matrix $\underset{\sim}{R}$:

$$R \begin{pmatrix} u_{Ax} \\ u_{Ay} \\ u_{Bx} \\ u_{By} \end{pmatrix} = \underset{\sim}{R} \begin{pmatrix} u_{Ax} \\ u_{Ay} \\ u_{Bx} \\ u_{By} \end{pmatrix}.$$

The traces of the matrices for the various operations in $G_o(\underset{\sim}{k})$ give the characters of the (reducible) representation of $G_o(\underset{\sim}{k})$ set by the displacement components. It can be easily seen that the trace of matrix $\underset{\sim}{R}$ is different from zero only for the operations E and C_2. If

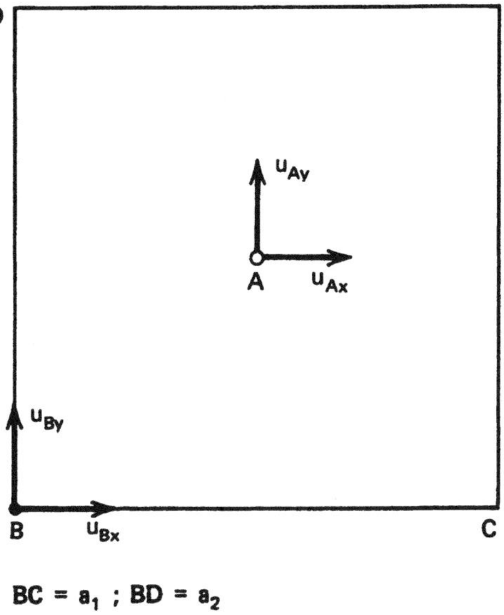

Fig. 6.8. Unit cell of a two-dimensional crystal with space group C_{4v}^1.

we call χ the trace of this matrix, we find

$$\chi(E;\underset{\sim}{k}) = 4,$$
$$\chi(C_2;\underset{\sim}{k}) = -2(1 + e^{-i\underset{\sim}{k}\cdot(\underset{\sim}{a}_1+\underset{\sim}{a}_2)}).$$

ORIGIN 0. $\underset{\sim}{k} = \underset{\sim}{0}$

At this point $K = G$ and $G_o(\underset{\sim}{k}) = G_o = C_{4v}$. Also

$$\underset{\sim}{D}\{\beta|\underset{\sim}{T}_n\} = \underset{\sim}{D}_{11}\{\beta|\underset{\sim}{T}_n\} = \underset{\sim}{\Gamma}_i(\beta),$$

where $\underset{\sim}{\Gamma}_i(\beta)$ is an irreducible representation of C_{4v}. The star $\{\underset{\sim}{k}\}$ has its maximum degeneracy ($q = 1$).

The displacement components set the following representation of $G_o(\underset{\sim}{k})$ which reduces as follows in terms of irreducible representation of C_{4v}:

E	C_4	C_2	$C_4{}^3$	σ_1	σ_2	σ_3	σ_4	Reduction
4	0	−4	0	0	0	0	0	$2\Gamma_5$

There are therefore two 2-fold degenerate eigenvalues of the frequency in correspondence to $\underset{\sim}{k} = \underset{\sim}{0}$.

Table 6.4. Group-theoretical properties of the dispersion curves.

Region	K	$G_o(\underset{\sim}{k})$	Dimension of Reps of G	Observations
$\underset{\sim}{k} = \underset{\sim}{0}$	$K = G$ $D_{11}\{\beta \vert \underset{\sim}{b}\} = \Gamma_i(\beta)$	$G_o(\underset{\sim}{k}) = G_o$	$dq \begin{cases} d = 1 \text{ or } >1 \\ d = 1 \end{cases}$	Irreducible representations of K given by irreducible representations of $G_o(\underset{\sim}{k})$. Displacement coordinates transform reducibly according to $G_o(\underset{\sim}{k})$.
$\underset{\sim}{k}$ inside BZ (no special symmetry)	$K = T$ $D_{11}\{\beta \vert \underset{\sim}{b}\} = e^{i\underset{\sim}{k}\cdot\underset{\sim}{b}}$	$G_o(\underset{\sim}{k}) = C_1 : E$	$dq \begin{cases} d = 1 \\ q = q_o \end{cases}$	Irreducible representations of K all one dimensional. 3J frequency eigenvalues for each $\underset{\sim}{k}$. (g_o = number of operation in G_o.)
$\underset{\sim}{k}$ inside BZ but special symmetry	$K : \{E \vert \underset{\sim}{T}_n\}, \ldots$	$G_o(\underset{\sim}{k}) : E, \ldots$	$dq \begin{cases} d = 1 \text{ or } >1 \\ 1 < q < g_0 \end{cases}$	Γ_i irreducible representations of $G_o(\underset{\sim}{k})$. In general, less than 3J frequency eigenvalues for each $\underset{\sim}{k}$.
$\underset{\sim}{k}$ on surface of BZ	$K : \{E \vert \underset{\sim}{T}_n\}, \ldots$	$G_o(\underset{\sim}{k}) : E, \ldots$	$dq \begin{cases} d = 1 \text{ or } >1 \\ 1 < q \leq g_0 \end{cases}$	
	If K symmorphic $D_{11}\{\beta \vert \underset{\sim}{b}\} = e^{i\underset{\sim}{k}\cdot\underset{\sim}{b}}\Gamma_i(\beta)$			Γ_i irreducible representations of $G_o(\underset{\sim}{k})$. In general, less than 3J frequency eigenvalues for each $\underset{\sim}{k}$.
	If K nonsymmorphic $D_{11}\{\beta \vert \underset{\sim}{b}\} \neq e^{i\underset{\sim}{k}\cdot\underset{\sim}{b}}\Gamma_i(\beta)$			Irreducible representations of K to be considered. In general, less than 3J frequency eigenvalues for each $\underset{\sim}{k}$.

POINT C. $\underset{\sim}{k} \equiv (\pi/a, \pi/a)$

At this point $K = G$ and $G_o(\underset{\sim}{k}) = G_o = C_{4v}$. Also, if $\underset{\sim}{T}_n = n_1\underset{\sim}{a}_1 + n_2\underset{\sim}{a}_2$,

$$\underset{\sim}{D}_{11}\{\beta|\underset{\sim}{T}_n\} = e^{i\underset{\sim}{k}\cdot\underset{\sim}{T}_n}\underset{\sim}{\Gamma}_i(\beta) = \pm\underset{\sim}{\Gamma}_i(\beta),$$

with plus if $(n_1 + n_2)$ even and minus if $(n_1 + n_2)$ odd and $\underset{\sim}{\Gamma}_i$ an irreducible representation of C_{4v}. The star $\{\underset{\sim}{k}\}$ has its maximum degeneracy $(q = 1)$.

The displacement components, shown in Fig. 6.8, set the following representation of $G_o(\underset{\sim}{k})$ which reduces as follows in terms of C_{4v}:

E	C_4	C_2	$C_4{}^3$	σ_1	σ_2	σ_3	σ_4	Reduction
4	0	−4	0	0	0	0	0	$2\Gamma_5$

In correspondence to the point C, we have then two 2-fold degenerate frequency eigenvalues.

POINTS A. $\underset{\sim}{k} \equiv (k_x, k_y)$

For a point inside the Section OBC of the Brillouin zone, $K = T$ and $G_o(\underset{\sim}{k}) = C_1$ contains only the operation E (identity). Also,

$$\underset{\sim}{D}_{11}\{\beta|\underset{\sim}{T}_n\} = \underset{\sim}{D}_{11}\{E|\underset{\sim}{T}_n\} = e^{i\underset{\sim}{k}\cdot\underset{\sim}{T}_n}.$$

The star $\{\underset{\sim}{k}\}$ has $q = 8$ arms and is nondegenerate.

The displacement components set a representation of $G_o(\underset{\sim}{k})$ that reduces as 4Γ, Γ being the only (one-dimensional) irreducible representation of C_1. For any point A, we have four nondegenerate frequency eigenvalues.

POINTS ON OB. $\underset{\sim}{k} \equiv (k_x, 0)$

For a point on OB, the group K consists of $\{E|\underset{\sim}{T}_n\}$ and $\{\sigma_1|\underset{\sim}{T}_n\}$. The group $G_o(\underset{\sim}{k}) = C_{1h}$ contains the operations E and σ_1. Also

$$\underset{\sim}{D}_{11}\{\beta|\underset{\sim}{T}_n\} = e^{i\underset{\sim}{k}\cdot\underset{\sim}{T}_n}\underset{\sim}{\Gamma}_i(\beta) = e^{in_1ak_x}\underset{\sim}{\Gamma}_i(\beta),$$

where $\underset{\sim}{\Gamma}_i$ is the irreducible representation of C_{1h}. The star $\{\underset{\sim}{k}\}$ has $q = 4$ arms.

The displacement components set the following representation of $G_o(\underset{\sim}{k})$ which reduces as follows in terms of irreducible representations of C_{1h}:

E	σ_1	Reduction
4	0	$2\Gamma_1 + 2\Gamma_2$

For any point on OB, we have then four nondegenerate frequency eigenvalues.

POINTS ON OC. $\underset{\sim}{k} \equiv (k_x, k_y = k_x)$

For a point on OC, the group *K* consists of $\{E|\underset{\sim}{T}_n\}$ and $\{\sigma_2|\underset{\sim}{T}_n\}$. The group $G_o(\underset{\sim}{k}) = C_{1h}$ contains the operations E and σ_2. Also

$$\underset{\sim}{D}_{11}\{\beta|\underset{\sim}{T}_n\} = e^{i\underset{\sim}{k}\cdot\underset{\sim}{T}_n}\underset{\sim}{\Gamma}_i(\beta) = e^{i(n_1+n_2)ak_x}\underset{\sim}{\Gamma}_i(\beta),$$

where $\underset{\sim}{\Gamma}_i$ is the irreducible representation of C_{1h}. The star $\{\underset{\sim}{k}\}$ has $q = 4$ arms.

The displacement components set a representation of $G_o(\underset{\sim}{k})$ that reduces as $2\Gamma_1 + 2\Gamma_2$ in terms of irreducible representations of C_{1h}. Therefore, for any point on OC, we have four nondegenerate frequency eigenvalues.

POINTS ON BC. $\underset{\sim}{k} \equiv (\pi/a, k_y)$

For a point on BC, the group *K* consists of $\{E|\underset{\sim}{T}_n\}$ and $\{\sigma_3|\underset{\sim}{T}_n\}$. The group $G_o(\underset{\sim}{k})$ contains the operations E and σ_3. Also

$$\underset{\sim}{D}_{11}\{\beta|\underset{\sim}{T}_n\} = e^{i\underset{\sim}{k}\cdot\underset{\sim}{T}_n}\underset{\sim}{\Gamma}_i(\beta) = \pm e^{in_2ak_y}\underset{\sim}{\Gamma}_i(\beta),$$

with plus if n_1 even and minus if n_1 odd and $\underset{\sim}{\Gamma}_i$ an irreducible representation of C_{1h}. The star $\{\underset{\sim}{k}\}$ has $q = 4$ arms.

The displacement components set a representation of $G_o(\underset{\sim}{k})$ that reduces in terms of irreducible representations of C_{1h} as $2\Gamma_1 + 2\Gamma_2$. Therefore, for the points on BC, we have four nondegenerate frequency eigenvalues.

POINT B. $\underset{\sim}{k} \equiv (\pi/a, 0)$

For this point, K consists of

$$\{E|\underset{\sim}{T}_n\}, \quad \{C_2|\underset{\sim}{T}_n\}, \quad \{\sigma_1|\underset{\sim}{T}_n\}, \quad \{\sigma_3|\underset{\sim}{T}_n\}$$

and the group $G_o(\underset{\sim}{k}) = C_{2v}$ contains the operations E, C_2, σ_1, and σ_3. Also

$$\underset{\sim}{D}_{11}\{\beta|\underset{\sim}{T}_n\} = e^{i\underset{\sim}{k}\cdot\underset{\sim}{T}_n}\underset{\sim}{\Gamma}_i(\beta) = \pm\underset{\sim}{\Gamma}_i(\beta),$$

with plus if n_1 even and minus if n_1 odd and $\underset{\sim}{\Gamma}_i$ an irreducible representation of C_{2v}. The star $\{\underset{\sim}{k}\}$ has $q = 2$ arms.

The displacement components set the following representation of $G_o(\underset{\sim}{k})$ which reduces as follows in terms of irreducible representations of C_{2v}:

E	C_2	σ_1	σ_3	Reduction
4	0	0	0	$\Gamma_1 + \Gamma_2 + \Gamma_3 + \Gamma_4$

In order to establish a correlation between the dispersion curves at different points of the Brillouin zone we establish *compatibility relations* among the representations of the various point groups in question. The concept on which these relations are based is simply this. Consider, for example, the point 0 of the Brillouin zone. A normal mode at this point is associated with an irreducible representation of the group of the $\underset{\sim}{k}$ vector K. As we move away from the point 0, say, along OB, the set of normal coordinates that form a basis for a representation of the group of the $\underset{\sim}{k} = Q$ vector must also form bases for irreducible representations of the group of the $\underset{\sim}{k}$ vector, with $\underset{\sim}{k}$ on OB. Let us call $(K0)$ and $K(\text{OB})$ the two groups of the $\underset{\sim}{k}$ vector above. A representation Γ_B of $K(\text{OB})$ is compatible with a representation Γ_A of $K(0)$ if the basis for Γ_B is included in the basis for Γ_A; this means that an irreducible representation Γ_B of $K(\text{OB})$ is compatible with an irreducible representation Γ_A of $K(0)$ if Γ_A, when reduced in terms of irreducible representations of $K(\text{OB})$, contains Γ_B.

The compatibility relations for the irreducible representations of the Group $G_o(\underset{\sim}{k})$ at different points of the Brillouin zone are presented in Table 6.5.

Table 6.5. Compatibility relations for the irreducible representations of the lattice vibrations of the crystal in Fig. 6.8

0	$\Gamma_1(0)$	$\Gamma_2(0)$	$\Gamma_3(0)$	$\Gamma_4(0)$	$\Gamma_5(0)$
OB	$\Gamma_1(\mathrm{OB})$	$\Gamma_2(\mathrm{OB})$	$\Gamma_1(\mathrm{OB})$	$\Gamma_2(\mathrm{OB})$	$\Gamma_1(\mathrm{OB})$, $\Gamma_2(\mathrm{OB})$
OC	$\Gamma_1(\mathrm{OC})$	$\Gamma_2(\mathrm{OC})$	$\Gamma_2(\mathrm{OC})$	$\Gamma_1(\mathrm{OC})$	$\Gamma_1(\mathrm{OC})$, $\Gamma_2(\mathrm{OC})$
B	$\Gamma_1(\mathrm{B})$	$\Gamma_2(\mathrm{B})$	$\Gamma_3(\mathrm{B})$	$\Gamma_4(\mathrm{B})$	
OB	$\Gamma_1(\mathrm{OB})$	$\Gamma_2(\mathrm{OB})$	$\Gamma_1(\mathrm{OB})$	$\Gamma_2(\mathrm{OB})$	
BC	$\Gamma_1(\mathrm{BC})$	$\Gamma_2(\mathrm{BC})$	$\Gamma_2(\mathrm{BC})$	$\Gamma_1(\mathrm{BC})$	
C	$\Gamma_1(\mathrm{C})$	$\Gamma_2(\mathrm{C})$	$\Gamma_3(\mathrm{C})$	$\Gamma_4(\mathrm{C})$	$\Gamma_5(\mathrm{C})$
OC	$\Gamma_1(\mathrm{OC})$	$\Gamma_2(\mathrm{OC})$	$\Gamma_2(\mathrm{OC})$	$\Gamma_1(\mathrm{OC})$	$\Gamma_1(\mathrm{OC})$, $\Gamma_2(\mathrm{OC})$
BC	$\Gamma_1(\mathrm{BC})$	$\Gamma_2(\mathrm{BC})$	$\Gamma_1(\mathrm{BC})$	$\Gamma_2(\mathrm{BC})$	$\Gamma_1(\mathrm{BC})$, $\Gamma_2(\mathrm{BC})$

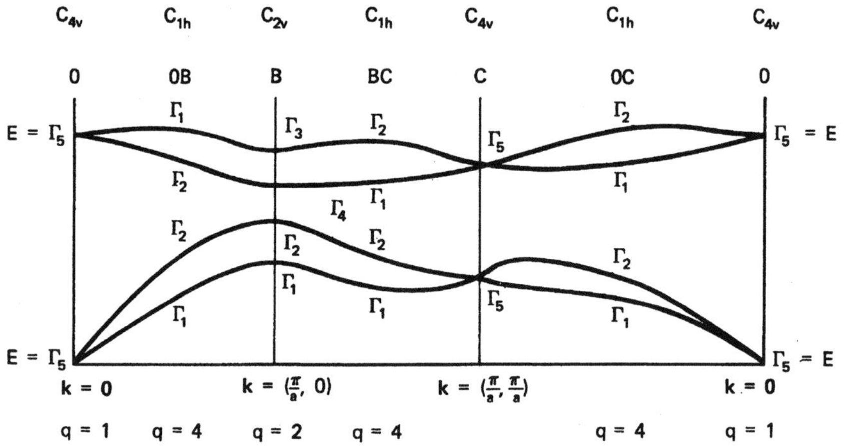

Fig. 6.9. Possible symmetry classification of the dispersion curves of the two-dimensional crystal in Fig. 6.8.

The compatibility relations can be used to set up the schematic diagram of a *possible symmetry* classification of the dispersion curves about OBC, in Fig. 6.9. Notice that group theory does not predict either the magnitude of the frequency eigenvalues, or the relative positions of the various branches, and therefore no significance should be attached to this diagram other than that it presents the group-theoretical classification of the dispersion curves.

Finally we need to consider the possibility of additional degeneracies due to time-reversal symmetry. Let us consider various points in the Section OBC of the Brillouin zone.

ORIGIN 0. $\underset{\sim}{k} = 0$

At this point every element of $G_o(\underset{\sim}{k}) = C_{4v}$ sends $\underset{\sim}{k}$ into $-\underset{\sim}{k}$. Therefore,

$$\sum_{\beta} \chi_j(\beta^2) = 6\chi_j(E) + 2\chi_j(C_2) = 8 = g_o{}',$$

where

$$\beta = E, C_4, C_4{}^3, C_2, \sigma_1, \sigma_2, \sigma_3, \sigma_4.$$

j ranges over the five irreducible representations of C_{4v} and $g_o{}'$ is the order of $G_o(\underset{\sim}{k})$. No time-reversal degeneracy is present.

POINT C. $\underset{\sim}{k} \equiv (\pi/a, \pi/a)$

As for 0, $G_o(\underset{\sim}{k}) = C_{4v}$ and every element of $G_o(\underset{\sim}{k})$ sends $\underset{\sim}{k}$ into $-k + \underset{\sim}{K}$:

$$\sum_{\beta} \chi_j(\beta^2) = 8 = g_o{}',$$

and no time-reversal degeneracy is present.

POINTS A. $\underset{\sim}{k} \equiv (k_x, k_y)$

For these points $G_o(\underset{\sim}{k}) = C_1$:E. Only the operation C_2 sends $\underset{\sim}{k}$ into $-\underset{\sim}{k}$. It is

$$\sum_{\gamma} \chi(\gamma^2) = \chi(C_2{}^2) = \chi(E) = 1.$$

No time-reversal degeneracy is present.

POINTS ON OB. $\underset{\sim}{k} \equiv (k_x, 0)$

For these points $G_o(\underset{\sim}{k}) = C_{1h} : E, \sigma_1$. The operations C_2 and σ_3 send $\underset{\sim}{k}$ into $-\underset{\sim}{k}$:

$$\sum_{\gamma} \chi_j(\gamma^2) = 2\chi_j(E) = 2 = g_o{}',$$

where

$$\gamma = C_2, \ \ \sigma_3; \ \ j = 1, 2,$$
$$g_o{}' = 2 = \text{order of } G_o(\underset{\sim}{k}).$$

No time-reversal degeneracy is present.

POINTS ON OC. $\underset{\sim}{k} \equiv (k_x, k_y = k_x)$

It is $G_o(\underset{\sim}{k}) = C_{1h} : E, \sigma_2$. The operations C_2 and σ_4 send $\underset{\sim}{k}$ into $-\underset{\sim}{k}$:

$$\sum_{\gamma} \chi_j(\gamma^2) = 2\chi_j(E) = 2 = g_o{}',$$

where

$$\gamma = C_2, \ \ \sigma_4; \ \ j = 1, 2.$$

No time-reversal degeneracy is present.

POINTS ON BC. $\underset{\sim}{k} \equiv (\pi/a, k_y)$

It is $G_o(\underset{\sim}{k}) = C_{1h} : E, \sigma_3$. The operations C_2 and σ_1 send $\underset{\sim}{k}$ into $-\underset{\sim}{k} + \underset{\sim}{K}$:

$$\sum_{\gamma} \chi_j(\gamma^2) = 2\chi_j(E) = 2 = g_o{}',$$

where

$$\gamma = C_2, \ \ \sigma_1; \ \ j = 1, 2.$$

No time-reversal symmetry is present.

POINT B. $\underset{\sim}{k} \equiv (\pi/a, 0)$

It is $G_o(\underset{\sim}{k}) = C_{2v} : E, C_2, \sigma_1, \sigma_3$. All these operations send $\underset{\sim}{k}$ into $-\underset{\sim}{k} + \underset{\sim}{K}$:

$$\sum_{\beta} \chi_j(\beta^2) = 4\chi_i(E) = 4 = g_o{}',$$

where

$$\beta = E, \ \ C_2, \ \ \sigma_1, \ \ \sigma_3; \ \ j = 1, 2, 3, 4.$$

No time-reversal symmetry is present.

Therefore for the present example no time-reversal symmetry is present for any value of $\underset{\sim}{k}$.

A summary of the irreducible representations of the lattice vibrations of the crystal in Fig. 6.8 is given in Table 6.6.

6.10. Example II. Lattice Vibrations of a Two-Dimensional Crystal with Symmetry C^2_{4v}

Let us consider the two-dimensional crystal whose unit cell is shown in Fig. 6.10. The space group of this crystal is C^2_{4v} and was treated in Section 4.8. We shall refer to the Brillouin zone in Fig. 4.10 and in particular to the Section OBC.

The displacement components can be visualized by attaching two vectors, one in the x and the other in the y direction to each of the 10 atoms in the unit cell? They are 20 in number. The effect of a generic operation of K on the displacement components can be described by a 20×20 matrix. The traces of these matrices in correspondence to the various operations of K give the characters of the (reducible) representation of K set by the displacement components.

It can be easily seen that the only matrices that have traces different from zero are the ones corresponding to the operations $\{E|\underset{\sim}{T}_n\}$ and $\{C_2|\underset{\sim}{T}_n\}$. Setting $\underset{\sim}{T}_n = 0$ and calling χ the trace of the matrix, we have

$$\chi(\{E|\underset{\sim}{0}\};\underset{\sim}{k}) = 20,$$
$$\chi(\{C_2|\underset{\sim}{0}\};\underset{\sim}{k}) = -2(1 + e^{-i\underset{\sim}{k}\cdot(\underset{\sim}{a}_1+\underset{\sim}{a}_2)}).$$

ORIGIN 0. $\underset{\sim}{k} = \underset{\sim}{0}$

At this point $K = G$ and $G_o(\underset{\sim}{k}) = G_o = C_{4v}$. Also,

$$\underset{\sim}{D}\{\beta|\underset{\sim}{b}\} = \underset{\sim}{D}_{11}\{\beta|\underset{\sim}{b}\} = \underset{\sim}{\Gamma}_i(\beta),$$

where $\underset{\sim}{\Gamma}_i(\beta)$ is an irreducible representation of C_{4v}. The star $\{\underset{\sim}{k}\}$ has its maximum degeneracy $(q = 1)$.

The displacement components set the following representations of $G_o(\underset{\sim}{k})$ which reduces as follows in terms of irreducible representations

Table 6.6. The irreducible representations of the lattice vibrations of the two-dimensional crystal in Fig. 6.8 ($\underset{\sim}{T}_n = n_1\underset{\sim}{a}_1 + n_2\underset{\sim}{a}_2$).

Region	K	$G_o(\underset{\sim}{k})$	Irreducible Reps of $G_o(\underset{\sim}{k})$	Number and Degeneracies of Eigenvalues	Dimension of Reps of G	Observations
O $\underset{\sim}{k}=\underset{\sim}{O}$	$K = G$	$G_o(\underset{\sim}{k}) = G_o = C_{4v}$	$2\Gamma_5$	Two 2-fold degenerate	$2\begin{cases} d=2 \\ q=1 \end{cases}$	$\underset{\sim}{D}_{11}\{\beta\|\underset{\sim}{T}_n\} = \underset{\sim}{\Gamma}_i(\beta)$
C $\underset{\sim}{k}\equiv(\pi/a,\pi/a)$	$K = G$	$G_o(\underset{\sim}{k}) = G_o = C_{4v}$	$2\Gamma_5$	Two 2-fold degenerate	$2\begin{cases} d=2 \\ q=1 \end{cases}$	$\underset{\sim}{D}_{11}\{\beta\|\underset{\sim}{T}_n\} = \pm\underset{\sim}{\Gamma}_i(\beta)$ $+ (n_1+n_2)$ even $- (n_1+n_2)$ odd
A $\underset{\sim}{k}\equiv(k_x,k_y)$	$K = T$	$G_o(\underset{\sim}{k}) = C_1 : E$	4Γ	Four nondegenerate	$8\begin{cases} d=1 \\ q=8 \end{cases}$	$\underset{\sim}{D}_{11}\{\beta\|\underset{\sim}{T}_n\} = e^{i\underset{\sim}{k}\cdot\underset{\sim}{T}_n}$
OB $\underset{\sim}{k}\equiv(k_x,O)$	$K : \{E\|\underset{\sim}{T}_n\}, \{\sigma_1\|\underset{\sim}{T}_n\}$	$G_o(\underset{\sim}{k}) = C_{1h} : E, \sigma_1$	$2\Gamma_1 + 2\Gamma_2$	Four nondegenerate	$4\begin{cases} d=1 \\ q=4 \end{cases}$	$\underset{\sim}{D}_{11}\{\beta\|\underset{\sim}{T}_n\}$ $= e^{in_1ak_x}\underset{\sim}{\Gamma}_i(\beta)$
OC $\underset{\sim}{k}\equiv(k_x,k_y=k_x)$	$K : \{E\|\underset{\sim}{T}_n\}, \{\sigma_2\|\underset{\sim}{T}_n\}$	$G_o(\underset{\sim}{k}) = C_{1h} : E, \sigma_2$	$2\Gamma_1 + 2\Gamma_2$	Four nondegenerate	$4\begin{cases} d=1 \\ q=4 \end{cases}$	$\underset{\sim}{D}_{11}\{\beta\|\underset{\sim}{T}_n\}$ $= e^{i(n_1+n_2)ak_x}\underset{\sim}{\Gamma}_i(\beta)$
BC $\underset{\sim}{k}\equiv(\pi/a,k_y)$	$K : \{E\|\underset{\sim}{T}_n\}, \{\sigma_3\|\underset{\sim}{T}_n\}$	$G_o(\underset{\sim}{k}) = C_{1h} : E, \sigma_3$	$2\Gamma_1 + 2\Gamma_2$	Four nondegenerate	$4\begin{cases} d=1 \\ q=4 \end{cases}$	$\underset{\sim}{D}_{11}\{\beta\|\underset{\sim}{T}_n\}$ $= \pm e^{in_2ak_y}\underset{\sim}{\Gamma}_i(\beta)$ $+ n_1$ even $- n_1$ odd
B $\underset{\sim}{k}\equiv(\pi/a,O)$	$K : \{E\|\underset{\sim}{T}_n\}, \{C_2\|\underset{\sim}{T}_n\}$ $\{\sigma_1\|\underset{\sim}{T}_n\}, \{\sigma_3\|\underset{\sim}{T}_n\}$	$G_o(\underset{\sim}{k}) = C_{2v}$: $E, C_2, \sigma_1, \sigma_3$	$\Gamma_1 + \Gamma_2 + \Gamma_3 + \Gamma_4$	Four nondegenerate	$2\begin{cases} d=1 \\ q=2 \end{cases}$	$\underset{\sim}{D}_{11}\{\beta\|\underset{\sim}{T}_n\} = \pm\underset{\sim}{\Gamma}_i(\beta)$ $+ n_1$ even $- n_1$ odd

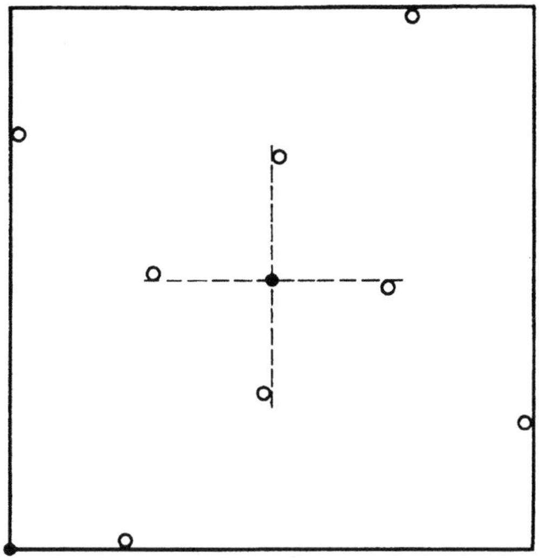

Fig. 6.10. Unit cell of a two-dimensional crystal with space group C^2_{4v}.

of C_{4v}:

E	C_4	C_2	${C_4}^3$	σ_1	σ_2	σ_3	σ_4	Reduction
20	0	−4	0	0	0	0	0	$2\Gamma_1 + 2\Gamma_2 + 2\Gamma_3 + 2\Gamma_4 + 6\Gamma_5$

Therefore in correspondence to $\underset{\sim}{k} = \underset{\sim}{0}$ we have eight nondegenerate and six 2-fold degenerate frequency eigenvalues.

POINTS A. $\underset{\sim}{k} \equiv (k_x, k_y)$

For a point A inside the Section OBC of the Brillouin zone, $K = T$ and $G_o(\underset{\sim}{k}) = C_1$ contains only the operation E. Also,

$$\underset{\sim}{D}_{11}\{\beta|\underset{\sim}{b}\} = \underset{\sim}{D}_{11}\{\beta|\underset{\sim}{T}_n\} = e^{i\underset{\sim}{k}\cdot\underset{\sim}{T}_n}.$$

The star $\{\underset{\sim}{k}\}$ has $q = 8$ arms and is nondegenerate.

The displacement components set a representation of $G_o(\underset{\sim}{k})$ that reduces as 20Γ, Γ being the only (one-dimensional) irreducible representation of C_1. For any point A we have then 20 nondegenerate frequency eigenvalues.

POINTS ON OB. $\underset{\sim}{k} \equiv (k_x, 0)$

For a point on OB, the group *K* consists of $\{E|\underset{\sim}{T}_n\}$ and $\{\sigma_1|\underset{\sim}{A}+\underset{\sim}{T}_n\}$. The group $G_o(\underset{\sim}{k}) = C_{1h}$ contains the operations E and σ_1. Also

$$\underset{\sim}{D}_{11}\{\beta|\underset{\sim}{b}\} = e^{i\underset{\sim}{k}\cdot\underset{\sim}{b}}\underset{\sim}{\Gamma}_i(\beta),$$

where $\underset{\sim}{\Gamma}_i$ is the irreducible representation of C_{1h}. The star $\{\underset{\sim}{k}\}$ has $q = 4$ arms.

The displacement components set the following representation of $G_o(\underset{\sim}{k})$ which reduces as follows in terms of irreducible representations of C_{1h}

E	σ_1	Reduction
20	0	$10\Gamma_1 + 10\Gamma_2$

For a point on OB we have then 20 nondegenerate frequency eigenvalues.

POINTS ON OC. $\underset{\sim}{k} \equiv (k_x, k_y = k_x)$

For a point on OC, the group *K* consists of $\{E|\underset{\sim}{T}_n\}$ and $\{\sigma_2|\underset{\sim}{A}+\underset{\sim}{T}_n\}$. The group $G_o(\underset{\sim}{k}) = C_{1h}$ contains the operations E and σ_2. Also

$$\underset{\sim}{D}_{11}\{\beta|\underset{\sim}{b}\} = e^{i\underset{\sim}{k}\cdot\underset{\sim}{b}}\underset{\sim}{\Gamma}_i(\beta),$$

where $\underset{\sim}{\Gamma}_i$ = irreducible representation of C_{1h}. The star $\{\underset{\sim}{k}\}$ has $q = 4$ arms.

The displacement components set the following representation of $G_o(\underset{\sim}{k})$ which reduces as follows in terms of irreducible representations of C_{1h}

E	σ_2	Reduction
20	0	$10\Gamma_1 + 10\Gamma_2$

For any point on OB we have then 20 nondegenerate frequency eigenvalues.

POINT C. $\underset{\sim}{k} \equiv (\pi/a, \pi/a)$

At this point K = G and $G_o(\underset{\sim}{k}) = G_o = C_{4v}$. The star $\{\underset{\sim}{k}\}$ has its maximum degeneracy (q = 1). Also, since K is nonsymmorphic the irreducible representations of K are derived by the method of the factor group. We have already found these representations and presented them in Table 4.16.

The displacement components set the following representation of K which reduces as follows in terms of irreducible representations of K reported in Table 4.16:

$\{E\|\underset{\sim}{a}_e\}$	$\{E\|\underset{\sim}{a}_o\}$	$\{C_2\|\underset{\sim}{a}_e\}$	$\{C_2\|\underset{\sim}{a}_o\}$	$\{\sigma_1\|\underset{\sim}{a}_e + \underset{\sim}{A}\}\cdots$		Reduction
20	−20	−4	4	0	0···	$3\Gamma_1 + 3\Gamma_2 + 3\Gamma_3 + 3\Gamma_4 + 4\Gamma_5$

where

$$\underset{\sim}{a}_e = 2n_1\underset{\sim}{a}_1 + 2n_2\underset{\sim}{a}_2 \quad \text{or} \quad (2n_1 + 1)\underset{\sim}{a}_1 + (2n_2 + 1)\underset{\sim}{a}_2,$$

$$\underset{\sim}{a}_o = 2n_1\underset{\sim}{a}_1 + (2n_2 + 1)\underset{\sim}{a}_2 \quad \text{or} \quad (2n_1 + 1)\underset{\sim}{a}_1 + 2n_2\underset{\sim}{a}_2.$$

We have then 12 nondegenerate and four 2-fold degenerate frequency eigenvalues in correspondence to the point C.

POINT B. $\underset{\sim}{k} \equiv (\pi/a, 0)$

At this point K contains the operations

$$\{E|\underset{\sim}{T}_n\}, \quad \{C_2|\underset{\sim}{T}_n\}, \quad \{\sigma_1|\underset{\sim}{A} + \underset{\sim}{T}_n\}, \quad \{\sigma_3|\underset{\sim}{A} + \underset{\sim}{T}_n\}.$$

The group $G_o(\underset{\sim}{k}) = C_{2v}$ contains the operations $E, C_2, \sigma_1, \sigma_3$. The star $\{\underset{\sim}{k}\}$ has q = 2 arms.

The irreducible representations of K were worked out by the factor group method and presented in Table 4.18.

The displacement components set the following representation of K which reduces as follows in terms of the (only) irreducible representation of K:

$\{E\|\underset{\sim}{a}_e\}$	$\{E\|\underset{\sim}{a}_o\}$	$\{C_2\|\underset{\sim}{a}_e\}$	$\{C_2\|\underset{\sim}{a}_o\}$	$\{\sigma_1\|\underset{\sim}{A} + \underset{\sim}{a}_e\}\cdots$		Reduction
20	−20	0	0	0	0···	10Γ

where

$$\underset{\sim}{a}_e = 2n_1\underset{\sim}{a}_1 + n_2\underset{\sim}{a}_2,$$
$$\underset{\sim}{a}_o = (2n_1 + 1)\underset{\sim}{a}_1 + n_2\underset{\sim}{a}_2.$$

We have then ten 2-fold degenerate frequency eigenvalues in correspondence to point C.

POINTS ON BC. $\underset{\sim}{k} \equiv (\pi/a, k_y)$

For a point on BC, the group *K* contains the operations $\{E|\underset{\sim}{T}_n\}$ and $\{\sigma_3|\underset{\sim}{A} + \underset{\sim}{T}_n\}$. The group $G_o(\underset{\sim}{k}) = C_{1h}$ contains the operations E and σ_3. The star $\{\underset{\sim}{k}\}$ has $q = 4$ arms.

The irreducible representations of *K* were worked out and presented in Table 4.20.

The displacement components set the following representation of *K* which reduces as follows in terms of the irreducible representations of *K*:

$\{E\|\underset{\sim}{0}\}$	$\{E\|n_1\underset{\sim}{a}_1 + n_2\underset{\sim}{a}_2\}$	$\{\sigma_3\|\underset{\sim}{A}\}$	Reduction
20	$20(-1)^{n_1}e^{in_2ak_y}$	0	$10\Gamma_1 + 10\Gamma_2$

For any point on BC we have then 20 nondegenerate frequency eigenvalues.

The compatibility relations for the irreducible representations of *K* at different points of the Brillouin zone are shown in Table 6.7. Figure 6.11 shows the schematic diagram of a possible symmetry classification of the dispersion curves.

Finally we need to consider the possibility of additional degeneracies due to time-reversal symmetry. Let us consider various points in the Section OBC of the Brillouin zone.

ORIGIN 0. $\underset{\sim}{k} = \underset{\sim}{0}$

At this point every element of *K* sends $\underset{\sim}{k}$ into $-\underset{\sim}{k}$. Therefore

$$\sum_{\{\beta|\underset{\sim}{b}\}} \chi_j(\{\beta|\underset{\sim}{b}\}^2) = 3\chi_j(\{E|\underset{\sim}{0}\} + 2\chi_j(\{C_2|\underset{\sim}{0}\}) + \chi_j(\{E|\underset{\sim}{a}_1\})$$
$$+ \chi_j(\{E|\underset{\sim}{a}_2\}) + \chi_j(\{E|\underset{\sim}{a}_1 + \underset{\sim}{a}_2\}) = 8.$$

Table 6.7. Compatibility relations for the irreducible representations of the lattice vibrations of the crystal in Fig. 6.10.

O	$\Gamma_1(O)$	$\Gamma_2(O)$	$\Gamma_3(O)$	$\Gamma_4(O)$	$\Gamma_5(O)$
OB	$\Gamma_1(OB)$	$\Gamma_2(OB)$	$\Gamma_1(OB)$	$\Gamma_2(OB)$	$\Gamma_1(OB)$, $\Gamma_2(OB)$
OC	$\Gamma_1(OC)$	$\Gamma_2(OC)$	$\Gamma_2(OC)$	$\Gamma_1(OC)$	$\Gamma_1(OC)$, $\Gamma_2(OC)$
C	$\Gamma_1(C)$	$\Gamma_2(C)$	$\Gamma_3(C)$	$\Gamma_4(C)$	$\Gamma_5(C)$
OC	$\Gamma_2(OC)$	$\Gamma_1(OC)$	$\Gamma_2(OC)$	$\Gamma_1(OC)$	$\Gamma_1(OC)$, $\Gamma_2(OC)$
BC	$\Gamma_1(BC)$	$\Gamma_2(BC)$	$\Gamma_2(BC)$	$\Gamma_1(BC)$	$\Gamma_1(BC)$, $\Gamma_2(BC)$
B	$\Gamma(B)$				
OB	$\Gamma_1(OB)$, $\Gamma_2(OB)$				
BC	$\Gamma_1(BC)$, $\Gamma_2(BC)$				

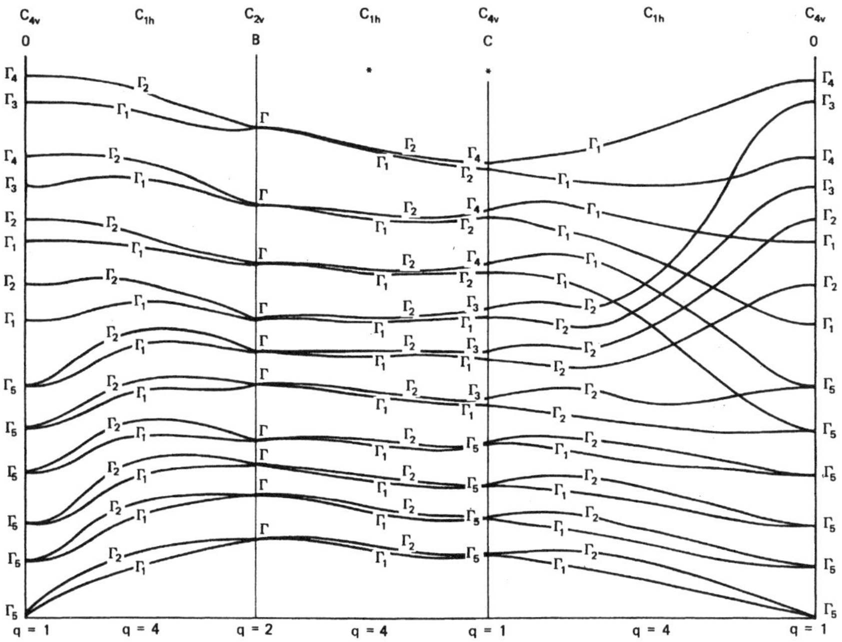

Fig. 6.11. Possible symmetry classification of the dispersion curves of the two-dimensional crystal in Fig. 6.10. Asterisks mark regions in which time-reversed degeneracy occurs.

where

$$\{\beta|\underset{\sim}{b}\} = \{E|\underset{\sim}{0}\}, \quad \{C_4|\underset{\sim}{0}\}, \quad \{C_2|\underset{\sim}{0}\}, \quad \{C_4{}^3|\underset{\sim}{0}\},$$
$$\{\sigma_1|\underset{\sim}{A}\}, \quad \{\sigma_2|\underset{\sim}{A}\}, \quad \{\sigma_3|\underset{\sim}{A}\}, \quad \{\sigma_4|\underset{\sim}{A}\};$$
$$j = 1, 2, \ldots, 5.$$

No time-reversal degeneracy is present.

POINTS A. $\underset{\sim}{k} = (k_x, k_y)$

At this point $\{C_2|\underset{\sim}{0}\}$ sends $\underset{\sim}{k}$ into $-\underset{\sim}{k}$. Therefore

$$\sum_{\{\gamma|\underset{\sim}{c}\}} \chi(\{\gamma|\underset{\sim}{c}\}^2) = \chi(\{C_2|\underset{\sim}{0}\}^2) = \chi(\{E|\underset{\sim}{0}\}) = 1$$

No time-reversal degeneracy is present.

POINTS ON OB. $\underset{\sim}{k} \equiv (k_x, 0)$

At these points $\{C_2|\underset{\sim}{0}\}$ and $\{\sigma_3|\underset{\sim}{A}\}$ send $\underset{\sim}{k}$ into $-\underset{\sim}{k}$. Therefore

$$\sum_{\{\gamma|\underset{\sim}{c}\}} \chi_j(\{\gamma|\underset{\sim}{c}\}^2) = \chi_j(\{C_2|\underset{\sim}{0}\}^2) + \chi_j(\{\sigma_3|\underset{\sim}{A}\}^2)$$
$$= \chi_j(\{E|\underset{\sim}{0}\}) + \chi_j(\{E|\underset{\sim}{a}_2\}) = 2$$

for both representations of *K*. No time-reversal degeneracy is present.

POINTS ON OC. $\underset{\sim}{k} \equiv (k_x, k_y = k_x)$

At these points $\{C_2|\underset{\sim}{0}\}$ and $\{\sigma_4|\underset{\sim}{A}\}$ send $\underset{\sim}{k}$ into $-\underset{\sim}{k}$. Therefore

$$\sum_{\{\gamma|\underset{\sim}{c}\}} \chi_j(\{\gamma|\underset{\sim}{c}\}^2) = \chi_j(\{C_2|\underset{\sim}{0}\}^2) + \chi_j(\{\sigma_4|\underset{\sim}{A}\}^2)$$
$$= 2\chi_j(\{E|\underset{\sim}{0}\}) = 2$$

for both representations of *K*. No time-reversal degeneracy is present.

POINT C. $\underset{\sim}{k} \equiv (\pi/a, \pi/a)$

At this point every element of K sends $\underset{\sim}{k}$ into $-\underset{\sim}{k} + \underset{\sim}{K}$. Therefore

$$\begin{aligned}\sum_{\{\beta|\underset{\sim}{b}\}} \chi_j(\{\beta|\underset{\sim}{b}\}^2) &= 3\chi_j(\{E|\underset{\sim}{0}\}) + 2\chi_j(\{C_2|\underset{\sim}{0}\}) + \chi_j(\{E|\underset{\sim}{a}_1\}) \\ &= \chi_j(\{E|\underset{\sim}{a}_2\}) + \chi_j(\{E|\underset{\sim}{a}_1 + \underset{\sim}{a}_2\}) \\ &= 4\chi_j(\{E|\underset{\sim}{a}_e\}) + 2\chi_j(\{E|\underset{\sim}{a}_o\}) + 2\chi_j(\{C_2|\underset{\sim}{0}\}) \\ &= \begin{cases} 0 & \text{for } j = 1,2,3,4 \\ 8 & \text{for } j = 5. \end{cases}\end{aligned}$$

Therefore there is no additional degeneracy for the four frequency eigenvalues that correspond to the four Γ_5 representations of K. There is degeneracy for the frequency eigenvalues corresponding to the $\Gamma_1, \Gamma_2, \Gamma_3$, and Γ_4 representations of K.

Applying the criterion (6.6.33) we find that time reversal must produce additional degeneracy between two representations T and $\bar{T}$ that are inequivalent (of the third kind). We may now determine which representations of K are related by time reversal. Let us choose $\{\psi_o|\underset{\sim}{t}\} = \{C_2|\underset{\sim}{0}\}$; then, from (6.6.32) and consulting Table 4.16,

$$\begin{aligned}\bar{T}(\{E|\underset{\sim}{T}_n\}) &= T(\{C_2|\underset{\sim}{0}\}\{E|\underset{\sim}{T}_n\}\{C_2|\underset{\sim}{0}\})^* \\ &= T(\{E|-\underset{\sim}{T}_n\})^* = T(\{E|\underset{\sim}{T}_n\})^*, \\ \bar{T}(\{C_4^n|\underset{\sim}{0}\}) &= T(\{C_2|\underset{\sim}{0}\}\{C_4^n|\underset{\sim}{0}\}\{C_2|\underset{\sim}{0}\})^* \\ &= T(\{C_4^n|\underset{\sim}{0}\})^*, \\ \bar{T}(\{\sigma_i|\underset{\sim}{A}\}) &= T(\{C_2|\underset{\sim}{0}\}\{\sigma_i|\underset{\sim}{A}\}\{C_2|\underset{\sim}{0}\})^* \\ &= T(\{\sigma_i|-\underset{\sim}{A}\})^* = T(\{\sigma_i|\underset{\sim}{A}\})^*.\end{aligned}$$

Thus for the point C, the time reversal degeneracy results in a representation and its complex conjugate corresponding to the same eigenvalue; this is the case for the two representations Γ_1 and Γ_3 and for the two representations Γ_2 and Γ_4.

POINT B. $\underset{\sim}{k} \equiv (\pi/a, 0)$

At this point every element of K sends $\underset{\sim}{k}$ into $-\underset{\sim}{k}+\underset{\sim}{K}$. Therefore

$$\sum_{\{\beta|\underset{\sim}{b}\}} \chi(\{\beta|\underset{\sim}{b}\}^2) = \chi(\{E|\underset{\sim}{0}\}^2) + \chi(\{C_2|\underset{\sim}{0}\}^2) + \chi(\{\sigma_1|\underset{\sim}{A}\}^2) + \chi(\{\sigma_3|\underset{\sim}{A}\}^2)$$
$$= 2\chi(\{E|\underset{\sim}{0}\}) + \chi(\{E|\underset{\sim}{a}_1\}) + \chi(\{E|\underset{\sim}{a}_2\})$$
$$= 3\chi(\{E|\underset{\sim}{a}_e\}) + \chi(\{E|\underset{\sim}{a}_o\}) = 4.$$

There is no additional degeneracy due to time reversal.

POINTS ON BC. $\underset{\sim}{k} = (\pi/a, k_y)$

At these points $\{C_2|\underset{\sim}{0}\}$ and $\{\sigma_1|\underset{\sim}{A}\}$ send $\underset{\sim}{k}$ into $-\underset{\sim}{k}+\underset{\sim}{K}$. Therefore

$$\sum_{\{\gamma|\underset{\sim}{c}\}} \chi_j(\{\gamma|\underset{\sim}{c}\}^2) = \chi_j(\{C_2|\underset{\sim}{0}\}^2) + \chi_j(\{\sigma_1|\underset{\sim}{A}\}^2)$$
$$= \chi_j(\{E|\underset{\sim}{0}\}) + \chi_j(\{E|\underset{\sim}{a}_1\}) = 1 - 1 = 0,$$

for both representations of K. Time-reversal degeneracy is present.

Applying again the criterion (6.6.33) we find that two degenerate representations T and $\bar{T}$ are inequivalent (of the third kind); they can be found by using (6.6.32) and consulting Table 4.20. Choosing again $\{\psi_o|\underset{\sim}{t}\} = \{C_2|\underset{\sim}{0}\}$, and setting $T = \Gamma_1$ we obtain

$$\begin{aligned}
\bar{T}(\{E|\underset{\sim}{T}_n\}) &= T(\{C_2|\underset{\sim}{0}\}\{E|\underset{\sim}{T}_n\}\{C_2|\underset{\sim}{0}\})^* \\
&= T(\{E|-\underset{\sim}{T}_n\})^* = [(-1)^{m_1} e^{-im_2k_ya}]^* \\
&= (-1)^{m_1} e^{im_2k_ya}, \\
\bar{T}(\{\sigma_3|\underset{\sim}{A}\}) &= T(\{C_2|\underset{\sim}{0}\}\{\sigma_3|\underset{\sim}{A}\}\{C_2|\underset{\sim}{0}\})^* = T(\{\sigma_3|-\underset{\sim}{A}\})^* \\
&= T(\{E|-(\underset{\sim}{a}_1+\underset{\sim}{a}_2)\}\{\sigma_3|\underset{\sim}{A}\})^* \\
&= [(-1)^{m_1} e^{-ik_ya} e^{ik_ya/2}]^* \\
&= -[e^{-ik_ya/2}]^* = -e^{ik_ya/2}.
\end{aligned}$$

Therefore $\bar{T} = \Gamma_2$; that is, the Γ_1 and Γ_2 representations are degenerate because of time reversal.

A summary of the irreducible representations of the lattice vibrations of the crystal in Fig. 6.10 is given in Table 6.8.

Table 6.8. The irreducible representations of the lattice vibrations of the two-dimensional crystal in Fig. 6.10.

Region	K	$G_o(\underset{\sim}{k})$	Irreducible reps of K	Number and degeneracies of eigenvalues	Dimension of reps of G	Observations
O $\underset{\sim}{k}=\underset{\sim}{O}$	$K=G$	$G_o(\underset{\sim}{k})=G_o=C_{4v}$	$2\Gamma_1+2\Gamma_2+2\Gamma_3+2\Gamma_4+6\Gamma_5$	Eight nondegenerate	$1\begin{cases}d=1\\q=1\end{cases}$	$\underset{\sim}{D}_{11}\{\beta\|\underset{\sim}{b}\}=\underset{\sim}{\Gamma}_i(\beta)$
				Six 2-fold degenerate	$2\begin{cases}d=1\\q=2\end{cases}$	
A $\underset{\sim}{k}\equiv(k_x,k_y)$	$K=T$	$G_o(\underset{\sim}{k})=C_1:E$	20Γ	20 nondegenerate	$8\begin{cases}d=1\\q=2\end{cases}$	$\underset{\sim}{D}_{11}\{\beta\|\underset{\sim}{T}_n\}=e^{i\underset{\sim}{k}\cdot\underset{\sim}{T}_n}$
OB $\underset{\sim}{k}\equiv(k_x,O)$	$K:\{E\|\underset{\sim}{T}_n\}$, $\{\sigma_1\|\underset{\sim}{A}+\underset{\sim}{T}_n\}$	$G_o(\underset{\sim}{k})=C_{1h}:E,\sigma_1$	$10\Gamma_1+10\Gamma_2$	20 nondegenerate	$4\begin{cases}d=1\\q=4\end{cases}$	$\underset{\sim}{D}_{11}\{\beta\|\underset{\sim}{b}\}=e^{i\underset{\sim}{k}\cdot\underset{\sim}{b}}\Gamma_i(\beta)$
OC $\underset{\sim}{k}\equiv(k_x,k_y=k_x)$	$K:\{E\|\underset{\sim}{T}_n\}$, $\{\sigma_2\|\underset{\sim}{A}+\underset{\sim}{T}_n\}$	$G_o(\underset{\sim}{k})=C_{1h}:E,\sigma_2$	$10\Gamma_1+10\Gamma_2$	20 nondegenerate	$4\begin{cases}d=1\\q=4\end{cases}$	$\underset{\sim}{D}_{11}\{\beta\|\underset{\sim}{b}\}=e^{i\underset{\sim}{k}\cdot\underset{\sim}{b}}\Gamma_i(\beta)$
C $\underset{\sim}{k}\equiv(\pi/a,\pi/a)$	$K=G$	$G_o(\underset{\sim}{k})=G_o=C_{4v}$	$3\Gamma_1+3\Gamma_2+3\Gamma_3+3\Gamma_4+4\Gamma_5$	Ten 2-fold degenerate	$1\begin{cases}d=1\\q=1\end{cases}$ $2\begin{cases}d=1\\q=1\end{cases}$	Time reversal increases degeneracy: Γ_1 and Γ_3 are degenerate, Γ_2 and Γ_4 are degenerate
B $\underset{\sim}{k}\equiv(\pi/a,O)$	$K:\{E\|\underset{\sim}{T}_n\},\{C_2\|\underset{\sim}{T}_n\}$ $\{\sigma_1\|\underset{\sim}{A}+\underset{\sim}{T}_n\}$, $\{\sigma_3\|\underset{\sim}{A}+\underset{\sim}{T}_n\}$	$G_o(\underset{\sim}{k})=C_{2v}:$ E,C_2,σ_1,σ_3	10Γ	Ten 2-fold degenerate	$4\begin{cases}d=2\\q=2\end{cases}$	
BC $k\equiv(\pi/a,k_y)$	$K:\{E\|\underset{\sim}{T}_n\}$, $\{\sigma_3\|\underset{\sim}{A}+\underset{\sim}{T}_n\}$	$G_o(\underset{\sim}{k})=C_{1h}:E,\sigma_3$	$10\Gamma_1+10\Gamma_2$	Ten 2-fold degenerate	$4\begin{cases}d=1\\q=4\end{cases}$	Time reversal doubles degeneracy: Γ_1 and Γ_2 are degenerate

References

1. A. A. Maradudin, E. W. Montroll and G. H. Weiss, "Theory of Lattice Dynamics in the Harmonic Approximation," in Suppl. 3 of *Solid State Physics*, Academic, New York, 1963.
2. C. Kittel, *Introduction to Solid State Physics*, Wiley, New York, 1971, p. 703.
3. W. A. Wall, "Selection Rules for Vibronic Transitions by Ion Impurities in Solids," Doctoral Dissertation, Boston College 1973, unpublished.
4. L. Van Hove, *Phys. Rev.* 89, 1189 (1953).
5. J. C. Phillips, *Phys. Rev.* 104, 1263 (1956).
6. B. Di Bartolo, *Optical Interactions in Solids*, Wiley, New York, 1968, p. 107.
7. A. A. Maradudin and S. H. Vosko, *Rev. Mod. Physics* 40, 1 (1968).
8. E. P. Wigner, *Group Theory and Its Application to the Quantum Mechanics of Atomic Spectra*, Academic, New York, 1959, p. 325.

Chapter 7

Thermodynamics of Lattice Vibrations

This chapter considers the relation of lattice vibrations with such physical properties as specific heat and melting temperature of solids. The Einstein and Debye theories of specific heat of solids are presented.

7.1. Thermodynamics of Specific Heats

Let us consider a homogeneous substance of volume V. The *heat capacity at constant volume* is defined as

$$C_V = \left(\frac{đQ}{dT}\right)_V = T\left(\frac{dS}{dT}\right)_V. \tag{7.1.1}$$

The heat capacity at constant pressure is defined as

$$C_p = \left(\frac{đQ}{dT}\right)_p = T\left(\frac{\partial S}{\partial T}\right)_p, \tag{7.1.2}$$

where đQ is the heat absorbed by the system. Using T and P as independent variables we can write

$$dS = \left(\frac{\partial S}{\partial T}\right)_p dT + \left(\frac{\partial S}{\partial p}\right)_T dp, \tag{7.1.3}$$

and

$$đQ = TdS = T\left(\frac{\partial S}{\partial T}\right)_p dT + T\left(\frac{\partial S}{\partial p}\right)_T dp = C_p dT + T\left(\frac{\partial S}{\partial p}\right)_T dp. \tag{7.1.4}$$

We can also write

$$dp = \left(\frac{\partial p}{\partial T}\right)_V dT + \left(\frac{\partial p}{\partial V}\right)_T dV, \tag{7.1.5}$$

and

$$đQ = TdS = C_p dT + T\left(\frac{\partial S}{\partial p}\right)_T \left[\left(\frac{\partial p}{\partial T}\right)_V dT + \left(\frac{\partial p}{\partial V}\right)_T dV\right]. \tag{7.1.6}$$

Therefore

$$C_V = \left(\frac{đQ}{dT}\right)_V = C_p + T\left(\frac{\partial S}{\partial p}\right)_T \left(\frac{\partial p}{\partial T}\right)_V. \tag{7.1.7}$$

On the other hand, a Maxwell relation gives

$$\left(\frac{\partial S}{\partial p}\right)_T = -\left(\frac{\partial V}{\partial T}\right)_p = -V\alpha, \tag{7.1.8}$$

where

$$\alpha = \frac{1}{V}\left(\frac{\partial V}{\partial T}\right)_p, \tag{7.1.9}$$

is defined as *volume coefficient of expansion*. We have also

$$dV = \left(\frac{\partial V}{\partial T}\right)_p dT + \left(\frac{\partial V}{\partial p}\right)_T dp, \tag{7.1.10}$$

and this for $dV = 0$ gives

$$\left(\frac{\partial p}{\partial T}\right)_V = -\frac{(\partial V/\partial T)_p}{(\partial V/\partial p)_T} = \frac{\alpha}{k}, \tag{7.1.11}$$

where

$$k = -\frac{1}{V}\left(\frac{\partial V}{\partial p}\right)_T, \tag{7.1.12}$$

is defined as *coefficient of isothermal compressibility*.

When we use (7.1.8) and (7.1.11) in (7.1.7) we obtain

$$C_p - C_V = VT\frac{\alpha^2}{k}, \tag{7.1.13}$$

and

$$c_p - c_v = vT\frac{\alpha^2}{k}, \tag{7.1.14}$$

where c_p, c_v, and v designate the specific heat at constant pressure, the specific heat at constant volume, and the specific volume, respectively.

It is relevant to investigate the properties of the specific heat near the absolute zero. The third law of thermodynamics states that as $T \to 0$, the entropy S tends to a constant value S_o, independent of all parameters of the system. Therefore, as T approaches zero, c_p and c_v also approach the value zero. Considering that as $T \to 0$, $(\partial S/\partial p)_T \to 0$, from (7.1.7) we derive the fact that as $T \to 0$, α and $c_p - c_v$ also tend to zero.

7.2. The Classical Theory of the Specific Heats of Solids

The lattice vibrations of a solid may be considered to be equivalent to a collection of independent harmonic oscillators, in the approximation of small oscillations. In the case of a crystal with N unit cells and J atoms per unit cell, the number of these oscillators is 3NJ, one for each degree of freedom.

The equipartition theorem of classical statistical mechanics states that each independent quadratic term contributes an energy of $\frac{1}{2}kT$ to the mean energy of the system. Therefore each harmonic oscillator contributes kT and the mean energy of the lattice vibrations at temperature T is given by

$$\langle E \rangle = 3NJkT, \tag{7.2.1}$$

where k is the Boltzmann constant, 1.38054×10^{-16} ergs/degree. Then the heat capacity at constant volume is given by

$$C_V = 3NJk. \tag{7.2.2}$$

If $J = 1$ (one atom per unit cell), the specific heat at constant volume per mole is given by

$$c_V = \frac{1}{n} C_V = \frac{3Nk}{n} = 3N_a k = 3R, \tag{7.2.3}$$

where

$$\begin{aligned} n &= \text{number of moles in the system}, \\ N_a &= \text{Avogadro's number} = 6.02252 \times 10^{23} \text{ mole}^{-1}, \\ R &= 8.3143 \times 10^7 \text{ ergs mole}^{-1} \text{ degree}^{-1}. \end{aligned}$$

The classical theory predicts a value for the specific heat independent of the temperature, contrary to the consequence of the third law and to the experimental observation of its tendency to zero as the temperature approaches the absolute zero. The relation (7.2.3) expresses the so-called *law of Dulong and Petit.*

7.3. The Einstein Theory of Specific Heat

The energy levels of a system consisting of the lattice vibrations of a crystal are given by (see 6.5.76)

$$E_{\{n_1 n_2 \ldots n_{3NJ}\}} = \sum_{\underset{\sim}{k}\lambda}^{3NJ} \left(n_{\underset{\sim}{k}}^{\lambda} + \frac{1}{2} \right) \hbar \omega_{\underset{\sim}{k}\lambda}, \tag{7.3.1}$$

where $n_{\underset{\sim}{k}}^{\lambda}$ corresponds to the degree of excitation of the $\underset{\sim}{k}\lambda$th oscillator.

According to Einstein's hypothesis[1] all the atoms vibrate with the same frequency, namely

$$\omega_{\underset{\sim}{k}\lambda} = \omega = \text{const.} \tag{7.3.2}$$

The energy levels of a single harmonic oscillator are given by

$$\varepsilon_n = \left(n + \frac{1}{2} \right) \hbar \omega. \tag{7.3.3}$$

The mean energy of an oscillator is then

$$\langle \varepsilon \rangle = \frac{\sum_{n=0}^{\infty} e^{-\beta \varepsilon_n} \varepsilon_n}{\sum_{n=0}^{\infty} e^{-\beta \varepsilon_n}} = -\frac{1}{Z} \frac{\partial Z}{\partial \beta} = -\frac{\partial}{\partial \beta} \ell n\, Z, \tag{7.3.4}$$

where $\beta = 1/kT$ and

$$Z = \sum_{n=0}^{\infty} e^{-\beta\varepsilon_n} = \text{partition sum}. \tag{7.3.5}$$

We can evaluate the partition sum as follows:

$$\begin{aligned} Z &= \sum_{n=0}^{\infty} e^{-\beta\varepsilon_n} = \sum_{n=0}^{\infty} e^{-\beta(n+\frac{1}{2})\hbar\omega} \\ &= e^{-(1/2)\beta\hbar\omega} \sum_{n=0}^{\infty} e^{-n\beta\hbar\omega} \\ &= e^{-(1/2)\beta\hbar\omega}(1 + e^{-\beta\hbar\omega} + e^{-2\beta\hbar\omega} + \cdots) \\ &= e^{-(1/2)\beta\hbar\omega} \frac{1}{1 - e^{-\beta\hbar\omega}}. \end{aligned} \tag{7.3.6}$$

Then

$$\ell n\, Z = -(1/2)\beta\hbar\omega - \ell n(1 - e^{-\beta\hbar\omega}),$$

and

$$\begin{aligned} \langle\varepsilon\rangle &= -\frac{\partial}{\partial\beta}(\ell n\, Z) = -\left(-\frac{1}{2}\hbar\omega - \frac{e^{-\beta\hbar\omega}\hbar\omega}{1 - e^{-\beta\hbar\omega}}\right) \\ &= \hbar\omega\left(\frac{1}{e^{\beta\hbar\omega} - 1} + \frac{1}{2}\right) = \hbar\omega(\bar{n} + 1/2), \end{aligned} \tag{7.3.7}$$

where

$$\bar{n} = \frac{1}{e^{\beta\hbar\omega} - 1}. \tag{7.3.8}$$

$\bar{n}$ represents the mean value of the degree of excitation of the oscillator.

At high temperatures, $\beta\hbar\omega = \hbar\omega/kT \ll 1$, and

$$\langle\varepsilon\rangle = \hbar\omega\left[\frac{1}{2} + \frac{1}{1 + \beta\hbar\omega + \cdots - 1}\right] \simeq \hbar\omega\left[\frac{1}{2} + \frac{1}{\beta\hbar\omega}\right] \simeq kT. \tag{7.3.9}$$

On the other hand, at low temperatures, $\beta\hbar\omega = \hbar\omega/kT \gg 1$ and

$$\langle\varepsilon\rangle = \hbar\omega\left(\frac{1}{2} + e^{-\beta\hbar\omega}\right). \tag{7.3.10}$$

The contribution to the specific heat due to one atom is then, in general, given by

$$
\begin{aligned}
C_V^{(1)} &= \left[\frac{\partial\langle\varepsilon\rangle}{\partial T}\right]_V = \left[\frac{\partial\langle\varepsilon\rangle}{\partial\beta}\right]_V \frac{\partial\beta}{\partial T} \\
&= -\frac{1}{kT^2}\left[\frac{\partial\langle\varepsilon\rangle}{\partial\beta}\right]_V = k\left[\frac{\hbar\omega}{kT}\right]^2 \frac{e^{\beta\hbar\omega}}{(e^{\beta\hbar\omega}-1)^2} \\
&= k(T_E/T)^2 \frac{e^{T_E/T}}{(e^{T_E/T}-1)^2},
\end{aligned} \tag{7.3.11}
$$

where

$$
T_E = \frac{\hbar\omega}{k} = \text{Einstein temperature.} \tag{7.3.12}
$$

At high temperature we find

$$
C_V^{(1)} = k, \tag{7.3.13}
$$

and at low temperatures

$$
\begin{aligned}
C_V^{(1)} &= \frac{\partial}{\partial T}\left[\hbar\omega\left(\frac{1}{2} + e^{-\beta\hbar\omega}\right)\right] \\
&= -\frac{1}{kT^2}\frac{\partial}{\partial\beta}\left[\hbar\omega\left(\frac{1}{2} + e^{-\beta\hbar\omega}\right)\right] \\
&= -\frac{1}{kT^2}\hbar\omega(-\hbar\omega)e^{-\beta\hbar\omega} = k\left[\frac{\hbar\omega}{kT}\right]^2 e^{-\beta\hbar\omega} \\
&= k\left[\frac{T_E}{T}\right]^2 e^{-T_E/T}.
\end{aligned} \tag{7.3.14}
$$

Considering now a crystal with one atom per unit cell the specific heat per mole is given, in general, by

$$
\begin{aligned}
c_V &= 3N_a C_V^{(1)} \\
&= 3N_a k\left[\frac{T_E}{T}\right]^2 \frac{e^{T_E/T}}{(e^{T_E/T}-1)^2} \\
&= 3R\left[\frac{T_E}{T}\right]^2 \frac{e^{T_E/T}}{(e^{T_E/T}-1)^2}.
\end{aligned} \tag{7.3.15}
$$

At high temperatures

$$c_V = 3N_a k = 3R, \tag{7.3.16}$$

and at low temperatures

$$c_V = 3N_a k \left[\frac{T_E}{T}\right]^2 e^{-T_E/T} = 3R \left[\frac{T_E}{T}\right]^2 e^{-T_E/T}. \tag{7.3.17}$$

According to (7.3.17) the specific heat should approach the zero value exponentially as $T \to 0$, contrary to the experimental observation that actually it approaches zero with a cubic law, namely $c_V \propto T^3$ as $T \to 0$. This discrepancy is due to the simplified assumption of Einstein's theory that all the atoms vibrate at the same frequency and independently. In particular, Einstein's theory does not take any account of the fact that some modes of oscillation may have such a low frequency ω that, even for small T, $\hbar\omega \ll kT$, and that therefore they contribute appreciably to the specific heat by preventing it from decreasing as rapidly as indicated by (7.3.17).

Consider now the expression for the Einstein temperature:

$$T_E = \frac{\hbar\omega}{k} = \frac{\hbar}{k}\sqrt{K_o/m}, \tag{7.3.18}$$

where K_o is the spring constant, and m is the mass of the atom.

If a solid has atoms of small mass m and if it is hard (namely, if it has a large spring constant K_o), its T_E is large. Most solids have a T_E of about 300°K; diamond, a hard solid consisting of the relatively light carbon atoms, has a T_E of about 1320°K. Notice that a T_E of about 300°K corresponds to $\omega/2\pi \simeq 6 \times 10^{12}$ cycles/sec and a wavelength of the electromagnetic spectrum of about 50μ.

Considering now the dependence of c_V on T in Fig. 7.1, we see that at $T \simeq T_E, c_V$ has almost reached its classical value. Therefore at room temperature most solids have a $c_V \simeq 3R$, but diamond has a c_V considerably smaller than 3R and about equal to 1.8R.

Despite its inadequacies, Einstein's theory of specific heat accounts for two different phenomena:

1. The falling of c_V below the classical value 3R and the approaching to the zero value as $T \to 0$,

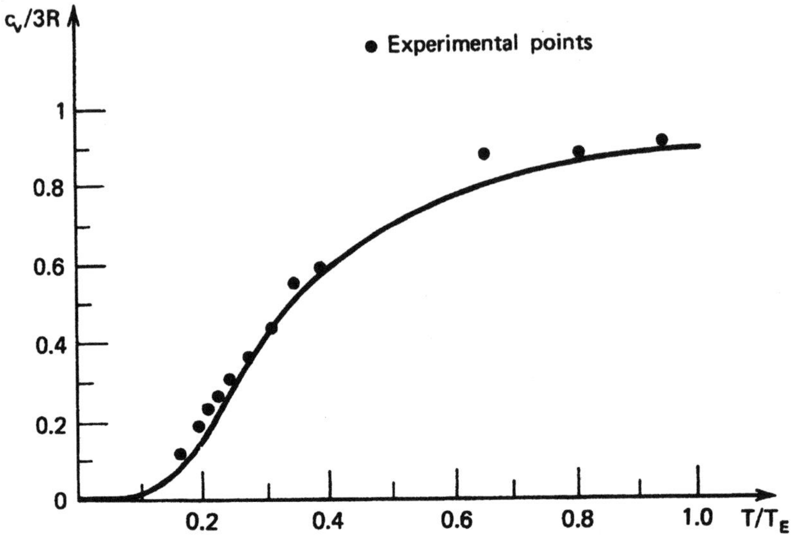

Fig. 7.1. Temperature dependence of c_V according to Einstein theory. The experimental points were obtained for diamond and the fit was done with $T_E = 1320°K$.[1]

2. The dependence of the approach to the zero value on such physical characteristics as hardness and mass of the atoms.

7.4. The Debye Theory of Specific Heat

7.4.1. *The Specific Heat of a Linear Crystal*

Let us consider a linear crystal with one atom per unit cell. The energy levels of the lattice vibrations are given by

$$E = \sum_{k=1}^{N} \left(n_k + \frac{1}{2} \right) \hbar\omega_k, \tag{7.4.1}$$

where N is the number of atoms in the crystal. The average energy of the kth oscillator is given by

$$\begin{aligned} \langle \varepsilon_k \rangle &= \left[\langle n_k \rangle + \frac{1}{2} \right] \hbar\omega_k \\ &= \left[\frac{1}{e^{\hbar\omega_k/kT} - 1} + \frac{1}{2} \right] \hbar\omega_k. \end{aligned} \tag{7.4.2}$$

The total average energy is given by

$$\langle E\rangle = \sum_{k=1}^{N}\langle \varepsilon_k\rangle = \sum_{k=1}^{N}\hbar\omega_k\left[\frac{1}{e^{\hbar\omega_k/kT}-1}+\frac{1}{2}\right]$$
$$\simeq \int_0^{\omega_{max}} g(\omega)d\omega\left[\frac{1}{e^{\hbar\omega/kT}-1}+\frac{1}{2}\right]\hbar\omega, \qquad (7.4.3)$$

where $g(\omega)d\omega$ is the number of modes with angular frequency in $(\omega, (\omega + d\omega)$. The number of modes with k in $(k, k + dk)$ is given, according to (6.2.5), by

$$dr = \frac{Na}{2\pi}dk, \qquad (7.4.4)$$

where a is the spacing between nearest-neighbor atoms. Then

$$g(\omega)d\omega = 2\frac{Na}{2\pi}\frac{dk}{d\omega}d\omega. \qquad (7.4.5)$$

The reason for the factor 2 in (7.4.5) is that two values of k correspond to each angular frequency.

Let us apply the above considerations to the case in which only the forces between nearest neighbors are non-negligible. In this case, according to (6.2.16),

$$\omega = 2\sqrt{\frac{\beta}{m}}\sin\frac{1}{2}ka, \qquad (7.4.6)$$

and

$$\frac{d\omega}{dk} = a\sqrt{\frac{\beta}{m}}\cos\frac{1}{2}ka. \qquad (7.4.7)$$

Then

$$g(\omega) = 2\frac{Na}{2\pi}\frac{1}{a\sqrt{\beta/m}\cos\frac{1}{2}ka}$$
$$= \frac{N}{\pi\sqrt{\beta/m}\left(1-\frac{m\omega^2}{4\beta}\right)^{\frac{1}{2}}}. \qquad (7.4.8)$$

We find, as expected, that

$$\int_o^{\omega_m} g(\omega)d\omega = \frac{N}{\pi}\sqrt{\frac{\beta}{m}}\int_o^{\omega_m} \frac{1}{\left[1-\frac{m\omega^2}{4\beta}\right]^{1/2}}d\omega$$

$$= \frac{N}{\pi}\sqrt{\frac{\beta}{m}}\int_o^{\omega_m} \frac{d\omega}{[1-(\omega/\omega_m)^2]^{1/2}}$$

$$= \frac{2N}{\pi}\int_o^1 \frac{dx}{\sqrt{1-x^2}} = N, \tag{7.4.9}$$

where

$$\omega_m = 2\sqrt{\frac{\beta}{m}}. \tag{7.4.10}$$

Now the integral in (7.4.3) becomes

$$\langle E\rangle = \frac{2N}{\pi\omega_m}\int_o^{\omega_m}\left[\frac{1}{e^{\hbar\omega/kT}-1}+\frac{1}{2}\right]\frac{\hbar\omega d\omega}{[1-(\omega/\omega_m)^2]^{1/2}}. \tag{7.4.11}$$

We have then, if n is the number of moles in the solid,

$$c_V = \frac{1}{n}\frac{\partial\langle E\rangle}{\partial T}$$
$$= \frac{2N_a k}{\pi\omega_m}\int_o^{\omega_m}\frac{(\hbar\omega/kT)^2}{[1-(\omega/\omega_m)^2]^{1/2}}\frac{e^{\hbar\omega/kT}d\omega}{(e^{\hbar\omega/kT}-1)^2}. \tag{7.4.12}$$

At high temperatures

$$e^{\hbar\omega/kT}-1 \approx \frac{\hbar\omega}{kT}, \quad e^{\hbar\omega/kT}\approx 1,$$

and

$$c_V = \frac{2N_a k}{\pi}\int_0^1\frac{dx}{\sqrt{1-x^2}} = N_a k = R. \tag{7.4.13}$$

At low temperatures

$$c_V = \frac{2N_a k}{\pi\omega_m}\int_o^{\omega_m}\frac{d\omega}{[1-(\omega/\omega_m)^2]^{1/2}}\left[\frac{\hbar\omega}{kT}\right]^2 e^{-\hbar\omega/kT}$$

$$= \frac{2N_a k}{\pi\omega_m}\int_o^{\infty}d\omega\left[\frac{\hbar\omega}{kT}\right]^2 e^{-\hbar\omega/kT}$$

$$= \frac{2N_a k}{\pi\omega_m}\frac{kT}{\hbar}\int_o^\infty y^2 e^{-y} dy$$

$$= R\frac{4}{\pi}\frac{kT}{\hbar\omega_m}. \tag{7.4.14}$$

The specific heat goes to zero as T.

7.4.2. *The Debye Theory Applied to a Linear Crystal*

The basic hypothesis of the Debye theory[2] consists in neglecting the fact that a crystal is a discrete array of atoms and in treating it as a continuous medium. This corresponds simply to setting

$$\frac{d\omega}{dk} = \frac{\omega}{k} = c. \tag{7.4.15}$$

The number of modes with angular frequency in $(\omega, \omega + d\omega)$ is then given by

$$g(\omega)d\omega = 2\frac{Na}{2\pi}\frac{1}{c}d\omega = \frac{L}{\pi c}d\omega, \tag{7.4.16}$$

where $L = Na$. The function $g(\omega)$ must be cut off at a frequency ω_D (called the *Debye frequency*) such that $g(\omega)$ yields the correct total number of N modes:

$$\int_o^{\omega_D} g(\omega)d\omega = N. \tag{7.4.17}$$

From (7.4.16) and (7.4.17) we derive

$$\omega_D = \frac{\pi c N}{L} = \frac{\pi c}{a}. \tag{7.4.18}$$

The average energy is given by

$$\begin{aligned}\langle E\rangle &= \int_o^{\omega_D} g(\omega)d\omega\left[\frac{1}{e^{\hbar\omega/kT}-1}+\frac{1}{2}\right]\hbar\omega \\ &= \frac{L}{\pi c}\int_o^{\omega_D} d\omega\left[\frac{1}{e^{\hbar\omega/kT}-1}+\frac{1}{2}\right]\hbar\omega,\end{aligned} \tag{7.4.19}$$

and the specific heat by

$$c_V = \frac{1}{n}\frac{\partial\langle E\rangle}{\partial T} = \frac{Lk}{n\pi c}\int_o^{\omega_D} d\omega\left(\frac{\hbar\omega}{kT}\right)^2\frac{e^{\hbar\omega/kT}}{(e^{\hbar\omega/kT}-1)^2}, \tag{7.4.20}$$

where n is the number of moles in the crystal. At high temperatures

$$e^{\hbar\omega/kT} - 1 \approx \frac{\hbar\omega}{kT}, \quad e^{\hbar\omega/kT} \approx 1,$$

and

$$\begin{aligned} c_V &= \frac{Lk}{n\pi c}\,\omega_D = \frac{Lk}{n\pi c}\frac{\pi c}{a} \\ &= \frac{Nk}{n} = N_a k = R. \end{aligned} \tag{7.4.21}$$

At low temperatures

$$\begin{aligned} c_V &= \frac{Lk}{n\pi c}\int_o^{\omega_D} d\omega \left[\frac{\hbar\omega}{kT}\right]^2 e^{-\hbar\omega/kT} \\ &\approx \frac{Lk}{n\pi c}\int_o^{\infty} d\omega \left[\frac{\hbar\omega}{kT}\right]^2 e^{-\hbar\omega/kT} \\ &= \frac{Lk}{n\pi c}\frac{kT}{\hbar}\int_o^{\infty} y^2 e^{-y} dy = \frac{Lk}{n\pi c}2\frac{kT}{\hbar} \\ &= \frac{Lk}{na}\frac{kT}{\hbar\omega_D}2 = 2N_a k\left(\frac{kT}{\hbar\omega_D}\right) = 2R\left(\frac{T}{T_D}\right), \end{aligned} \tag{7.4.22}$$

where

$$T_D = \frac{\hbar\omega_D}{k} = \text{Debye temperature.} \tag{7.4.23}$$

7.4.3. *The Debye Theory Applied to a Three-Dimensional Crystal*

Let us consider a three-dimensional crystal with N identical atoms. The energy levels of the lattice vibrations are given by

$$E_{\{n_1 n_2 \cdots n_{3N}\}} = \sum_{\underset{\sim}{k}\lambda}^{3N}\left(n_{\underset{\sim}{k}}^{\lambda} + \frac{1}{2}\right)\hbar\omega_{\underset{\sim}{k}\lambda}, \tag{7.4.24}$$

where $n_{\underset{\sim}{k}}^{\lambda}$ indicates the excitation level of the $\underset{\sim}{k}\lambda$th oscillator. The partition sum of the system is given by

$$
\begin{aligned}
z &= \sum_{n_1 n_2 \cdots n_{3N}} \exp\left[-\beta \sum_{\underset{\sim}{k}\lambda}^{3N}\left(\frac{1}{2}\hbar\omega_{k\lambda} + n_{\underset{\sim}{k}}^{\lambda}\hbar\omega_{k\lambda}\right)\right] \\
&= e^{-\beta E_o} \sum_{n_1 n_2 \cdots n_{3N}} e^{-\beta(n_1\hbar\omega_1 + n_2\hbar\omega_2 + \cdots + n_{3N}\hbar\omega_{3N})} \\
&= e^{-\beta E_o} \sum_{n_1=0}^{\infty} e^{-\beta n_1\hbar\omega_1} \sum_{n_2=0}^{\infty} e^{-\beta n_2\hbar\omega_2} \cdots \sum_{n_{3N}=0}^{\infty} e^{-\beta n_{3N}\hbar\omega_{3N}} \\
&= e^{-\beta E_o} \frac{1}{1-e^{-\beta\hbar\omega_1}} \frac{1}{1-e^{-\beta\hbar\omega_2}} \cdots \frac{1}{1-e^{-\beta\hbar\omega_{3N}}},
\end{aligned}
\tag{7.4.25}
$$

where

$$E_o = \text{zero point energy} = \sum_{\underset{\sim}{k}\lambda} \frac{1}{2}\hbar\omega_{\underset{\sim}{k}\lambda},$$

$$\beta = \frac{1}{kT}.$$

Then

$$
\begin{aligned}
\ell n\, Z &= -\beta E_o - \sum_{\underset{\sim}{k}\lambda}^{3N} \ell n(1 - e^{-\beta\hbar\omega_{\underset{\sim}{k}\lambda}}) \\
&= -\beta E_o - \int_0^{\infty} [\ell n(1 - e^{-\beta\hbar\omega})] g(\omega) d\omega,
\end{aligned}
\tag{7.4.26}
$$

where $g(\omega)d\omega$ is the number of modes with angular frequency in $(\omega, \omega + d\omega)$. The mean energy of the system is given by

$$\langle E\rangle = -\frac{\partial \ell nz}{\partial\beta} = E_o + \int_0^{\infty} \frac{\hbar\omega}{e^{\beta\hbar\omega} - 1} g(\omega) d\omega. \tag{7.4.27}$$

The heat capacity is given by

$$
\begin{aligned}
c_V &= \frac{\partial\langle E\rangle}{\partial T} = \left(\frac{\partial\langle E\rangle}{\partial\beta}\right)_V \frac{\partial\beta}{\partial T} \\
&= -k\beta^2 \left(\frac{\partial\langle E\rangle}{\partial\beta}\right)_V
\end{aligned}
$$

$$= k\beta^2 \int_0^\infty \frac{(\hbar\omega)^2}{(e^{\beta\hbar\omega}-1)^2} e^{\beta\hbar\omega} g(\omega)d\omega$$

$$= k \int_0^\infty \frac{e^{\beta\hbar\omega}}{(e^{\beta\hbar\omega}-1)^2} (\beta\hbar\omega)^2 g(\omega)d\omega. \tag{7.4.28}$$

We may find the high-temperature limit for C_v irrespective of the actual shape of $g(\omega)$. For T very large

$$C_V = k \int_0^\infty \frac{1}{(\beta\hbar\omega)^2} (\beta\hbar\omega)^2 g(\omega)d\omega$$

$$= \int_0^\infty g(\omega)d\omega = 3Nk, \tag{7.4.29}$$

and the specific heat is given by

$$c_V = \frac{1}{n} C_V = \frac{3N}{n} k = 3N_a k = 3R. \tag{7.4.30}$$

The relation above is simply the classical result derived in Section 7.2.

Let us apply now the Debye theory to the present case. The following assumptions have to be made.

1. $\omega = ck$, for the entire range of k. In particular, for each of the three branches $\omega_\lambda = c_\lambda k$.
2. Each velocity c_λ is independent of the direction of propagation of the waves.
3. In agreement with elasticity theory, for each $\underset{\sim}{k}$ vector there are two transverse modes with a common value of velocity of propagation c_t and one longitudinal mode with propagation velocity c_ℓ. The number of modes with k_x in $(k_x, k_x + dk_x)$, k_y in $(k_y, k_y + dk_y)$, and k_z in $(k_z, k_z + dk_z)$ is given, according to (5.5.29), by

$$dr_1 dr_2 dr_3 = \frac{V}{(2\pi)^3} dk_x dk_y dk_z, \tag{7.4.31}$$

where V is the volume of the crystal. The number of modes with k in $(k, k + dk)$ is given by

$$\frac{V}{(2\pi)^3} 4\pi k^2 dk = \frac{V}{2\pi^2} k^2 \frac{dk}{d\omega} d\omega = g(\omega)d\omega. \tag{7.4.32}$$

According to the Debye theory the number of transverse modes in the frequency interval $(\omega, \omega + d\omega)$ is given by

$$g_t(\omega)d\omega = 2\frac{V}{2\pi^2 c_t^3}\omega^2 d\omega, \tag{7.4.33}$$

where the factor 2 takes into account the fact that there are two transverse modes for each ω. Similarly the number of longitudinal modes in the frequency interval $(\omega, \omega + d\omega)$ is given by

$$g_\ell(\omega)d\omega = \frac{V}{2\pi^2 c_\ell^3}\omega^2 d\omega. \tag{7.4.34}$$

The total number of modes in the frequency interval $(\omega, \omega + d\omega)$ is given by

$$g(\omega)d\omega = 3\frac{V}{2\pi^2 c_s^3}\omega^2 d\omega, \tag{7.4.35}$$

where

$$\frac{3}{c_s^3} = \frac{2}{c_t^3} + \frac{1}{c_\ell^3}. \tag{7.4.36}$$

The cut-off frequency for $g(\omega)$, called Debye frequency ω_D, must be so chosen as to yield the correct number of modes 3N:

$$\int_0^{\omega_D} g(\omega)d\omega = \frac{3V}{2\pi^2 c_s^3}\frac{\omega_D^3}{3} = 3N, \tag{7.4.37}$$

or

$$\omega_D = c_s\left(6\pi^2\frac{N}{V}\right)^{1/3}. \tag{7.4.38}$$

The Debye frequency depends on the velocity of sound c_s in the solid and on the number of atoms per unit volume. Putting the approximate values $c_s \approx 5 \times 10^5$ cm/sec and $(V/N)^{1/3} \approx$ interatomic distance $\approx 10^{-8}$ cm in (6.4.38) we find $\omega_D \approx 10^{14}\ \text{sec}^{-1}$.

The number of modes in the unit frequency range is given, according to the Debye approximation, by

$$g(\omega) = \begin{cases} \dfrac{3V}{2\pi^2 c_s^3}\omega^2 = \dfrac{9N}{\omega_D^3}\omega^2 & 0 \le \omega \le \omega_D \\ 0 & \omega > \omega_D \end{cases}. \tag{7.4.39}$$

We can now calculate the average energy of the system. From (7.4.27)

$$\begin{aligned}\langle E\rangle &= E_o + \int_0^{\omega_D} \frac{\hbar\omega}{e^{\beta\hbar\omega}-1} g(\omega)d\omega \\ &= E_o + \frac{9N}{\omega_D^3}\int_0^{\omega_D} \frac{\hbar\omega^3}{e^{\beta\hbar\omega}-1} d\omega \\ &= E_o + \frac{9N}{\omega_D^3}\frac{1}{\beta^4\hbar^3}\int_0^{T_D/T} \frac{x^3dx}{e^x-1} \\ &= E_o + 9NkT\left(\frac{T}{T_D}\right)^3\int_0^{T_D/T}\frac{x^3dx}{e^x-1},\end{aligned} \tag{7.4.40}$$

where

$$T_D = \frac{\hbar\omega_D}{k} = \text{Debye temperature.} \tag{7.4.41}$$

The heat capacity, following (7.4.28), is given by

$$\begin{aligned}C_V = \frac{\partial\langle E\rangle}{\partial T} &= k\int_0^{\omega_D}\frac{e^{\beta\hbar\omega}}{(e^{\beta\hbar\omega}-1)^2}(\beta\hbar\omega)^2 g(\omega)d\omega \\ &= \frac{9Nk}{\omega_D^3}\int_0^{\omega_D}\frac{e^{\beta\hbar\omega}(\beta\hbar\omega)^2}{(e^{\beta\hbar\omega}-1)^2}\omega^2 d\omega \\ &= \frac{9Nk}{\omega_D^3}\frac{1}{\beta^3\hbar^3}\int_0^{T_D/T}\frac{x^4e^x}{(e^x-1)^2}dx \\ &= 9Nk\left(\frac{T}{T_D}\right)^3\int_0^{T_D/T}\frac{x^4e^x}{(e^x-1)^2}dx \\ &= 3Nk\, f_D\left(\frac{T_D}{T}\right),\end{aligned} \tag{7.4.42}$$

where f_D (y), the *Debye function*, is defined as

$$f_D(y) = \frac{3}{y^3}\int_0^y \frac{x^4e^x}{(e^x-1)^2}dx. \tag{7.4.43}$$

Let us examine now the Debye function in the limit of $y \ll 1$ and $y \gg 1$. For $y \ll 1$

$$f_D(y) \approx \frac{3}{y^3}\int_0^Y x^2 dx = 1. \tag{7.4.44}$$

For $y \gg 1$, we can replace the upper limit y in the integral in (7.4.43) by ∞ and

$$f_D(y) \approx \frac{3}{y^3}\int_0^\infty \frac{x^4 e^x}{(e^x - 1)^2} dx = \frac{4\pi^4}{5}\frac{1}{y^3}. \tag{7.4.45}$$

We can now calculate the high- and low-temperature values of the heat capacity C_V. At $T \gg T_D$ or $T_D/T \ll 1$,

$$C_V = 3Nk, \tag{7.4.46}$$

and the specific heat is given by

$$c_V = \frac{C_V}{n} = 3R, \tag{7.4.47}$$

where n = number of moles in the solid. At $T \ll T_D$ or $T_D/T \gg 1$,

$$C_V = \frac{12\pi^4}{5} Nk \left(\frac{T}{T_D}\right)^3, \tag{7.4.48}$$

and

$$c_V = \frac{12\pi^4}{5} R \left(\frac{T}{T_D}\right)^3. \tag{7.4.49}$$

The Debye approximation gives at low temperatures a temperature dependence for the specific heat which is in agreement with the experimental results. The Debye theory provides also a temperature dependence over the entire temperature range which is in many cases in fairly good agreement with the experimental results.

We notice also from (7.4.38) that ω_D can be computed from the velocities of sound c_t and c_ℓ, which, on the other hand, can be derived from known elastic moduli. The agreement between ω_D derived from specific heat measurements and ω_D derived from elastic constants is good in many cases.

The Debye theory was modified by Born, who proposed a different method for the cut-off provision of the frequency spectrum.

In essence his procedure consists in assigning to the longitudinal and transverse modes a minimum common value of the wavelength rather than a maximum value of their frequencies:

$$\lambda_{\min} = \frac{2\pi c_\ell}{\omega_{m\ell}} = \frac{2\pi c_t}{\omega_{mt}}, \tag{7.4.50}$$

where $\omega_{m\ell}$ and ω_{mt} are the cut-off frequencies for the longitudinal and transverse modes, respectively. The number of modes in the frequency interval $(\omega, \omega + d\omega)$ is given by

$$g(\omega)d\omega = \frac{V}{2\pi^2 c_\ell^3}\omega^2 d\omega + 2\frac{V}{2\pi^2 c_t^3}\omega^2 d\omega. \tag{7.4.51}$$

Then

$$\frac{V}{2\pi^2 c_\ell^3}\frac{\omega_{m\ell}^2}{3} + 2\frac{V}{2\pi^2 c_t^3}\frac{\omega_{mt}^3}{3} = 3N. \tag{7.4.52}$$

Taking (7.4.50) into account the above relation gives

$$\lambda_{\min} = \left(\frac{4\pi V}{3N}\right)^{1/3}. \tag{7.4.53}$$

The number of longitudinal and transverse modes in the unit frequency interval can now be expressed as follows:

$$g_\ell(\omega) = \begin{cases} \dfrac{V}{2\pi^2 c_\ell^3}\omega^2 = \dfrac{3N}{\omega_{m\ell}^3}\omega^2 & 0 \le \omega \le \omega_\ell \\ 0 & \omega > \omega_\ell \end{cases},$$
$$g_t(\omega) = \begin{cases} 2\dfrac{V}{2\pi^2 c_t^3}\omega^2 = 2\dfrac{3N}{\omega_{mt}^3}\omega^2 & 0 \le \omega \le \omega_t \\ 0 & \omega > \omega_t \end{cases}. \tag{7.4.54}$$

The average energy of the system is then given by

$$\begin{aligned}\langle E\rangle &= E_o + \int_0^{\omega_{m\ell}} \frac{\hbar\omega}{e^{\beta\hbar\omega} - 1} g_\ell(\omega)d\omega + \int_0^{\omega_{mt}} \frac{\hbar\omega}{e^{\beta\hbar\omega} - 1} g_t(\omega)d\omega \\ &= E_o + \frac{3N}{\omega_{m\ell}^3}\frac{1}{\beta^4\hbar^3}\int_0^{T_\ell/T}\frac{x^3dx}{e^x - 1} + 2\frac{3N}{\omega_{mt}^3}\int_0^{T_t/T}\frac{x^3dx}{e^x - 1}\end{aligned}$$

$$= E_o + 3NkT\left[\left(\frac{T}{T_\ell}\right)^3 \int_0^{T_\ell/T} \frac{x^3 dx}{e^x - 1} + 2\left(\frac{T}{T_t}\right)^3 \int_0^{T_t/T} \frac{x^3 dx}{e^x - 1}\right], \tag{7.4.55}$$

where

$$T_\ell = \frac{\hbar\omega_{m\ell}}{k}, \qquad T_t = \frac{\hbar\omega_{mt}}{k}. \tag{7.4.56}$$

The heat capacity is given by

$$\begin{aligned} C_V &= \frac{\partial\langle E\rangle}{\partial T} \\ &= \frac{3Nk}{\omega_{m\ell}^3}\int_0^{\omega_{m\ell}} \frac{e^{\beta\hbar\omega}(\beta\hbar\omega)^2}{(e^{\beta\hbar\omega}-1)^2}\omega^2 d\omega + 2\frac{3N}{\omega_{mt}^3}\int_0^{\omega_{mt}} \frac{e^{\beta\hbar\omega}(\beta\hbar\omega)^2}{(e^{\beta\hbar\omega}-1)^2}\omega^2 d\omega \\ &= \frac{3Nk}{\beta^3\hbar^3}\left[\frac{1}{\omega_{m\ell}^3}\int_0^{T_\ell/T} \frac{x^4 e^x}{(e^x-1)^2} dx + \frac{2}{\omega_{mt}^3}\int_0^{T_t/T} \frac{x^4 e^x}{(e^x-1)^2}\right] \\ &= 3Nk\left[\left(\frac{T}{T_\ell}\right)^3 \int_0^{T_\ell/T} \frac{x^4 e^x}{(e^x-1)^2} dx + 2\left(\frac{T}{T_\ell}\right)^3 \int_0^{T_\ell/T} \frac{x^4 e^x}{e^x-1} dx\right]. \end{aligned} \tag{7.4.57}$$

The specific heat is then given by

$$c_V = R\left[f_D\left(\frac{T_\ell}{T}\right) + 2f_D\left(\frac{T_t}{T}\right)\right]. \tag{7.4.58}$$

We have considered until now crystals with one atom per unit cell. The Debye theory can, however, be applied also to the acoustic branches of the phonon spectra of polyatomic crystals. A simple approximation may be used in these cases to take into account the presence of the optical branches. This approximation consists of using the Einstein model for the optical modes, while retaining the Debye approximation for the acoustical branches. The justification for this approximation derives from the relatively small spread in frequency of the optical branches. This fact is particularly evident in the case we treated in the example of Section 5.4 of diatomic linear crystals. If the mass ratio $M/m \gg 1$, and the crystal has 2N atoms, we can consider the N optical modes as equivalent to N harmonic oscillators of frequency [see (6.4.21)]

$$\omega = 2\beta\left(\frac{1}{M}+\frac{1}{m}\right)^{1/2} \approx \left(\frac{2\beta}{m}\right)^{1/2} \tag{7.4.59}$$

7.5. Temperature Dependence of the Amplitude of Vibrations Solids. The Lindemann Law of Melting

It is interesting to investigate the temperature dependence of the thermal vibrations in solids. In the present treatment the vibrations of each atom are considered to be harmonic and therefore the time average of the displacement of any atom even at high temperature is zero. Therefore it is more meaningful to consider the square of the displacement as a function of temperature.

In particular, we shall consider the following quantity:

$$\begin{aligned}\langle[u_n(t)]^2\rangle &= \frac{1}{N}\sum_s\langle[u_s(t)]^2\rangle \\ &= \frac{1}{N}\sum_s\langle\psi|\underset{\sim}{u}_s\cdot\underset{\sim}{u}_s|\psi\rangle,\end{aligned} \tag{7.5.1}$$

where ψ is the wavefunction representing the system "lattice vibrations," which may be represented by

$$\psi = \prod_{\underset{\sim}{k}\lambda}|n_{\underset{\sim}{k}}^{\lambda}\rangle. \tag{7.5.2}$$

Considering now a crystal with one atom per unit cell, from (6.5.77)

$$\underset{\sim}{u}_s = \sum_{\underset{\sim}{k}\lambda}\left(\frac{\hbar}{2M\omega_{\underset{\sim}{k}\lambda}}\right)^{1/2}[\underset{\sim}{V}(\underset{\sim}{k},\lambda)e^{i\underset{\sim}{k}\cdot\underset{\sim}{R}_s}a_{\underset{\sim}{k}}^{\lambda}+\underset{\sim}{V}(\underset{\sim}{k},\lambda)^*e^{-i\underset{\sim}{k}\cdot\underset{\sim}{R}_s}a_{\underset{\sim}{k}}^{\lambda+}]. \tag{7.5.3}$$

From (6.5.48) we have also

$$\sum_{s}[\underset{\sim}{V}(\underset{\sim}{k},\lambda)^{*}e^{-i\underset{\sim}{k}\cdot\underset{\sim}{R}_s}\cdot\underset{\sim}{V}(\underset{\sim}{k}',\lambda')e^{i\underset{\sim}{k}'\cdot\underset{\sim}{R}_s}]=N\delta_{\underset{\sim}{k}\underset{\sim}{k}'}\delta_{\lambda\lambda}. \tag{7.5.4}$$

Taking the above relation into account and considering only the terms that contribute to the eigenvalue in (7.5.1) we obtain

$$\begin{aligned}\langle[u_n(t)]^2\rangle &= \frac{1}{Nm}\sum_{\underset{\sim}{k}\lambda}\left[\frac{\hbar}{2\omega_{\underset{\sim}{k}\lambda}}\langle\psi|a^{\lambda}_{\underset{\sim}{k}}a^{\lambda+}_{\underset{\sim}{k}}+a^{\lambda+}_{\underset{\sim}{k}}a^{\lambda}_{\underset{\sim}{k}}|\psi\rangle\right]\\ &= \frac{1}{Nm}\sum_{\underset{\sim}{k}\lambda}\left[\frac{1}{\omega^2_{\underset{\sim}{k}\lambda}}\langle\psi|H_{\underset{\sim}{k}\lambda}|\psi\rangle\right],\end{aligned} \tag{7.5.5}$$

where

$$H_{\underset{\sim}{k}\lambda}=\frac{\hbar\omega_{\underset{\sim}{k}\lambda}}{2}(a^{\lambda}_{\underset{\sim}{k}}a^{\lambda+}_{\underset{\sim}{k}}+a^{\lambda+}_{\underset{\sim}{k}}a^{\lambda}_{\underset{\sim}{k}}). \tag{7.5.6}$$

Therefore

$$\langle[u_n(t)]^2\rangle=\frac{1}{Nm}\sum_{\underset{\sim}{k}\lambda}\frac{1}{\omega^2_{\underset{\sim}{k}\lambda}}\hbar\omega_{\underset{\sim}{k}\lambda}\left(\frac{1}{2}+n^{\lambda}_{\underset{\sim}{k}}\right). \tag{7.5.7}$$

Taking the ensemble average,

$$\begin{aligned}\overline{\langle[u_n(t)]^2\rangle} &= \frac{1}{Nm}\sum_{\underset{\sim}{k}\lambda}\frac{1}{\omega^2_{\underset{\sim}{k}\lambda}}\hbar\omega_{\underset{\sim}{k}\lambda}\left(\frac{1}{2}+\bar{n}^{\lambda}_{k}\right)\\ &= \frac{1}{Nm}\sum_{\underset{\sim}{k}\lambda}\frac{1}{\omega^2_{\underset{\sim}{k}\lambda}}\hbar\omega_{\underset{\sim}{k}\lambda}\left(\frac{1}{2}+\frac{1}{e^{\hbar\omega_{\underset{\sim}{k}\lambda}/kT}-1}\right).\end{aligned} \tag{7.5.8}$$

At low temperature, in the Debye approximation

$$\begin{aligned}\overline{\langle[u_n(t)]^2\rangle} &\approx \frac{1}{2Nm}\sum_{\underset{\sim}{k}\lambda}\frac{\hbar}{\omega_{\underset{\sim}{k}\lambda}}=\frac{\hbar}{2Nm}\int_0^{\omega_D}\frac{g(\omega)}{\omega}d\omega\\ &= \frac{\hbar}{2Nm}\frac{9N}{\omega_D^3}\int_0^{\omega_D}\omega_d d\omega\\ &= \frac{9\hbar}{2m\omega_D^3}\frac{\omega_D^2}{2}=\frac{9\hbar}{4m\omega_D}\end{aligned}$$

$$= \frac{9\hbar}{4m}\frac{1}{c_s}\left(\frac{V}{N}\right)^{1/3}\frac{1}{(6\pi^2)^{1/3}}$$

$$= \left(\frac{3}{4\pi}\right)^{2/3}\frac{\hbar}{2m}\left(\frac{V}{N}\right)^{1/3}\frac{3}{c_s}. \tag{7.5.9}$$

At high temperature, in the Debye approximation

$$\begin{aligned}
\overline{\langle[u_n(t)]^2\rangle} &\approx \frac{1}{Nm}\sum_{\underset{\sim}{k}\lambda}\left[\frac{1}{\omega_{\underset{\sim}{k}\lambda}^2}\hbar\omega_{\underset{\sim}{k}\lambda}\frac{kT}{\hbar\omega_{\underset{\sim}{k}\lambda}}\right] \\
&= \frac{kT}{Nm}\sum_{\underset{\sim}{k}\lambda}\frac{1}{\omega_{\underset{\sim}{k}\lambda}^2} \\
&= \frac{kT}{Nm}\int_0^{\omega_D}\frac{g(\omega)}{\omega^2}d\omega \\
&= \frac{kT}{Nm}\frac{9N}{\omega_D^3}\omega_D = \frac{9kT}{m}\frac{1}{\omega_D^2} \\
&= \frac{9\hbar^2}{km}\frac{T}{T_D^2}.
\end{aligned} \tag{7.5.10}$$

The above formula allows us to relate the Debye temperature to the melting temperature of a solid. Melting may actually result when the vibration amplitude reaches a value comparable with the interatomic distance R_o, or when the temperature is such that

$$\frac{\overline{\langle[u_n(t)]^2\rangle}}{R_o^2} = \lambda, \tag{7.5.11}$$

where the critical value λ is found experimentally to be approximately 0.1. From (7.5.10) and (7.5.11) we find

$$\frac{9\hbar^2}{km}\frac{T_{melt}}{T_D^2} = \lambda R_o^2, \tag{7.5.12}$$

or

$$T_D = \left(\frac{9\hbar^2}{k\lambda}\frac{m}{R_o^2}T_{melt}\right)^{1/2}. \tag{7.5.13}$$

It is, on the other hand,

$$\frac{4}{3}\pi R_o^3 = v_A = \text{atomic volume}$$

$$m = \frac{\text{Atomic Weight}}{\text{Avogadro's Number}} = \frac{\text{At.W.}}{N_a}.$$

Therefore (7.5.13) becomes

$$T_D = \left[\frac{9\hbar^2(4\pi)^{2/3}}{3^{2/3}k\,\lambda}\frac{T_{melt}}{(\text{At.W.})(v_A)^{2/3}}\right]^{1/2}$$

$$= \text{const}\left[\frac{T_{melt}}{(\text{At.W.})(v_A)^{2/3}}\right]^{1/2}. \qquad (7.5.14)$$

The constant in (7.5.14) has been found to be approximately the same in metal solids. The relation (7.5.14) expresses the *Lindemann law of melting.*

References

1. A. Einstein, *Ann. Physik* 22, 186 (1907).
2. P. Debye, *Ann. Physik* 39, 789 (1912).

Chapter 8

Effect of Lattice Vibrations on X-ray Scattering and Neutron Scattering

X-ray scattering was introduced in Chapter 5. In this chapter we investigate how this technique is affected by the lattice vibrations of solids. Then we introduce the technique of neutron scattering and examine its application to the study of lattice vibrations.

8.1. Effect of Lattice Vibrations on the Intensity of the Scattered Radiation

8.1.1. *The Intensity of the Scattered Radiation*

We have already established the fact that reinforcements of the scattered wave take place in correspondence to directions $\underset{\sim}{k}'$ such that $\underset{\sim}{k} - \underset{\sim}{k}' = \underset{\sim}{K}$ respects the Laue equations (5.5.7). The scattered wave (5.5.2) contains, however, other factors such as the structure amplitude

$$S = \sum_{i=1}^{J} e^{i\underset{\sim}{K}\cdot\underset{\sim}{r}_i} f_{ai}(\theta), \tag{8.1.1}$$

which will determine the intensity of the reinforced "spots." For the first-order diffraction,

$$\underset{\sim}{K} = s\underset{\sim}{b}_1 + t\underset{\sim}{b}_2 + u\underset{\sim}{b}_3. \tag{8.1.2}$$

The position of the i-th atom in the unit cell may be expressed as

$$\underset{\sim}{r}_i = r_{i1}\underset{\sim}{a}_1 + r_{i2}\underset{\sim}{a}_2 + r_{i3}\underset{\sim}{a}_3. \tag{8.1.3}$$

Then

$$S = \sum_{i=1}^{J} e^{i2\pi(r_{i1}s + r_{i2}t + r_{i3}u)} f_{ai}. \tag{8.1.4}$$

The intensity of the spots depends on the positions of the atoms in the unit cell. In other words the position of the spots tells us something in regard to the planes of reflection, and the intensity tells us something in regard to the location of the atoms in the unit cell.

The arguments above do not tell the entire story in regard to the intensity of the spot. The lattice vibrations, which have been ignored so far, are indeed present and have an effect on the intensity of the spots.

8.1.2. *The Effect of Lattice Vibrations: Einstein Model*

We make the following assumptions:

1. The crystal has one atom per unit cell.
2. The lattice vibrations can be described by the Einstein model.

The scattered wave is given by

$$\begin{aligned} \underset{\sim}{E}(\underset{\sim}{r}, t) &= \hat{1}_\phi \frac{e^2 E_o}{mc^2} \sin\phi \frac{e^{i(\underset{\sim}{k}' \cdot \underset{\sim}{r} - \omega t)}}{r} f_a \sum_n e^{i\underset{\sim}{K} \cdot \underset{\sim}{d}_n} \\ &= \underset{\sim}{A} f_a \sum_n e^{i\underset{\sim}{K} \cdot \underset{\sim}{d}_n}, \end{aligned} \tag{8.1.5}$$

where

$$\underset{\sim}{A} = \hat{1}_\phi \frac{e^2 E_o}{mc^2} \sin\phi \frac{e^{i(\underset{\sim}{k}' \cdot \underset{\sim}{r} - \omega t)}}{r}, \tag{8.1.5$'$}$$

and

$$\underset{\sim}{d}_n = \underset{\sim}{R}_n + \underset{\sim}{u}_n,$$

is the position of the n-th atom, $\underset{\sim}{u}_n$ is the displacement of the nth atom, and $\underset{\sim}{R}_n$ is the equilibrium position of the nth atom. Then

$$\begin{aligned} |\underset{\sim}{E}|^2 &= |\underset{\sim}{A}|^2 f_a^2 \sum_n \sum_m e^{i\underset{\sim}{K} \cdot (\underset{\sim}{d}_n - \underset{\sim}{d}_m)} \\ &= |\underset{\sim}{A}|^2 f_a^2 \sum_n \sum_m e^{i\underset{\sim}{K} \cdot (\underset{\sim}{R}_n - \underset{\sim}{R}_m)} e^{i\underset{\sim}{K} \cdot (\underset{\sim}{u}_n - \underset{\sim}{u}_m)}, \end{aligned} \tag{8.1.6}$$

and, taking the thermal average,

$$\overline{|E|^2} = |\underset{\sim}{A}|^2 f_a^2 \sum_n \sum_m e^{i\underset{\sim}{K}\cdot(\underset{\sim}{R}_n - \underset{\sim}{R}_m)} \overline{e^{i\underset{\sim}{K}\cdot(\underset{\sim}{u}_n - \underset{\sim}{u}_m)}}. \tag{8.1.7}$$

Set

$$\underset{\sim}{K} \cdot (\underset{\sim}{u}_n - \underset{\sim}{u}_m) = \eta_{nm}. \tag{8.1.8}$$

Then

$$\begin{aligned} \overline{e^{i\underset{\sim}{K}\cdot(\underset{\sim}{u}_n - \underset{\sim}{u}_m)}} &= \overline{e^{i\eta_{nm}}} \\ &= \overline{1 + i\frac{\eta_{nm}}{1!} - \frac{\eta_{nm}^2}{2!} - i\frac{\eta_{nm}^3}{3!} + \frac{\eta_{nm}^4}{4!} + \cdots} \\ &\simeq 1 - \frac{\overline{\eta_{nm}^2}}{2!} \simeq e^{-\overline{\eta_{nm}^2}/2}. \end{aligned} \tag{8.1.9}$$

Also

$$\begin{aligned} \overline{\eta_{nm}^2} &= \overline{[\underset{\sim}{K} \cdot (\underset{\sim}{u}_n - \underset{\sim}{u}_m)][\underset{\sim}{K} \cdot (\underset{\sim}{u}_n - \underset{\sim}{u}_m)]} \\ &= K^2 \overline{[\underset{\sim}{u}_n - \underset{\sim}{u}_m]^2} = K^2 [\overline{u_n^2} + \overline{u_m^2} - \overline{2\underset{\sim}{u}_n \cdot \underset{\sim}{u}_m}] \\ &= 2K^2 \overline{u_n^2}, \end{aligned} \tag{8.1.10}$$

where we have disregarded any correlation between the motions of different atoms. Therefore

$$\overline{e^{i\underset{\sim}{K}\cdot(\underset{\sim}{u}_n - \underset{\sim}{u}_m)}} = e^{-\overline{\eta_{nm}^2}/2} = e^{-2M}, \tag{8.1.11}$$

where

$$\begin{aligned} 2M = \frac{\overline{\eta_{nm}^2}}{2} = K^2 \overline{u_n^2} = \frac{16\pi^2}{\lambda^2} \left(\sin^2 \frac{1}{2}\theta \right) \overline{u_n^2} \\ = \text{Debye–Waller factor.} \end{aligned} \tag{8.1.12}$$

We can write

$$\overline{|\underset{\sim}{E}|^2} = |\underset{\sim}{A}|^2 f_a^2 \left\{ N + \left[\sum_{\substack{n,m \\ n \neq m}} e^{i\underset{\sim}{K}\cdot(\underset{\sim}{R}_n - \underset{\sim}{R}_m)} \right] e^{-2M} \right\}, \tag{8.1.13}$$

where N is the number of unit cells in the crystal. From (5.5.3) and (5.5.5)

$$\sum_n \sum_m e^{i\underset{\sim}{K}\cdot(\underset{\sim}{R}_n - \underset{\sim}{R}_m)} = I_o$$

$$= N + \sum_{\substack{n,m \\ n\neq m}} e^{i\underset{\sim}{K}\cdot(\underset{\sim}{R}_n - \underset{\sim}{R}_m)}, \qquad (8.1.14)$$

so that

$$\sum_{\substack{n,m \\ n\neq m}} e^{i\underset{\sim}{K}\cdot(\underset{\sim}{R}_n - \underset{\sim}{R}_m)} = I_o - N, \qquad (8.1.15)$$

and

$$\begin{aligned} \overline{|\underset{\sim}{E}|^2} &= |\underset{\sim}{A}|^2 f_a^2 \{N + [I_o - N]e^{-2M}\} \\ &= |\underset{\sim}{A}|^2 f_a^2 \{N(1 - e^{-2M}) + I_o e^{-2M}\}. \end{aligned} \qquad (8.1.16)$$

In absence of lattice vibrations, $\overline{u_n^2} = 0$ and $M = 0$; in this case (5.8.16) reduces to

$$\overline{|\underset{\sim}{E}|^2} = |\underset{\sim}{A}|^2 f_a^2 I_o, \qquad (8.1.17)$$

as expected.

We still have to calculate $\overline{u_n^2}$. The average energy of a harmonic oscillator is given by

$$\bar{E} = 3\hbar\omega\left(\frac{1}{e^{\hbar\omega/kT} - 1} + \frac{1}{2}\right), \qquad (8.1.18)$$

having assumed the oscillator to be isotropic. On the other hand, because of the virial theorem

$$\bar{E} = 3\left[\frac{1}{2m}\overline{p^2} + \frac{1}{2}m\omega^2\overline{u^2}\right] = 3[m\omega^2\overline{u^2}],$$

or

$$\begin{aligned} \overline{u^2} &= \frac{\bar{E}}{3m\omega^2} = \frac{3\hbar\omega}{3m\omega^2}\left(\frac{1}{e^{\hbar\omega/kT} - 1} + \frac{1}{2}\right) \\ &= \frac{\hbar\omega}{m\omega^2}\left(\frac{1}{e^{\hbar\omega/kT} - 1} + \frac{1}{2}\right). \end{aligned} \qquad (8.1.19)$$

We can then write in conclusion:

$$\overline{|\underset{\sim}{E}|^2} = |\underset{\sim}{A}|^2 f_a^2 [N(1 - e^{-2M}) + I_o(\underset{\sim}{K}) e^{-2M}], \tag{8.1.20}$$

where

$$2M = \frac{16\pi^2}{\lambda^2}\left(\sin\frac{1}{2}\theta\right)\frac{\hbar\omega}{m\omega^2}\left(\frac{1}{e^{\hbar\omega/kT} - 1} + \frac{1}{2}\right), \tag{8.1.21}$$

and θ is the angle between $\underset{\sim}{k}$ and $\underset{\sim}{k}'$ $(0 < \theta < \pi)$, λ is the wavelength of x-radiation, ω is the frequency of vibration of atoms, and 2M is the Debye-Waller factor.

In the expression for $\overline{|\underset{\sim}{E}|^2}$,

1. M is due to the presence of lattice vibrations. It is approximately 0 in the absence of lattice vibrations.
2. $M = 0$ if $\theta = 0$ or 2π; going from 0 to π, M increases. The function $N(1 - e^{-2M})$ varies slowly with the angle θ.
3. The real angle dependence is still with I_o. The directions of maximum scattering intensity and the width of the spots are the same.
4. The factor e^{-2M} makes the term $I_o e^{-2M}$ maximum at $\theta = 0$, 2π and minimum at $\theta = \pi$.

In tabular form,

θ	M	e^{-2M}	$N(1 - e^{-2M})$	$I_o e^{-2M}$
0	0	1	0	I_o
π	max	min	max	$I_o e^{-2M} < I_o$
2π	0	1	0	I_o

8.1.3. *The Effect of Lattice Vibrations: Normal Mode Treatment*

We have already found for a crystal with one atom per unit cell

$$\overline{|\underset{\sim}{E}|^2} = |\underset{\sim}{A}|^2 f_a^2 \sum_n \sum_m e^{i\underset{\sim}{K}\cdot(\underset{\sim}{R}_n - \underset{\sim}{R}_m)} \overline{e^{i\underset{\sim}{K}\cdot(\underset{\sim}{u}_n - \underset{\sim}{u}_m)}}. \tag{8.1.22}$$

We can write

$$e^{i\underset{\sim}{K}\cdot(\underset{\sim}{u}_n - \underset{\sim}{u}_m)} \simeq 1 - \frac{1}{2}[\underset{\sim}{K}\cdot(\underset{\sim}{u}_n - \underset{\sim}{u}_m)]^2, \tag{8.1.23}$$

disregarding terms with power higher than two. It is (remember, $\underset{\sim}{u}_n, \underset{\sim}{u}_m$ are real operators)

$$\begin{aligned}\langle[\underset{\sim}{K}\cdot(\underset{\sim}{u}_n-\underset{\sim}{u}_m)]^2\rangle &= \langle K^2[\hat{1}_{\underset{\sim}{K}}\cdot(\underset{\sim}{u}_n-\underset{\sim}{u}_m)][\hat{1}_{\underset{\sim}{K}}\cdot(\underset{\sim}{u}_n^+-\underset{\sim}{u}_m^+)]\rangle \\ &= \langle K^2(\hat{1}_{\underset{\sim}{K}}\cdot\underset{\sim}{u}_n-\hat{1}_{\underset{\sim}{K}}\cdot\underset{\sim}{u}_m)(\hat{1}_{\underset{\sim}{K}}\cdot\underset{\sim}{u}_n^+-\hat{1}_{\underset{\sim}{K}}\cdot\underset{\sim}{u}_m^+)\rangle \\ &= \langle K^2[(\hat{1}_{\underset{\sim}{K}}\cdot\underset{\sim}{u}_n)(\hat{1}_{\underset{\sim}{K}}\cdot\underset{\sim}{u}_n^+)-(\hat{1}_{\underset{\sim}{K}}\cdot\underset{\sim}{u}_n)(\hat{1}_{\underset{\sim}{K}}\cdot\underset{\sim}{u}_m^+) \\ &\quad -(\hat{1}_{\underset{\sim}{K}}\cdot\underset{\sim}{u}_m)(\hat{1}_{\underset{\sim}{K}}\cdot\underset{\sim}{u}_n^+)+(\hat{1}_{\underset{\sim}{K}}\cdot\underset{\sim}{u}_m)(\hat{1}_{\underset{\sim}{K}}\cdot\underset{\sim}{u}_m^+)]\rangle,\end{aligned} \tag{8.1.24}$$

where angular brackets indicate a quantum-mechanical average. From (6.5.77)

$$\underset{\sim}{u}_n = \sum_{\underset{\sim}{\chi}\lambda}\left(\frac{\hbar}{2M\omega_{\underset{\sim}{\chi}\lambda}}\right)^{\frac{1}{2}}[\underset{\sim}{V}(\underset{\sim}{\chi},\lambda)e^{i\underset{\sim}{\chi}\cdot\underset{\sim}{R}_n}a_{\underset{\sim}{\chi}}^{\lambda}+\underset{\sim}{V}(\underset{\sim}{\chi},\lambda)^*e^{-i\underset{\sim}{\chi}\cdot\underset{\sim}{R}_n}a_{\underset{\sim}{\chi}}^{\lambda+}], \tag{8.1.25}$$

where $\underset{\sim}{\chi}$ is the generic $\underset{\sim}{k}$ vector of the phonon and M is the total mass of the crystal. We also know that [see (6.3.54)]

$$\begin{cases} a_{\underset{\sim}{\chi}}^{\lambda}|n_{\underset{\sim}{\chi}}^{\lambda}\rangle = \sqrt{n_{\underset{\sim}{\chi}}^{\lambda}}|n_{\underset{\sim}{\chi}}^{\lambda}-1\rangle \\ a_{\underset{\sim}{\chi}}^{\lambda}|n_{\underset{\sim}{\chi}}^{\lambda}\rangle = \sqrt{n_{\underset{\sim}{\chi}}^{\lambda}+1}|n_{\underset{\sim}{\chi}}^{\lambda}+1\rangle. \end{cases} \tag{8.1.26}$$

Also

$$\begin{cases} \langle|a_{\underset{\sim}{\chi}}^{\lambda}a_{\underset{\sim}{\chi}'}^{\lambda'}|\rangle = 0 \\ \langle|a_{\underset{\sim}{\chi}}^{\lambda+}a_{\underset{\sim}{\chi}'}^{\lambda'+}|\rangle = 0 \\ \langle|a_{\underset{\sim}{\chi}}^{\lambda}a_{\underset{\sim}{\chi}'}^{\lambda'+}|\rangle = \langle|a_{\underset{\sim}{\chi}}^{\lambda}|\rangle\langle|a_{\underset{\sim}{\chi}'}^{\lambda'+}|\rangle = n_{\underset{\sim}{\chi}}^{\lambda}+1, \quad \text{if } \underset{\sim}{\chi}=\underset{\sim}{\chi}',\ \lambda=\lambda' \\ \langle|a_{\underset{\sim}{\chi}}^{\lambda+}a_{\underset{\sim}{\chi}'}^{\lambda'}|\rangle = \langle|a_{\underset{\sim}{\chi}}^{\lambda+}|\rangle\langle|a_{\underset{\sim}{\chi}'}^{\lambda'}|\rangle = n_{\underset{\sim}{\chi}}^{\lambda}, \quad \text{if } \underset{\sim}{\chi}=\underset{\sim}{\chi}',\ \lambda=\lambda'. \end{cases} \tag{8.1.27}$$

Now

$$\langle|\,|\hat{1}_{\underset{\sim}{K}}\cdot\underset{\sim}{u}_n|^2|\rangle$$

$$= \left\langle\left|\sum_{\underset{\sim}{\chi}\lambda}\sum_{\underset{\sim}{\chi}'\lambda'}\left(\frac{\hbar}{2M\omega_{\underset{\sim}{\chi}\lambda}}\right)^{\frac{1}{2}}\left(\frac{\hbar}{2M\omega_{\underset{\sim}{\chi}'\lambda'}}\right)^{\frac{1}{2}}\right.\right.$$

$$\times[\hat{1}_{\underset{\sim}{K}} \cdot \underset{\sim}{V}(\underset{\sim}{\chi}, \lambda)e^{i\underset{\sim}{\chi}\cdot\underset{\sim}{R}_n}a^{\lambda}_{\underset{\sim}{\chi}} + \hat{1}_{\underset{\sim}{K}} \cdot \underset{\sim}{V}(\underset{\sim}{\chi}, \lambda)^* e^{-i\underset{\sim}{\chi}\cdot\underset{\sim}{R}_n}a^{\lambda+}_{\underset{\sim}{\chi}}]$$

$$\times[\hat{1}_{\underset{\sim}{K}} \cdot \underset{\sim}{V}(\underset{\sim}{\chi}', \lambda')^* e^{-i\underset{\sim}{\chi}'\cdot\underset{\sim}{R}_n}a^{\lambda'+}_{\underset{\sim}{\chi}'} + \hat{1}_{\underset{\sim}{K}} \cdot \underset{\sim}{V}(\underset{\sim}{\chi}', \lambda')e^{-i\underset{\sim}{\chi}'\cdot\underset{\sim}{R}_n}a^{\lambda'}_{\underset{\sim}{\chi}'}]\Bigg|\Bigg\rangle$$

$$= \sum_{\underset{\sim}{\chi}\lambda} \langle | \, |\hat{1}_{\underset{\sim}{K}} \cdot \underset{\sim}{V}(\underset{\sim}{\chi}, \lambda)|^2 (a^{\lambda}_{\underset{\sim}{\chi}} a^{\lambda+}_{\underset{\sim}{\chi}} + a^{\lambda+}_{\underset{\sim}{\chi}} a^{\lambda}_{\underset{\sim}{\chi}})|\rangle \frac{\hbar}{2M\omega_{\underset{\sim}{\chi}\lambda}}$$

$$= \sum_{\underset{\sim}{\chi}\lambda} |\hat{1}_{\underset{\sim}{K}} \cdot \underset{\sim}{V}(\underset{\sim}{\chi}, \lambda)|^2 \langle | a^{\lambda}_{\underset{\sim}{\chi}} a^{\lambda+}_{\underset{\sim}{\chi}} + a^{\lambda+}_{\underset{\sim}{\chi}} a^{\lambda}_{\underset{\sim}{\chi}} |\rangle \frac{\hbar}{2M\omega_{\underset{\sim}{\chi}\lambda}}$$

$$= \sum_{\underset{\sim}{\chi}\lambda} |\hat{1}_{\underset{\sim}{K}} \cdot \underset{\sim}{V}(\underset{\sim}{\chi}, \lambda)|^2 \frac{\hbar}{2M\omega_{\underset{\sim}{\chi}\lambda}} (n^{\lambda}_{\underset{\sim}{\chi}} + 1 + n^{\lambda}_{\underset{\sim}{\chi}})$$

$$= \sum_{\underset{\sim}{\chi}\lambda} |\hat{1}_{\underset{\sim}{K}} \cdot \underset{\sim}{V}(\underset{\sim}{\chi}, \lambda)|^2 \frac{\hbar}{M\omega_{\underset{\sim}{\chi}\lambda}} \left(n^{\lambda}_{\underset{\sim}{\chi}} + \frac{1}{2}\right). \tag{8.1.28}$$

Also

$$\langle |(\hat{1}_{\underset{\sim}{K}} \cdot \underset{\sim}{u}_n)(\hat{1}_{\underset{\sim}{K}} \cdot \underset{\sim}{u}^+_m)|\rangle$$

$$= \Bigg\langle \Bigg| \sum_{\underset{\sim}{\chi}\lambda} \sum_{\underset{\sim}{\chi}'\lambda'} \left(\frac{\hbar}{2M\omega_{\underset{\sim}{\chi}\lambda}}\right)^{\frac{1}{2}} \left(\frac{\hbar}{2M\omega_{\underset{\sim}{\chi}'\lambda'}}\right)^{\frac{1}{2}}$$

$$\times[\hat{1}_{\underset{\sim}{K}} \cdot \underset{\sim}{V}(\underset{\sim}{\chi}, \lambda)e^{i\underset{\sim}{\chi}\cdot\underset{\sim}{R}_n}a^{\lambda}_{\underset{\sim}{\chi}} + \hat{1}_{\underset{\sim}{K}} \cdot \underset{\sim}{V}(\underset{\sim}{\chi}, \lambda)^* e^{-i\underset{\sim}{\chi}\cdot\underset{\sim}{R}_n}a^{\lambda+}_{\underset{\sim}{\chi}}]$$

$$\times[\hat{1}_{\underset{\sim}{K}} \cdot \underset{\sim}{V}(\underset{\sim}{\chi}', \lambda')^* e^{-i\underset{\sim}{\chi}'\cdot\underset{\sim}{R}_m}a^{\lambda'+}_{\underset{\sim}{\chi}'} + \hat{1}_{\underset{\sim}{K}} \cdot \underset{\sim}{V}(\underset{\sim}{\chi}', \lambda')e^{-i\underset{\sim}{\chi}'\cdot\underset{\sim}{R}_m}a^{\lambda'}_{\underset{\sim}{\chi}'}]\Bigg|\Bigg\rangle$$

$$= \sum_{\underset{\sim}{\chi}\lambda} |\hat{1}_{\underset{\sim}{K}} \cdot \underset{\sim}{V}(\underset{\sim}{\chi}, \lambda)|^2 \langle | a^{\lambda}_{\underset{\sim}{\chi}} a^{\lambda+}_{\underset{\sim}{\chi}} e^{i\underset{\sim}{\chi}\cdot(\underset{\sim}{R}_n - \underset{\sim}{R}_m)} + a^{\lambda+}_{\underset{\sim}{\chi}} a^{\lambda}_{\underset{\sim}{\chi}} e^{-i\underset{\sim}{\chi}\cdot(\underset{\sim}{R}_n - \underset{\sim}{R}_m)} |\rangle$$

$$\times \frac{\hbar}{2M\omega_{\underset{\sim}{\chi}\lambda}}. \tag{8.1.29}$$

Changing $\underset{\sim}{\chi}$ into $-\underset{\sim}{\chi}$ in the second term we obtain

$$\langle |(\hat{1}_{\underset{\sim}{K}} \cdot \underset{\sim}{u}_n)(\hat{1}_{\underset{\sim}{K}} \cdot \underset{\sim}{u}^+_m)|\rangle$$

$$= \sum_{\underset{\sim}{\chi}\lambda} \frac{\hbar}{2M\omega_{\underset{\sim}{\chi}\lambda}} |\hat{1}_{\underset{\sim}{K}} \cdot \underset{\sim}{V}(\underset{\sim}{\chi}, \lambda)|^2 e^{i\underset{\sim}{\chi}\cdot(\underset{\sim}{R}_n - \underset{\sim}{R}_m)} \langle |a^{\lambda}_{\underset{\sim}{\chi}} a^{\lambda+}_{\underset{\sim}{\chi}} + a^{\lambda+}_{-\underset{\sim}{\chi}} a^{\lambda}_{-\underset{\sim}{\chi}}|\rangle$$

$$= \sum_{\underset{\sim}{\chi}\lambda} \frac{\hbar}{2M\omega_{\underset{\sim}{\chi}\lambda}} |\hat{1}_{\underset{\sim}{K}} \cdot \underset{\sim}{V}(\underset{\sim}{\chi}, \lambda)|^2 e^{i\underset{\sim}{\chi}\cdot(\underset{\sim}{R}_n - \underset{\sim}{R}_m)} (n^{\lambda}_{\underset{\sim}{\chi}} + 1 + n^{\lambda}_{-\underset{\sim}{\chi}}). \tag{8.1.30}$$

Therefore, taking the thermal average

$$\overline{\langle [\underset{\sim}{K} \cdot (\underset{\sim}{u}_n - \underset{\sim}{u}_m)]^2 \rangle}$$

$$= 2K^2 \overline{\langle |\hat{1}_{\underset{\sim}{K}} \cdot \underset{\sim}{u}_n|^2 - (\hat{1}_{\underset{\sim}{K}} \cdot \underset{\sim}{u}_n)(\hat{1}_{\underset{\sim}{K}} \cdot \underset{\sim}{u}^{+}_m) \rangle}$$

$$= 2K^2 \sum_{\underset{\sim}{\chi}\lambda} \frac{\hbar}{M\omega_{\underset{\sim}{\chi}\lambda}} |\hat{1}_{\underset{\sim}{K}} \cdot \underset{\sim}{V}(\underset{\sim}{\chi}, \lambda)|^2 \overline{\left(n^{\lambda}_{\underset{\sim}{\chi}} + \frac{1}{2} \right)} [1 - e^{i\underset{\sim}{\chi}\cdot(\underset{\sim}{R}_n - \underset{\sim}{R}_m)}], \tag{8.1.31}$$

and

$$\overline{e^{i\underset{\sim}{K}\cdot(\underset{\sim}{u}_n - \underset{\sim}{u}_m)}} \simeq 1 - \frac{\overline{[\underset{\sim}{K} \cdot (\underset{\sim}{u}_n - \underset{\sim}{u}_m)]^2}}{2}$$

$$= 1 - K^2 \sum_{\underset{\sim}{\chi}\lambda} \frac{\hbar}{M\omega_{\underset{\sim}{\chi}\lambda}} |\hat{1}_K \cdot \underset{\sim}{V}(\underset{\sim}{\chi}, \lambda)|^2$$

$$\times [1 - e^{i\underset{\sim}{\chi}\cdot(\underset{\sim}{R}_n - \underset{\sim}{R}_m)}] \overline{\left(n^{\lambda}_{\underset{\sim}{\chi}} + \frac{1}{2} \right)}$$

$$= 1 - K^2 \sum_{\underset{\sim}{\chi}\lambda} \frac{|\hat{1}_{\underset{\sim}{K}} \cdot \underset{\sim}{V}(\underset{\sim}{\chi}, \lambda)|^2}{M\omega^2_{\underset{\sim}{\chi}\lambda}} \hbar\omega_{\underset{\sim}{\chi}\lambda} \left(\frac{1}{2} + \frac{1}{e^{\hbar\omega_{\chi\lambda}/kT} - 1} \right)$$

$$\times [1 - e^{i\underset{\sim}{\chi}\cdot(\underset{\sim}{R}_n - \underset{\sim}{R}_m)}]$$

$$= 1 - \sum_{\underset{\sim}{\chi}\lambda} G^{\lambda}_{\underset{\sim}{\chi}} [1 - e^{i\underset{\sim}{\chi}\cdot(\underset{\sim}{R}_n - \underset{\sim}{R}_m)}]$$

$$\simeq e^{-\sum_{\underset{\sim}{\chi}\lambda} G^{\lambda}_{\underset{\sim}{\chi}}} + \sum_{\underset{\sim}{\chi}\lambda} e^{i\underset{\sim}{\chi}\cdot(\underset{\sim}{R}_n - \underset{\sim}{R}_m)} G^{\lambda}_{\underset{\sim}{\chi}}$$

$$= e^{-2M} + \sum_{\underset{\sim}{\chi}\lambda} e^{i\underset{\sim}{\chi}\cdot(\underset{\sim}{R}_n - \underset{\sim}{R}_m)} G^{\lambda}_{\underset{\sim}{\chi}}, \tag{8.1.32}$$

where

$$2\mathrm{M} = \sum_{\underset{\sim}{\chi}\lambda} \mathrm{G}^{\lambda}_{\underset{\sim}{\chi}} = \text{Debye–Waller factor} \tag{8.1.33}$$

$$\begin{aligned}\mathrm{G}^{\lambda}_{\underset{\sim}{\chi}} &= \mathrm{K}^2 \frac{|\hat{1}_{\underset{\sim}{\mathrm{K}}} \cdot \underset{\sim}{\mathrm{V}}(\underset{\sim}{\chi}, \lambda)|^2}{\mathrm{M}\omega^2_{\underset{\sim}{\chi}\lambda}} \hbar\omega_{\underset{\sim}{\chi}\lambda} \left(\frac{1}{2} + \frac{1}{e^{\hbar\omega_{\underset{\sim}{\chi}\lambda}/\mathrm{kT}} - 1}\right) \\ &= \mathrm{K}^2 \frac{|\hat{1}_{\underset{\sim}{\mathrm{K}}} \cdot \underset{\sim}{\mathrm{V}}(\underset{\sim}{\chi}, \lambda)|^2}{2\mathrm{M}\omega^2_{\underset{\sim}{\chi}\lambda}} \hbar\omega_{\underset{\sim}{\chi}\lambda} \coth \frac{\hbar\omega_{\underset{\sim}{\chi}\lambda}}{2\mathrm{kT}}.\end{aligned} \tag{8.1.34}$$

Then

$$\begin{aligned}\overline{|\underset{\sim}{\mathrm{E}}|^2} &= |\underset{\sim}{\mathrm{A}}|^2 \mathrm{f}_a^2 \sum_{nm} e^{i\underset{\sim}{\mathrm{K}}\cdot(\underset{\sim}{\mathrm{R}}_n - \underset{\sim}{\mathrm{R}}_m)} \overline{e^{i\underset{\sim}{\mathrm{K}}\cdot(\underset{\sim}{u}_n - \underset{\sim}{u}_m)}} \\ &= |\underset{\sim}{\mathrm{A}}|^2 \mathrm{f}_a^2 \sum_{nm} e^{i\underset{\sim}{\mathrm{K}}\cdot(\underset{\sim}{\mathrm{R}}_n - \underset{\sim}{\mathrm{R}}_m)} \left[e^{-2\mathrm{M}} + \sum_{\underset{\sim}{\chi}\lambda} e^{i\underset{\sim}{\chi}\cdot(\underset{\sim}{\mathrm{R}}_n - \underset{\sim}{\mathrm{R}}_m)} \mathrm{G}^{\lambda}_{\underset{\sim}{\chi}}\right] \\ &= |\underset{\sim}{\mathrm{A}}|^2 \mathrm{f}_a^2 \left[\sum_{nm} e^{i\underset{\sim}{\mathrm{K}}\cdot(\underset{\sim}{\mathrm{R}}_n - \underset{\sim}{\mathrm{R}}_m)} e^{-2\mathrm{M}} + \sum_{nm} \sum_{\underset{\sim}{\chi}\lambda} e^{i(\underset{\sim}{\mathrm{K}}+\underset{\sim}{\chi})\cdot(\underset{\sim}{\mathrm{R}}_n - \underset{\sim}{\mathrm{R}}_m)} \mathrm{G}^{\lambda}_{\underset{\sim}{\chi}}\right],\end{aligned} \tag{8.1.35}$$

or

$$\overline{|\underset{\sim}{\mathrm{E}}|^2} = |\underset{\sim}{\mathrm{A}}|^2 \mathrm{f}_a^2 \mathrm{I}_o(\underset{\sim}{\mathrm{K}}) e^{-2\mathrm{M}} + \mathrm{I}_2(\underset{\sim}{\mathrm{K}}), \tag{8.1.36}$$

where

$$\mathrm{I}_o(\underset{\sim}{\mathrm{K}}) = \sum_{nm} e^{i\underset{\sim}{\mathrm{K}}\cdot(\underset{\sim}{\mathrm{R}}_n - \underset{\sim}{\mathrm{R}}_m)} \tag{8.1.37}$$

$$\mathrm{I}_2(\underset{\sim}{\mathrm{K}}) = \sum_{\underset{\sim}{\chi}\lambda} \mathrm{G}^{\lambda}_{\underset{\sim}{\chi}} \mathrm{I}_o(\underset{\sim}{\mathrm{K}} + \underset{\sim}{\chi}) \tag{8.1.38}$$

$$= \frac{\hbar}{\mathrm{M}} \sum_{\underset{\sim}{\chi}\lambda} \frac{|\underset{\sim}{\mathrm{K}} \cdot \underset{\sim}{\mathrm{V}}(\underset{\sim}{\chi}, \lambda)|^2}{2\omega_{\underset{\sim}{\chi}\lambda}} \coth \frac{\hbar\omega_{\underset{\sim}{\chi}\lambda}}{2\mathrm{kT}} \sum_{n,m} e^{i(\underset{\sim}{\mathrm{K}}+\underset{\sim}{\chi})\cdot(\underset{\sim}{\mathrm{R}}_n - \underset{\sim}{\mathrm{R}}_m)} \tag{8.1.39}$$

$$
\begin{aligned}
2M &= \sum_{\underset{\sim}{\chi}\lambda} G^{\lambda}_{\underset{\sim}{\chi}} \\
&= \frac{\hbar}{M} \sum_{\underset{\sim}{\chi}\lambda} \frac{|\underset{\sim}{K} \cdot \underset{\sim}{V}(\underset{\sim}{\chi}, \lambda)|^2}{2\omega_{\underset{\sim}{\chi}\lambda}} \coth \frac{\hbar\omega_{\underset{\sim}{\chi}\lambda}}{2kT}.
\end{aligned} \tag{8.1.40}
$$

We can now make the following observations:

1. The predominant term in the spot intensity is the term $I_o(\underset{\sim}{K})e^{-2M}$. There will be a very bright spot in correspondence to $\underset{\sim}{K}$ being a vector of the reciprocal lattice. The intensity will be proportional to $\sim$N.
2. The intensity of the spot falls off exponentially as e^{-2M}; as θ increases from 0 to π, M increases and the intensity decreases.
3. In correspondence to a certain scattering direction the decrease in intensity is given by

$$
\begin{aligned}
I_o(1 - e^{-2M}) &\simeq I_o 2M \\
&= \sum_{nm} e^{i\underset{\sim}{K}\cdot(\underset{\sim}{R}_n - \underset{\sim}{R}_m)} \frac{\hbar}{\omega} \sum_{\underset{\sim}{\chi}\lambda} \frac{|\underset{\sim}{K} \cdot \underset{\sim}{V}(\underset{\sim}{\chi}, \lambda)|^2}{2\omega_{\underset{\sim}{\chi}\lambda}} \coth \frac{\hbar\omega_{\underset{\sim}{\chi}\lambda}}{2kT}.
\end{aligned} \tag{8.1.41}
$$

4. $I_2(\underset{\sim}{K})$ also contributes to the intensity. It is different from zero if $\underset{\sim}{K}+\underset{\sim}{\chi}$ is a vector of the reciprocal lattice; since $\underset{\sim}{\chi}$ can assume N discrete but very close values (allowed by the boundary conditions), $\underset{\sim}{K}$ will be allowed to range over practically a continuum. This fact produces a continuous background for the scattering spot.
5. The background produced by $I_2(\underset{\sim}{K})$ is not uniform. Assume that we have a certain primitive vector of the reciprocal lattice $\underset{\sim}{K}_\eta$ such that

$$
\underset{\sim}{K} + \underset{\sim}{\chi} = \underset{\sim}{K}_\eta. \tag{8.1.42}
$$

We may regard $\underset{\sim}{\chi}$ as a function of $\underset{\sim}{K}$ through the above relation. The background intensity is then proportional to

$$
\sum_{\lambda} |\underset{\sim}{K} \cdot \underset{\sim}{V}(\underset{\sim}{\chi}, \lambda)|^2 \frac{1}{\omega_{\underset{\sim}{\chi}\lambda}} \coth \frac{\hbar\omega_{\underset{\sim}{\chi}\lambda}}{2kT}. \tag{8.1.43}
$$

We notice the following:

a. The first factor varies very slowly with $\underset{\sim}{\chi}$. The last two factors are together proportional to ω^{-2} at *high temperatures*; the intensity is greatest at $\underset{\sim}{\chi} = 0$ and, since for small $\chi, \omega \propto \chi$, it will vary as $\underset{\sim}{\chi}^{-2}$. The background intensity is then inversely proportional to the square of the deviation in reciprocal space from the nearest reciprocal lattice point; the proportionality factor depends on the direction of this deviation.

b. At high temperatures the background intensity is proportional to T.

c. At $T = 0$, $n^{\lambda}_{\underset{\sim}{\chi}} = 0$ and

$$\coth \frac{\hbar\omega_{\underset{\sim}{\chi}\lambda}}{2kT} \simeq 1, \tag{8.1.44}$$

and only emission of phonons contributes to the background. The intensity in this case varies as χ^{-1}.

6. Finally, comparing (5.7.41) with (5.7.39) we see that the loss of intensity from the regular X-ray scattering is equal to the total intensity of the background.

8.2. Theory of Neutron Scattering

Since high-flux reactors have become available, the interaction of neutrons with solids has been an important area of research in physics. Neutron scattering provides a powerful tool for determining crystal structures, studying lattice dynamics, investigating the magnetic properties of materials, and elucidating energy loss mechanisms for neutrons passing through solids. In this chapter we outline the basic concepts of neutron scattering by crystals and their application to the study of lattice dynamics. For discussions on other applications of neutron scattering the reader is referred to the review articles by Ringo,[1] Shull and Wollan,[2] and Bacon and Lonsdale[3] on crystal structure, by Jacrot and Riste[4] on magnetic inelastic scattering, and by Kothari and Singwi[5] on the slowing down of neutrons in solids.

The following important facts characterize thermal neutrons vis-a-vis x rays:

1. Typical *x-ray* wavelengths are about 1 Å; in x-ray scattering momentum transfer values are comparable with the momenta of phonons. *Thermal neutrons* have De Broglie wavelengths also of the order of 1 Å, that is, of the same order of magnitude of the interatomic distance in crystals. From this point of view x-rays and neutrons are similar.
2. Typical x rays have energies of the order 10 keV, which is much greater than the order of magnitude of phonon energies, 10 meV. Thermal neutrons obtained from a reactor have energies of about 40 meV, that is, of the same order of magnitude of phonon energies. Thermal neutrons have both momenta and energies comparable with those of phonons; therefore in a single scattering event we may expect significant changes in both wave vector and energies.
3. The sources of neutrons are considerably less intense than the sources of x rays or, for that matter, the optical sources. As a consequence, the resolutions achievable with neutrons are not as good as those achievable with x rays. Moreover, the equipment used for neutron scattering is very complex and bulky and has to be located near a nuclear reactor.
4. X rays are scattered by the electrons of the atoms. Neutrons are scattered by the nuclei. An atom, such as hydrogen, which is a poor x-ray scatterer may be an efficient neutron scatterer.
5. For both x rays and neutrons the conservation of momentum and energy give

$$\underset{\sim}{Q} = \underset{\sim}{k}_i - \underset{\sim}{k}_s = \underset{\sim}{K}_\eta \pm \underset{\sim}{\chi}, \tag{8.2.1}$$

$$\Delta E = \hbar\omega_j(\underset{\sim}{Q}), \tag{8.2.2}$$

where

$$\Delta E = \hbar c(k_i - k_s) \quad \text{for x rays,}$$

$$\Delta E = \frac{\hbar^2}{2m}(k_i^2 - k_s^2) \quad \text{for neutrons,}$$

and $\underset{\sim}{K}_\eta$ is the primitive vector of the reciprocal lattice, $\underset{\sim}{\chi}$ is the wave vector of the created or annihilated phonon, $\underset{\sim}{k}_i$ is the incident wave vector, and $\underset{\sim}{k}_s$ is the scattered wave vector.

In a typical neutron scattering experiment a monochromatic beam of neutrons is scattered by a crystal; as a result we obtain:

1. Neutrons scattered in various directions without change of energy, said to be *elastically* scattered.
2. Neutrons scattered in various directions with a change in energy, said to be *inelastically* scattered.

It will be necessary to treat the problem of neutron scattering quantum mechanically.

The following assumptions will be made:

1. The neutron absorption by the scattering material is negligible,
2. Each neutron undergoes a single scattering event,
3. The target nuclei are spinless.

We consider the system that consists of the neutron and the target crystal. The Hamiltonian of such a system, neglecting the interaction between the neutrons and the crystal, is given by

$$\mathrm{H} = \mathrm{H_n} + \mathrm{H_v}, \tag{8.2.3}$$

where

$$\mathrm{H_n} = \frac{\mathrm{p}^2}{2\mathrm{m}} = \frac{\hbar^2\mathrm{k}^2}{2\mathrm{m}}, \tag{8.2.4}$$

$$\mathrm{H_v} = \sum_{\underset{\sim}{\chi}\lambda}^{3\mathrm{NJ}} \hbar\omega_{\underset{\sim}{\chi}\lambda}\left(\mathrm{a}_{\underset{\sim}{\chi}}^{\lambda+}\mathrm{a}_{\underset{\sim}{\chi}}^{\lambda} + \frac{1}{2}\right). \tag{8.2.5}$$

The eigenvalue equation

$$(\mathrm{H_n} + \mathrm{H_v})\psi = \mathrm{E}\psi, \tag{8.2.6}$$

has the eigenvalues

$$\mathrm{E} = \frac{\hbar^2\mathrm{k}^2}{2\mathrm{m}} + \sum_{\underset{\sim}{\chi}\lambda}^{3\mathrm{NJ}}\left(\mathrm{n}_{\underset{\sim}{\chi}}^{\lambda} + \frac{1}{2}\right)\hbar\omega_{\underset{\sim}{\chi}\lambda}, \tag{8.2.7}$$

and the eigenfunctions

$$\psi = \psi_\mathrm{n}\psi_\mathrm{v}, \tag{8.2.8}$$

where

$$\psi_n = \frac{1}{L^{3/2}} e^{i\underset{\sim}{k}\cdot\underset{\sim}{r}} = \frac{1}{L^{3/2}} |\underset{\sim}{k}\rangle, \tag{8.2.9}$$

$$\psi_v = \prod_{\underset{\sim}{\chi}\lambda} |n_{\underset{\sim}{\chi}}^{\lambda}\rangle. \tag{8.2.10}$$

We shall use for ψ the following short notation:

$$\psi = \frac{1}{L^{3/2}} |\underset{\sim}{k}; \{n_{\underset{\sim}{\chi}}^{\lambda}\}\rangle. \tag{8.2.11}$$

The initial and final states of the system are given by

$$\psi_i = \frac{1}{L^{3/2}} |\underset{\sim}{k}_i; \{n_{\underset{\sim}{\chi}}^{\lambda}\}\rangle, \tag{8.2.12}$$

$$\psi_f = \frac{1}{L^{3/2}} |\underset{\sim}{k}_s; \{n_{\underset{\sim}{\chi}'}^{\lambda'}\}\rangle, \tag{8.2.13}$$

respectively.

The interaction Hamiltonian is simply given by the potential $V(\underset{\sim}{r})$ between the neutron and the target crystal. The transition probability per unit time is given by

$$\begin{aligned} W &= \frac{2\pi}{\hbar} \left| \int \psi_f^* V(\underset{\sim}{r}) \psi_i d\tau \right|^2 \rho(E) \\ &= \frac{2\pi}{\hbar} \frac{1}{L^3} |\langle \underset{\sim}{k}_s; \{n_{\underset{\sim}{\chi}'}^{\lambda'}\} | V(\underset{\sim}{r}) | \underset{\sim}{k}_i; \{n_{\underset{\sim}{\chi}}^{\lambda}\}\rangle|^2 \rho(E). \end{aligned} \tag{8.2.14}$$

The first task in evaluating W is to find the density of final states. It is

$$\begin{aligned} \rho(E_f = E_i) &= \rho(E_f)\delta(E_f - E_i)dE_f \\ &= \rho(E_{fn})\delta[(E_{fn} + E_{fv}) - (E_{in} + E_{iv})]dE_{fn}, \end{aligned} \tag{8.2.15}$$

$$\rho(\underset{\sim}{k}_s)d^2\underset{\sim}{k}_s = \frac{L^3}{8\pi^3} k_s^2 dk_s d\Omega, \tag{8.2.16}$$

$$\begin{aligned} \rho(E_{fn})dE_{fn} &= \frac{L^3}{8\pi^3} k_s^2 \frac{dk_s}{dE_{fn}} dE_{fn} d\Omega \\ &= \frac{L^3}{8\pi^3} k_s^2 \frac{m}{\hbar^2 k_s} dE_{fn} d\Omega = \frac{L^3}{8\pi^3} \frac{mk_s}{\hbar^2} dE_{fn} d\Omega. \end{aligned} \tag{8.2.17}$$

The transition probability is then given by

$$\begin{aligned} W &= \frac{2\pi}{\hbar}\frac{1}{L^6}|\langle \underset{\sim}{k}_s; \{n_{\underset{\sim}{\chi}'}^{\lambda'}\}|V(\underset{\sim}{r})|\underset{\sim}{k}_i; \{n_{\underset{\sim}{\chi}}^{\lambda}\}\rangle|^2 \frac{L^3}{8\pi^3}\frac{mk_s}{\hbar^3}d\Omega \\ &= \frac{m}{4\pi^2\hbar^3L^3}|\langle \underset{\sim}{k}_s; \{n_{\underset{\sim}{\chi}'}^{\lambda'}\}|V(\underset{\sim}{r})|\underset{\sim}{k}_i; \{n_{\underset{\sim}{\chi}}^{\lambda}\}\rangle|^2 k_s d\Omega \ [\sec^{-1}], \end{aligned} \tag{8.2.18}$$

subject to the condition

$$\frac{\hbar^2k_i^2}{2m} + \sum_{\underset{\sim}{\chi}\lambda} \hbar\omega_{\underset{\sim}{\chi}\lambda}\left(n_{\underset{\sim}{\chi}}^{\lambda} + \frac{1}{2}\right) = \frac{\hbar^2k_s^2}{2m} + \sum_{\underset{\sim}{\chi}'\lambda'} \hbar\omega_{\underset{\sim}{\chi}'\lambda'}\left(n_{\underset{\sim}{\chi}'}^{\lambda'} + \frac{1}{2}\right). \tag{8.2.19}$$

The incident flux of neutrons corresponding to one neutron in the volume L^3 is given by

$$\begin{aligned} F &= \frac{\text{velocity of incident neutrons}}{L^3} \\ &= \frac{\hbar k_i}{mL^3}. \end{aligned} \tag{8.2.20}$$

The number of particles scattered in the unit time in the solid angle $d\Omega$ with wave vector $\underset{\sim}{k}_s$ is given by

$$dN = Fd\sigma = W, \tag{8.2.21}$$

where $d\sigma$ is the differential scattering cross section. Therefore

$$\begin{aligned} d\sigma &= \frac{W}{F} = \frac{mL^3}{\hbar k_i}W \\ &= \frac{mL^3}{\hbar k_i}\frac{m}{4\pi^2\hbar^3L^3}|\langle \underset{\sim}{k}_s; \{n_{\underset{\sim}{\chi}'}^{\lambda'}\}|V(\underset{\sim}{r})|\underset{\sim}{k}_i; \{n_{\underset{\sim}{\chi}}^{\lambda}\}\rangle|^2 k_s d\Omega \\ &= \frac{k_s}{k_i}\left(\frac{m}{2\pi\hbar^2}\right)^2 |\langle \underset{\sim}{k}_s; \{n_{\underset{\sim}{\chi}'}^{\lambda'}\}|V(\underset{\sim}{r})|\underset{\sim}{k}_i; \{n_{\underset{\sim}{\chi}}^{\lambda}\}\rangle|^2 d\Omega, \end{aligned} \tag{8.2.22}$$

and

$$\frac{d\sigma}{d\Omega} = \frac{k_s}{k_i}\left(\frac{m}{2\pi\hbar^2}\right)^2 |\langle \underset{\sim}{k}_s; \{n_{\underset{\sim}{\chi}'}^{\lambda'}\}|V(\underset{\sim}{r})|\underset{\sim}{k}_i; \{n_{\underset{\sim}{\chi}}^{\lambda}\}\rangle|^2, \tag{8.2.23}$$

subject to the condition (8.2.19). The expression (8.2.23) is the first *Born approximation* to the scattering cross section.

Indeed the first Born approximation is essentially derived by perturbation theory; this theory does not seem applicable if one considers the fact that the neutron-nucleus interaction potential has a very short range and is strong.

A formal artifice was introduced by Fermi, who replaced the actual potential by means of a pseudopotential[6–8]

$$V(\underset{\sim}{r}) = \frac{2\pi\hbar^2}{m} a\delta(\underset{\sim}{r} - \underset{\sim}{R}), \qquad (8.2.24)$$

where a is the scattering length of the nucleus ($\approx 10^{-12}$ cm), $\underset{\sim}{r}$ is the position of the neutron, and $\underset{\sim}{R}$ is the position of the nucleus.

When using the pseudopotential and the Born approximation we get the correct result for s-wave scattering. For an unbound atom ($k_s = k_i$)

$$\frac{d\sigma}{d\Omega} = \left(\frac{m}{2\pi\hbar^2}\right)^2 \left| \int \frac{2\pi\hbar^2}{m} a\delta(\underset{\sim}{r} - \underset{\sim}{R}) d^3\underset{\sim}{r} \right|^2 = a^2, \qquad (8.2.25)$$

and the total cross section is

$$\sigma = 4\pi a^2. \qquad (8.2.26)$$

For a monochromatic beam of neutrons interacting with a crystal,

$$V(\underset{\sim}{r}) = \sum_{s\nu} \frac{2\pi\hbar^2}{m} a_\nu \delta(\underset{\sim}{r} - \underset{\sim}{R}_{s\nu} - \underset{\sim}{u}_{s\nu}), \qquad (8.2.27)$$

where $\underset{\sim}{r}$ is the position of the neutron, $\underset{\sim}{R}_{s\nu}$ is the equilibrium position of the ν-th atom in the s-th unit cell, $\underset{\sim}{u}_{s\nu}$ is the displacement of sν-th atom, a_ν is the scattering length of ν-th atom in the unit cell.

The relevant matrix element is given by

$$\begin{aligned}
&\langle \underset{\sim}{k}_s; \{n^{\lambda'}_{\underset{\sim}{\chi}'}\} | V(\underset{\sim}{r}) | \underset{\sim}{k}_i; \{n^{\lambda}_{\underset{\sim}{\chi}}\}\rangle \\
&\quad = \langle \{n^{\lambda'}_{\underset{\sim}{\chi}'}\} | e^{i(\underset{\sim}{k}_i - \underset{\sim}{k}_s)\cdot \underset{\sim}{r}} \sum_{s\nu} \frac{2\pi\hbar^2}{m} a_\nu \delta(\underset{\sim}{r} - \underset{\sim}{R}_{s\nu} - \underset{\sim}{u}_{s\nu}) | \{n^{\lambda}_{\underset{\sim}{\chi}}\}\rangle \\
&\quad = \frac{2\pi\hbar^2}{m} \sum_{s\nu} a_\nu e^{i\underset{\sim}{Q}\cdot \underset{\sim}{R}_s} \langle \{n^{\lambda'}_{\underset{\sim}{\chi}'}\} | e^{i\underset{\sim}{Q}\cdot \underset{\sim}{u}_{s\nu}} | \{n^{\lambda}_{\underset{\sim}{\chi}}\}\rangle, \qquad (8.2.28)
\end{aligned}$$

where

$$\underset{\sim}{Q} = \underset{\sim}{k}_i - \underset{\sim}{k}_s,$$

and where, of course, energy is conserved. In the above formula $\underset{\sim}{u}_{s\nu}$ is, according to (6.5.77), given by

$$\underset{\sim}{u}_{s\nu} = \sum_{\underset{\sim}{\chi}\lambda} \left(\frac{\hbar}{2M\omega_{\underset{\sim}{\chi}\lambda}}\right)^{1/2} [\underset{\sim}{V}_\nu(\underset{\sim}{\chi},\lambda)e^{i\underset{\sim}{\chi}\cdot\underset{\sim}{R}_s}a^\lambda_{\underset{\sim}{\chi}} + \underset{\sim}{V}_\nu(\underset{\sim}{\chi},\lambda)^* e^{-i\underset{\sim}{\chi}\cdot\underset{\sim}{R}_s}a^{\lambda+}_{\underset{\sim}{\chi}}]. \tag{8.2.29}$$

Using (8.2.29) in (8.2.28) we find

$$\begin{aligned}
&\langle\underset{\sim}{k}_s;\{n^{\lambda'}_{\underset{\sim}{\chi}'}\}|V(\underset{\sim}{r})|\underset{\sim}{k}_i;\{n^\lambda_{\underset{\sim}{\chi}}\}\rangle \\
&\quad = \frac{2\pi\hbar^2}{m}\sum_{s\nu} a_\nu e^{i\underset{\sim}{Q}\cdot\underset{\sim}{R}_{s\nu}}\langle\{n^{\lambda'}_{\underset{\sim}{\chi}'}\}\rangle \\
&\qquad \times\left|\exp\left[i\sum_{\underset{\sim}{\chi}\lambda}\left(U^{\underset{\sim}{\chi}\lambda}_{s\nu}a^\lambda_{\underset{\sim}{\chi}} + U^{\underset{\sim}{\chi}\lambda*}_{s\nu}a^{\lambda+}_{\underset{\sim}{\chi}}\right)\right]\right|\{n^\lambda_{\underset{\sim}{\chi}}\}\rangle \\
&\quad = \frac{2\pi\hbar^2}{m}\sum_{s\nu} a_\nu e^{i\underset{\sim}{Q}\cdot\underset{\sim}{R}_s}\prod_{\underset{\sim}{\chi}\lambda}\langle n^{\lambda'}_{\underset{\sim}{\chi}'}|\exp[i(U^{\underset{\sim}{\chi}\lambda}_{s\nu}a^\lambda_{\underset{\sim}{\chi}} + U^{\underset{\sim}{\chi}\lambda*}_{s\nu}a^{\lambda+}_{\underset{\sim}{\chi}})]|n_{\underset{\sim}{\chi}\lambda}\rangle,
\end{aligned} \tag{8.2.30}$$

where

$$U^{\underset{\sim}{\chi}\lambda}_{s\nu} = \left(\frac{\hbar}{2M\omega_{\underset{\sim}{\chi}\lambda}}\right)^{1/2} e^{i\underset{\sim}{\chi}\cdot\underset{\sim}{R}_s}\underset{\sim}{V}_\nu(\underset{\sim}{\chi},\lambda)\cdot\underset{\sim}{Q}. \tag{8.2.31}$$

It may be useful to calculate the order of magnitude of the coefficients U. It is

$$U \approx \frac{1}{\sqrt{N}}\left(\frac{\hbar\omega}{m\omega^2}\right)^{1/2} Q.$$

We may use the typical values $\hbar\omega \simeq 3\times 10^{-14}$ ergs; $\omega = \frac{3\times 10^{-14}}{10^{-27}} = 3\times 10^{13}$; $m \approx 1$ amu $= 1.66\times 10^{-24}$ g; $Q \approx 10^8$. We find

$$U \approx \frac{1}{\sqrt{N}}\left(\frac{3\times 10^{-14}\times 10^{16}}{1.66\times 10^{-24}\times 9\times 10^{26}}\right)^{1/2} \approx \frac{1}{\sqrt{N}}\left(\frac{10^2}{10^3}\right)^{1/2} \approx \frac{1}{\sqrt{N}}.$$

8.3. Elastic Neutron Scattering

In this case $k_i = k_s$ and $\{n_{\underset{\sim}{\chi}}^{\lambda}\} = \{n_{\underset{\sim}{\chi}'}^{\lambda'}\}$. The relevant matrix element is

$$\langle \underset{\sim}{k}_s; \{n_{\underset{\sim}{\chi}}^{\lambda}\}|V(\underset{\sim}{r})|\underset{\sim}{k}_i; \{n_{\underset{\sim}{\chi}}^{\lambda}\}\rangle$$

$$= \frac{2\pi\hbar^2}{m}\sum_{s\nu} a_\nu e^{i\underset{\sim}{Q}\cdot\underset{\sim}{R}_{s\nu}} \prod_{\underset{\sim}{\chi}\lambda}^{N} \langle n_{\underset{\sim}{\chi}}^{\lambda}|\exp[i(U_{s\nu}^{\underset{\sim}{\chi}\lambda} a_{\underset{\sim}{\chi}}^{\lambda} + U_{s\nu}^{\underset{\sim}{\chi}\lambda *} a_{\underset{\sim}{\chi}}^{\lambda +})]|n_{\underset{\sim}{\chi}}^{\lambda}\rangle. \tag{8.3.1}$$

It is a simple matter to calculate the quantity

$$\langle n|e^{i(Ua+U^*a^+)}|n\rangle \simeq \langle n|1 + i(Ua + U^*a^+) - \frac{1}{2}|U|^2(aa^+ + a^+a)|n\rangle$$
$$= 1 - \frac{1}{2}|U|^2(2n+1) = e^{-1/2}|U|^2(2n+1). \tag{8.3.2}$$

Therefore

$$\langle \underset{\sim}{k}_s; \{n_{\underset{\sim}{\chi}}^{\lambda}\}|V(\underset{\sim}{r})|\underset{\sim}{k}_i; \{n_{\underset{\sim}{\chi}}^{\lambda}\}\rangle$$

$$= \frac{2\pi\hbar^2}{m}\sum_{s\nu} a_\nu e^{i\underset{\sim}{Q}\cdot\underset{\sim}{R}_{s\nu}} \prod_{\underset{\sim}{\chi}\lambda} \exp\left[-\frac{1}{2}|U_{s\nu}^{\underset{\sim}{\chi}\lambda}|^2(2n_{\underset{\sim}{\chi}}^{\lambda}+1)\right]. \tag{8.3.3}$$

The differential scattering cross section, according to (8.2.23), is given by

$$\frac{d\sigma}{s\Omega} = \left(\frac{m}{2\pi\hbar^2}\right)^2 |\langle k_s; \{n_{\underset{\sim}{\chi}}^{\lambda}\}|V(\underset{\sim}{r})|k_i; \{n_{\underset{\sim}{\chi}}^{\lambda}\}\rangle|^2$$
$$= \sum_{s\nu}\sum_{s'\nu'} a_\nu a_{\nu'} e^{i\underset{\sim}{Q}\cdot(\underset{\sim}{R}_{s\nu} - \underset{\sim}{R}_{s'\nu'})}$$
$$\times \prod_{\underset{\sim}{\chi}\lambda} \exp\left[-\frac{1}{2}(|U_{s\nu}^{\underset{\sim}{\chi}\lambda}|^2 + |U_{s'\nu'}^{\underset{\sim}{\chi}\lambda}|^2)(2n_{\underset{\sim}{\chi}}^{\lambda}+1)\right], \tag{8.3.4}$$

where

$$U_{s\nu}^{\underset{\sim}{\chi}\lambda} = \left(\frac{\hbar}{2M\omega_{\underset{\sim}{\chi}\lambda}}\right)^{1/2} \underset{\sim}{V}_\nu(\underset{\sim}{\chi}, \lambda)e^{i\underset{\sim}{\chi}\cdot\underset{\sim}{R}_s} \cdot \underset{\sim}{Q}, \tag{8.3.5}$$

and

$$|U_{\underset{\sim}{s\nu}}^{\chi\lambda}|^2 = |U_{\underset{\sim}{s\nu}}^{\chi\lambda}|^2, \text{ independent of s.} \tag{8.3.6}$$

Taking the thermal average of (8.3.4) and taking (8.3.6) into account we may write

$$\overline{\frac{d\sigma}{d\Omega}} = \sum_{s\nu}\sum_{s'\nu'} a_\nu a_{\nu'} e^{i\underset{\sim}{Q}\cdot(\underset{\sim}{R}_{s\nu}-\underset{\sim}{R}_{s'\nu'})}$$

$$\times \prod_{\underset{\sim}{\chi}\lambda} \exp\left[-\frac{1}{2}(|U_{\underset{\sim}{\nu}}^{\chi\lambda}|^2 + |U_{\underset{\sim}{\nu'}}^{\chi\lambda}|^2)\coth\frac{\hbar\omega_{\underset{\sim}{\chi}\lambda}}{2kT}\right]. \tag{8.3.7}$$

Let us define the *Debye–Waller factor* as follows:

$$2W_\nu| = \sum_{\underset{\sim}{\chi}\lambda} |U_{\underset{\sim}{\nu}}^{\chi\lambda}|^2 \coth\frac{\hbar\omega_{\underset{\sim}{\chi}\lambda}}{2kT}$$

$$= \sum_{\underset{\sim}{\chi}\lambda} \frac{\hbar}{2M\omega_{\underset{\sim}{\chi}\lambda}} |\underset{\sim}{Q}\cdot\underset{\sim}{V}_\nu(\underset{\sim}{\chi},\lambda)|^2 \coth\frac{\hbar\omega_{\underset{\sim}{\chi}\lambda}}{2kT}. \tag{8.3.8}$$

Then

$$\overline{\frac{d\sigma}{d\Omega}} = \sum_{s\nu}\sum_{s'\nu'} a_\nu a_{\nu'} e^{i\underset{\sim}{Q}\cdot(\underset{\sim}{R}_{s\nu}-\underset{\sim}{R}_{s'\nu'})} e^{-(W_\nu+W_{\nu'})}. \tag{8.3.9}$$

If the crystal contains different and randomly distributed isotopes of the various atoms we may write

$$\overline{\frac{d\sigma}{d\Omega}} = \sum_{s\nu}\sum_{s'\nu'} a_{s\nu} a_{s'\nu'} e^{i\underset{\sim}{Q}\cdot(\underset{\sim}{R}_{s\nu}-\underset{\sim}{R}_{s'\nu'})} e^{-(W_\nu+W_{\nu'})}$$

$$= \sum_{s\nu} a_{s\nu}^2\, e^{-2W_\nu} + \sum_{s\nu}\sum_{\substack{s'\nu' \\ (s\nu)\neq(s'\nu')}} a_{s\nu}\, a_{s'\nu'} e^{i\underset{\sim}{Q}\cdot(\underset{\sim}{R}_{s\nu}-\underset{\sim}{R}_{s'\nu'})} e^{-(W_\nu+W_{\nu'})}$$

$$\simeq N\sum_\nu \overline{a_\nu^2} e^{-2W_\nu} + \sum_{s\nu}\sum_{\substack{s'\nu' \\ (s\nu)\neq(s'\nu')}} \bar{a}_\nu\, \bar{a}_{\nu'} e^{i\underset{\sim}{Q}\cdot(\underset{\sim}{R}_{s\nu}-\underset{\sim}{R}_{s'\nu'})} x^{-(W_\nu+W_{\nu'})}$$

$$= N\sum_{\nu}\overline{a_\nu^2}e^{-2W_\nu} - \sum_{s\nu}(\overline{a_\nu})^2e^{-2W_\nu}$$

$$+\sum_{s\nu}\sum_{s'\nu'} \bar{a}_\nu\bar{a}_{\nu'}e^{i\underset{\sim}{Q}\cdot(\underset{\sim}{R}_{s\nu}-\underset{\sim}{R}_{s'\nu'})}e^{-(W_\nu+W_{\nu'})}$$

$$= N\sum_{\nu}\overline{a_\nu^2}e^{-2W_\nu} - \sum_{\nu}N(\overline{a_\nu})^2e^{-2W_\nu}$$

$$+\sum_{s\nu}\sum_{s'\nu'} \bar{a}_\nu\bar{a}_{\nu'}\, e^{i\underset{\sim}{Q}\cdot(\underset{\sim}{R}_{s\nu}-\underset{\sim}{R}_{s'\nu'})}e^{-(W_\nu+W_{\nu'})}$$

$$= N\sum_{\nu}[\overline{a_\nu^2} - (\overline{a_\nu})^2]e^{-2W_\nu}$$

$$+\sum_{s\nu}\sum_{s'\nu'} \bar{a}_\nu\bar{a}_{\nu'}e^{i\underset{\sim}{Q}\cdot(\underset{\sim}{R}_{s\nu}-\underset{\sim}{R}_{s'\nu'})}e^{-(W_\nu+W_{\nu'})}, \tag{8.3.10}$$

where we have performed an average over the scattering lengths, and neglected mass differences. But

$$\begin{aligned} \underset{\sim}{R}_{s\nu} &= \underset{\sim}{R}_s + \underset{\sim}{r}_\nu, \\ \underset{\sim}{R}_{s'\nu'} &= \underset{\sim}{R}_{s'} + \underset{\sim}{r}_{\nu'}. \end{aligned} \tag{8.3.11}$$

Then

$$\overline{\frac{d\sigma}{d\Omega}} = N\sum_{\nu}[\overline{a_\nu^2} - (\bar{a}_\nu)^2]e^{-2W_\nu}$$

$$+\sum_{ss'}e^{i\underset{\sim}{Q}\cdot(\underset{\sim}{R}_s-\underset{\sim}{R}_{s'})}\sum_{\nu\nu'}e^{i\underset{\sim}{Q}\cdot(\underset{\sim}{r}_\nu-\underset{\sim}{r}_{\nu'})}\bar{a}_\nu\bar{a}_{\nu'}e^{-(W_\nu+W_{\nu'})}. \tag{8.3.12}$$

We may want to examine the sum over ss′ appearing in (8.3.12). It is

$$\sum_{ss'}e^{i\underset{\sim}{Q}\cdot(\underset{\sim}{R}_s-\underset{\sim}{R}_{s'})} = \left|\sum_{s}e^{i\underset{\sim}{Q}\cdot\underset{\sim}{R}_s}\right|^2$$

$$= \frac{8\pi^3N}{\gamma}\sum_{\underset{\sim}{K}_\eta}\delta(\underset{\sim}{K}_\eta - \underset{\sim}{Q}), \tag{8.3.13}$$

where γ is the volume of the unit cell, N is the number of unit cells, and $\underset{\sim}{K}_\eta$ is the primitive vector of the reciprocal space.

Therefore

$$\overline{\frac{d\sigma}{d\Omega}} = N\sum_{\nu}[\overline{a_\nu^2} - (\overline{a_\nu})^2]e^{-2W_\nu}$$

$$+\frac{8\pi^3 N}{\gamma}\sum_{\underset{\sim}{K}_\eta}\delta(\underset{\sim}{Q} - \underset{\sim}{K}_\eta)\left|\sum_{\nu}\overline{a_\nu}e^{i\underset{\sim}{Q}\cdot\underset{\sim}{r}_\nu}e^{-W_\nu}\right|^2$$

$$= \overline{\frac{d\sigma}{d\Omega}}\bigg|_{\text{incoh}} + \overline{\frac{d\sigma}{d\Omega}}\bigg|_{\text{coh}}, \tag{8.3.14}$$

where

$$\overline{\frac{d\sigma}{d\Omega}}\bigg|_{\text{incoh}} = N\sum_{\nu}[\overline{a_\nu^2} - (\overline{a_\nu})^2]e^{-2W_\nu}, \tag{8.3.15}$$

is the *incoherent elastic differential cross section*, and

$$\overline{\frac{d\sigma}{d\Omega}}\bigg|_{\text{coh}} = \frac{8\pi^3 N}{\gamma}\sum_{\underset{\sim}{K}_\eta}\delta(\underset{\sim}{Q} - \underset{\sim}{K}_\eta)\left|\sum_{\nu}\bar{a}_\nu e^{i\underset{\sim}{Q}\cdot\underset{\sim}{R}_\nu}e^{-2W_\nu}\right|^2, \tag{8.3.16}$$

is the *coherent elastic differential cross section.*

The incoherent differential scattering cross section depends on the variance of the scattering length of the various isotopes of the atoms in the crystal; it is zero if only one isotope per atom is present. The coherent differential scattering cross section is the result of interference effects of the waves scattered by the nuclei in the crystal. The term

$$\left|\sum_{\nu}\bar{a}_\nu e^{i\underset{\sim}{Q}\cdot\underset{\sim}{r}_\nu}e^{-W_\nu}\right|^2 \tag{8.3.17}$$

is called the *elastic structure factor.* It is clear from (8.3.16) that if $\underset{\sim}{K}_{\text{min}}$ is the shortest possible primitive vector in the reciprocal lattice, no coherent elastic scattering can take place if

$$k_i < \frac{1}{2}K_{\text{min}}. \tag{8.3.18}$$

8.4. Inelastic Neutron Scattering

We wish now to calculate the cross section for the scattering process in which *one* phonon is created or destroyed. The formula to use is (8.2.21) where we sum over all the possible final states:

$$\left(\frac{d^2\sigma}{d\Omega dE}\right)_1 = \frac{k_s}{k_i} \sum_{\{n_{\underset{\sim}{\chi}'}^{\lambda'}\}} |\langle \underset{\sim}{k}_s; \{n_{\underset{\sim}{\chi}'}^{\lambda'}\}|V(\underset{\sim}{r})|\underset{\sim}{k}_i; \{n_{\underset{\sim}{\chi}}^{\lambda}\}\rangle|^2$$

$$\times \left(\frac{m}{2\pi\hbar^2}\right)^2 \delta\left[\frac{\hbar k_i^2}{2m} - \frac{\hbar^2 k_s^2}{2m} \mp \hbar\omega_{\underset{\sim}{\chi}\lambda}\right], \quad (8.4.1)$$

where the subscript 1 indicates that the process involves one phonon; the minus and plus signs indicate the creation and destruction of a phonon, respectively.

Let us consider the scattering process in which one phonon is destroyed. The relevant matrix element [see (8.2.26)] is in this case

$$\langle \underset{\sim}{k}_s; \{n_{\underset{\sim}{\chi}}^{\lambda}\}|V(\underset{\sim}{r})|\underset{\sim}{k}_i; \{n_{\underset{\sim}{\chi}}^{\lambda}\}\rangle$$

$$= \frac{2\pi\hbar^2}{m} \sum_{s\nu} a_\nu e^{i\underset{\sim}{Q}\cdot\underset{\sim}{R}_{s\nu}} \langle n_1^1 - 1|$$

$$\times \prod_{\substack{\underset{\sim}{\chi}\lambda \\ (\underset{\sim}{\chi}\lambda)\neq(11)}} \langle n_{\underset{\sim}{\chi}}^{\lambda}| e^{i\underset{\sim}{Q}\cdot\underset{\sim}{u}_{s\nu}} \prod_{\substack{\underset{\sim}{\chi}\lambda \\ (\underset{\sim}{\chi}\lambda)\neq(11)}} |n_{\underset{\sim}{\chi}}^{\lambda}\rangle |n_1^1\rangle$$

$$= \frac{2\pi\hbar^2}{m} \sum_{s\nu} a_\nu e^{i\underset{\sim}{Q}\cdot\underset{\sim}{R}_{s\nu}} \langle n_1^1 - 1| \exp[i(U_{s\nu}^{11} a_1^1 + U_{s\nu}^{11^*} a_1^{1+})]|n_1^1\rangle$$

$$\times \prod_{\substack{\underset{\sim}{\chi}\lambda \\ (\underset{\sim}{\chi}\lambda)\neq(11)}} \langle n_{\underset{\sim}{\chi}}^{\lambda}| \exp[i(U_{s\nu}^{\underset{\sim}{\chi}\lambda} a_{\underset{\sim}{\chi}}^{\lambda} + U_{s\nu}^{\underset{\sim}{\chi}\lambda^*} a_{\underset{\sim}{\chi}}^{\lambda+})]|n_{\underset{\sim}{\chi}}^{\lambda}\rangle. \quad (8.4.2)$$

The first matrix element in the above relation can be approximated as

$$\langle n-1|e^{i(Ua+U^*a^+)}|n\rangle = \langle n-1|1 + iUa|n\rangle + O\left(\frac{1}{N^{3/2}}\right)$$

$$\simeq i\sqrt{n}U\langle n|e^{i(Ua+U^*a^+)}|n\rangle, \quad (8.4.3)$$

because

$$\langle n|e^{i(Ua+U^*a^+)}|n\rangle \approx 1 + O\left(\frac{1}{N}\right). \tag{8.4.4}$$

Therefore

$$\begin{aligned}\langle \underset{\sim}{k}_s; \{\underset{\sim}{n}^{\lambda'}_{\underset{\sim}{\chi}'}\}|V(\underset{\sim}{r})|\underset{\sim}{k}_i; \{\underset{\sim}{n}^{\lambda}_{\underset{\sim}{\chi}}\}\rangle &= \frac{2\pi\hbar^2}{m}\sum_{s\nu} a_\nu e^{i\underset{\sim}{Q}\cdot\underset{\sim}{R}_{s\nu}} i\sqrt{n^1_1}\, U^{11}_{s\nu} \\ &\quad \times \prod_{\underset{\sim}{\chi}\lambda} \langle n^{\lambda}_{\underset{\sim}{\chi}} \exp[i(U^{\underset{\sim}{\chi}\lambda}_{s\nu} a^{\lambda}_{\underset{\sim}{\chi}} + U^{\underset{\sim}{\chi}\lambda}_{s\nu} a^{\lambda+}_{\underset{\sim}{\chi}})]|n^{\lambda}_{\underset{\sim}{\chi}}\rangle \\ &= \frac{2\pi\hbar^2}{m}\sum_{s\nu} a_\nu e^{i\underset{\sim}{Q}\cdot\underset{\sim}{R}_{s\nu}} i\sqrt{n^1_1}\, U^{11}_{s\nu} \\ &\quad \times \prod_{\underset{\sim}{\chi}\lambda} \exp\left[-\frac{1}{2}|U^{\underset{\sim}{\chi}\lambda}_{s\nu}|^2(2n^{\lambda}_{\underset{\sim}{\chi}} + 1)\right]. \end{aligned} \tag{8.4.5}$$

It is then

$$\begin{aligned}&|\langle \underset{\sim}{k}_s; \{\underset{\sim}{n}^{\lambda'}_{\underset{\sim}{\chi}'}\}|V(\underset{\sim}{r})|\underset{\sim}{k}_i; \{\underset{\sim}{n}^{\lambda}_{\underset{\sim}{\chi}}\}\rangle|^2 \\ &= \left(\frac{2\pi\hbar^2}{m}\right)^2 \sum_{s\nu}\sum_{s'\nu'} a_\nu a_{\nu'} e^{i\underset{\sim}{Q}\cdot(\underset{\sim}{R}_{s\nu} - \underset{\sim}{R}_{s'\nu'})} \\ &\qquad \times n^1_1 U^{11}_{s\nu} U^{11*}_{s'\nu'} \prod_{\underset{\sim}{\chi}\lambda} \exp\left[-\frac{1}{2}(|U^{\underset{\sim}{\chi}\lambda}_{s\nu}|^2 + |U^{\underset{\sim}{\chi}\lambda}_{s'\nu'}|^2)(2n^{\lambda}_{\underset{\sim}{\chi}} + 1)\right], \end{aligned} \tag{8.4.6}$$

and taking the thermal average,

$$\begin{aligned}&\overline{|\langle \underset{\sim}{k}_s; \{\underset{\sim}{n}^{\lambda'}_{\underset{\sim}{\chi}'}\}|V(\underset{\sim}{r})|\underset{\sim}{k}_i; \{\underset{\sim}{n}^{\lambda}_{\underset{\sim}{\chi}}\}\rangle|^2} \\ &= \left(\frac{2\pi\hbar^2}{m}\right)^2 \sum_{s\nu}\sum_{s'\nu'} a_\nu a_{\nu'} e^{i\underset{\sim}{Q}\cdot(\underset{\sim}{R}_{s\nu} - \underset{\sim}{R}_{s'\nu'})} \\ &\qquad \times U^{11}_{s\nu} U^{11*}_{s'\nu'} \overline{n^1_1} \prod_{\underset{\sim}{\chi}\lambda} \exp\left[-\frac{1}{2}(|U^{\underset{\sim}{\chi}\lambda}_{s\nu}|^2 + |U^{\underset{\sim}{\chi}\lambda}_{s'\nu'}|^2)(2\overline{n^{\lambda}_{\underset{\sim}{\chi}}} + 1)\right] \end{aligned}$$

$$= \left(\frac{2\pi\hbar^2}{m}\right)^2 \sum_{s\nu}\sum_{s'\nu'} a_\nu a_{\nu'} e^{i\underset{\sim}{Q}\cdot(\underset{\sim}{R}_{s\nu}-\underset{\sim}{R}_{s'\nu'})} U_{s\nu}^{11} U_{s'\nu'}^{11*} \frac{1}{e^{\hbar\omega_{11}/kT}-1}$$

$$\times \prod_{\underset{\sim}{\chi}\lambda} \exp\left[-\frac{1}{2}(|U_{s\nu}^{\underset{\sim}{\chi}\lambda}|^2 + |U_{s'\nu'}^{\underset{\sim}{\chi}\lambda}|^2)\coth\frac{\hbar\omega_{\underset{\sim}{\chi}\lambda}}{2kT}\right]$$

$$= \left(\frac{2\pi\hbar^2}{m}\right)^2 \sum_{s\nu}\sum_{s'\nu'} a_\nu a_{\nu'} e^{i\underset{\sim}{Q}\cdot(\underset{\sim}{R}_{s\nu}-\underset{\sim}{R}_{s'\nu'})}$$

$$\times U_{s\nu}^{11} U_{s'\nu'}^{11*} \frac{1}{e^{\hbar\omega_{11}/kT}-1} e^{-(W_\nu+W_{\nu'})}. \tag{8.4.7}$$

The differential scattering cross section is then given by summing over all the possible final states:

$$\overline{\left(\frac{d^2\sigma}{d\Omega dE}\right)_1} = \frac{k_s}{k_i}\sum_{s\nu}\sum_{s'\nu'} a_\nu a_{\nu'} e^{i\underset{\sim}{Q}\cdot(\underset{\sim}{R}_{s\nu}-\underset{\sim}{R}_{s'\nu'})}$$

$$\times e^{-(W_\nu+W_{\nu'})} \sum_{\underset{\sim}{\chi}\lambda} U_{s\nu}^{\underset{\sim}{\chi}\lambda} U_{s'\nu'}^{\underset{\sim}{\chi}\lambda *} \frac{1}{e^{\hbar\omega_{\underset{\sim}{\chi}\lambda}/kT}-1}$$

$$\times \delta\left(\frac{\hbar^2 k_i^2}{2m} - \frac{\hbar^2 k_s^2}{2m} + \hbar\omega_{\underset{\sim}{\chi}\lambda}\right). \tag{8.4.8}$$

But

$$U_{s\nu}^{\underset{\sim}{\chi}\lambda} = \left(\frac{\hbar}{2M\omega_{\underset{\sim}{\chi}\lambda}}\right)^{\frac{1}{2}} e^{i\underset{\sim}{\chi}\cdot\underset{\sim}{R}_s} \underset{\sim}{V}_\nu(\underset{\sim}{\chi},\lambda)\cdot\underset{\sim}{Q}. \tag{8.4.9}$$

Using (8.4.9) in (8.4.8) and taking into account the presence of isotopes,

$$\left(\frac{d^2\sigma_1}{d\Omega dE}\right)_1 = \frac{k_s}{k_i}\sum_{s\nu}\sum_{s'\nu'} a_{s\nu} a_{s'\nu'} e^{i\underset{\sim}{Q}\cdot(\underset{\sim}{R}_{s\nu}-\underset{\sim}{R}_{s'\nu'})} e^{-(W_\nu+W_{\nu'})}$$

$$\times \sum_{\underset{\sim}{\chi}\lambda} \frac{\hbar}{2M\omega_{\underset{\sim}{\chi}\lambda}(e^{\hbar\omega_{\underset{\sim}{\chi}\lambda}/kT}-1)} e^{i\underset{\sim}{\chi}\cdot(\underset{\sim}{R}_s-\underset{\sim}{R}_{s'})}$$

$$\times [\underset{\sim}{V}_\nu(\underset{\sim}{\chi},\lambda)\cdot\underset{\sim}{Q}][\underset{\sim}{V}_\nu(\underset{\sim}{\chi},\lambda)\cdot\underset{\sim}{Q}]^*\delta\left(\frac{\hbar^2 k_i^2}{2m} - \frac{\hbar^2 k_s^2}{2m} + \hbar\omega_{\underset{\sim}{\chi}\lambda}\right)$$

$$= \frac{k_s}{k_i} \sum_{\underset{\sim}{\chi}\lambda} \frac{\hbar}{2M\omega_{\underset{\sim}{\chi}\lambda}(e^{\hbar\omega_{\underset{\sim}{\chi}\lambda}/kT} - 1)} \left\{ \sum_{s\nu} a_{s\nu}^2 e^{-2W_\nu} |\underset{\sim}{V}_\nu(\underset{\sim}{\chi}, \lambda) \cdot \underset{\sim}{Q}|^2 \right.$$

$$+ \sum_{s\nu} \sum_{\substack{s'\nu' \\ (s\nu) \neq (s'\nu')}} a_{s\nu} a_{s'\nu'} e^{i\underset{\sim}{Q}\cdot(\underset{\sim}{R}_{s\nu} - \underset{\sim}{R}_{s'\nu'})} e^{-(W_\nu + W_{\nu'})}$$

$$\left. \times [\underset{\sim}{V}_\nu(\underset{\sim}{\chi}, \lambda) \cdot \underset{\sim}{Q}][\underset{\sim}{V}_{\nu'}(\underset{\sim}{\chi}, \lambda) \cdot \underset{\sim}{Q}]^* \right\}$$

$$\times e^{i\underset{\sim}{\chi}\cdot(\underset{\sim}{R}_s - \underset{\sim}{R}_{s'})} \delta\left(\frac{\hbar^2 k_i^2}{2m} - \frac{\hbar^2 k_s^2}{2m} + \hbar\omega_{\underset{\sim}{\chi}\lambda}\right)$$

$$= \frac{k_s}{k_i} \sum_{\underset{\sim}{\chi}\lambda} \frac{\hbar}{2M\omega_{\underset{\sim}{\chi}\lambda}(e^{\hbar\omega_{\underset{\sim}{\chi}\lambda}/kT} - 1)}$$

$$\times \left\{ N \sum_{\nu} [\overline{a_\nu^2} - (\bar{a}_\nu)^2] e^{-2W_\nu} |\underset{\sim}{V}_\nu(\underset{\sim}{\chi}, \lambda) \cdot \underset{\sim}{Q}|^2 \right.$$

$$+ \sum_{ss'} e^{i(\underset{\sim}{Q}+\underset{\sim}{\chi})\cdot(\underset{\sim}{R}_s - \underset{\sim}{R}_{s'})} \sum_{\nu\nu'} \bar{a}_\nu \bar{a}_{\nu'} e^{i\underset{\sim}{Q}\cdot(\underset{\sim}{r}_\nu - \underset{\sim}{r}_{\nu'})} e^{-(W_\nu + W_{\nu'})}$$

$$\left. \times [\underset{\sim}{V}_\nu(\underset{\sim}{\chi}, \lambda) \cdot \underset{\sim}{Q}][\underset{\sim}{V}_{\nu'}(\underset{\sim}{\chi}, \lambda) \cdot \underset{\sim}{Q}]^* \right\} \delta\left(\frac{\hbar^2 k_i^2}{2m} - \frac{\hbar^2 k_s^2}{2m} + \hbar\omega_{\underset{\sim}{\chi}\lambda}\right)$$

$$= \frac{k_s}{k_i} \sum_{\underset{\sim}{\chi}\lambda} \frac{\hbar}{2M\omega_{\underset{\sim}{\chi}\lambda}(e^{\hbar\omega_{\underset{\sim}{\chi}\lambda}/kT} - 1)}$$

$$\times \left\{ N \sum_{\nu} [\overline{a_\nu^2} - (\bar{a}_\nu)^2] e^{-2W_\nu} |\underset{\sim}{V}_\nu(\underset{\sim}{\chi}, \lambda) \cdot \underset{\sim}{Q}|^2 \right.$$

$$+\frac{8\pi^3 N}{\gamma}\sum_{\underset{\sim}{K}_\eta}\delta(\underset{\sim}{Q}+\underset{\sim}{\chi}-\underset{\sim}{K}_\eta)$$

$$\times\left.\left|\sum_\nu \bar{a}_\nu e^{i\underset{\sim}{Q}\cdot\underset{\sim}{r}_\nu}e^{-W_\nu}\underset{\sim}{V}_\nu(\underset{\sim}{\chi},\lambda)\cdot\underset{\sim}{Q}\right|^2\right\}$$

$$\times\,\delta\left(\frac{\hbar^2k_i^2}{2m}-\frac{\hbar^2k_s^2}{2m}+\hbar\omega_{\underset{\sim}{\chi}\lambda}\right), \tag{8.4.10}$$

where γ is the volume of the primitive cell in reciprocal space.

$$\left(\overline{\frac{d^2\sigma}{d\Omega dE}}\right)_1=\left(\overline{\frac{d^2\sigma}{d\Omega dE}}\right)_{\text{incoh}}+\left(\overline{\frac{d^2\sigma}{d\Omega dE}}\right)_{\text{coh}}, \tag{8.4.11}$$

where

$$\left(\overline{\frac{d^2\sigma}{d\Omega dE}}\right)_{\text{incoh}}=\frac{k_s}{k_i}\sum_{\underset{\sim}{\chi}\lambda}\frac{N\hbar\left(\overline{n^\lambda_{\underset{\sim}{\chi}}}+\frac{1}{2}\pm\frac{1}{2}\right)}{2M\omega_{\underset{\sim}{\chi}\lambda}}\sum_\nu[\overline{a_\nu^2}-(\bar{a}_\nu)^2]e^{-2W_\nu}$$

$$\times|\underset{\sim}{V}_\nu(\underset{\sim}{\chi},\lambda)\cdot\underset{\sim}{Q}|^2\delta\left(\frac{\hbar^2k_i^2}{2m}-\frac{\hbar^2k_s^2}{2m}\mp\hbar\omega_{\underset{\sim}{\chi}\lambda}\right), \tag{8.4.12}$$

is the *incoherent one-phonon double differential cross section* and

$$\left(\frac{d^2\sigma}{d\Omega dE}\right)_{\text{coh}}=\frac{k_s}{k_i}\sum_{\underset{\sim}{\chi}\lambda}\sum_{\underset{\sim}{K}_\eta}\frac{8\pi^3}{\gamma}\delta(\underset{\sim}{Q}\mp\underset{\sim}{\chi}-\underset{\sim}{K}_\eta)$$

$$\times\left|\sum_\nu \bar{a}_\nu e^{i\underset{\sim}{Q}\cdot\underset{\sim}{r}_\nu}e^{-W_\nu}\underset{\sim}{V}_\nu(\underset{\sim}{\chi},\lambda)\cdot\underset{\sim}{Q}\right|^2$$

$$\times\frac{N\hbar\left(\overline{n^\lambda_{\underset{\sim}{\chi}}}+\frac{1}{2}\pm\frac{1}{2}\right)}{2M\omega_{\underset{\sim}{\chi}\lambda}}\delta\left(\frac{\hbar^2k_i^2}{2m}-\frac{\hbar^2k_s^2}{2m}\mp\hbar\omega_{\underset{\sim}{\chi}\lambda}\right), \tag{8.4.13}$$

is the *coherent one-phonon double differential cross section.*

In the above formulas the upper sign corresponds to the creation of one phonon and the lower sign to the destruction of one phonon.

The term

$$\left|\sum_{\nu} \bar{a}_\nu e^{i\underset{\sim}{Q}\cdot\underset{\sim}{r}_\nu} e^{-W_\nu} \underset{\sim}{V}_\nu(\underset{\sim}{\chi},\lambda)\cdot\underset{\sim}{Q}\right|^2 = |S(\underset{\sim}{Q},\underset{\sim}{\chi}\lambda)|^2,$$

is called *inelastic structure factor* and determines the way in which the scattering intensity varies for different momentum transfers. We notice that the coherent part of the cross section satisfies the condition for conservation of energy and wave vector; the incoherent part satisfies only the condition for conservation of energy.

8.5. Application of Neutron Scattering to the Study of Lattice Vibrations

For the study of phonons in crystals one must rely primarily on one-phonon inelastic coherent scattering whose differential scattering cross section is given in (8.4.13). It is possible to eliminate completely the elastic scattering by using such slow neutrons that their wavelength is larger than the wavelength for Bragg cut-off given by

$$\lambda_c = \frac{2\pi}{\frac{1}{2}|\underset{\sim}{K}_\eta|_{min}}, \qquad (8.5.1)$$

where $|\underset{\sim}{K}_\eta|_{min}$ is the magnitude of the smallest primitive vector of the reciprocal lattice.

In the actual experimental set up, the target is maintained at a low temperature and the detector is placed at a particular position in space and arranged so as to detect only neutrons of a certain energy. In these conditions the double differential cross section is determined.

Two basic techniques are commonly used at the present time for the determination of phonon spectra:

1. The triple-axis crystal spectrometer developed at Chalk River by Brookhouse.[9] In this apparatus the neutrons from the reactor are Bragg reflected by a crystal (A) and made monochromatic; the monochromatic beam is then sent to the sample (S) and scattered off by it. The energy of the neutrons scattered by S is determined by reflection from an analyzing crystal (B) and sent to a detector. The entire process consists of three scattering events; the counting rate upon arrival at the detector is rather low, typically

10 neutrons are recorded in the counter per minute for a flux in the center of the reactor of 2×10^4 neutrons cm^{-2} sec^{-1}. The name "triple axis" derives from the fact that A, S, and B are mounted on revolving drums.

2. Time-of-flight machines are also available for phonon studies. In these machines a monochromatic beam of neutrons is pulsed and made to fall on the sample crystal. The scattered neutrons are then detected by several detectors arranged around the sample and recording the time of arrivals of the neutrons. The knowledge of the time elapsed between the arrival of the neutrons at the sample and of the sample-detector distance allows the determination of the velocity, and therefore of the wavevector and energy of the scattered neutrons. Pulsed monochromatic beams are produced by either a succession of rotors, phased in such a way that only neutrons of a certain velocity are let through or by a spinning single crystal.

We are reminded at this point that incoherently scattered neutrons obey energy conservation, whereas coherently scattered neutrons obey energy and momentum conservation. Coherent scattering peaks are more well defined than incoherent scattering peaks; incoherent scattering gives a fairly constant counting rate with occasional peaks resembling those of the coherent scattering. Neutrons scattered incoherently and those that interact with more than one phonon form in general a continuous background to the inelastic coherent scattering. Occasional incoherent scattering peaks can be sorted out by changing $\underset{\sim}{Q}$ slightly and maintaining the energy of the detectable neutrons constant; if the peaks persist the scattering is incoherent.

The study of phonons in crystals may be directed towards the evaluation of (a) dispersion curves and (b) phonon frequency distribution. Incoherent scattering provides under certain conditions a direct measure of the phonon frequency distribution for a limited class of crystals; however, in most cases, experiments are designed in order to maximize the coherent scattering, and for this reason the crystals have to be as perfect as possible.

Conservation of energy and momentum in coherent inelastic scattering allows, in principle, the determination of the dispersion curves at any point of the Brillouin zone. The double differential one-phonon inelastic coherent cross section is given by

$$\left(\overline{\frac{d^2\sigma}{d\Omega dE}}\right)_{\text{coh}} = \frac{k_s}{k_i}\frac{8\pi^3}{\gamma}\sum_{\underset{\sim}{\chi}\lambda} N\left|\sum_{\nu} \bar{a}_\nu e^{i\underset{\sim}{Q}\cdot\underset{\sim}{r}_\nu} e^{-W_\nu} \underset{\sim}{V}_\nu(\underset{\sim}{\chi},\lambda)\cdot\underset{\sim}{Q}\right|^2$$
$$\times \frac{\hbar}{2M\omega_{\underset{\sim}{\chi}\lambda}}\left(\overline{n^\lambda_{\underset{\sim}{\chi}}}+\frac{1}{2}\pm\frac{1}{2}\right)$$
$$\times\, \delta(\underset{\sim}{Q}\mp\underset{\sim}{\chi}-\underset{\sim}{K}_\eta)\,\delta\left(\frac{\hbar^2k_i^2}{2m}-\frac{\hbar^2k_s^2}{2m}+\hbar\omega_{\underset{\sim}{\chi}\lambda}\right) \tag{8.5.2}$$

where the upper (lower) sign indicates the creation (destruction) of a phonon. It depends on the following factors:

1. k_s/k_i, the ratio of the scattered to the incident wavevector magnitudes.
2. $\underset{\sim}{Q} = \underset{\sim}{k}_i - \underset{\sim}{k}_s \cdot \underset{\sim}{Q}$ can be in principle in the first, second or any other Brillouin zone.
3. The frequency of the normal mode $\omega_{\underset{\sim}{\chi}\lambda}$. Since $\omega_{\underset{\sim}{\chi}\lambda}$ is periodical in the reciprocal lattice, the energy transfer that takes place in the scattering event does not depend on $\underset{\sim}{Q}$ being in the first, second, or any other Brillouin zone.
4. The inelastic scattering factor, which determines the way in which the intensity of the scattering varies for different momentum transfers. It is to be noted that this factor contains the product $\underset{\sim}{Q}\cdot\underset{\sim}{V}_\nu(\underset{\sim}{\chi},\lambda)$.

On the basis of the above observations one can understand what type of information is obtainable by neutron scattering measurements.

The experimenter can vary the magnitudes and directions of both $\underset{\sim}{k}_i$ and $\underset{\sim}{k}_s$. One commonly used method, called the *constant Q method*, consists of varying these parameters in such a way that $\underset{\sim}{Q}$ remains constant: this may be accomplished by varying the magnitude of both $\underset{\sim}{k}_i$ and $\underset{\sim}{k}_s$ and the direction of $\underset{\sim}{k}_s$. The peaks observed with such a scan represent the energies of phonons in different branches at the point $\underset{\sim}{\chi}$ of the Brillouin zone corresponding to the chosen $\underset{\sim}{Q}$. Repeating this for many values of $\underset{\sim}{Q}$, one can obtain the dispersion curves along the desired directions of the Brillouin zone: by restricting $\underset{\sim}{Q}$ to the first Brillouin zone one can drop $\underset{\sim}{K}_\eta$ in (8.5.2).

The conservation of momentum in (8.5.2) allows a number of choices for $\underset{\sim}{Q}$, since $\underset{\sim}{K}_\eta$ can be any primitive vector of the reciprocal lattice. The possibility of this multiple choice may be used to extract information about the polarization vectors. Polarization vectors of normal modes of vibration obey orthogonality relations; moreover, group theory provides information about their degeneracies and orientation (but certainly not about the frequency magnitudes). Taking advantage of the various allowed vectors $\underset{\sim}{Q}$, it is possible to use various values and, more importantly, various directions of $\underset{\sim}{Q}$, in order to obtain information about the polarization vectors for points $\underset{\sim}{\chi}$ of high symmetry. This is most easily seen considering cubic crystals; in such crystals there are certain directions for which the polarization vectors are purely longitudinal or purely transverse. For longitudinal modes the cross section is maximized by choosing $\underset{\sim}{Q}$ parallel to $\underset{\sim}{\chi}$; for transverse modes the cross section is maximized by choosing $\underset{\sim}{Q}$ perpendicular to $\underset{\sim}{\chi}$. By properly using these different geometries it is possible to distinguish between longitudinal and transverse modes.

As reported above, most efforts are generally directed to maximize coherent scattering. However, incoherent scattering may provide a direct measure of phonon frequency distribution for simple cubic crystals; it is left to the reader as an exercise to show that the inelastic incoherent one-phonon double differential cross section is proportional to the density of phonon states for crystals of cubic symmetry, with one atom per unit cell. If the nuclei in such crystals have large incoherent cross sections the direct determination of the phonon frequency distribution from incoherent scattering measurements is possible.[10]

References

1. G. R. Ringo, "Neutron Diffraction and Interference," *Handbuch der Physik*, Edited by S. Flugge, Vol. 32, Springer-Verlag, Berlin, 1957, p. 552.
2. C. G. Shull and E. O. Wollan, "Applications of Neutron Diffraction to Solid State Problems," in *Solid State Physics*, Edited by F. Seitz and D. Turnbull, Vol. 2, Academic, New York, 1956, p. 137.

3. G. E. Bacon and K. Lonsdale, *Rept. Prog. Phys.* 16, 1 (1953).
4. B. Jacrot and T. Riste, "Magnetic Inelastic Scattering of Neutrons," in *Thermal Neutron Scattering*, Edited by P. A. Egelstaff, Academic, New York, 1965, p. 251.
5. L. S. Kothari and K. S. Singwi, "Interaction of Thermal Neutrons with Solids," in *Solid State Physics*, Edited by F. Seitz and D. Turnbull, Vol. 8, Academic, New York, 1959, p. 109.
6. G. Breit, *Phys. Rev.* 71, 215 (1947).
7. J. M. Blatt and V. F. Weisskopf, *Theoretical Nuclear Physics*, Wiley, New York, 1952, p. 71.
8. E. Fermi, *Ricerca Scientifica* 7, 13 (1936).
9. B. N. Brookhouse, in *Phonons in Perfect Lattices and in Lattices with Imperfections*, Edited by R. W. H. Stevenson, Plenum, New York, 1966, p. 110.
10. A. A. Maradudin, E. W. Montroll, and G. H. Weiss, *Theory of Lattice Dynamics in the Harmonic Approximation*, Academic, New York, 1963, p. 264.

Part III

Optical Spectroscopy of Crystals

The third part of this book deals with what happens when a beam of light is focused onto a crystal. It can be transmitted, scattered or absorbed. It can produce fluorescence emission and laser action. Depending on the wavelength of the light, it can interact with electronic transitions or vibrational states of the crystal. If the crystal is doped with impurity ions, the light can interact with the electronic and vibrational energy levels of these ions.

This field of physics is called *optical spectroscopy*. The information gained through optical spectroscopy has been critically important in the development of solid state laser systems. This book does not attempt to cover the broad, general field of optical spectroscopy, but rather focuses on the aspects of the field important to solid state lasers. Both crystal symmetry and lattice vibrations play important roles in determining the spectroscopic properties of ions in solids. Thus the material covered in PARTS I and II of this book provide an important background for understanding the material covered in this part.

PART III is divided into five chapters. Chapter 9 develops the quantum mechanical formalism for describing the interaction of electromagnetic radiation with atoms in crystals. The next two chapters focus on the optical spectra of impurity ions in crystals. The first of these describes the effects of lattice vibrations on spectra involving the absorption or emission of photons including, including processes involving the scattering, absorption or emission of phonons. The second of these chapters discusses the relationship of the observed spectra to the strength of the electron-phonon interaction. Chapter 12 describes the use of optical spectroscopy to directly probe the lattice vibrations of the crystal. This includes the spectrum of resonant absorption of infrared photons and the inelastic scattering of optical photons during which energy is exchanged with phonons. Finally, in Chapter 13 the optical spectroscopy of solid state laser materials is discussed. The central focus of this chapter is how lattice vibrations affect the operational characteristics of the laser system.

Chapter 9

Interaction of Radiation with Matter

The classical description of electromagnetic radiation developed in the late 1800's is described by Maxwell's equations and the wave equation. In the early part of the 1900's the quantum theory of radiation was developed. Electromagnetic radiation was shown to have a duel wave and particle nature. The particle nature was especially important in explaining some the experimental observations of the interaction of radiation with matter.

The basic concept of the quantum theory of radiation is that electromagnetic energy is described as a beam of particles called *photons*. The amount of energy carried by each photon is proportional to the frequency of the electromagnetic wave. An electromagnetic wave interacts with matter through the absorption, emission, or scattering of photons with discrete amounts of energy depending on the frequency of the radiation.

This chapter describes how an electromagnetic field made up of photons can be modeled as an ensemble of harmonic oscillators. The properties of the field depend on the *occupation number* of photons in different harmonic oscillator modes, each having specific frequencies and polarizations. This approach is called *second quantization* and is especially useful in describing the interaction of radiation with matter. After developing this formalism, the rest of the chapter shows how it can be used to derive the transition matrix elements for absorption, emission, and scattering processes involved in the interaction of radiation with matter. The application of these concepts to impurity ions in solids, lattice vibrations of solids, and solid state lasers is covered in subsequent chapters.

9.1. The Classical Radiative Field

An electromagnetic field in the presence of currents and charges is given by the four Maxwell equations

$$\underset{\sim}{\nabla} \cdot \underset{\sim}{E} = 4\pi\rho, \tag{9.1.1}$$

$$\underset{\sim}{\nabla} \cdot \underset{\sim}{B} = 0, \tag{9.1.2}$$

$$\underset{\sim}{\nabla} \times \underset{\sim}{E} + \frac{1}{c}\frac{\partial \underset{\sim}{B}}{\partial t} = 0, \tag{9.1.3}$$

$$\underset{\sim}{\nabla} \times \underset{\sim}{B} - \frac{1}{c}\frac{\partial \underset{\sim}{E}}{\partial t} = \frac{4\pi}{c}\underset{\sim}{j}. \tag{9.1.4}$$

Let us now consider the second and third (homogeneous) equations above. We know that

$$\underset{\sim}{\nabla} \cdot \underset{\sim}{\nabla} \times \underset{\sim}{u} = 0. \tag{9.1.5}$$

Therefore we can set, from (9.1.2),

$$\underset{\sim}{B} = \underset{\sim}{\nabla} \times \underset{\sim}{A}, \tag{9.1.6}$$

where $\underset{\sim}{A}$ is defined as *vector potential.*

Replacing this expression of $\underset{\sim}{B}$ in (9.1.3) we obtain

$$\underset{\sim}{\nabla} \times \underset{\sim}{E} = -\frac{1}{c}\frac{\partial}{\partial t}(\underset{\sim}{\nabla} \times \underset{\sim}{A}). \tag{9.1.7}$$

We know also that

$$\underset{\sim}{\nabla} \times \underset{\sim}{\nabla}\phi = 0. \tag{9.1.8}$$

Therefore we can set

$$\underset{\sim}{E} = -\underset{\sim}{\nabla}\phi - \frac{1}{c}\frac{\partial \underset{\sim}{A}}{\partial t}, \tag{9.1.9}$$

where ϕ is defined as *scalar potential.* Let us now consider the first and fourth (inhomogeneous) Maxwell equations. Replacing $\underset{\sim}{E}$ and $\underset{\sim}{B}$

with the expressions (9.1.9) and (9.1.6), respectively, we find

$$\underset{\sim}{\nabla} \cdot \left(-\underset{\sim}{\nabla} \phi - \frac{1}{c} \frac{\partial \underset{\sim}{A}}{\partial t} \right) = 4\pi \rho, \tag{9.1.10}$$

$$\underset{\sim}{\nabla} \times (\underset{\sim}{\nabla} \times \underset{\sim}{A}) - \frac{1}{c} \frac{\partial}{\partial t} \left(-\underset{\sim}{\nabla} \phi - \frac{1}{c} \frac{\partial \underset{\sim}{A}}{\partial t} \right) = \frac{4\pi}{c} \underset{\sim}{j}. \tag{9.1.11}$$

These two expressions become

$$\nabla^2 \phi + \frac{1}{c} \frac{\partial}{\partial t} (\underset{\sim}{\nabla} \cdot \underset{\sim}{A}) = -4\pi \rho, \tag{9.1.12}$$

$$\left(\nabla^2 \underset{\sim}{A} - \frac{1}{c^2} \frac{\partial^2 \underset{\sim}{A}}{\partial t^2} \right) - \underset{\sim}{\nabla} \left(\underset{\sim}{\nabla} \cdot \underset{\sim}{A} + \frac{1}{c} \frac{\partial \phi}{\partial t} \right) = -\frac{4\pi}{c} \underset{\sim}{j}, \tag{9.1.13}$$

since

$$\underset{\sim}{\nabla} \times (\underset{\sim}{\nabla} \times \underset{\sim}{A}) = \underset{\sim}{\nabla} (\underset{\sim}{\nabla} \cdot \underset{\sim}{A}) - \nabla^2 \underset{\sim}{A}. \tag{9.1.14}$$

Note that ϕ and $\underset{\sim}{A}$ are not uniquely determined. In fact, if we set

$$\phi' = \phi - \frac{1}{c} \frac{\partial f}{\partial t}, \tag{9.1.15}$$

$$\underset{\sim}{A}' = \underset{\sim}{A} + \underset{\sim}{\nabla} f, \tag{9.1.16}$$

f being any function of position and time, we find

$$\begin{aligned} \underset{\sim}{E} &= -\frac{1}{c} \frac{\partial \underset{\sim}{A}}{\partial t} - \underset{\sim}{\nabla} \phi \\ &= -\frac{1}{c} \frac{\partial}{\partial t} (\underset{\sim}{A}' - \underset{\sim}{\nabla} f) - \underset{\sim}{\nabla} \left(\phi' + \frac{1}{c} \frac{\partial f}{\partial t} \right) \\ &= -\frac{1}{c} \frac{\partial \underset{\sim}{A}'}{\partial t} + \frac{1}{c} \frac{\partial}{\partial t} (\underset{\sim}{\nabla} f) - \underset{\sim}{\nabla} \phi' - \frac{1}{c} \frac{\partial}{\partial t} (\underset{\sim}{\nabla} f) \\ &= -\frac{1}{c} \frac{\partial \underset{\sim}{A}'}{\partial t} - \underset{\sim}{\nabla} \phi', \end{aligned} \tag{9.1.17}$$

$$\underset{\sim}{B} = \underset{\sim}{\nabla} \times \underset{\sim}{A} = \underset{\sim}{\nabla} \times (\underset{\sim}{A}' - \underset{\sim}{\nabla} f) = \underset{\sim}{\nabla} \times \underset{\sim}{A}'. \tag{9.1.18}$$

It is easy to show that (9.1.12) and (9.1.13) are also valid if we replace ϕ and $\underset{\sim}{A}$ with ϕ' and $\underset{\sim}{A}'$, respectively.

The indeterminacy of ϕ and $\underset{\sim}{A}$ is removed by imposing an additional condition, which we choose to be

$$\underset{\sim}{\nabla} \cdot \underset{\sim}{A} = 0. \tag{9.1.19}$$

With this condition (9.1.12) and (9.1.13) become

$$\nabla^2 \phi = -4\pi\rho \quad \text{(Poisson's equation)}, \tag{9.1.20}$$

$$\nabla^2 \underset{\sim}{A} - \frac{1}{c^2}\frac{\partial^2 \underset{\sim}{A}}{\partial t^2} = -\frac{4\pi}{c}\underset{\sim}{j} + \frac{1}{c}\frac{\partial}{\partial t}(\underset{\sim}{\nabla}\phi). \tag{9.1.21}$$

The equation (9.1.20) can be integrated by using *Green's theorem*, which is expressed by the equality

$$\int d\tau (G\nabla^2\phi - \phi\nabla^2 G) = \int ds \left[G\frac{\partial\phi}{\partial n} - \phi\frac{\partial G}{\partial n}\right], \tag{9.1.22}$$

where the integral in the left-hand side is over a volume and the integral on the right-hand side is over a surface enclosing the volume, and $\partial\phi/\partial n$ is the derivative of ϕ in the direction perpendicular to the surface. If we set

$$G(\underset{\sim}{r}, \underset{\sim}{r}') = \frac{1}{|\underset{\sim}{r} - \underset{\sim}{r}'|}, \tag{9.1.23}$$

we obtain

$$\nabla_r^2 G\,(\underset{\sim}{r}, \underset{\sim}{r}') = -4\pi\delta(\underset{\sim}{r} - \underset{\sim}{r}'), \tag{9.1.24}$$

and

$$\int d^3\underset{\sim}{r}'(G\nabla^2\phi - \phi\nabla^2 G)$$

$$= \int d^3\underset{\sim}{r}' \left[\frac{-4\pi\rho(\underset{\sim}{r}', t)}{|\underset{\sim}{r} - \underset{\sim}{r}'|} + \phi(\underset{\sim}{r}', t)4\pi\delta(\underset{\sim}{r} - \underset{\sim}{r}')\right] = 0, \tag{9.1.25}$$

since

$$\phi(\infty) = G(\infty) = 0. \tag{9.1.26}$$

Then

$$\phi(\underset{\sim}{r}, t) = \int \frac{\rho(\underset{\sim}{r}', t)}{|\underset{\sim}{r} - \underset{\sim}{r}'|}\, d^3\underset{\sim}{r}'. \tag{9.1.27}$$

ϕ is a function only of the charge distribution; if $\rho = 0$, $\phi = 0$.

On the other hand, if $\rho = \underset{\sim}{j} = 0$ then (9.1.21) becomes

$$\nabla^2 \underset{\sim}{A}(\underset{\sim}{r}, t) - \frac{1}{c^2}\frac{\partial^2}{\partial t^2}\underset{\sim}{A}(\underset{\sim}{r}, t) = 0, \tag{9.1.28}$$

which is called *field equation.* This, together with

$$\underset{\sim}{\nabla} \cdot \underset{\sim}{A} = 0, \tag{9.1.29}$$

$$\underset{\sim}{E} = -\frac{1}{c}\frac{\partial \underset{\sim}{A}}{\partial t}, \tag{9.1.30}$$

$$\underset{\sim}{B} = \underset{\sim}{\nabla} \times \underset{\sim}{A}, \tag{9.1.31}$$

defines the *radiative field.* We assume that the solutions of the field equation are of the form

$$\underset{\sim}{A}(\underset{\sim}{r}, t) = q(t)\underset{\sim}{\mathcal{A}}(\underset{\sim}{r}). \tag{9.1.32}$$

The field equation becomes then

$$q\nabla^2 \underset{\sim}{\mathcal{A}} = \frac{1}{c^2}\underset{\sim}{\mathcal{A}}\ddot{q},$$

or

$$\frac{\ddot{q}}{q} = \frac{c^2\nabla^2 \underset{\sim}{\mathcal{A}}}{\underset{\sim}{\mathcal{A}}} = -\omega^2, \tag{9.1.33}$$

where ω^2 is a constant. From (9.1.33) we obtain

$$\ddot{q} + \omega^2 q = 0, \tag{9.1.34}$$

$$\nabla^2 \underset{\sim}{\mathcal{A}} + \frac{\omega^2}{c^2}\underset{\sim}{\mathcal{A}} = 0, \tag{9.1.35}$$

with the solutions

$$q(t) = |q|e^{-i\omega t}, \tag{9.1.36}$$

$$\underset{\sim}{\mathcal{A}}(\underset{\sim}{r}) = \underset{\sim}{\pi}\left(\frac{4\pi c^2}{V}\right)^{\frac{1}{2}} e^{i\underset{\sim}{k}\cdot\underset{\sim}{r}}, \tag{9.1.37}$$

where $\underset{\sim}{\pi}$ is the unit vector in the direction of polarization, $V = L_x L_y L_z$ is the volume of the space in which the field is confined, and $k = \omega/c$. Therefore

$$\underset{\sim}{A}(\underset{\sim}{r}, t) = |q| \left(\frac{4\pi c^2}{V}\right)^{\frac{1}{2}} \underset{\sim}{\pi} e^{i(\underset{\sim}{k}\cdot\underset{\sim}{r}-\omega t)}, \tag{9.1.38}$$

which represents a plane wave of wavelength $\lambda = 2\pi/k$. Because of the condition $\underset{\sim}{\nabla} \cdot \underset{\sim}{A} = 0$, it is

$$\pi_x k_x + \pi_y k_y + \pi_z k_z = \underset{\sim}{\pi} \cdot \underset{\sim}{k} = 0, \tag{9.1.39}$$

that is, the directions of propagation and polarization are perpendicular to each other.

Generalizing the expression for $\underset{\sim}{A}(\underset{\sim}{r}, t)$ to include allowed polarizations and allowed $\underset{\sim}{k}$ and $\underset{\sim}{\omega}$, we may write

$$\underset{\sim}{A}(\underset{\sim}{r}, t) = \sum_{\alpha} \sum_{\lambda} [q_{\alpha}^{\lambda}(t) \underset{\sim}{A}_{\alpha}^{\lambda}(\underset{\sim}{r}) + q_{\alpha}^{\lambda}(t)^* \underset{\sim}{A}_{\alpha}^{\lambda}(\underset{\sim}{r})^*], \tag{9.1.40}$$

where λ ranges over the two possible polarizations and the values of α can be determined by the periodic boundary conditions

$$\begin{aligned} \underset{\sim}{A}(\underset{\sim}{r} + L_x \underset{\sim}{1}_x, t) &= \underset{\sim}{A}(\underset{\sim}{r}, t), \\ \underset{\sim}{A}(\underset{\sim}{r} + L_y \underset{\sim}{1}_y, t) &= \underset{\sim}{A}(\underset{\sim}{r}, t), \\ \underset{\sim}{A}(\underset{\sim}{r} + L_z \underset{\sim}{1}_z, t) &= \underset{\sim}{A}(\underset{\sim}{r}, t). \end{aligned} \tag{9.1.41}$$

Here $\underset{\sim}{1}_x$, $\underset{\sim}{1}_y$, and $\underset{\sim}{1}_z$ are the unit vectors in the x, y, and z directions, respectively. It must then be

$$\begin{aligned} k_x L_x &= 2\pi n_x, \\ k_y L_y &= 2\pi n_y, \\ k_z L_z &= 2\pi n_z, \end{aligned} \tag{9.1.42}$$

or

$$\begin{aligned} k_x &= \frac{2\pi}{L_x} n_x, \\ k_y &= \frac{2\pi}{L_y} n_y, \\ k_z &= \frac{2\pi}{L_z} n_z, \end{aligned} \tag{9.1.43}$$

with n_x, n_y and $n_z = 0, \pm 1, \pm 2, \pm 3, \ldots$. In (9.1.40), α stands for the three numbers n_x, n_y, and n_z. By specifying these three numbers we specify $\underset{\sim}{k}$ and $\underset{\sim}{\omega}$. We can then express $\underset{\sim}{A}$ as

$$\underset{\sim}{A}(\underset{\sim}{r}, t) = \sum_{\alpha} \sum_{\lambda} \underset{\sim}{\pi}_{\alpha}^{\lambda} \left(\frac{4\pi c^2}{V} \right)^{\frac{1}{2}} |q_{\alpha}^{\lambda}|$$
$$\times \left\{ \exp \left[i \left(\frac{2\pi}{L_x} n_{x\alpha} x + \frac{2\pi}{L_y} n_{y\alpha} y + \frac{2\pi}{L_z} n_{z\alpha} z - \omega_{\alpha} t \right) \right] \right.$$
$$\left. + \text{complex conjugate} \right\}. \tag{9.1.44}$$

In what follows we shall drop for simplicity of notation the subscript λ. We note that

$$\begin{aligned} \underset{\sim}{k}_{-\alpha} &= -\underset{\sim}{k}_{\alpha}, \\ \underset{\sim}{A}_{-\alpha} &= \underset{\sim}{A}_{\alpha}^{*}, \\ \underset{\sim}{\pi}_{-\alpha} &= \underset{\sim}{\pi}_{\alpha}, \end{aligned} \tag{9.1.45}$$

and that, since

$$\underset{\sim}{A}_{\alpha}(\underset{\sim}{r}) = \underset{\sim}{\pi}_{\alpha} \left(\frac{4\pi c^2}{V} \right)^{\frac{1}{2}} e^{i\underset{\sim}{k}_{\alpha} \cdot \underset{\sim}{r}}, \tag{9.1.46}$$

it is therefore

$$\int \underset{\sim}{A}_{\alpha} \cdot \underset{\sim}{A}_{\alpha'}\, d^3\underset{\sim}{r} = 4\pi c^2\, \delta_{\alpha,-\alpha'}, \tag{9.1.47}$$

$$\int \underset{\sim}{A}_{\alpha} \cdot \underset{\sim}{A}_{\alpha}\, d^3\underset{\sim}{r} = 0 = \int \underset{\sim}{A}_{\alpha}^{*} \cdot \underset{\sim}{A}_{\alpha}^{*} d^3\underset{\sim}{r}, \tag{9.1.48}$$

$$\int \underset{\sim}{A}_{\alpha}^{*} \cdot \underset{\sim}{A}_{\alpha}\, d^3\underset{\sim}{r} = 4\pi c^2. \tag{9.1.49}$$

Also, since

$$\underset{\sim}{A}_{\alpha}(\underset{\sim}{r}, t) = q_{\alpha}(t) \underset{\sim}{A}_{\alpha}(\underset{\sim}{r}) + q_{\alpha}(t)^{*} \underset{\sim}{A}_{\alpha}(\underset{\sim}{r})^{*}, \tag{9.1.50}$$

it is

$$\underset{\sim}{E}_{\alpha} = -\frac{1}{c} \frac{\partial \underset{\sim}{A}_{\alpha}}{\partial t} = \frac{i\omega_{\alpha}}{c} (q_{\alpha} \underset{\sim}{A}_{\alpha} - q_{\alpha}^{*} \underset{\sim}{A}_{\alpha}^{*}), \tag{9.1.51}$$

and

$$\underset{\sim}{E} = \sum_\alpha \frac{i\omega_\alpha}{c}(q_\alpha \underset{\sim}{A}_\alpha - q_\alpha^* \underset{\sim}{A}_\alpha^*). \tag{9.1.52}$$

Therefore

$$(\underset{\sim}{E})^2 = \sum_\alpha \sum_{\alpha'} \left[\frac{i\omega_\alpha}{c}(q_\alpha \underset{\sim}{A}_\alpha - q_\alpha^* \underset{\sim}{A}_\alpha^*)\right] \cdot \left[\frac{i\omega_{\alpha'}}{c}(q_{\alpha'} \underset{\sim}{A}_{\alpha'} - q_{\alpha'}^* \underset{\sim}{A}_{\alpha'}^{*'})\right]$$
$$= \sum_\alpha \sum_{\alpha'} \left(-\frac{\omega_\alpha \omega_{\alpha'}}{c^2}\right) [q_\alpha q_{\alpha'} \underset{\sim}{A}_\alpha \cdot \underset{\sim}{A}_{\alpha'} + q_\alpha^* q_{\alpha'}^* \underset{\sim}{A}_\alpha^* \cdot \underset{\sim}{A}_{\alpha'}^*$$
$$- q_\alpha^* q_{\alpha'} \underset{\sim}{A}_\alpha^* \cdot \underset{\sim}{A}_{\alpha'} - q_\alpha q_{\alpha'}^* \underset{\sim}{A}_\alpha \cdot \underset{\sim}{A}_{\alpha'}^*]. \tag{9.1.53}$$

By integrating over space,

$$\int (\underset{\sim}{E})^2 d^3\underset{\sim}{r} = \sum_\alpha \sum_{\alpha'} \left(-\frac{\omega_\alpha \omega_{\alpha'}}{c^2}\right) [q_\alpha q_{\alpha'} 4\pi c^2 \delta_{\alpha,-\alpha'} + q_\alpha^* q_{\alpha'}^* 4\pi c^2 \delta_{\alpha-\alpha'}$$
$$- q_\alpha^* q_{\alpha'} 4\pi c^2 \delta_{\alpha,\alpha'} - q_\alpha q_{\alpha'}^* 4\pi c^2 \delta_{\alpha,\alpha'}]$$
$$= 4\pi \sum_\alpha (-\omega_\alpha^2)[q_\alpha q_{-\alpha} + q_\alpha^* q_{-\alpha}^* - q_\alpha^* q_\alpha - q_\alpha q_\alpha^*]$$
$$= 4\pi \sum_\alpha [\omega_\alpha^2 (q_\alpha q_\alpha^* + q_\alpha^* q_\alpha) - \omega_\alpha^2 (q_\alpha q_{-\alpha} + q_\alpha^* q_{-\alpha}^*]. \tag{9.1.54}$$

On the other hand

$$\underset{\sim}{B} = \underset{\sim}{\nabla} \times \underset{\sim}{A} = \underset{\sim}{\nabla} \times \left[\sum_\alpha (q_\alpha \underset{\sim}{A}_\alpha + q_\alpha^* \underset{\sim}{A}_\alpha^*)\right]$$
$$= \sum_\alpha [q_\alpha (\underset{\sim}{\nabla} \times \underset{\sim}{A}_\alpha) + q_\alpha^* (\underset{\sim}{\nabla} \times \underset{\sim}{A}_\alpha^*)], \tag{9.1.55}$$

and

$$(\underset{\sim}{B})^2 = (\underset{\sim}{\nabla} \times \underset{\sim}{A}) \cdot (\underset{\sim}{\nabla} \cdot \underset{\sim}{A})$$
$$= \sum_\alpha \sum_{\alpha'} [q_\alpha (\underset{\sim}{\nabla} \times \underset{\sim}{A}_\alpha) + q_\alpha^* (\underset{\sim}{\nabla} \times \underset{\sim}{A}_\alpha^*)]$$

$$\cdot [q_{\alpha'}(\underset{\sim}{\nabla} \times \underset{\sim}{A}_{\alpha'}) + q^*_{\alpha'}(\underset{\sim}{\nabla} \times \underset{\sim}{A}^*_{\alpha'})]$$

$$= \sum_{\alpha}\sum_{\alpha'}[q_\alpha q_{\alpha'}(\underset{\sim}{\nabla} \times \underset{\sim}{A}_\alpha) \cdot (\underset{\sim}{\nabla} \times \underset{\sim}{A}_{\alpha'})$$

$$+ q^*_\alpha q^*_{\alpha'}(\underset{\sim}{\nabla} \times \underset{\sim}{A}^*_\alpha) \cdot (\underset{\sim}{\nabla} \times \underset{\sim}{A}^*_{\alpha'}) + q^*_\alpha q_{\alpha'}(\underset{\sim}{\nabla} \times \underset{\sim}{A}^*_\alpha) \cdot (\underset{\sim}{\nabla} \times \underset{\sim}{A}_{\alpha'})$$

$$+ q_\alpha q^*_{\alpha'}(\underset{\sim}{\nabla} \times \underset{\sim}{A}_\alpha) \cdot (\underset{\sim}{\nabla} \times \underset{\sim}{A}^*_{\alpha'}). \tag{9.1.56}$$

From vector analysis we know that, given two vectors $\underset{\sim}{C}$ and $\underset{\sim}{D}$, it is

$$\underset{\sim}{\nabla} \cdot (\underset{\sim}{C} \times \underset{\sim}{D}) = \underset{\sim}{D} \cdot (\underset{\sim}{\nabla} \times \underset{\sim}{C}) - \underset{\sim}{C} \cdot (\underset{\sim}{\nabla} \times \underset{\sim}{D}), \tag{9.1.57}$$

or

$$\underset{\sim}{D} \cdot (\underset{\sim}{\nabla} \times \underset{\sim}{C}) = \underset{\sim}{\nabla} \cdot (\underset{\sim}{C} \times \underset{\sim}{D}) + \underset{\sim}{C} \cdot (\underset{\sim}{\nabla} \times \underset{\sim}{D}). \tag{9.1.58}$$

It is then

$$\begin{aligned}(\underset{\sim}{\nabla} \times \underset{\sim}{A}_\alpha) \cdot (\underset{\sim}{\nabla} \times \underset{\sim}{A}_{\alpha'}) &= \underset{\sim}{\nabla} \cdot \underset{\sim}{A}_\alpha \times (\underset{\sim}{\nabla} \times \underset{\sim}{A}_{\alpha'}) + \underset{\sim}{A}_\alpha \cdot \underset{\sim}{\nabla} \times (\underset{\sim}{\nabla} \times \underset{\sim}{A}_{\alpha'})\\
&= \underset{\sim}{\nabla} \cdot \underset{\sim}{A}_\alpha \times (\underset{\sim}{\nabla} \times \underset{\sim}{A}_{\alpha'}) + \underset{\sim}{A}_\alpha \cdot [\underset{\sim}{\nabla}(\underset{\sim}{\nabla} \cdot \underset{\sim}{A}_{\alpha'}) - \nabla^2 \underset{\sim}{A}_{\alpha'}]\\
&= \underset{\sim}{\nabla} \cdot \underset{\sim}{A}_\alpha \times (\underset{\sim}{\nabla} \times \underset{\sim}{A}_{\alpha'}) - \underset{\sim}{A}\alpha \cdot \nabla^2 \underset{\sim}{A}_{\alpha'}\\
&= \underset{\sim}{\nabla} \cdot \underset{\sim}{A}_\alpha \times (\underset{\sim}{\nabla} \times \underset{\sim}{A}_{\alpha'}) + k^2_{\alpha'} \underset{\sim}{A}_\alpha \cdot \underset{\sim}{A}_{\alpha'},\end{aligned} \tag{9.1.59}$$

where we have taken into account the fact that $\underset{\sim}{\nabla} \cdot \underset{\sim}{A}_{\alpha'} = 0$. Then

$$\begin{aligned}&\int (\underset{\sim}{\nabla} \times \underset{\sim}{A}_\alpha) \cdot (\underset{\sim}{\nabla} \times \underset{\sim}{A}_{\alpha'}) d^3\underset{\sim}{r}\\
&\quad = \int dS[\underset{\sim}{A}_\alpha \times (\underset{\sim}{\nabla} \times \underset{\sim}{A}_{\alpha'})]_n + \frac{\omega^2_{\alpha'}}{c^2} \int \underset{\sim}{A}_\alpha \cdot \underset{\sim}{A}_{\alpha'} d^3\underset{\sim}{r}\\
&\quad = \frac{\omega^2_{\alpha'}}{c^2} 4\pi c^2 \delta_{\alpha,-\alpha'} = \omega^2_{\alpha'} 4\pi \delta_{\alpha,-\alpha'},\end{aligned} \tag{9.1.60}$$

where the subscript n above indicates the component of the vector $\underset{\sim}{A}_\alpha \times (\underset{\sim}{\nabla} \times \underset{\sim}{A}_{\alpha'})$ normal to the boundary surface. The integral over the boundary surface vanishes because of the periodic boundary conditions. We can now write

$$\int (\underset{\sim}{B})^2 \, d^3r$$

$$= \sum_{\alpha} \sum_{\alpha'} [q_\alpha q_{\alpha'} \omega_{\alpha'}^2 4\pi \delta_{\alpha,-\alpha'} + q_\alpha^* q_{\alpha'}^* \omega_{\alpha'}^2 4\pi \delta_{\alpha,-\alpha'}$$

$$+ q_\alpha^* q_{\alpha'} \omega_{\alpha'}^2 4\pi \delta_{\alpha\alpha'} + q_\alpha q_{\alpha'}^* \omega_\alpha^2 4\pi \delta_{\alpha\alpha'}]$$

$$= 4\pi \sum_\alpha [\omega_\alpha^2 (q_\alpha q_{-\alpha} + q_\alpha^* q_{-\alpha}^*) + \omega_\alpha^2 (q_\alpha q_\alpha^* + q_\alpha^* q_\alpha)]. \quad (9.1.61)$$

The energy of the radiative field is then given by

$$\frac{1}{8\pi} \int (\underset{\sim}{E})^3 d^3\underset{\sim}{r} + \frac{1}{8\pi} \int (\underset{\sim}{B})^3 d^3\underset{\sim}{r} = \sum_\alpha \omega_\alpha^2 (q_\alpha q_\alpha^* + q_\alpha^* q_\alpha). \quad (9.1.62)$$

The Hamiltonian of the radiative field can then be expressed as follows:

$$H = \sum_\alpha \omega_\alpha^2 (q_\alpha q_\alpha^* + q_\alpha^* q_\alpha). \quad (9.1.63)$$

Two new variables can now be introduced for each α:

$$\begin{aligned} Q_\alpha &= q_\alpha + q_\alpha^*, \\ P_\alpha &= -i\omega_\alpha (q_\alpha - q_\alpha^*). \end{aligned} \quad (9.1.64)$$

Since

$$\begin{aligned} q_\alpha &= \frac{1}{2}\left(Q_\alpha - \frac{1}{i\omega_\alpha} P_\alpha\right), \\ q_\alpha^* &= \frac{1}{2}\left(Q_\alpha + \frac{1}{i\omega_\alpha} P_\alpha\right). \end{aligned} \quad (9.1.65)$$

the Hamiltonian becomes

$$H = \sum_\alpha \omega_\alpha^2 (q_\alpha q_\alpha^* + q_\alpha^* q_\alpha)$$

$$= \sum_\alpha \left(\frac{1}{2}\omega_\alpha^2 Q_\alpha^2 + \frac{1}{2} P_\alpha^2\right) = \sum_\alpha H_\alpha, \quad (9.1.66)$$

where

$$H_\alpha = \frac{1}{2}\omega_\alpha^2 Q_\alpha^2 + \frac{1}{2}P_\alpha^2. \tag{9.1.67}$$

Q_α and P_α are canonical (and real) variables. In fact, since for each α,

$$\ddot{q}_\alpha + \omega_\alpha^2 q_\alpha = 0, \tag{9.1.68}$$

it must also be

$$\ddot{Q}_\alpha + \omega_\alpha^2 Q_\alpha = 0. \tag{9.1.69}$$

This relation can be deduced from the Hamiltonian equation

$$\begin{aligned} \frac{\partial H_\alpha}{\partial P_\alpha} &= \dot{Q}_\alpha = P_\alpha, \\ \frac{\partial H_\alpha}{\partial Q_\alpha} &= -\dot{P}_\alpha = \omega_\alpha^2 Q_\alpha, \end{aligned} \tag{9.1.70}$$

and therefore Q_α and P_α are canonical variables.

9.2. The Quantum Theory of the Radiative Field

Since Q_α and P_α are canonical variables we can go over to quantum mechanics by setting

$$\begin{aligned} [Q_\alpha, P_{\alpha'}] &= i\hbar\delta_{\alpha\alpha'}, \\ [Q_\alpha, Q_{\alpha'}] &= [P_\alpha, P_{\alpha'}] = 0, \end{aligned} \tag{9.2.1}$$

from which we can derive

$$[q_\alpha, q_{\alpha'}^+] = \frac{\hbar}{2\omega_\alpha}\delta_{\alpha\alpha'}, \tag{9.2.2}$$

q_α and q_α^+ are now two operators, replacing q_α and q_α^*. These two operators can be expressed in terms of the dimensionless operators

a_α and a_α^+ called *creation* and *annihilation* operators, and given by

$$a_\alpha = \sqrt{\frac{2\omega_\alpha}{\hbar}} q_\alpha,$$

$$a_\alpha^+ = \sqrt{\frac{2\omega_\alpha}{\hbar}} q_\alpha^+, \tag{9.2.3}$$

respectively. It is also

$$[a_\alpha, a_{\alpha'}^+] = \delta_{\alpha\alpha'},$$

$$[a_\alpha, a_{\alpha'}] = [a_\alpha^+, a_{\alpha'}^+] = 0. \tag{9.2.4}$$

The Hamiltonian of the radiative field now takes the form

$$H = \sum_\alpha \omega_\alpha^2 \frac{\hbar}{2\omega_\alpha}(a_\alpha a_\alpha^+ + a_\alpha^+ a_\alpha)$$

$$= \sum_\alpha \frac{\hbar\omega_\alpha}{2}(a_\alpha a_\alpha^+ + a_\alpha^+ a_\alpha)$$

$$= \sum_\alpha \hbar\omega_\alpha \left(a_\alpha^+ a_\alpha + \frac{1}{2}\right),$$

or, reintroducing the superscript λ,

$$H = \sum_\alpha \sum_\lambda \hbar\omega_\alpha \left(a_\alpha^{\lambda+} a_\alpha^\lambda + \frac{1}{2}\right). \tag{9.2.5}$$

The eigenfunction of the radiative field is simply given by

$$\psi_{\substack{\lambda_1\lambda_2 \\ n_1 n_2 \ldots}} = \underset{\alpha\lambda}{\pi} |n_\alpha^\lambda\rangle, \tag{9.2.6}$$

and the energy levels by

$$E_{\substack{\lambda_1\lambda_2 \\ n_1 n_2 \ldots}} = \sum_\alpha \sum_\lambda \hbar\omega_\alpha \left(n_\alpha^\lambda + \frac{1}{2}\right). \tag{9.2.7}$$

The vector potential can now be expressed in terms of the operators a_α^λ and $a_\alpha^{\lambda+}$ as follows:

$$\begin{aligned}\underset{\sim}{A} &= \sum_\alpha \sum_\lambda [\underset{\sim}{A}_\alpha^\lambda(\underset{\sim}{r}) q_\alpha^\lambda + \underset{\sim}{A}_\alpha^\lambda(\underset{\sim}{r})^* q_\alpha^{\lambda+}] \\ &= \sum_\alpha \sum_\lambda \sqrt{\frac{4\pi c^2}{V}} \sqrt{\frac{\hbar}{2\omega_\alpha}}\, \underset{\sim}{\pi}_\alpha^\lambda (e^{i\underset{\sim}{k}_\alpha \cdot \underset{\sim}{r}} a_\alpha^\lambda + e^{-i\underset{\sim}{k}_\alpha \cdot \underset{\sim}{r}} a_\alpha^{\lambda+}) \\ &= \sum_\alpha \sum_\lambda \sqrt{\frac{hc^2}{\omega_\alpha V}}\, \underset{\sim}{\pi}_\alpha^\lambda (e^{i\underset{\sim}{k}_\alpha \cdot \underset{\sim}{r}} a_\alpha^\lambda + e^{-i\underset{\sim}{k}_\alpha \cdot \underset{\sim}{r}} a_\alpha^{\lambda+}). \qquad (9.2.8)\end{aligned}$$

9.3. The Hamiltonian of a Charged Particle in an Electromagnetic Field

Let us consider a particle with a charge q and mass m in an electromagnetic field defined by the scalar potential ϕ and the vector potential $\underset{\sim}{A}$. The Lagrangian of this particle is given by

$$L = T - q\phi + \frac{q}{c}\underset{\sim}{v} \cdot \underset{\sim}{A}, \qquad (9.3.1)$$

where T is the kinetic energy of the particle and $\underset{\sim}{v}$ is the velocity of the particle. In fact, considering, for example, the x component of the particle's position we have

$$\frac{d}{dt}\frac{\partial L}{\partial \dot{x}} = m\ddot{x} + \frac{q}{c}\frac{dA_x}{dt}, \qquad (9.3.2)$$

$$\frac{\partial L}{\partial x} = -q\frac{\partial \phi}{\partial x} + \frac{\partial}{\partial x}\left(\frac{q}{c}\underset{\sim}{v} \cdot \underset{\sim}{A}\right), \qquad (9.3.3)$$

and the equation of motion in the x direction is given by

$$m\ddot{x} + \frac{q}{c}\frac{dA_x}{dt} = -q\frac{\partial \phi}{\partial x} + \frac{\partial}{\partial x}\left(\frac{q}{c}\underset{\sim}{v} \cdot \underset{\sim}{A}\right). \qquad (9.3.4)$$

Therefore we may write

$$m\ddot{\underset{\sim}{r}} = q\left[-\underset{\sim}{\nabla}\phi + \underset{\sim}{\nabla}\left(\frac{1}{c}\underset{\sim}{v} \cdot \underset{\sim}{A}\right) - \frac{1}{c}\frac{d\underset{\sim}{A}}{dt}\right]. \qquad (9.3.5)$$

But

$$\underset{\sim}{v} \times (\underset{\sim}{\nabla} \times \underset{\sim}{A}) = \underset{\sim}{\nabla}(\underset{\sim}{v} \cdot \underset{\sim}{A}) - (\underset{\sim}{v} \cdot \underset{\sim}{\nabla})A, \tag{9.3.6}$$

and

$$\frac{d\underset{\sim}{A}}{dt} = \frac{\partial \underset{\sim}{A}}{\partial t} + (\underset{\sim}{v} \cdot \underset{\sim}{\nabla})\underset{\sim}{A}. \tag{9.3.7}$$

Then

$$\begin{aligned} m\ddot{\underset{\sim}{r}} &= q\left\{-\underset{\sim}{\nabla}\phi + \frac{1}{c}[\underset{\sim}{v} \times (\underset{\sim}{\nabla} \times \underset{\sim}{A}) + (\underset{\sim}{v} \cdot \underset{\sim}{\nabla})\underset{\sim}{A}] - \frac{1}{c}\frac{\partial \underset{\sim}{A}}{\partial t} - \frac{1}{c}(\underset{\sim}{v} \cdot \underset{\sim}{\nabla})\underset{\sim}{A}\right\} \\ &= q\left\{-\underset{\sim}{\nabla}\phi - \frac{1}{c}\frac{\partial \underset{\sim}{A}}{\partial t} + \frac{1}{c}\underset{\sim}{v} \times (\underset{\sim}{\nabla} \times \underset{\sim}{A})\right\} \\ &= q\left\{\underset{\sim}{E} + \frac{1}{c}(\underset{\sim}{v} \times \underset{\sim}{B})\right\}, \end{aligned} \tag{9.3.8}$$

since

$$\underset{\sim}{E} = -\underset{\sim}{\nabla}\phi - \frac{1}{c}\frac{\partial A}{\partial t}, \tag{9.3.9}$$

$$\underset{\sim}{B} = \underset{\sim}{\nabla} \times \underset{\sim}{A}. \tag{9.3.10}$$

Since the right side of (9.3.8) expresses the Lorentz force on the charge, that L in (9.3.1) is the correct Lagrangian is now proved.

The generalized momentum can be derived as follows:

$$\begin{aligned} p_i = \frac{\partial L}{\partial \dot{x}_i} &= \frac{\partial T}{\partial \dot{x}_i} + \frac{q}{c}\frac{\partial}{\partial \dot{x}_i}(\underset{\sim}{v} \cdot \underset{\sim}{A}) \\ &= \frac{\partial T}{\partial \dot{x}_i} + \frac{q}{c}A_i = mv_i + \frac{q}{c}A_i, \end{aligned} \tag{9.3.11}$$

where $x_i = x, y, z$. The Hamiltonian is then given by

$$\begin{aligned} H = \sum_i (p_i\dot{x}_i - L) &= \sum_i \left[\frac{\partial L}{\partial \dot{x}_i}x_i - \left(T - q\phi + \frac{q}{c}\underset{\sim}{v} \cdot \underset{\sim}{A}\right)\right] \\ &= \sum_i \left[\left(\frac{\partial T}{\partial \dot{x}_i} + \frac{q}{c}A_i\right)\dot{x}_i\right] - \left(T - q\phi + \frac{q}{c}\underset{\sim}{v} \cdot \underset{\sim}{A}\right) \\ &= \sum_i \frac{\partial T}{\partial \dot{x}_i}\dot{x}_i - T + q\phi = T + q\phi = \frac{1}{2}mv^2 + q\phi. \end{aligned} \tag{9.3.12}$$

From (9.3.11)

$$\underset{\sim}{p} = m\underset{\sim}{v} + \frac{q}{c}\underset{\sim}{A}, \tag{9.3.13}$$

and

$$\underset{\sim}{v} = \frac{\underset{\sim}{p} - q/c\underset{\sim}{A}}{m}. \tag{9.3.14}$$

Therefore we can write

$$H = \frac{[\underset{\sim}{p} - (q/c)\,\underset{\sim}{A}]^2}{2m} + q\phi. \tag{9.3.15}$$

9.4. The Interaction Between a Charged Particle and a Radiative Field

Let us assume that a charged particle of mass m and charge q is in some bound state and imbedded in a radiative field. We shall consider the total system, which consists of the bound particle and of the radiative field; the Hamiltonian of this system is given by

$$H = \frac{1}{2m}\left(\underset{\sim}{p} - \frac{q}{c}\underset{\sim}{A}\right)^2 + q\phi + \frac{1}{8\pi}\int (E^2 + B^2)d^3\underset{\sim}{r}, \tag{9.4.1}$$

where $\underset{\sim}{p}$ is the momentum of the particle, ϕ is the scalar potential, and $\underset{\sim}{A}$ is the vector potential. We can write the Hamiltonian as follows:

$$\begin{aligned} H &= \frac{\underset{\sim}{p}^2}{2m} + q\phi + \frac{1}{8\pi}\int (E^2 + B^2)d^3\underset{\sim}{r} \\ &\quad - \frac{q}{2mc}(\underset{\sim}{p}\cdot\underset{\sim}{A} + \underset{\sim}{A}\cdot\underset{\sim}{p}) + \frac{q^2}{2mc^2}(\underset{\sim}{A})^2. \end{aligned} \tag{9.4.2}$$

But

$$\underset{\sim}{p}\cdot\underset{\sim}{A} = \underset{\sim}{A}\cdot\underset{\sim}{p} + [\underset{\sim}{p}, \underset{\sim}{A}]. \tag{9.4.3}$$

Considering the x component of the operator $[\underset{\sim}{p}, \underset{\sim}{A}]$ operating on some function ϕ,

$$\begin{aligned} [p_x, A_x]\phi &= p_x A_x \phi - A_x p_x \phi \\ &= -i\hbar\frac{\partial}{\partial x}(A_x\phi) + A_x\, i\hbar\frac{\partial\phi}{\partial x} \end{aligned}$$

$$= -i\hbar \left[\frac{\partial A_x}{\partial x}\right] \phi - i\hbar\, A_x \frac{\partial \phi}{\partial x} + A_x i\hbar \frac{\partial \phi}{\partial x}$$

$$= -i\hbar \left[\frac{\partial A_x}{\partial x}\right] \phi. \tag{9.4.4}$$

Therefore

$$[\underset{\sim}{p}, \underset{\sim}{A}] = -i\hbar(\underset{\sim}{\nabla} \cdot \underset{\sim}{A}) = 0, \tag{9.4.5}$$

because of the Coulomb gauge $\underset{\sim}{\nabla} \cdot \underset{\sim}{A} = 0$. We can then write

$$\underset{\sim}{p} \cdot \underset{\sim}{A} = \underset{\sim}{A} \cdot \underset{\sim}{p}, \tag{9.4.6}$$

and

$$\underset{\sim}{p} \cdot \underset{\sim}{A} + \underset{\sim}{A} \cdot \underset{\sim}{p} = 2\underset{\sim}{A} \cdot \underset{\sim}{p}. \tag{9.4.7}$$

The Hamiltonian (9.4.2) can now be written

$$H = H_0 + H_1 + H_2, \tag{9.4.8}$$

where

$$H_0 = \frac{\underset{\sim}{p}^2}{2m} + q\phi + \frac{1}{8\pi} \int (E^2 + B^2) d^3\underset{\sim}{r}$$

$$= -\frac{\hbar^2}{2m}\nabla^2 + q\phi + \sum_{\alpha\lambda} \hbar\omega_\alpha \left(a_\alpha^{\lambda +} a_\alpha^\lambda + \frac{1}{2} \right), \tag{9.4.9}$$

$$H_1 = -\frac{q}{mc}\underset{\sim}{A} \cdot \underset{\sim}{p}$$

$$= -\frac{q}{m} \sum_{\alpha\lambda} \sqrt{\frac{h}{\omega_\alpha V}} (a_\alpha^\lambda e^{i\underset{\sim}{k}_\alpha \cdot \underset{\sim}{r}} + a_\alpha^{\lambda +} e^{-i\underset{\sim}{k}_\alpha \cdot \underset{\sim}{r}}) \underset{\sim}{\pi}_\alpha^\lambda \cdot p, \tag{9.4.10}$$

$$H_2 = -\frac{q^2}{2mc^2}(\underset{\sim}{A})^2$$

$$= \frac{q^2}{2m}\frac{h}{V}\sum_{\alpha\lambda}\sum_{\alpha'\lambda'}\frac{1}{\sqrt{\omega_\alpha\omega_{\alpha'}}}(\underset{\sim}{\pi}_\alpha^\lambda \cdot \underset{\sim}{\pi}_{\alpha'}^{\lambda'})$$

$$\times\,(a_\alpha^\lambda e^{i\underset{\sim}{k}_\alpha\cdot\underset{\sim}{r}} + a_\alpha^{\lambda+}e^{-i\underset{\sim}{k}_\alpha\cdot\underset{\sim}{r}})(a_{\alpha'}^{\lambda'}e^{i\underset{\sim}{k}_\alpha\cdot\underset{\sim}{r}} + a_{\alpha'}^{\lambda'+}e^{-i\underset{\sim}{k}_{\alpha'}\cdot\underset{\sim}{r}}). \qquad (9.4.11)$$

The terms H_1 and H_2 of the Hamiltonian represent the interaction between the charged particle and the radiative field. We notice that this interaction is *time dependent.* The eigenfunctions of the system, neglecting the interaction between particle and radiative field, that is, the eigenfunctions of H_0 are given by

$$\psi_{e;n_1,n_2,\ldots} = \psi^e \prod_{\alpha\lambda} |n_\alpha^\lambda\rangle$$

$$= \psi^e|n_1^1\rangle|n_1^2\rangle|n_2^1\rangle|n_2^2\rangle\cdots, \qquad (9.4.12)$$

where ψ^e is the eigenfunction of the particle and $\prod_{\alpha\lambda}|n_\alpha^\lambda\rangle$ is the eigenfunction of the radiative field. The energies of such eigenstates are given by

$$E_{e;n_1,n_2,\ldots} = E^e + \sum_{\alpha\lambda}\hbar\omega_\alpha\left(n_\alpha^\lambda + \frac{1}{2}\right), \qquad (9.4.13)$$

where E^e is the energy of the particle. The interaction Hamiltonian given by $H_1 + H_2$ can produce transitions between states described by the eigenfunctions ψ in (9.4.12) provided energy is conserved in the process. Such transitions or processes can be classified as follows:

1. *First-Order Processes.* They are due to H_1 in first order.
   1a. Annihilation of a photon of frequency ω_α, called *absorption.*
   1b. Creation of a photon of frequency ω_α, called *emission.*
2. *Second-Order Processes.* They are due to H_1 in second order and/or to H_2 in first order.
   2a. Photon scattering, consisting in the simultaneous absorption of a photon of frequency ω_α and emission of a photon of frequency to ω_α, $(\omega_\alpha \gtrless \omega_{\alpha'})$.
   2b. Two-photon absorption.
   2c. Two-photon emission.

Higher-order processes are possible but are much less probable.

9.5. First-Order Processes: Absorption and Emission of Radiation

As reported in the Section 9.4, first-order processes require the use of the interaction Hamiltonian H_1 in first order. H_1 is given by

$$
\begin{aligned}
H_1 &= -\frac{q}{mc}\underset{\sim}{A}\cdot\underset{\sim}{p} \\
&= -\frac{q}{m}\sum_{\alpha\lambda}\left(\frac{h}{\omega_\alpha V}\right)^{\frac{1}{2}}(a_\alpha^\lambda e^{i\underset{\sim}{k}_\alpha\cdot\underset{\sim}{r}} + a_\alpha^{\lambda+}e^{-i\underset{\sim}{k}_\alpha\cdot\underset{\sim}{r}})\underset{\sim}{\pi}_\alpha^\lambda\cdot\underset{\sim}{p} \\
&\simeq -\frac{q}{m}\sum_{\alpha\lambda}\left(\frac{h}{\omega_\alpha V}\right)^{\frac{1}{2}}\{a_\alpha^\lambda[1+i(\underset{\sim}{k}_\alpha\cdot r)]\underset{\sim}{\pi}_\alpha^\lambda\cdot\underset{\sim}{p}\} \\
&\quad -\frac{q}{m}\sum_{\alpha\lambda}\left(\frac{h}{\omega_\alpha V}\right)^{\frac{1}{2}}\{a_\alpha^{\lambda+}[1-i(\underset{\sim}{k}_\alpha\cdot r)]\underset{\sim}{\pi}_\alpha^\lambda\cdot\underset{\sim}{p}\} \\
&= -\frac{q}{m}\sum_{\alpha\lambda}\left(\frac{h}{\omega_\alpha V}\right)^{\frac{1}{2}}a_\alpha^\lambda\underset{\sim}{\pi}_\alpha^\lambda\cdot\underset{\sim}{p} - \frac{q}{m}\sum_{\alpha\lambda}\left(\frac{h}{\omega_\alpha V}\right)^{\frac{1}{2}}a_\alpha^{\lambda+}\underset{\sim}{\pi}_\alpha^\lambda\cdot\underset{\sim}{p} \\
&\quad -i\frac{q}{m}\sum_{\alpha\lambda}\left(\frac{h}{\omega_\alpha V}\right)^{\frac{1}{2}}a_\alpha^\lambda(\underset{\sim}{k}_\alpha\cdot\underset{\sim}{r})\underset{\sim}{\pi}_\alpha^\lambda\cdot\underset{\sim}{p} \\
&\quad +i\frac{q}{m}\sum_{\alpha\lambda}\left(\frac{h}{\omega_\alpha V}\right)^{\frac{1}{2}}a_\alpha^{\lambda+}(\underset{\sim}{k}_\alpha\cdot\underset{\sim}{r})\underset{\sim}{\pi}_\alpha^\lambda\cdot\underset{\sim}{p}. \qquad (9.5.1)
\end{aligned}
$$

For any function of coordinates, momenta and time we have classically

$$
\begin{aligned}
\frac{dF(x_i,p_i,t)}{dt} &= \frac{\partial F}{\partial t} + \sum_i\left(\frac{\partial F}{\partial x_i}\frac{\partial x_i}{\partial t} + \frac{\partial F}{\partial p_i}\frac{\partial p_i}{\partial t}\right) \\
&= \frac{\partial F}{\partial t} + \sum_i\left(\frac{\partial F}{\partial x_i}\frac{\partial H}{\partial p_i} - \frac{\partial F}{\partial p_i}\frac{\partial H}{\partial x_i}\right) \\
&= \frac{\partial F}{\partial t} + \{F, H\}, \qquad (9.5.2)
\end{aligned}
$$

where the curly brackets indicate a Poisson bracket. If F is not an explicit function of time, the term $\partial F/\partial t$ is zero. Quantum mechanically, if F is an operator with no explicit dependence on time, we can write, replacing in (9.5.2) the Poisson bracket by $(-i/\hbar)$ times the commutator $[F, H]$,

$$\frac{dF}{dt} = -\frac{i}{\hbar}[F, H] = \frac{i}{\hbar}[H, F]. \tag{9.5.3}$$

If we set $F = x$

$$\dot{x} = \frac{i}{\hbar}[H, x], \tag{9.5.4}$$

and

$$p_x = m\dot{x} = i\frac{m}{\hbar}[H, x]. \tag{9.5.5}$$

Taking a matrix element of $[H, x]$ between two eigenfunctions of the particle we find

$$\begin{aligned}\langle \psi_f^e | [H, x] | \psi_i^e \rangle &= \langle \psi_f^e | Hx - xH | \psi_i^e \rangle \\ &= (E_f^e - E_i^e)\langle \psi_f^e | x | \psi_i^e \rangle = \hbar\omega \langle \psi_f^e | x | \psi_i^e \rangle,\end{aligned} \tag{9.5.6}$$

and

$$\langle \psi_f^e | \underset{\sim}{p} | \psi_i^e \rangle = im\omega \langle \psi_f^e | \underset{\sim}{r} | \psi_i^e \rangle. \tag{9.5.7}$$

Therefore we can write

$$\begin{aligned}&-\frac{q}{m}\sum_{\alpha\lambda}\left(\frac{h}{\omega_\alpha V}\right)^{\frac{1}{2}} a_\alpha^\lambda \underset{\sim}{\pi}_\alpha^\lambda \cdot \underset{\sim}{p} - \frac{q}{m}\sum_{\alpha\lambda}\left(\frac{h}{\omega_\alpha V}\right)^{\frac{1}{2}} a_\alpha^{\lambda+} \underset{\sim}{\pi}_\alpha^\lambda \cdot \underset{\sim}{p} \\ &\quad = -i\sum_{\alpha\lambda}\left(\frac{h\omega_\alpha}{V}\right)^{\frac{1}{2}} a_\alpha^\lambda \underset{\sim}{\pi}_\alpha^\lambda \cdot (q\underset{\sim}{r}) + i\sum_{\alpha\lambda}\left(\frac{h\omega_\alpha}{V}\right)^{\frac{1}{2}} a_\alpha^{\lambda+} \underset{\sim}{\pi}_\alpha^\lambda \cdot (q\underset{\sim}{r}).\end{aligned} \tag{9.5.8}$$

Consider now the quantity $(\underset{\sim}{k} \cdot \underset{\sim}{r})\underset{\sim}{p}$. We can write

$$(\underset{\sim}{k} \cdot \underset{\sim}{r})\underset{\sim}{p} = \frac{1}{2}\{(\underset{\sim}{k} \cdot \underset{\sim}{r})\underset{\sim}{p} - (\underset{\sim}{k} \cdot \underset{\sim}{p})\underset{\sim}{r}\} + \frac{1}{2}\{(\underset{\sim}{k} \cdot \underset{\sim}{r})\underset{\sim}{p} + (\underset{\sim}{k} \cdot \underset{\sim}{p})\underset{\sim}{r}\}. \tag{9.5.9}$$

Also

$$\begin{aligned}\frac{1}{2}\{(\mathbf{k}\cdot\mathbf{r})\mathbf{p}-(\mathbf{k}\cdot\mathbf{p})\mathbf{r}\} &= -\frac{1}{2}\{\mathbf{k}\times(\mathbf{r}\times\mathbf{p})\}\\ &= -\frac{\omega}{2c}\{\mathbf{l}_{\mathbf{k}}\times\mathbf{L}\},\end{aligned}\tag{9.5.10}$$

where $\mathbf{L}$ is the angular momentum of the particle, and $\mathbf{l}_{\mathbf{k}}$ is the unit vector in the $\mathbf{k}$ direction. The second term in (9.5.9) can be expressed as follows:

$$\begin{aligned}\frac{1}{2}\{(\mathbf{k}\cdot\mathbf{r})\mathbf{p}+(\mathbf{k}\cdot\mathbf{p})\mathbf{r}\} &= \frac{1}{2}m\{(\mathbf{k}\cdot\mathbf{r})\dot{\mathbf{r}}+(\mathbf{k}\cdot\dot{\mathbf{r}})\mathbf{r}\}\\ &= \frac{m}{2}\frac{d}{dt}[(\mathbf{k}\cdot\mathbf{r})\mathbf{r}] = \frac{im\omega}{2}[(\mathbf{k}\cdot\mathbf{r})\mathbf{r}].\end{aligned}\tag{9.5.11}$$

Therefore we can write

$$\begin{aligned}&-i\frac{q}{m}\sum_{\alpha\lambda}\left(\frac{h}{\omega_\alpha V}\right)^{\frac{1}{2}} a_\alpha^\lambda(\mathbf{k}_\alpha\cdot\mathbf{r})\mathbf{p}\cdot\boldsymbol{\pi}_\alpha^\lambda + i\frac{q}{m}\sum_{\alpha\lambda}\left(\frac{h}{\omega_\alpha V}\right)^{\frac{1}{2}} a_\alpha^{\lambda+}(\mathbf{k}_\alpha\cdot\mathbf{r})\mathbf{p}\cdot\boldsymbol{\pi}_\alpha^\lambda\\ &= i\sum_{\alpha\lambda}\left(\frac{h\omega_\alpha}{V}\right)^{\frac{1}{2}}\left[\left(\mathbf{l}_{\mathbf{k}_\alpha}\times\frac{q}{2mc}\mathbf{L}\right)\cdot\boldsymbol{\pi}_\alpha^\lambda\right]a_\alpha^\lambda\\ &\quad + i\sum_{\alpha\lambda}\left(\frac{h\omega_\alpha}{V}\right)^{\frac{1}{2}}\left[\left(\mathbf{l}_{\mathbf{k}_\alpha}\times\frac{q}{2mc}\mathbf{L}\right)\cdot\boldsymbol{\pi}_\alpha^\lambda\right]a_\alpha^{\lambda+}\\ &\quad + \sum_{\alpha\lambda}\left(\frac{h\omega_\alpha}{V}\right)^{\frac{1}{2}}\frac{1}{2}(\mathbf{k}_\alpha\cdot\mathbf{r})(q\mathbf{r}\cdot\boldsymbol{\pi}_\alpha^\lambda)a_\alpha^\lambda\\ &\quad + \sum_{\alpha\lambda}\left(\frac{h\omega_\alpha}{V}\right)\frac{1}{2}(\mathbf{k}_\alpha\cdot\mathbf{r})(q\mathbf{r}\cdot\boldsymbol{\pi}_\alpha^\lambda)a_\alpha^{\lambda+}.\end{aligned}\tag{9.5.12}$$

The interaction Hamiltonian can now be written as

$$H_1 = H_1(E1) + H_1(M1) + H_1(E2),\tag{9.5.13}$$

where the electric dipole interaction, $H_1(E1)$, is given by

$$H_1(E1) = -i\sum_{\alpha\lambda}\left(\frac{h\omega_\alpha}{V}\right)^{\frac{1}{2}}(q\underset{\sim}{r}\cdot\underset{\sim}{\pi}_\alpha^\lambda)a_\alpha^\lambda + i\sum_{\alpha\lambda}\left(\frac{h\omega_\alpha}{V}\right)^{\frac{1}{2}}(q\underset{\sim}{r}\cdot\underset{\sim}{\pi}_\alpha^\lambda)a_\alpha^{\lambda+}, \tag{9.5.14}$$

and the magnetic dipole interaction, $H_1(M1)$ is given by

$$\begin{aligned} H_1(M1) = {} & i\sum_{\alpha\lambda}\left(\frac{h\omega_\alpha}{V}\right)^{\frac{1}{2}}\left[\left(\underset{\sim}{1}_{k_\alpha}\times\frac{q}{2mc}\underset{\sim}{L}\right)\cdot\underset{\sim}{\pi}_\alpha^\lambda\right]a_\alpha^\lambda \\ & + i\sum_{\alpha\lambda}\left(\frac{h\omega_\alpha}{V}\right)^{\frac{1}{2}}\left[\left(\underset{\sim}{1}_{k_\alpha}\times\frac{q}{2mc}\underset{\sim}{L}\right)\cdot\underset{\sim}{\pi}_\alpha^{\lambda+}\right]a_\alpha^{\lambda+}, \end{aligned} \tag{9.5.15}$$

and the electric quadrupole interaction, $H(E2)$, is given by

$$\begin{aligned} H(E2) = {} & \sum_{\alpha\lambda}\left(\frac{h\omega_\alpha}{V}\right)^{\frac{1}{2}}\frac{1}{2}(\underset{\sim}{k}_\alpha\cdot\underset{\sim}{r})(q\underset{\sim}{r}\cdot\underset{\sim}{\pi}_\alpha^\lambda)a_\alpha^\lambda \\ & + \sum_{\alpha\lambda}\left(\frac{h\omega_\alpha}{V}\right)^{\frac{1}{2}}\frac{1}{2}(\underset{\sim}{k}_\alpha\cdot\underset{\sim}{r})(q\underset{\sim}{r}\cdot\underset{\sim}{\pi}_\alpha^\lambda)a_\alpha^{\lambda+}. \end{aligned} \tag{9.5.16}$$

These three terms produce *electric dipole, magnetic dipole,* and *electric quadrupole* transitions, respectively.

If the wavelength of the light is much greater than the size of the system the light is interacting with, that is, if

$$|\underset{\sim}{k}\cdot\underset{\sim}{r}| \simeq \frac{r}{\lambda} \ll 1, \tag{9.5.17}$$

a condition fulfilled in free atoms, then electric dipole transitions are predominant.

Considering now the process of absorption or emission of one photon, the probability per unit time that the system (charged particle and radiative field) is found with one less or one more photon of frequency ω_α and polarization $\underset{\sim}{\pi}_\alpha^\lambda$ in the direction $(\Omega_\alpha, \Omega_\alpha + d\Omega_\alpha)$

is given by

$$P_\alpha^\lambda d\Omega_\alpha = \frac{2\pi}{\hbar^2}|M_\alpha^\lambda|^2\, g(\omega_\alpha), \tag{9.5.18}$$

where M_α^λ is the relevant matrix element and $g(\omega_\alpha)$ can be found as follows:

$$\begin{aligned} g(\underset{\sim}{k}_\alpha)d^3\underset{\sim}{k}_\alpha &= \frac{V}{8\pi^3}d^3\underset{\sim}{k}_\alpha = \frac{V}{8\pi^3}k_\alpha^2\, dk_\alpha\, d\Omega_\alpha \\ &= \frac{V}{8\pi^3}k_\alpha^2\frac{dk_\alpha}{d\Omega_\alpha}d\omega_\alpha\, d\Omega_\alpha = \frac{V\omega_\alpha^2}{8\pi^3c^3}d\omega_\alpha\, d\Omega_\alpha, \end{aligned} \tag{9.5.19}$$

and

$$g(\omega_\alpha) = \frac{V\omega_\alpha^2}{8\pi^3c^3}d\Omega_\alpha. \tag{9.5.20}$$

Also

$$|M_\alpha^\lambda|^2 = \frac{q^2}{m^2}\frac{h}{\omega_\alpha V}\left|\left\langle \psi_f^e \left| \begin{matrix} e^{i\underset{\sim}{k}_\alpha\cdot\underset{\sim}{r}} & \underset{\sim}{\pi}_\alpha^\lambda\cdot\underset{\sim}{p} \\ e^{-i\underset{\sim}{k}_\alpha\cdot\underset{\sim}{r}} & \underset{\sim}{\pi}_\alpha^\lambda\cdot\underset{\sim}{p} \end{matrix} \right| \psi_i^e \right\rangle\right|^2 \begin{pmatrix} n_\alpha^\lambda \\ n_\alpha^\lambda+1 \end{pmatrix}, \tag{9.5.21}$$

and

$$\begin{aligned} P_\alpha^\lambda\, d\Omega_\alpha &= \frac{V\omega_\alpha^2}{4\pi^2c^3h^2}\frac{q^2}{m^2}\frac{h}{\omega_\alpha V}\left|\left\langle \psi_f^e \left| \begin{matrix} e^{i\underset{\sim}{k}_\alpha\cdot\underset{\sim}{r}} & \underset{\sim}{\pi}_\alpha^\lambda\cdot\underset{\sim}{p} \\ e^{-i\underset{\sim}{k}_\alpha\cdot\underset{\sim}{r}} & \underset{\sim}{\pi}_\alpha^\lambda\cdot\underset{\sim}{p} \end{matrix} \right| \psi_i \right\rangle\right|^2 \begin{bmatrix} n_\alpha^\lambda \\ n_\alpha^\lambda+1 \end{bmatrix} d\Omega_\alpha \\ &= \frac{\omega_\alpha q^2}{hc^3m^2}\left|\left\langle \psi_f \left| \begin{matrix} e^{i\underset{\sim}{k}_\alpha\cdot\underset{\sim}{r}} & \underset{\sim}{\pi}_\alpha^\lambda\cdot\underset{\sim}{p} \\ e^{-i\underset{\sim}{k}_\alpha\cdot\underset{\sim}{r}} & \underset{\sim}{\pi}_\alpha^\lambda\cdot\underset{\sim}{p} \end{matrix} \right| \psi_i \right\rangle\right|^2 \begin{bmatrix} n_\alpha^\lambda \\ n_\alpha^\lambda+1 \end{bmatrix} d\Omega_\alpha, \end{aligned} \tag{9.5.22}$$

where the upper (lower) row corresponds to the process of absorption (emission) of one photon.

We note that, since $\underset{\sim}{\pi}_\alpha^\lambda \cdot \underset{\sim}{k} = 0$, we can interchange in (9.5.22) $\underset{\sim}{\pi}_\alpha^\lambda \cdot \underset{\sim}{p}$ with $e^{-\underset{\sim}{k}_\alpha\cdot\underset{\sim}{r}}$. Therefore we may write

$$\begin{aligned} |\langle\psi_f^e|e^{-i\underset{\sim}{k}\cdot\underset{\sim}{r}}(\underset{\sim}{\pi}_\alpha^\lambda\cdot\underset{\sim}{p})|\psi_i^e\rangle|^2 &= |\langle\psi_f^e|(\underset{\sim}{\pi}_\alpha^\lambda\cdot\underset{\sim}{p})e^{-i\underset{\sim}{k}\cdot\underset{\sim}{r}}|\psi_i^e\rangle|^2 \\ &= |\langle(\underset{\sim}{\pi}_\alpha^\lambda\cdot\underset{\sim}{p})\psi_f^e|e^{-i\underset{\sim}{k}\cdot\underset{\sim}{r}}|\psi_i^e\rangle|^2 \\ &= |\langle\psi_i^e|e^{i\underset{\sim}{k}\cdot\underset{\sim}{r}}(\underset{\sim}{\pi}_\alpha^\lambda\cdot\underset{\sim}{p})|\psi_f^e\rangle|^2, \end{aligned} \tag{9.5.23}$$

and (9.5.22) becomes

$$P_\alpha^\lambda \, d\Omega_\alpha = \frac{\omega_\alpha q^2}{hc^3 m^2} |\langle \psi_f^e | e^{i\underset{\sim}{k}_\alpha \cdot \underset{\sim}{r}} \underset{\sim}{\pi}_\alpha^\lambda \cdot \underset{\sim}{p} | \psi_i^e \rangle|^2 \begin{pmatrix} n_\alpha^\lambda \\ n_\alpha^\lambda + 1 \end{pmatrix} d\Omega_\alpha. \tag{9.5.24}$$

If more than one charged particle is present

$$P_\alpha^\lambda \, d\Omega_\alpha = \frac{\omega_\alpha}{hc^3} \left| \left\langle \sum_\ell \frac{q_\ell}{m_\ell} (\underset{\sim}{\pi}_\alpha^\lambda \cdot \underset{\sim}{p}_\ell) e^{i\underset{\sim}{k}_\alpha \cdot \underset{\sim}{r}_\ell} \right\rangle_{fi} \right|^2 \begin{pmatrix} n_\alpha^\lambda \\ n_\alpha^\lambda + 1 \end{pmatrix} d\Omega_\alpha. \tag{9.5.25}$$

The probability per unit time of spontaneous emission of a photon of frequency ω_α and polarization $\underset{\sim}{\pi}_\alpha^\lambda$ in the solid angle $d\Omega_\alpha$ is given by

$$P_\alpha^\lambda(sp) d\Omega_\alpha = \frac{\omega_\alpha}{hc^3} \left| \left\langle \sum_\ell \frac{q_\ell}{m_\ell} (\underset{\sim}{\pi}_\alpha^\lambda \cdot \underset{\sim}{p}_\ell) e^{i\underset{\sim}{k}_\alpha \cdot \underset{\sim}{r}_\ell} \right\rangle_{fi} \right|^2 d\Omega_\alpha. \tag{9.5.26}$$

The probability per unit time of induced absorption or emission of a photon of frequency ω_α and polarization $\underset{\sim}{\pi}_\alpha^\lambda$ in the solid angle $d\Omega_\alpha$ is given by

$$P_\alpha^\lambda(abs; emi) d\Omega_\alpha = \frac{\omega_\alpha}{hc^3} \left| \left\langle \sum_\ell \frac{q_\ell}{m_\ell} (\underset{\sim}{\pi}_\alpha^\lambda \cdot \underset{\sim}{p}_\ell) e^{i\underset{\sim}{k}_\alpha \cdot \underset{\sim}{r}_\ell} \right\rangle_{fi} \right|^2 n_\alpha^\lambda \, d\Omega_\alpha. \tag{9.5.27}$$

The probability per unit time of spontaneous emission of a photon of frequency ω_α is given by

$$\sum_\lambda \int P_\alpha^\lambda(sp) d\Omega_\alpha = \frac{\omega_\alpha}{hc^3} \sum_\lambda \int d\Omega_\alpha \left| \left\langle \sum_\ell \frac{q_\ell}{m_\ell} (\underset{\sim}{\pi}_\alpha^\lambda \cdot \underset{\sim}{p}_\ell) e^{i\underset{\sim}{k}_\alpha \cdot \underset{\sim}{r}_\ell} \right\rangle_{fi} \right|^2. \tag{9.5.28}$$

The probability per unit time of induced absorption or emission of a photon of frequency ω_α is given by

$$\frac{\omega_\alpha}{hc^3} \sum_\ell \int d\Omega_\alpha \left| \left\langle \sum_\lambda \frac{q_\ell}{m_\ell} (\underset{\sim}{\pi}_\alpha^\lambda \cdot \underset{\sim}{p}_\ell) e^{i\underset{\sim}{k}_\alpha \cdot \underset{\sim}{r}_\ell} \right\rangle_{fi} \right|^2 n_\alpha^\lambda. \tag{9.5.29}$$

We consider now the case in which (a) the relevant matrix element is not polarization or direction dependent, a case always occurring in free atoms, and (b) the intensity of the field at a certain frequency is equal in the two polarizations. In this case the probability per unit time of spontaneous emission is given by

$$A = \frac{2\omega_\alpha}{3hc^3} 4\pi \left| \left\langle \sum_\ell \frac{q_\ell}{m_\ell} \underset{\sim}{p}_\ell e^{i\underset{\sim}{k}_\alpha \cdot \underset{\sim}{r}_\ell} \right\rangle_{fi} \right|^2, \tag{9.5.30}$$

and the probability of induced emission or absorption by

$$B\rho(\nu_\alpha) = An_\alpha, \tag{9.5.31}$$

where A and B are called *Einstein's coefficients* and $\rho(\nu_\alpha)$ is the energy density per unit frequency range. It is

$$\rho(\nu_\alpha)d\nu_\alpha = \rho(\omega_\alpha)d\omega_\alpha,$$
$$\rho(\nu_\alpha) = 2\pi\ \rho(\omega_\alpha).$$

But

$$\rho(\omega_\alpha) = 2\frac{4\pi}{V}\frac{V\omega_\alpha^2}{8\pi^3c^3}\hbar\omega_\alpha\ n_\alpha = \frac{\omega_\alpha^3\hbar}{\pi^2c^3}n_\alpha. \tag{9.5.32}$$

Then

$$\rho(\nu_\alpha) = \frac{\omega_\alpha^3 h}{\pi^2c^3}n_\alpha = \frac{8\pi h\nu_\alpha^3}{c^3}n_\alpha, \tag{9.5.33}$$

and

$$\frac{A}{B} = \frac{8\pi h\nu_\alpha^3}{c^3}. \tag{9.5.34}$$

In the electric dipole approximation the formula (9.5.25) becomes

$$P_\alpha^\lambda d\Omega_\alpha = \frac{\omega_\alpha}{hc^3}\left|\left\langle \sum_\ell \frac{q_\ell}{m_\ell}(\underset{\sim}{\pi}_\alpha^\lambda \cdot \underset{\sim}{p}_\ell)\right\rangle_{fi}\right|^2 \begin{bmatrix} n_\alpha^\lambda \\ n_\alpha^\lambda + 1 \end{bmatrix} d\Omega_\alpha$$
$$= \frac{\omega_\alpha^3}{hc^3}|\langle \underset{\sim}{\pi}_\alpha^\lambda \cdot \underset{\sim}{M}\rangle_{fi}|^2 \begin{bmatrix} n_\alpha^\lambda \\ n_\alpha^\lambda + 1 \end{bmatrix} d\Omega_\alpha, \tag{9.5.35}$$

where

$$\underset{\sim}{M} = \sum_\ell q_\ell \underset{\sim}{r}_\ell, \tag{9.5.36}$$

is the electric dipole operator.

The radiation density per unit frequency range at $\omega = \omega_\alpha$ and polarization $\underset{\sim}{\pi}_\alpha^\lambda$ and per unit solid angle is given by

$$\frac{g(\omega_\alpha)\hbar\omega_\alpha n_\alpha^\lambda}{V d\Omega_\alpha} = \frac{\omega_\alpha^2}{8\pi^3 c^3}\hbar\omega_\alpha\, n_\alpha^\lambda = \frac{\hbar\omega_\alpha^3}{8\pi^3 c^3} n_\alpha^\lambda, \tag{9.5.37}$$

and the radiation intensity per unit frequency range at $\omega = \omega_\alpha$ and polarization $\underset{\sim}{\pi}_\alpha^\lambda$ and per unit solid angle is given by

$$I(\omega_\alpha, \lambda) = \frac{\hbar\omega_\alpha^3}{8\pi^3 c^2} n_\alpha^\lambda. \tag{9.5.38}$$

It is

$$\frac{4\pi^2 I}{\hbar^2 c} = \frac{4\pi^2}{\hbar^2 c}\frac{\hbar\omega_\alpha^3}{8\pi^3 c^2} n_\alpha^\lambda = \frac{\omega_\alpha^3}{hc^3} n_\alpha^\lambda. \tag{9.5.39}$$

Therefore we can write the formula (9.5.35) as

$$P_\alpha^\lambda = \begin{cases} \dfrac{4\pi^2 I(\omega_\alpha;\lambda)}{\hbar^2 c}|M_{fi}|^2 \\[2ex] \left\{\dfrac{4\pi^2 I(\omega_\alpha;\lambda)}{\hbar^2 c} + \dfrac{\omega_\alpha^3}{hc^3}\right\}|M_{fi}|^2, \end{cases} \tag{9.5.40}$$

where the upper (lower) row corresponds to absorption (emission) and where

$$|M_{fi}|^2 = |\langle \underset{\sim}{\pi}_\alpha^\lambda \cdot \underset{\sim}{M} \rangle_{fi}|^2.$$

9.6. Second-Order Processes

In examining second-order processes we restrict ourselves to the consideration of the process of photon scattering. The system as in the previous sections consists of the bound charged particle and the radiative field; the unperturbed Hamiltonian of this system is given by

$$H_o = -\frac{\hbar^2}{2m}\nabla^2 + q\phi + \sum_\alpha \hbar\omega_\alpha \left(a_\alpha^+ a_\alpha + \frac{1}{2}\right). \tag{9.6.1}$$

In the formula above and in the following ones we shall drop for simplicity of notation the polarization label λ. We shall reintroduce this index at the proper time.

The eigenfunctions of the systems are given by

$$\psi_{e;n_1,n_2,\ldots} = \psi^e \prod_\alpha |n_\alpha\rangle, \tag{9.6.2}$$

and the energies of such eigenstates are given by

$$E_{e;n_1,n_2,\ldots} = E^e + \sum_\alpha \hbar\omega_\alpha \left(n_\alpha + \frac{1}{2}\right). \tag{9.6.3}$$

The interaction between charged particle and radiative field is represented by $H_1 + H_2$. H_1 and H_2 are given in (9.4.10) and (9.4.11), respectively.

In a scattering process the initial state of the system is given by

$$\psi_i = |\psi_i^e; n_1, n_2, \ldots, n_f, n_i, \ldots\rangle, \tag{9.6.4}$$

and the final state by

$$\psi_f = |\psi_f^e; n_1, n_2, \ldots, n_f + 1, n_i - 1, \ldots\rangle. \tag{9.6.5}$$

The energies of these states are

$$E_i = E_i^e + \sum_\alpha \hbar\omega_\alpha \left(n_\alpha + \frac{1}{2}\right) \tag{9.6.6}$$

$$E_f = E_f^e + \sum_{\alpha \neq f,i} \hbar\omega_\alpha \left(n_\alpha + \frac{1}{2}\right)$$
$$+ \hbar\omega_f \left(n_f + \frac{3}{2}\right) + \hbar\omega_i \left(n_i - \frac{1}{2}\right), \tag{9.6.7}$$

respectively. Since energy must be conserved in the process ($E_i = E_f$), it must be

$$E_f^e = E_i^e + \hbar(\omega_i - \omega_f). \tag{9.6.8}$$

Second-order processes can be brought about by the perturbation Hamiltonian H_1 when used in second order and by perturbation H_2 when used in first order. We shall consider now the two contributions separately.

9.6.1. Matrix Element Due to $\mathbf{H_1}$

When using H_1 in second order the effective matrix element is given by

$$\sum_j \frac{\langle\psi_f|H_1|\psi_j\rangle\langle\psi_j|H_1|\psi_i\rangle}{E_i - E_j}, \tag{9.6.9}$$

where ψ_i and ψ_f are given by (9.6.4) and (9.6.5), respectively. The intermediate states can be of two types:

1. Intermediate state that consists of the charged particle in some state ψ_j^e and the radiative field with one photon of wave vector $\underset{\sim}{k}_i$ less

$$\psi_j = |\psi_j^e; n_1, n_2, \ldots, n_i - 1, n_f, \ldots\rangle. \tag{9.6.10}$$

We have in this case

$$E_i - E_j = E_i^e - (E_j^e - \hbar\omega_i) = E_i^e - E_j^e + \hbar\omega_i. \tag{9.6.11}$$

The relevant matrix elements are now

$$\langle\psi_j|H_1|\psi_i\rangle = -\frac{q}{m}\sqrt{\frac{h}{\omega_i V}}\langle j|e^{i\underset{\sim}{k}_i\cdot\underset{\sim}{r}}\,\underset{\sim}{p}\cdot\underset{\sim}{\pi}_i|i\rangle\sqrt{n_i}, \tag{9.6.12}$$

$$\langle\psi_f|H_1|\psi_j\rangle = -\frac{q}{m}\sqrt{\frac{h}{\omega_f V}}\langle f|e^{-i\underset{\sim}{k}_f\cdot\underset{\sim}{r}}\,\underset{\sim}{p}\cdot\underset{\sim}{\pi}_f|j\rangle\sqrt{n_f + 1}, \tag{9.6.13}$$

where the bras and kets indicate states of the charged particle alone. The contribution of intermediate states of this type to the sum in (9.6.9) is then given by

$$\frac{q^2}{m}\frac{h}{V}\frac{1}{\sqrt{\omega_f\omega_i}}\left[\frac{1}{m}\frac{\langle f|e^{-i\underset{\sim}{k}_f\cdot\underset{\sim}{r}}\,\underset{\sim}{p}\cdot\underset{\sim}{\pi}_f|j\rangle\langle j|e^{i\underset{\sim}{k}_i\cdot\underset{\sim}{r}}\,\underset{\sim}{p}\cdot\underset{\sim}{\pi}_i|i\rangle}{E_i^e - E_j^e + \hbar\omega_i}\right]\sqrt{n_i(n_f + 1)}. \tag{9.6.14}$$

2. Intermediate state that consists of the charged particle in some state ψ_j^e and the radiative field with one photon of wave vector $\underset{\sim}{k}_f$

more

$$\psi_j = |\psi_j^e; n_1, n_2, \ldots, n_i, n_f + 1, \ldots\rangle. \tag{9.6.15}$$

We have in this case

$$E_i - E_j = E_i^e - (E_j^e + \hbar\omega_f) = E_i^e - E_j^e - \hbar\omega_f. \tag{9.6.16}$$

The relevant matrix elements are now

$$\langle\psi_j|H_1|\psi_i\rangle = -\frac{q}{m}\sqrt{\frac{h}{\omega V}}\langle j|e^{-i\underset{\sim}{k}_f\cdot\underset{\sim}{r}}\,\underset{\sim}{p}\cdot\underset{\sim}{\pi}_f|i\rangle\sqrt{n_f+1}, \tag{9.6.17}$$

$$\langle\psi_f|H_1|\psi_j\rangle = -\frac{q}{m}\sqrt{\frac{h}{\omega V}}\langle f|e^{i\underset{\sim}{k}_i\cdot\underset{\sim}{r}}\,\underset{\sim}{p}\cdot\underset{\sim}{\pi}_i|j\rangle\sqrt{n_i}, \tag{9.6.18}$$

where the bras and the kets indicate states of the charged particle alone. The contribution of intermediate states of this type to the sum in (9.6.9) is then

$$\frac{q^2}{m}\frac{h}{V}\frac{1}{\sqrt{\omega_f\omega_i}}\left[\frac{1}{m}\frac{\langle f|e^{i\underset{\sim}{k}_i\cdot\underset{\sim}{r}}\,\underset{\sim}{p}\cdot\underset{\sim}{\pi}_i|j\rangle\langle j|e^{-i\underset{\sim}{k}_f\cdot\underset{\sim}{r}}\,\underset{\sim}{p}\cdot\underset{\sim}{\pi}_f|i\rangle}{E_i^e - E_j^e - \hbar\omega_f}\right]\sqrt{n_i(n_f+1)}. \tag{9.6.19}$$

9.6.2. *Matrix Elements Due to* $\mathbf{H_2}$

When using H_2 in first order the only term that can give a matrix element different from zero between the states ψ_f and ψ_i, is

$$\frac{q^2}{m}\frac{h}{v}\frac{1}{\sqrt{\omega_i\omega_f}}\underset{\sim}{\pi}_i\cdot\underset{\sim}{\pi}_f\, a_i a_f^+ e^{i(\underset{\sim}{k}_i-\underset{\sim}{k}_f)\cdot\underset{\sim}{r}}. \tag{9.6.20}$$

This matrix element is

$$\frac{q^2}{m}\frac{h}{v}\frac{1}{\sqrt{\omega_i\omega_f}}(\underset{\sim}{\pi}_i\cdot\underset{\sim}{\pi}_f)\langle f|e^{i(\underset{\sim}{k}_i-\underset{\sim}{k}_f)\cdot\underset{\sim}{r}}|i\rangle\sqrt{n_i(n_f+1)}. \tag{9.6.21}$$

9.6.3. *Effective Matrix Element*

Putting everything together, the effective matrix element for a photon scattering process is given by

$$\frac{q^2 h}{mV}\left[\frac{n_i(n_f+1)}{\omega_f\omega_i}\right]^{\frac{1}{2}}\left\{\frac{1}{m}\sum_j\left[\frac{\langle f|e^{-i\underset{\sim}{k}_f\cdot\underset{\sim}{r}}\,\underset{\sim}{p}\cdot\underset{\sim}{\pi}_f|j\rangle\langle j|e^{i\underset{\sim}{k}_i\cdot\underset{\sim}{r}}\,\underset{\sim}{p}\cdot\underset{\sim}{\pi}_i|i\rangle}{E^e_i-E^e_j-\hbar\omega_i}\right.\right.$$

$$\left.\left.+\frac{\langle f|e^{i\underset{\sim}{k}_i\cdot\underset{\sim}{r}}\,\underset{\sim}{p}\cdot\underset{\sim}{\pi}_i|j\rangle\langle j|e^{-i\underset{\sim}{k}_f\cdot\underset{\sim}{r}}\underset{\sim}{\pi}_f\cdot\underset{\sim}{p}|i\rangle}{E^e_i-E^e_j-\hbar\omega_f}\right]+\underset{\sim}{\pi}_i\cdot\underset{\sim}{\pi}_f\langle f|e^{i(\underset{\sim}{k}_i-\underset{\sim}{k}_f)\cdot\underset{\sim}{r}}|i\rangle\right\}. \tag{9.6.22}$$

Before proceeding with further elaborations we point out some physical phenomena predicted by the formula above:

1. If

$$\omega_f=\omega_i\gg\left|\frac{E^e_i-E^e_f}{\hbar}\right|, \tag{9.6.23}$$

   the sum of the first two terms in the curly brackets in (9.6.22) is negligible with respect to the third term in the brackets. The formula (9.6.22) so simplified is used in evaluating the relevant matrix element in *Thompson scattering*, in which x-ray photons are scattered by free or loosely bound electrons.
2. If the initial and final states of the charged particle are the same we have the *Rayleigh or coherent scattering.*
3. If the initial and final states of the charged particle are different, we have the *Raman scattering.*

We now make the following simplifying assumptions:

1. We assume with no loss of generality that $n_i = 1$ and $n_f = 0$.
2. We set all the exponential factors equal to one, using as in (9.5.14), the electric dipole approximation. This assumption is somewhat restrictive because it implies that the wavelengths of the absorbed and scattered photons are much larger than the system with which the light interacts.

With these assumptions we obtain

$$\frac{q^2 h}{mV}(\omega_f \omega_i)^{-\frac{1}{2}} \left\{ \frac{1}{m} \sum_j \left[\frac{\langle f|\underset{\sim}{p} \cdot \underset{\sim}{\pi}_f|j\rangle\langle j|\underset{\sim}{p} \cdot \underset{\sim}{\pi}_i|i\rangle}{E_i^e - E_j^e + \hbar\omega_i} \right.\right.$$

$$\left.\left. + \frac{\langle f|\underset{\sim}{p} \cdot \underset{\sim}{\pi}_i|j\rangle\langle j|\underset{\sim}{p} \cdot \underset{\sim}{\pi}_f|i\rangle}{E_i^e - E_j^e - \hbar\omega_f} \right] + (\underset{\sim}{\pi}_i \cdot \underset{\sim}{\pi}_f)\delta_{if} \right\}. \qquad (9.6.24)$$

If we set i = f in (9.6.24), as is the case for coherent scattering, we obtain the *dispersion formula* first derived by Kramers and Heisenberg from the classical theory by the application of the correspondence principle[1,2]; Waller[3] showed the existence of the terms $(\underset{\sim}{\pi}_i \cdot \underset{\sim}{\pi}_f)$ in the dispersion formula.

We now take into account the result (9.5.7):

$$\langle j|\underset{\sim}{p}|i\rangle = \frac{im}{\hbar} E_{ji}^e \langle j|\underset{\sim}{r}|i\rangle, \qquad (9.6.25)$$

$$\langle f|\underset{\sim}{p}|j\rangle = \frac{im}{\hbar} E_{fj}^e \langle f|\underset{\sim}{r}|j\rangle, \qquad (9.6.26)$$

where

$$\begin{aligned} E_{ji}^e &= E_j^e - E_i^e, \\ E_{fj}^e &= E_f^e - E_j^e. \end{aligned} \qquad (9.6.27)$$

The matrix element (9.6.24) becomes then

$$\frac{q^2 h}{V}(\omega_f \omega_i)^{-\frac{1}{2}} \left\{ -\frac{1}{\hbar^2} \sum_j E_{fi}^e E_{ji}^e \left[\frac{\langle f|\underset{\sim}{r} \cdot \underset{\sim}{\pi}_f|j\rangle\langle j|\underset{\sim}{r} \cdot \underset{\sim}{\pi}_i|i\rangle}{-E_{ji}^e + \hbar\omega_i} \right.\right.$$

$$\left.\left. - \frac{\langle f|\underset{\sim}{r} \cdot \underset{\sim}{\pi}_i|j\rangle\langle j|\underset{\sim}{r} \cdot \underset{\sim}{\pi}_f|i\rangle}{E_{ji}^e + \hbar\omega_f} \right] + \frac{\underset{\sim}{\pi}_i \cdot \underset{\sim}{\pi}_f}{m}\delta_{if} \right\}. \qquad (9.6.28)$$

We shall make use of the following commutation relations[2]:

$$(\underset{\sim}{\pi}_f \cdot \underset{\sim}{r})(\underset{\sim}{\pi}_i \cdot \underset{\sim}{p}) - (\underset{\sim}{\pi}_i \cdot \underset{\sim}{p})(\underset{\sim}{\pi}_f \cdot \underset{\sim}{r}) = i\hbar(\underset{\sim}{\pi}_f \cdot \underset{\sim}{\pi}_i), \qquad (9.6.29)$$

$$(\underset{\sim}{\pi}_f \cdot \underset{\sim}{r})(\underset{\sim}{\pi}_i \cdot \underset{\sim}{r}) - (\underset{\sim}{\pi}_i \cdot \underset{\sim}{r})(\underset{\sim}{\pi}_f \cdot \underset{\sim}{r}) = 0. \qquad (9.6.30)$$

Using (9.6.29) we find

$$
\begin{aligned}
&\langle f|(\underset{\sim}{\pi}_f \cdot \underset{\sim}{r})(\underset{\sim}{\pi}_i \cdot \underset{\sim}{p})|i\rangle - \langle f|(\underset{\sim}{\pi}_i \cdot \underset{\sim}{p})(\underset{\sim}{\pi}_f \cdot \underset{\sim}{r})|i\rangle \\
&\quad = \sum_j \langle f|\underset{\sim}{\pi}_f \cdot \underset{\sim}{r}|j\rangle\langle j|\underset{\sim}{\pi}_i \cdot \underset{\sim}{p}|i\rangle - \sum_j \langle f|\underset{\sim}{\pi}_i \cdot \underset{\sim}{p}|j\rangle\langle j|\underset{\sim}{\pi}_f \cdot \underset{\sim}{r}|i\rangle \\
&\quad = \sum_j \langle f|\underset{\sim}{\pi}_f \cdot \underset{\sim}{r}|j\rangle\langle j|\underset{\sim}{\pi}_i \cdot \underset{\sim}{p}|i\rangle \frac{im}{\hbar} E^e_{ji} \\
&\qquad - \sum_j \langle f|\underset{\sim}{\pi}_i \cdot \underset{\sim}{r}|j\rangle\langle j|\underset{\sim}{\pi}_f \cdot \underset{\sim}{r}|i\rangle \frac{im}{\hbar} E^e_{fj} = i\hbar(\underset{\sim}{\pi}_f \cdot \underset{\sim}{\pi}_i)\delta_{if},
\end{aligned}
\tag{9.6.31}
$$

or

$$
\begin{aligned}
&\sum_j \frac{1}{\hbar^2}[\langle f|\underset{\sim}{\pi}_f \cdot \underset{\sim}{r}|j\rangle\langle j|\underset{\sim}{\pi}_i \cdot \underset{\sim}{r}|i\rangle E^e_{ji} \\
&\qquad - \langle f|\underset{\sim}{\pi}_i \cdot \underset{\sim}{r}|j\rangle\langle j|\underset{\sim}{\pi}_f \cdot \underset{\sim}{r}|i\rangle E^e_{fj}] = \frac{\underset{\sim}{\pi}_f \cdot \underset{\sim}{\pi}_i}{m}\delta_{if}.
\end{aligned}
\tag{9.6.32}
$$

Using now (9.6.30)

$$
\begin{aligned}
&\frac{\omega_f}{\hbar}[\langle f|(\underset{\sim}{\pi}_f \cdot \underset{\sim}{r})(\underset{\sim}{\pi}_i \cdot \underset{\sim}{r})|i\rangle - \langle f|(\underset{\sim}{\pi}_i \cdot \underset{\sim}{r})(\underset{\sim}{\pi}_f \cdot \underset{\sim}{r})|i\rangle] \\
&\quad = \frac{1}{\hbar^2}\sum_j [\hbar\omega_f \langle f|\underset{\sim}{\pi}_f \cdot \underset{\sim}{r}|j\rangle\langle j|\underset{\sim}{\pi}_i \cdot \underset{\sim}{r}|i\rangle \\
&\qquad - \hbar\omega_f \langle f|\underset{\sim}{\pi}_i \cdot \underset{\sim}{r}|j\rangle\langle j|\underset{\sim}{\pi}_f \cdot \underset{\sim}{r}|i\rangle] = 0.
\end{aligned}
\tag{9.6.33}
$$

We add now (9.6.32) and (9.6.33) and find

$$
\begin{aligned}
&\frac{1}{\hbar^2}\sum_j [(\hbar\omega_f + E^e_{ji})\langle f|\underset{\sim}{\pi}_f \cdot \underset{\sim}{r}|j\rangle\langle j|\underset{\sim}{\pi}_i \cdot \underset{\sim}{r}|i\rangle \\
&\qquad - (\hbar\omega_f + E^e_{fj})\langle f|\underset{\sim}{\pi}_i \cdot \underset{\sim}{r}|j\rangle\langle j|\underset{\sim}{\pi}_f \cdot \underset{\sim}{r}|i\rangle] = \frac{\underset{\sim}{\pi}_f \cdot \underset{\sim}{\pi}_i}{m}\,\delta_{if}.
\end{aligned}
\tag{9.6.34}
$$

Using this result in (9.6.28) we obtain for the expression in curly brackets

$$\frac{1}{\hbar^2}\sum_j\left[\frac{E^e_{fj}E^e_{ji}}{E^e_{ji}-\hbar\omega_i}+(\hbar\omega_f+E^e_{ji})\right]\langle f|\underset{\sim}{\pi}_f\cdot\underset{\sim}{r}|j\rangle\langle j|\underset{\sim}{r}_i\cdot\underset{\sim}{r}|i\rangle$$

$$+\frac{1}{\hbar^2}\sum_j\left[\frac{E^e_{fj}E^e_{ji}}{E^e_{ji}+\hbar\omega_f}-(\hbar\omega_f+E^e_{ji})\right]\langle f|\underset{\sim}{\pi}_i\cdot\underset{\sim}{r}|j\rangle\langle j|\underset{\sim}{\pi}_f\cdot\underset{\sim}{r}|i\rangle. \tag{9.6.35}$$

But

$$E^e_{ji}+E^e_{fj}=E^e_f-E^e_i=\hbar\omega_i-\hbar\omega_f. \tag{9.6.36}$$

Then

$$\begin{aligned}
&\frac{E^e_{fj}E^e_{ji}}{E^e_{ji}-\hbar\omega_i}+(\hbar\omega_f+E^e_{ji})\\
&\quad=(E^e_{ji}-\hbar\omega_i)^{-1}[E^e_{fi}E^e_{ji}+(\hbar\omega_f+E^e_{ji})(E^e_{ji}-\hbar\omega_i)]\\
&\quad=(E^e_{ji}-\hbar\omega_i)^{-1}[E^e_{fj}E^e_{ji}+\hbar\omega_f E^e_{ji}-\hbar^2\omega_f\omega_i+(E^e_{ji})^2-\hbar\omega_i E^e_{ji}]\\
&\quad=(E^e_{ji}-\hbar\omega_i)^{-1}[E^e_{fj}E^e_{ji}+(\hbar\omega_f-\hbar\omega_i)E^e_{ji}-(E^e_{ji})^2-\hbar^2\omega_f\omega_i]\\
&\quad=(E^e_{ji}-\hbar\omega_i)^{-1}[E^e_{fj}E^e_{ji}+(E^e_{ji}+E^e_{fj})E^e_{ji}-(E^e_{ji})^2-\hbar^2\omega_f\omega_i]\\
&\quad=-\frac{\hbar^2\omega_f\omega_i}{E^e_{ji}-\hbar\omega_i}.
\end{aligned} \tag{9.6.37}$$

Similarly

$$\begin{aligned}
&\frac{E^e_{fj}E^e_{ji}}{E^e_{ji}+\hbar\omega_f}-(E^e_{fj}+\hbar\omega_f)\\
&\quad=(E^e_{ji}+\hbar\omega_f)^{-1}[E^e_{fj}E^e_{ji}-(E^e_{fj}+\hbar\omega_f)(E^e_{ji}+\hbar\omega_f)]\\
&\quad=(E^e_{ji}+\hbar\omega_f)^{-1}[E^e_{fj}E^e_{ji}-(E^e_{fj}E^e_{ji}+\hbar\omega_f E^e_{ji}+\hbar\omega_f E^e_{fj}+\hbar^2\omega_f^2)]
\end{aligned}$$

$$= (E^e_{ji} + \hbar\omega_f)^{-1}[-\hbar\omega_f E^e_{ji} - \hbar\omega_f E^e_{fj} - \hbar\omega_f(\hbar\omega_i - E^e_{ji} - E^e_{fj})]$$

$$= -\frac{\hbar^2\omega_f\omega_i}{E^e_{ji} + \hbar\omega_f}. \tag{9.6.38}$$

The expression in curly brackets of (9.6.28), which has been found to be equal to (9.6.35), is, because of the last results, given by

$$\omega_i\omega_f \sum_j \left[\frac{\langle f|\underset{\sim}{\pi}_f \cdot \underset{\sim}{r}|j\rangle\langle j|\underset{\sim}{\pi}_i \cdot \underset{\sim}{r}|i\rangle}{E^e_i - E^e_j + \hbar\omega_i} - \frac{\langle f|\underset{\sim}{\pi}_i \cdot \underset{\sim}{r}|j\rangle\langle j|\underset{\sim}{\pi}_i \cdot \underset{\sim}{r}|i\rangle}{E^e_j - E^e_i + \hbar\omega_f}\right]. \tag{9.6.39}$$

The matrix element (9.6.28) is now given by

$$\frac{q^2h}{V}(\omega_f\omega_i)^{-\frac{1}{2}} \sum_j \left[\frac{\langle f|\underset{\sim}{\pi}_f \cdot \underset{\sim}{r}|j\rangle\langle j|\underset{\sim}{\pi}_i \cdot \underset{\sim}{r}|i\rangle}{E^e_i - E^e_j + \hbar\omega_i} + \frac{\langle f|\underset{\sim}{\pi}_i \cdot \underset{\sim}{r}|j\rangle\langle j|\underset{\sim}{\pi}_i \cdot \underset{\sim}{r}|i\rangle}{E^e_i - E^e_j - \hbar\omega_f}\right]. \tag{9.6.40}$$

9.6.4. *Transition Rates of Scattering Processes*

Having found the relevant matrix element, it is easy to calculate the transition rate for second-order scattering processes by using the Fermi golden rule. The probability per unit time that the system absorbs a photon of frequency ω_i, wave vector $\underset{\sim}{k}_i$, and polarization $\underset{\sim}{\pi}^{\lambda}_i$ and scatters photon of frequency ω_f, wave vector $\underset{\sim}{k}_f$, and polarization $\underset{\sim}{\pi}^{\lambda}_f$ into the solid angle $d\Omega_f$ is given by

$$P_{f\lambda';i\lambda}d\Omega_f = \frac{2\pi}{\hbar^2}|M_{f\lambda';i\lambda}|^2 g(\omega_f), \tag{9.6.41}$$

where

$$M_{f\lambda';i\lambda} = \frac{q^2h}{V}(\omega_f\omega_i)^{\frac{1}{2}} \sum_j \left[\frac{\langle f|\underset{\sim}{\pi}^{\lambda'}_f \cdot \underset{\sim}{r}|j\rangle\langle j|\underset{\sim}{\pi}^{\lambda}_i \cdot \underset{\sim}{r}|i\rangle}{E^e_i - E^e_j + \hbar\omega_i} + \frac{\langle f|\underset{\sim}{\pi}^{\lambda}_i \cdot \underset{\sim}{r}|j\rangle\langle j|\underset{\sim}{\pi}^{\lambda'}_f \cdot \underset{\sim}{r}|i\rangle}{E^e_i - E^e_j + \hbar\omega_f}\right], \tag{9.6.42}$$

and

$$g(\omega_f) = \frac{V\omega_f^2}{8\pi^3c^3}d\Omega_f. \tag{9.6.43}$$

Using (9.6.42) and (9.6.43) in (9.6.41) we find

$$\mathrm{P}_{\mathrm{f}\lambda';\mathrm{i}\lambda}\mathrm{d}\Omega_{\mathrm{f}} = \frac{\mathrm{q}^4}{\mathrm{V}\mathrm{c}^3}\omega_{\mathrm{f}}^3\omega_{\mathrm{i}}\left|\sum_{\mathrm{j}}\left[\frac{\langle \mathrm{f}|\underset{\sim}{\pi}_{\mathrm{f}}^{\lambda'}\cdot\underset{\sim}{\mathrm{r}}|\mathrm{j}\rangle\langle \mathrm{j}|\underset{\sim}{\pi}_{\mathrm{i}}^{\lambda}\cdot\underset{\sim}{\mathrm{r}}|\mathrm{i}\rangle}{\mathrm{E}_{\mathrm{i}}^{\mathrm{e}}-\mathrm{E}_{\mathrm{j}}^{\mathrm{e}}+\hbar\omega_{\mathrm{i}}} + \frac{\langle \mathrm{f}|\underset{\sim}{\pi}_{\mathrm{f}}^{\lambda}\cdot\underset{\sim}{\mathrm{r}}|\mathrm{j}\rangle\langle \mathrm{j}|\underset{\sim}{\pi}_{\mathrm{f}}^{\lambda'}\cdot\underset{\sim}{\mathrm{r}}|\mathrm{i}\rangle}{\mathrm{E}_{\mathrm{i}}^{\mathrm{e}}-\mathrm{E}_{\mathrm{j}}^{\mathrm{e}}-\hbar\omega_{\mathrm{f}}}\right]\right|^2 \mathrm{d}\Omega_{\mathrm{f}}. \tag{9.6.44}$$

We can see that the transition probability per unit time of the scattering process is found to be dependent on V. In order to find a more appropriate quantity to describe the scattering process we consider the notion of *differential scattering cross section.* The differential cross section is defined as the ratio the light energy scattered into the solid angle $\mathrm{d}\Omega_{\mathrm{f}}$ per unit time to the intensity of the light incident upon the scatterer:

$$\mathrm{d}\sigma = \frac{\text{Light energy into } \mathrm{d}\Omega_{\mathrm{f}}/\text{unit time}}{\text{Incident light intensity}}. \tag{9.6.45}$$

The light energy scattered into the solid angle $\mathrm{d}\Omega_{\mathrm{f}}$ per unit time is given by

$$\mathrm{P}_{\mathrm{f}\lambda';\mathrm{i}\lambda}\ \mathrm{d}\Omega_{\mathrm{f}}\ \hbar\omega_{\mathrm{f}}. \tag{9.6.46}$$

The intensity of the incident light is, on the other hand, given by $(\hbar\omega_{\mathrm{i}}/\mathrm{V})\mathrm{c}$. Therefore

$$(\mathrm{d}\sigma)_{\mathrm{f}\lambda';\mathrm{i}\lambda} = \frac{\mathrm{P}_{\mathrm{f}\lambda';\mathrm{i}\lambda}\mathrm{d}\Omega_{\mathrm{f}}\hbar\omega_{\mathrm{f}}}{(\hbar\omega_{\mathrm{i}}/\mathrm{V})\mathrm{c}} = \mathrm{P}_{\mathrm{f}\lambda';\mathrm{i}\lambda}\frac{\mathrm{V}\omega_{\mathrm{f}}}{\mathrm{c}\omega_{\mathrm{i}}}\mathrm{d}\Omega_{\mathrm{f}}, \tag{9.6.47}$$

or

$$(\mathrm{d}\sigma)_{\mathrm{f}\lambda';\mathrm{i}\lambda} = \frac{\mathrm{q}^4\omega_{\mathrm{f}}^4}{\mathrm{c}^4}\left|\sum_{\mathrm{j}}\left[\frac{\langle \mathrm{f}|\underset{\sim}{\pi}_{\mathrm{f}}^{\lambda'}\cdot\underset{\sim}{\mathrm{r}}|\mathrm{j}\rangle\langle \mathrm{j}|\underset{\sim}{\pi}_{\mathrm{i}}^{\lambda}\cdot\underset{\sim}{\mathrm{r}}|\mathrm{i}\rangle}{\mathrm{E}_{\mathrm{i}}^{\mathrm{e}}-\mathrm{E}_{\mathrm{j}}^{\mathrm{e}}+\hbar\omega_{\mathrm{i}}} + \frac{\langle \mathrm{f}|\underset{\sim}{\pi}_{\mathrm{i}}^{\lambda}\cdot\underset{\sim}{\mathrm{r}}|\mathrm{j}\rangle\langle \mathrm{j}|\underset{\sim}{\pi}_{\mathrm{f}}^{\lambda'}\cdot\underset{\sim}{\mathrm{r}}|\mathrm{i}\rangle}{\mathrm{E}_{\mathrm{i}}^{\mathrm{e}}-\mathrm{E}_{\mathrm{j}}^{\mathrm{e}}-\hbar\omega_{\mathrm{f}}}\right]\right|^2 \mathrm{d}\Omega_{\mathrm{f}}. \tag{9.6.48}$$

The above treatment breaks down if the frequency of the incoming light is such that for some state j of the system

$$\hbar\omega_i = E_j^e - E_i^e. \tag{9.6.49}$$

In this case in (9.6.48) one of the denominations vanishes and the scattering cross section blows up. This case is related to the phenomenon of *resonance fluorescence*.[1]

References

1. W. Heitler, *Quantum Theory of Radiation*, Oxford University Press, London, 1944.
2. H. A. Kramers and W. Heisenberg, *Z. Phys.* 31, 681 (1925).
3. I. Waller, *Z. Phys.* 51, 213 (1928).
4. G. N. Fowler, "Noncovariant Quantum Theory of Radiation" in *Quantum Theory*, edited by D. R. Bates, Vol. III, Academic, New York, 1962, p. 47, and references therein.

Chapter 10

Optical Spectra of Impurities in Solids I

It is possible to alter the properties of a material by *doping*. This refers to replacing a small amount of one of the naturally occurring ions of the material by a different type of ion called an *impurity ion*. This produces a *point defect* in a lattice structure of a crystal. It has been found that doping with only very small amounts of impurity ions can make significant changes in the electrical, optical, and mechanical properties of materials. For example, doping semiconductor crystals provides the technology to make transistors and other devices that form the basis of the field of microelectronics. Similarly, doping transparent insulator crystals provides the technology for solid state lasers.

Impurity ions destroy the translational symmetry of the lattice discussed in Part I of this book. Their physical properties are associated with their properties as a free ion as modified by the environment they have in the crystal. The symmetry of the system is described by the point group symmetry at the site of the impurity ion as presented in Part I. The electrostatic interaction with neighboring ions changes the ability of the impurity ion to absorb and emit electromagnetic radiation. The local vibrational modes of impurity ion and its neighbors can have thermal effects on its properties as described in Part II of this book. Using the formalism developed in Chapter 9, this chapter describes the optical spectroscopy of impurity ions in solids including both purely radiative and phonon-assisted transitions.

10.1. Impurities in Crystals

An impurity in a crystal may be defined as a localized perturbation of the ordered array of atoms in the crystal.[1] It may consist of an extraneous atom or ion, or of any other point defect. The reader is referred to solid-state physics textbooks such as Kittel's[2] for a classification of the various types of impurities that may occur in crystals. For the purposes of our treatment we shall assume the following:

1. The impurity concentration is very low (no impurity-impurity interaction).
2. The electronic eigenstates of the impurity are localized, that is they are at different energies than the electronic bands or states of the crystal.

We may notice at this point the following effects of the presence of the impurity:

1. It destroys the translational symmetry of the crystal. The crystal is no longer invariant under the operations of a "space group." The relevant symmetry is now the one of the impurity *site* (see Section 4.2.3).
2. The presence of the impurity perturbs the lattice vibrations by giving rise to "localized vibrations" not existing in the perfect crystal. The impurity center will, in general, interact with both "band" vibrations and localized vibrations.

10.2. Review of the Theory of Small Vibrations (Classical)

This section and Section 10.4 follow basically the treatment due to K. K. Rebane.[1] Some of these concepts have been discussed in Section 5.5. Consider a system of N material points (atoms) held together by some attractive forces. We may expand the potential energy in a Taylor series about the equilibrium positions of the atoms:

$$\mathrm{V} = \frac{1}{2}\sum_{\mathrm{s}=1}^{3\mathrm{N}}\sum_{\mathrm{s}'=1}^{3\mathrm{N}} \mathrm{A}_{\mathrm{ss}'}\mathrm{u}_{\mathrm{s}}\mathrm{u}_{\mathrm{s}'} + \mathrm{V}_{\mathrm{o}}, \qquad (10.2.1)$$

where

$$A_{ss'} = \left[\frac{\partial^2 v}{\partial u_s \partial u_{s'}}\right]_o . \tag{10.2.2}$$

Notice that $A_{ss'} = A_{s's}$. A component of the force acting on an atom is given by

$$F_s = -\frac{\partial V}{\partial u_s} = -\frac{1}{2}\left(2\sum_{s'} A_{ss'}u_{s'}\right) = -\sum_{s'} A_{ss'}u_{s'},$$

and the 3N equations of motion are given by

$$m_s\ddot{u}_s + \sum_{s'=1}^{3N} A_{ss'}u_{s'} = 0 \quad (s = 1, 2, \ldots, 3N), \tag{10.2.3}$$

which can be written as

$$\begin{aligned}
&m_1\ddot{u}_1 + A_{11}u_1 + A_{12}u_2 + \cdots + A_{1,3N}u_{3N} = 0,\\
&m_2\ddot{u}_2 + A_{22}u_1 + A_{22}u_2 + \cdots + A_{2,3N}u_{3N} = 0,\\
&\vdots\\
&m_{3N}\ddot{u}_{3N} + A_{3N,1}u_1 + A_{3N,2}u_2 + \cdots + A_{3N,3N}u_{3N} = 0.
\end{aligned} \tag{10.2.4}$$

In order to decouple the equations of motion we look for solutions of the type

$$u_s = \frac{a_s}{\sqrt{m_s}}e^{i\omega t}. \tag{10.2.5}$$

From this we derive

$$\ddot{u}_s = -\omega^2\frac{a_s}{\sqrt{m_s}}e^{i\omega t}. \tag{10.2.6}$$

Equations (10.2.6) give a system of 3N homogeneous algebraic equations

$$-m_s\omega^2\frac{a_s}{\sqrt{m_s}}e^{i\omega t} + \sum_{s'=1}^{3N}\frac{A_{ss'}}{\sqrt{m_s}}a_{s'}e^{i\omega t} = 0,$$

or

$$-\omega^2 a_s + \sum_{s'=1}^{3N}\frac{A_{ss'}}{\sqrt{m_s m_{s'}}}a_{s'} = 0, \tag{10.2.7}$$

which can be written in expanded form as

$$\left(\frac{A_{11}}{m_1} - \omega^2\right) a_1 + \frac{A_{12}}{\sqrt{m_1 m_2}} a_2 + \cdots + \frac{A_{1,3N}}{\sqrt{m_1 m_{3N}}} a_{3N} = 0,$$

$$\frac{A_{21}}{\sqrt{m_2 m_1}} a_1 + \left(\frac{A_{22}}{m_2} - \omega^2\right) a_2 + \cdots + \frac{A_{2,3N}}{\sqrt{m_2 m_3 N}} a_{3N} = 0, \quad (10.2.8)$$

$$\vdots$$

$$\frac{A_{3N,1}}{\sqrt{m_{3N} m_1}} a_1 + \frac{A_{3N,2}}{\sqrt{m_{3N} m_2}} a_2 \cdots + \left(\frac{A_{3N,3N}}{m_{3N}} - \omega^2\right) a_{3N} = 0.$$

These equations imply that

$$\begin{vmatrix} \frac{A_{11}}{m_1} - \omega & \frac{A_{12}}{\sqrt{m_1 m_2}} & \cdots & \frac{A_{1,3N}}{\sqrt{m_1 m_{3N}}} \\ \frac{A_{22}}{\sqrt{m_2 m_1}} & \frac{A_{22}}{m_2} - \omega^2 & \cdots & \frac{A_{2,3N}}{\sqrt{m_2 m_{3N}}} \\ \vdots & & & \\ \frac{A_{3N,1}}{\sqrt{m_{3N} m_1}} & \frac{A_{3N,2}}{\sqrt{m_{3N} m_2}} & \cdots & \frac{A_{3N,3N}}{m_{3N}} - \omega^2 \end{vmatrix} = 0. \quad (10.2.9)$$

This equation determines the 3N solutions for ω^2 some of which may be degenerate. We shall assume for the moment that these solutions are not degenerate.

Equations (10.2.8) are 3N in number but only $3N-1$ of them are independent. The knowledge of the values of ω^2 does not determine uniquely the values of the constants $a_i (i = 1, 2, \ldots, 3N)$; rather we have to use one of these 3N constants, say a_{3N}, as arbitrary and express each other constant as follows:

$$a_i' = \frac{a_i}{a_{3N}}. \quad (10.2.10)$$

We then obtain, by substitution of (10.2.10) into (10.2.8) and neglecting the last equation, the following system of 3N − 1 independent equations whose determinant is *not* equal to zero:

$$
\begin{aligned}
&\left(\frac{A_{11}}{m_1}-\omega^2\right)a_1'+\frac{A_{12}}{\sqrt{m_1 m_2}}a_2'+\cdots+\frac{A_{1,3N-1}}{\sqrt{m_1 m_{3N-1}}}a_{3N-1}' = -\frac{A_{1,3N}}{\sqrt{m_1 m_{3N}}},\\
&\frac{A_{21}}{\sqrt{m_1 m_2}}a_1'+\left(\frac{A_{22}}{m_2}-\omega^2\right)a_2'+\cdots+\frac{A_{2,3N-1}}{\sqrt{m_2 m_{3N-1}}}a_{3N-1}' = -\frac{A_{2,3N}}{\sqrt{m_2 m_{3N}}},\\
&\quad\vdots\\
&\frac{A_{3N-1,1}}{\sqrt{m_{3N-1} m_1}}a_1'+\frac{A_{3N-1,2}}{\sqrt{m_{3N-1} m_2}}a_2'+\cdots+\left(\frac{A_{3N-1,3N-1}}{m_{3N-1}}-\omega^2\right)a_{3N-1}'\\
&\qquad = -\frac{A_{3N-1,3N}}{\sqrt{m_{3N-1} m_{3N}}}.
\end{aligned}
\tag{10.2.11}
$$

The above equations give the solutions

$$
a_s' = \frac{D_s(\omega_k^2)}{D_{3N}(\omega_k^2)}, \tag{10.2.12}
$$

where

$$
D_{3N}(\omega_k^2) = \begin{vmatrix}
\frac{A_{11}}{m_1}-\omega_k^2 & \frac{A_{12}}{\sqrt{m_1 m_2}} & \cdots & \frac{A_{1,3N-1}}{\sqrt{m_1 m_{3N-1}}}\\
\frac{A_{21}}{\sqrt{m_2 m_1}} & \frac{A_{22}}{m_2}-\omega_k^2 & \cdots & \frac{A_{2,3N-1}}{\sqrt{m_2 m_{3N-1}}}\\
\vdots & \vdots & \vdots & \vdots\\
\frac{A_{3N-1,1}}{\sqrt{m_{3N-1} m_1}} & \frac{A_{3N-1,2}}{\sqrt{m_{3N-1} m_2}} & \cdots & \frac{A_{3N-1,3N-1}}{m_{3N-1}}-\omega_k^2
\end{vmatrix}, \tag{10.2.13}
$$

$$D_s(\omega_k^2) = \begin{vmatrix} \frac{A_{11}}{m_1} - \omega_k^2 & \cdots & \frac{A_{1,S-1}}{\sqrt{m_1 m_{s-1}}} & \frac{-A_{1,3N}}{\sqrt{m_1 m_{3N}}} & \cdots & \frac{A_{1,3N-1}}{\sqrt{m_1 m_{3N-1}}} \\ \frac{A_{21}}{\sqrt{m_2 m_1}} & \cdots & \frac{A_{2,S-1}}{\sqrt{m_2 m_{s-1}}} & \frac{-A_{2,3N}}{\sqrt{m_2 m_{3N}}} & \cdots & \frac{A_{2,3N-1}}{\sqrt{m_2 m_{3N-1}}} \\ \vdots & & \vdots & \vdots & & \vdots \\ \frac{A_{3N-1,1}}{\sqrt{m_{3N-1} m_1}} & \cdots & \frac{A_{3N-1,S-1}}{\sqrt{m_{3N-1} m_{s-1}}} & \frac{-A_{3N-1,3N}}{\sqrt{m_{3N-1} m_{3N}}} & \cdots & \frac{A_{3N-1,3N-1}}{m_{3N-1}} - \omega_k^2 \end{vmatrix}. \tag{10.2.14}$$

Also,

$$a_s = \frac{a_{3N}}{D_{3N}(\omega_k^2)} D_s(\omega_k^2) = b_k D_s(\omega_k^2), \tag{10.2.15}$$

where

$$b_k = \frac{a_{3N}}{D_{3N}(\omega_k^2)}. \tag{10.2.16}$$

The relations (10.2.5) can now be written as

$$\sqrt{m_s} U_s(t) = \left(b_k e^{i\omega_k t} + b_k^* e^{-i\omega_k t}\right) D_s(\omega_k^2). \tag{10.2.17}$$

We can set

$$b_k = \frac{1}{2} c_k e^{i\alpha_k}, \tag{10.2.18}$$

and find

$$\sqrt{m_s} U_s(t) = D_s(\omega_k^2) c_k \cos(\omega_k t + \alpha_k). \tag{10.2.19}$$

Summing over the different frequencies we obtain

$$\begin{aligned} \sqrt{m_s} U_s(t) &= \sum_{k=1}^{3N} D_s(\omega_k^2) c_k \cos(\omega_k t + \alpha_k) \\ &= \sum_{k=1}^{3N} D_s(\omega_k^2) q_k(t), \end{aligned} \tag{10.2.20}$$

which contains 6N arbitrary constants c_k and α_k to be determined by the use of the initial conditions. The above relations express the fact that displacements are linear combinations of the *normal coordinates*

$$q_k(t) = c_k \cos(\omega_k t + \alpha_k). \tag{10.2.21}$$

The coefficients of the linear combinations $D_s(\omega_k^2)$ are independent of time. It is interesting to consider some properties of these coefficients. From (10.2.8) and (10.2.15) we derive

$$\sum_{s'=1}^{3N}\left(\frac{A_{ss'}}{\sqrt{m_s m_{s'}}} - \delta_{ss'}\omega_k^2\right) a'_s = \sum_{s'=1}^{3N}\left(\frac{A_{ss'}}{\sqrt{m_s m_{s'}}} - \delta_{ss'}\omega_k^2\right) b_k D_{s'}(\omega_k^2) = 0, \tag{10.2.22}$$

or

$$\sum_{s'=1}^{3N}\left(\frac{A_{ss'}}{\sqrt{m_s m_{s'}}} - \delta_{ss'}\omega_{k'}^2\right) D_{s'}(\omega_k^2) = 0. \tag{10.2.23}$$

The above relations give a system of 3N equations; multiplying the first relation by $D_1(\omega_{k'}^2)$, the second by $D_2(\omega_{k'}^2)$, and so on, and summing we obtain

$$\sum_s\sum_{s'}\left(\frac{A_{ss'}}{\sqrt{m_s m_{s'}}} - \delta_{ss'}\omega_k^2\right) D_{s'}(\omega_k^2) D_s(\omega_{k'}^2) = 0. \tag{10.2.24}$$

Interchanging the indices s and s′ we can write

$$\sum_s\sum_{s'}\left(\frac{A_{ss'}}{\sqrt{m_s m_{s'}}} - \delta_{ss'}\omega_k^2\right) D_s(\omega_k^2) D_{s'}(\omega_{k'}^2) = 0. \tag{10.2.25}$$

On the other hand, interchanging the indices k and k′ in (10.2.24) we find

$$\sum_s\sum_{s'}\left(\frac{A_{ss'}}{\sqrt{m_s m_{s'}}} - \delta_{ss'}\omega_{k'}^2\right) D_s(\omega_k^2) D_{s'}(\omega_{k'}^2) = 0. \tag{10.2.26}$$

Subtracting (10.2.25) from (10.2.26)

$$(\omega_k^2 - \omega_{k'}^2) \sum_s \sum_{s'} \delta_{ss'} D_s(\omega_k^2) D_{s'}(\omega_{k'}^2)$$
$$= (\omega_k^2 - \omega_{k'}^2) \sum_s D_s(\omega_k^2) D_s(\omega_{k'}^2) = 0, \qquad (10.2.27)$$

$$\sum_s D_s(\omega_k^2) D_s(\omega_{k'}^2) = 0, \quad \text{if } k \neq k'. \qquad (10.2.28)$$

Consider now (10.2.25); taking (10.2.28) into account we find that if $k \neq k'$ then

$$\sum_s \sum_{s'} \frac{A_{ss'}}{\sqrt{m_s m_{s'}}} D_s(\omega_k^2) D_{s'}(\omega_{k'}^2) - \sum_s \sum_{s'} \omega_k^2 \delta_{ss'} D_s(\omega_k^2) D_{s'}(\omega_{k'}^2)$$
$$= \sum_s \sum_{s'} \frac{A_{ss'}}{\sqrt{m_s m_{s'}}} D_s(\omega_k^2) D_{s'}(\omega_{k'}^2) - \omega_k^2 \sum_s D_s(\omega_k^2) D_{s'}(\omega_{k'}^2)$$
$$= \sum_s \sum_{s'} \frac{A_{ss'}}{\sqrt{m_s m_{s'}}} D_s(\omega_k^2) D_{s'}(\omega_{k'}^2) = 0. \qquad (10.2.29)$$

We now want to express the potential and kinetic energy in terms of normal coordinates. The potential energy is given by

$$V = V_o + \frac{1}{2} \sum_{s=1}^{3N} \sum_{s'=1}^{3N} A_{ss'} u_s u_{s'}$$
$$= V_o + \frac{1}{2} \sum_{s=1}^{3N} \sum_{s'=1}^{3N} \frac{A_{ss'}}{\sqrt{m_s m_{s'}}} \sum_{k=1}^{3N} D_s(\omega_k^2) q_k \sum_{k'=1}^{3N} D_s(\omega_{k'}^2) q_{k'}$$
$$= V_o + \frac{1}{2} \sum_{k=1}^{3N} \sum_{k'=1}^{3N} q_k q_{k'} \sum_{s=1}^{3N} \sum_{s'=1}^{3N} \frac{A_{ss'}}{\sqrt{m_s m_{s'}}} D_s(\omega_k^2) D_{s'}(\omega_{k'}^2)$$
$$= V_o + \frac{1}{2} \sum_{k=1}^{3N} q_k^2 \sum_{s=1}^{3N} \sum_{s'=1}^{3N} \frac{A_{ss'}}{\sqrt{m_s m_{s'}}} D_s(\omega_k^2) D_{s'}(\omega_k^2), \qquad (10.2.30)$$

where we have taken (10.2.29) into account. The kinetic energy is given by

$$\begin{aligned}
T &= \frac{1}{2}\sum_{s=1}^{3N} m_s \dot{u}_s^2 \\
&= \frac{1}{2}\sum_{s=1}^{3N}\sum_{k=1}^{3N}\sum_{k'=1}^{3N} D_s(\omega_k^2)D_s(\omega_{k'}^2)\dot{q}_k\dot{q}_{k'} \\
&= \frac{1}{2}\sum_{k=1}^{3N}\sum_{k'=1}^{3N} \dot{q}_k\dot{q}_{k'}\sum_s D_s(\omega_k^2)D_s(\omega_{k'}^2) \\
&= \frac{1}{2}\sum_{k=1}^{3N} \dot{q}_k^2 \sum_s D_s(\omega_k^2),
\end{aligned} \tag{10.2.31}$$

where we have taken (10.2.28) into account. The energy of the vibrations is then given by

$$\begin{aligned}
E &= \frac{1}{2}\sum_{k=1}^{3N}\left\{\sum_{s=1}^{3N}\sum_{s'=1}^{3N}\frac{A_{ss'}}{\sqrt{m_s m_{s'}}}D_s(\omega_k^2)D_{s'}(\omega_k^2)\right\}q_k^2 \\
&+ \frac{1}{2}\sum_{k=1}^{3N}\left\{\sum_{s=1}^{3N}D_s^2(\omega_k^2)\right\}\dot{q}_k^2,
\end{aligned} \tag{10.2.32}$$

and the energy of the k-th oscillator by

$$\begin{aligned}
E_k &= \frac{1}{2}\left\{\sum_{s=1}^{3N}\sum_{s'=1}^{3N}\frac{A_{ss'}}{\sqrt{m_s m_{s'}}}D_s(\omega_k^2)D_{s'}(\omega_k^2)\right\}q_k^2 \\
&+ \frac{1}{2}\left\{\sum_{s=1}^{3N}D_s^2(\omega_k^2)\right\}\dot{q}_k^2.
\end{aligned} \tag{10.2.33}$$

Differentiating with respect to time,

$$\ddot{q}_k + \frac{\sum_{s=1}^{3N}\sum_{s'=1}^{3N}(A_{ss'}/\sqrt{m_s m_{s'}})D_s(\omega_k^2)D_{s'}(\omega_k^2)}{\sum_{s=1}^{3N}D_s^2(\omega_k^2)}q_k = 0. \tag{10.2.34}$$

On the other hand the relation (10.2.24), setting $k = k'$, becomes

$$\sum_{s}\sum_{s'}\left(\frac{A_{ss'}}{\sqrt{m_s m_{s'}}} - \delta_{ss'}\omega_k^2\right) D_s(\omega_k^2) D_{s'}(\omega_k^2) = 0,$$

or

$$\sum_{s}\sum_{s'}\frac{A_{ss'}}{\sqrt{m_s m_{s'}}} D_s(\omega_k^2) D_{s'}(\omega_k^2) = \omega_k^2 \sum_{s} D_s^2(\omega_k^2). \tag{10.2.35}$$

Therefore (10.2.34) becomes

$$\ddot{q}_k + \omega_k^2 q_k = 0, \tag{10.2.36}$$

as expected.

Until now we have assumed that the frequencies of vibration are not degenerate. If there are two frequencies that are degenerate ($\omega_k = \omega_j$), and all the other frequencies are nondegenerate, then only $3N - 2$ of the $3N$ equations (10.2.8) are independent and two constants, say a_{3N} and a_{3N-1}, have to be considered arbitrary for $\omega_k = \omega_j$; correspondingly, two normal coordinates q_k and q_j are related to the same frequency and linear combinations of them are also acceptable as normal coordinates. The actual number of degenerate frequencies of vibration is related to the symmetry of the crystal.

Now that we have described the vibrations let us ask ourselves the question: How could we have the system vibrate only in one normal mode? The displacement component of an atom in correspondence to the k-th mode is given by

$$\sqrt{m_s}\, U_s(t) = D_s(\omega_k^2) c_k \cos(\omega_k t + \alpha_k), \tag{10.2.37}$$

where $D_s(\omega_k^2)$ depends on s, and the coefficient c_k and the phase angle α_k are the same for all s. We can, in principle, excite the k-th mode by displacing each atom by the amount $C m_s^{-\frac{1}{2}} D_s(\omega_k^2)$ and then releasing the atoms; in this case the atoms start vibrating with a phase $\alpha_k = 0$ and an amplitude $c_k = C$ in the k-th mode and continue to vibrate in this mode without any effect due to the other normal modes. Such a pattern of motion could actually be set in the system in the absence of anharmonic effects; anharmonicity constitutes a mechanism by means of which energy is transferred from one mode to the other (the entire energy remaining constant if the system is isolated). Realistically we must then expect that, having excited a particular mode of vibration, the corresponding amplitude decreases in time because of anharmonicity until the energy is distributed over all the modes of vibrations according to the proper statistics of the situation; in other words, anharmonicity is responsible for the achievement of thermal equilibrium. A certain characteristic time τ_k may be associated with the process of transfer of energy from a single mode k to all other modes of the solid on account of the anharmonicity; this in turn causes a broadening $\Delta\omega_k$ of the frequency distribution associated with a particular mode; in the absence of anharmonic interactions this distribution could simply be represented by a delta function $\delta(\omega - \omega_k)$.

10.3. Harmonic and Anharmonic Relaxation

Consider two pendula coupled by a spring as in Fig. 10.1. The most general pattern of motion of the two pendula is described by the following equations:

$$
\begin{aligned}
x_1 &= A\cos(\omega_1 t + \phi_1) + B\cos(\omega_2 t + \phi_2),\\
x_2 &= A\cos(\omega_1 t + \phi_1) - B\cos(\omega_2 t + \phi_2),
\end{aligned}
\tag{10.3.1}
$$

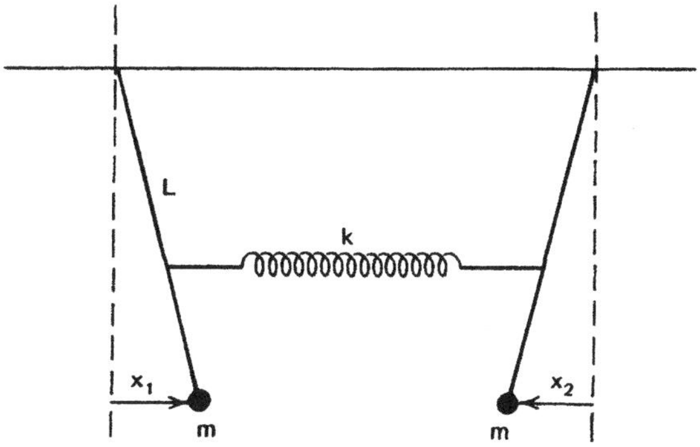

Fig. 10.1. Two coupled pendula.

where

$$\omega_1 = \sqrt{\frac{q}{L}},$$
$$\omega_2 = \sqrt{\frac{q}{L} + \frac{2k}{m}}, \tag{10.3.2}$$

and A, B, ϕ_1, and ϕ_2 are four constants that can be determined by the initial conditions at the time, say, $t = 0$. Let us assume that at time $t = 0$ one pendulum is displaced by a certain amount C and the other pendulum is at its equilibrium position and that the velocity of each pendulum is zero. The initial conditions give

$$\begin{aligned} C &= A\cos\phi_1 + B\cos\phi_2, \\ 0 &= A\cos\phi_1 - B\cos\phi_2, \\ 0 &= -\omega_1 A\sin\phi_1 - \omega_2 B\sin\phi_2, \\ 0 &= -\omega_1 A\sin\phi_1 + \omega_2 B\sin\phi_2. \end{aligned} \tag{10.3.3}$$

From this we derive

$$\begin{aligned} A = B &= \frac{C}{2}, \\ \phi_1 = \phi_2 &= 0. \end{aligned} \tag{10.3.4}$$

Introducing the values of these constants in (10.3.1) gives

$$\begin{aligned} x_1 &= \frac{C}{2}(\cos\omega_1 t + \cos\omega_2 t), \\ x_2 &= \frac{C}{2}(\cos\omega_1 t - \cos\omega_2 t). \end{aligned} \tag{10.3.5}$$

The above equations describe the motion of the pendula following the initial situation of one pendulum displaced by C and the other at its equilibrium position with both velocities equal to zero.

We notice the following:

1. A "localized" excitation of pendulum 1 is obtained by the constructive interference of *both* modes of vibration.
2. A "relaxation" of the excitation energy, that is, a transfer of energy from one pendulum to the other, follows the initial

situation. No energy is lost in the process; the energy is conserved even within each mode. This relaxation is *spatial* in nature and is called *harmonic.*

3. In due time the original situation is recreated in the sense that again we shall find the two pendula in conditions identical to the initial ones. This will occur at a time

$$T = 2\pi \frac{n_1}{\omega_1} = 2\pi \frac{n_2}{\omega_2} \quad (n_1, n_2 \text{ positive integers}),$$

that is, when two multiples of the periods of the normal oscillations coincide.

4. In the case of more than two coupled pendula one can create a similar situation of "localized" excitation by displacing only one pendulum and keeping the others at their equilibrium positions. In the case of three pendula it is to be expected that the initial localized excitation relaxes as the two other pendula become involved in the motion and that in due time the original situation is recreated. This time is obviously going to be longer for three pendula than for two, since it requires that

$$\frac{n_1}{\omega_1} = \frac{n_2}{\omega_2} = \frac{n_3}{\omega_3}.$$

In general, the time to recreate the initial situation is greater the greater is the number of coupled pendula.

5. Anharmonic distortions may affect the patterns of motion in the simple case of two pendula or in the more complicated multipendulum system. Anharmonicity produces transfer of energy from one mode to the other at a rate that, in any case, has to be compared to the period T of the relaxation. If such rate is slow, for a two-pendulum system there is the possibility of recreating after a certain time a situation close to the original one; in the case of a multipendulum system the possibility may not exist anymore because of the greater time necessary to go back to the original situation. In the latter case the relaxation of the localized excitation is an irreversible process; the relaxation is in this case called *anharmonic.*

Let us turn our attention now to localized excitations in solids. Initial conditions corresponding to a localized excitation can be created by displacing one atom and leaving all the others in their equilibrium positions. We note here that:

1. A localized excitation of such type is created by the interference of many modes of vibration; all the normal modes participating in the constructive interference at the site of the localized excitation form essentially a *wave packet.* The sharper the wave packet the wider its frequency spectrum. On the contrary, for a completely delocalized oscillation the frequency spectrum may be very sharp; in particular, a single frequency corresponds to a normal mode excitation.
2. In absence of anharmonic forces a harmonic relaxation follows the establishment of a localized excitation and the motion spreads rapidly to all the other atoms, each normal mode conserving at any time its initial energy. The rate with which relaxation occurs, that is, the wave packet spreads to the whole crystal, is about equal to the half-width $\Delta\omega$ of the frequency distribution of the normal modes contributing to the wave packet:

   highly localized excitation $\equiv$ large $\Delta\omega \equiv$ large relaxation rate

   A normal mode excitation, for which $\Delta\omega = 0$, extends to the entire crystal and never relaxes into anything.
3. In the harmonic approximation the time T it takes to restore the initial localized excitation, must be exceedingly large, given the large number of interacting particles.
4. The effects of anharmonicity on the relaxation process can be considered by examining two cases; (a) $\Delta\omega_k < \Delta\omega$ and (b) $\Delta\omega_k > \Delta\omega_k$, where $\Delta\omega_k$ is the frequency broadening of one mode caused by anharmonicity and $\Delta\omega$ is the half-width of the distribution of the frequencies in the packet. In case (a) anharmonicity does not initially influence the relaxation process, but affects it rather when the wave packet has already spread out. In case (b) anharmonicity affects the relaxation process even at the initial state.
5. In any case if $\Delta\omega_k > T^{-1}$, T being the period of the relaxation, anharmonicity makes the relaxation process irreversible.

10.4. Review of the Theory of Small Vibrations (Quantum Mechanical)

The potential and kinetic energies are given by

$$V = \frac{1}{2}\sum_{s}\sum_{s'} A_{ss'} u_s u_{s'} + V_o, \tag{10.4.1}$$

and

$$T = \frac{1}{2}\sum_{s} m_s \dot{u}_s^2, \tag{10.4.2}$$

respectively. We set

$$e_s = \sqrt{m_s} u_s. \tag{10.4.3}$$

Then

$$\begin{aligned} V &= \frac{1}{2}\sum_{s}\sum_{s'} \frac{A_{ss'}}{\sqrt{m_s m_{s'}}} e_s e_{s'} \\ &= \frac{1}{2}\sum_{s}\sum_{s'} D_{ss'} e_s e_{s'}, \end{aligned} \tag{10.4.4}$$

where

$$D_{ss'} = \frac{A_{ss'}}{\sqrt{m_s m_{s'}}}, \tag{10.4.5}$$

and

$$T = \frac{1}{2}\sum_{s} \dot{e}_s^2. \tag{10.4.6}$$

The Hamiltonian of the system is then given by

$$H = T + V + \frac{1}{2}\sum_{s} \dot{e}_s^2 + \frac{1}{2}\sum_{s}\sum_{s'} D_{ss'} e_s e_{s'}. \tag{10.4.7}$$

We introduce a new set of coordinates

$$q_k = \sum_{s} x_{sk} e_s. \tag{10.4.8}$$

The inverse transformation must be such that

$$
\begin{aligned}
e_s &= \sum_k y_{ks} q_k = \sum_k y_{ks} \sum_{s'} x_{s'k} e_{s'} \\
&= \sum_{s'} \left[\sum_k y_{ks} x_{s'k} \right] e_{s'} = \sum_{s'} \delta_{ss'} e_{s'}, \qquad (10.4.9)
\end{aligned}
$$

that is, it must be

$$
\sum_k x_{s'k} y_{ks} = \delta_{ss'}, \qquad (10.4.10)
$$

which means that the two $3N \times 3N$ matrices of the coefficients x and y are one the inverse of the other. We have to choose the coefficients x and y in such a way as to eliminate the cross terms $q_k q_{k'}$ in the Hamiltonian. In particular, the potential energy expressed in terms of the new coordinates should give

$$
\begin{aligned}
V &= \frac{1}{2} \sum_s \sum_{s'} D_{ss'} e_s e_{s'} \\
&= \frac{1}{2} \sum_s \sum_{s'} D_{ss'} \sum_k y_{ks} q_k \sum_{k'} y_{k's'} q_{k'} \\
&= \frac{1}{2} \sum_{k'} \left[\sum_s \sum_{s'} \sum_k D_{ss'} y_{ks} y_{k's'} q_k q_{k'} \right] \\
&= \frac{1}{2} \sum_{k'} \omega_{k'}^2 q_{k'}^2, \qquad (10.4.11)
\end{aligned}
$$

or

$$
\sum_s \sum_{s'} \sum_k D_{ss'}\, y_{ks} y_{k's'}\, q_k q_{k'} = \omega_{k'}^2 q_{k'}^2. \qquad (10.4.12)
$$

The above condition is satisfied if we set

$$
\sum_{s'} D_{ss'}\, y_{k's'} = \omega_{k'}^2 y_{k's}, \qquad (10.4.13)
$$

$$
\sum_s y_{k's}\, y_{ks} = \delta_{kk'}. \qquad (10.4.14)
$$

In fact (10.4.12), taking the two above conditions into account, becomes

$$\begin{aligned}\sum_k q_k q_{k'} \sum_s \sum_{s'} [D_{ss'} y_{k's'}] y_{ks} &= \sum_k q_k q_{k'} \sum_s \omega_{k'}^2 y_{k's} y_{ks} \\ &= \sum_k \omega_{k'}^2 q_k q_{k'} \sum_s y_{k's} y_{ks} \\ &= \sum_k \omega_{k'}^2 q_k q_{k'} \delta_{kk'} = \omega_{k'}^2 q_{k'}^2. \end{aligned} \tag{10.4.15}$$

Comparing (10.4.10) with (10.4.14) we find

$$y_{ks} = x_{sk}, \tag{10.4.16}$$

which means that the $3N \times 3N$ matrix for the inverse transformation (10.4.9) is the transpose of the $3N \times 3N$ matrix for the direct transformation (10.4.8). The conditions (10.4.13) and (10.4.14) can now be rewritten together with condition (10.4.10) as follows:

$$\sum_{s'} D_{ss'} x_{s'k} = \omega_k^2 x_{sk}, \tag{10.4.17}$$

$$\sum_s x_{sk} x_{sk'} = \delta_{kk'}, \tag{10.4.18}$$

$$\sum_k x_{sk} x_{s'k} = \delta_{ss'}. \tag{10.4.19}$$

From (10.4.18) we can further deduce that both transformation matrices are *unitary.*

The equations (10.4.17) can be written out as follows

$$\begin{aligned} D_{11} x_{1k} + D_{12} x_{2k} + \cdots + D_{1,3N} x_{3N,k} &= \omega_k^2 x_{1k}, \\ D_{21} x_{1k} + D_{22} x_{2k} + \cdots + D_{2,3N} x_{3N,k} &= \omega_k^2 x_{2k}, \\ D_{3N,1} x_{1k} + D_{3N,2} x_{2k} + \cdots + D_{3N,3N} x_{3N,k} &= \omega_k^2 x_{3N,k}, \end{aligned} \tag{10.4.20}$$

which imply

$$\begin{vmatrix} D_{11}-\omega^2 & D_{12} & \cdots & D_{1,3N} \\ D_{21} & D_{22}-\omega^2 & \cdots & D_{2,3N} \\ D_{3N,1} & D_{3N,2} & \cdots & D_{3N,3N}-\omega^2 \end{vmatrix} = 0, \qquad (10.4.21)$$

which coincides with the determinantal equation (10.2.9) obtained in the classical treatment.

Now that we have found the proper coordinates q_k we want to verify that the kinetic-energy operator also appears as a sum of squares when expressed in terms of them. Assume that we have a function

$$f[q_1(e_1, e_2, \ldots, e_{3N}); q_2(e_1, e_2, \ldots, e_{3N}); \ldots q_{3N}(e_1, e_2, \ldots, e_{3N})], \qquad (10.4.22)$$

and let the kinetic energy operator operate on it:

$$\begin{aligned} Tf &= -\frac{\hbar^2}{2}\sum_s \frac{\partial^2 f}{\partial e_s^2} = -\frac{\hbar^2}{2}\sum_s \left[\sum_k \frac{\partial}{\partial q_k}\left(\sum_{k'} \frac{\partial f}{\partial q_{k'}}\frac{\partial q_{k'}}{\partial e_s}\right)\frac{\partial q_k}{\partial e_s}\right] \\ &= -\frac{\hbar^2}{2}\sum_s\sum_k \frac{\partial}{\partial q_k}\left(\sum_{k'} \frac{\partial f}{\partial q_{k'}} x_{sk'}\right) x_{sk} \\ &= -\frac{\hbar^2}{2}\sum_k\sum_{k'} \frac{\partial^2 f}{\partial q_k \partial q_{k'}} \sum_s x_{sk'} x_{sk} \\ &= -\frac{\hbar^2}{2}\sum_k\sum_{k'} \frac{\partial^2 f}{\partial q_k \partial q_{k'}} \delta_{kk'} = -\frac{\hbar^2}{2}\sum_k \frac{\partial^2 f}{\partial q_k^2}. \end{aligned} \qquad (10.4.23)$$

The Hamiltonian operator can now be expressed as follows:

$$H = \frac{1}{2}\sum_k \left(-\hbar^2 \frac{\partial^2}{\partial q_k^2} + \omega_k^2 q_k^2\right) = \sum_k H_k, \qquad (10.4.24)$$

where

$$H_k = -\frac{\hbar^2}{2}\frac{\partial^2}{\partial q_k^2} + \frac{1}{2}\omega_k^2 q_k^2. \qquad (10.4.25)$$

The Schröedinger equation

$$H_k\phi(q_k) = -\frac{\hbar^2}{2}\frac{\partial^2}{\partial q_k}\phi(q_k) + \frac{1}{2}\omega_k^2 q_k^2\phi(q_k), \qquad (10.4.26)$$

has the solution

$$\phi_{n_k}(q_k) = (\bar{q}_k)^{-1/2} e^{-\xi_k^2/2} H_{n_k}(\xi_k), \qquad (10.4.27)$$

where

$$\xi_k = \frac{q_k}{\bar{q}_k} = \frac{q_k}{\sqrt{\hbar/\omega_k}}, \qquad (10.4.28)$$

and

$$H_{n_k}(\xi_k) = \frac{(-1)^{n_k}}{\sqrt{2^{n_k} n_k : \sqrt{\pi}}} e^{\xi_{k^2}} \frac{d_k^n}{d{\xi_k}^{n_k}} e^{-\xi_{k^2}}. \qquad (10.4.29)$$

The eigenvalues of H_k are given by

$$E_k = \left(n_k + \frac{1}{2}\right)\hbar\omega_k. \qquad (10.4.30)$$

The eigenfunctions and the eigenvalues of the total Hamiltonian H are given by

$$\begin{aligned}\phi_{\{n\}}(q_1, q_2, \ldots, q_{3N}) &= \phi_{n_1}(q_1)\phi_{n_2}(q_2)\cdots\phi_{n_{3N}}(q_{3N}) \\ &= \prod_{k=1}^{3N}\phi_{n_k}(q_k), \end{aligned} \qquad (10.4.31)$$

and

$$E_{\{n\}} = \sum_{k=1}^{3N} E_k(n_k), \qquad (10.4.32)$$

respectively. In summary, Cartesian coordinates and normal coordinates are related as follows:

$$q_k = \sum_s x_{sk} e_s, \qquad (10.4.33)$$

$$e_s = \sum_k x_{sk} q_k, \qquad (10.4.34)$$

with

$$\sum_{s=1}^{3N} x_{sk} x_{sk'} = \delta_{kk'}, \tag{10.4.35}$$

$$\sum_{k=1}^{3N} x_{sk} x_{s'k} = \delta_{ss'}. \tag{10.4.36}$$

Setting $k = k'$ in (10.4.35) or $s = s'$ in (10.4.36) we find

$$x_{sk}^2 \approx \frac{1}{N},$$

and

$$x_{sk} \approx \frac{1}{\sqrt{N}}. \tag{10.4.37}$$

The Cartesian coordinates and the normal coordinates are related by a unitary transformation whose coefficients are of the order of magnitude $1/\sqrt{N}$; for this reason the displacements of some atoms confined in a small region of the solid do not affect almost at all the normal coordinates. Conversely the change of a small number of normal coordinates does not affect appreciably the actual Cartesian displacements of the atoms.

10.5. The Effect of Impurities on Lattice Vibrations

In the simple case of a diatomic molecule the vibrational motion is performed while the molecule is in a potential well whose shape is determined by the electronic energy $\varepsilon_k(R)$, as given in the adiabatic approximation by the solution of the Schrödinger equation (1.1.15). When the molecule is electronically excited the shape of the potential changes and the molecule in general vibrates with a different frequency; in particular, if the potential energy has no minimum, the vibrational frequencies will be in a continuum, that is, the molecule dissociates.

In a crystal (or molecule) with N atoms the potential energy is a function of 3N Cartesian or normal coordinates. Since the normal coordinates are independent, it is possible to represent in a diagram a potential curve for each normal coordinate. In representing the potential this way one should not put in the same diagram other potential curves corresponding to electronic excited states because the system of normal coordinates changes when the electronic state of the system changes.

In particular, in the case of an impurity in a solid, only the electronic states of the impurity are of interest. Potential curves may be introduced, one in correspondence to each normal coordinate, and only when neglecting the scrambling of the normal coordinates, can the same diagrams contain curves corresponding to ground and excited electronic states.

In order to consider the effect of the presence of an impurity on the vibrations of a solid, it is instructive to consider the case of a molecular impurity with certain natural (intramolecular) frequencies of vibration. These frequencies will be somewhat affected by the presence of the molecule in the crystal due to the coupling of the molecule with other atoms. In what follows we shall indicate as ω_m a frequency of vibration of the molecule in the crystal. Two different situations may arise:

1. The molecular frequency ω_m falls in an allowed frequency band of the perfect crystal. The resonance between the molecular and the crystal frequencies allows energy to be transferred from the molecular impurity to the crystal. When equilibrium is reached the atoms of the impurity molecule participate in the vibration corresponding to the normal mode of frequency ω_m. The energy of the mode is in this case equally distributed over all the atoms of the crystal, the energy of each atom being $\sim 1/N$ of the total energy of the normal mode. The vibrations so described are called *band vibrations.*
2. The molecular frequency ω_m falls in a frequency region that is forbidden in the perfect crystal. In this case the energy remains essentially localized at the impurity and immediate surroundings. The coefficients x_{sk} of the $3N \times 3N$ transformation matrix are now related as

$$\sum_{k=1}^{3N} x_{sk}^2 = \sum_{k=1}^{3N}{}' x_{sk}^2 + x_{s\lambda}^2 = 1, \qquad (10.5.1)$$

where λ refers to the *localized vibration* and the sum $\sum'$ is exclusive of the localized mode of vibration. The above relation may be written

$$\sum' x_{sk}^2 = 1 - x_{s\lambda}^2, \tag{10.5.2}$$

which means that for an atom at the s-th position the more it participates in the localized λ vibration, the less it participates in the band vibrations. In other words band vibrations are practically excluded from the region of the crystal in which a localized vibration is active; physically this is understandable since the lack of resonance between band modes and localized mode(s) makes it impossible to the band modes to propagate to the region of the crystal in which localized vibrations are active.

It is important to examine at this point the effects that a change in the electronic state of an impurity may have on band and localized modes. A change in the electronic state of the impurity produces essentially two physical effects:

1. A change in the equilibrium position of a small number of atoms including the impurity and its immediate neighbors.
2. A change in the force constants between the impurity atom and its immediate neighbors.

Correspondingly the following changes occur in the adiabatic potential:

1. A shift in the minimum of the potential curve.
2. A change in the curvature of the potential curve, that is, a change in the frequency of vibration.

A logical consequence of the above changes is the fact that when the impurity is in an excited state a new system of normal coordinates should be introduced to describe exactly the situation. Let us examine now these two effects for localized and band modes:

1. *Shifts in the Equilibrium Position*

Band Modes. The shifts in the equilibrium position of very few atoms (which, if localized vibrations are active, do not participate in the band vibrations) has very small effect on the normal coordinates. A variation in the position of the minimum of the potential of the order of $1/\sqrt{N}$ times the Cartesian shifts of the atoms above can be expected.

Localized Modes. The shifts in the equilibrium positions of the impurity and its immediate neighbors will produce a change in the position of the potential minimum of the same order of magnitude of the Cartesian shifts of the atoms above. We note here also that the shifts in the potential minima are independent of N.

2. *Changes in the Frequency of Vibration Band Modes.* Considering the relation (10.4.17)

$$\sum_{s'} D_{ss'} x_{s'k} = \sum_{s'} \frac{A_{ss'}}{\sqrt{m_s m_s'}} x_{s'k} = \omega_k^2 x_{sk'}, \tag{10.5.3}$$

the change of very few terms in the sum above will affect very little the frequency. For a simple harmonic oscillator

$$\omega = \sqrt{\frac{k}{m}}, \tag{10.5.4}$$

and

$$\frac{\Delta\omega}{\omega} = \frac{\Delta(\sqrt{k/m})}{\sqrt{k/m}}. \tag{10.5.5}$$

In a band vibration the frequency will change by

$$\frac{\Delta\omega}{\omega} \sim \frac{\Delta(\sqrt{D})}{\sqrt{D}} \frac{1}{N}, \tag{10.5.6}$$

that is, by 1/N x the relative change in some particular $\sqrt{D}$.

Localized Modes. Here the sum (10.5.3) above (s corresponds now to an atom in the region of the localized vibration) contains only few terms and for each of these terms D changes. The change in the frequency are now of the order

$$\frac{\Delta\omega}{\omega} \sim \frac{\Delta(\sqrt{D})}{\sqrt{D}}. \tag{10.5.7}$$

We have considered the impurity as consisting of a molecule with internal degrees of freedom. A particular case of this general situation is the one most common in laser crystals in which the impurity is represented by an atom or ion that replaces at certain sites the natural component of the perfect crystal. For simple crystals the occurrence of forbidden gaps is a rare occurrence because of the strong curvatures and the several intersections of the dispersion curves. For simple crystals with, say, two atoms per unit cell a look at Fig. 5.5 will give the reader the feeling that only for crystals like alkali halides[3] in which the two constituents have very different masses may one expect gaps in the phonon spectrum.

Summarizing the above results we can say the following for a crystal containing an impurity and presenting both band and localized vibrations:

1. For a band vibration the coefficients x_{sk} are in general all approximately equal to $1/\sqrt{N}$ if the s-th atom is outside a locally vibrating region and ≈ 0 if the s-th atom is in a locally vibrating region. Small shifts ($\sim 1/\sqrt{N}$ of the Cartesian shifts of the impurity and surrounding atoms) of the potential minima and small frequency changes ($\sim (1/N)[\Delta(\sqrt{D})/\sqrt{D}]$) are expected when the impurity goes from the ground to an excited electronic state; these effects are even smaller if a localized vibration is present.
2. For localized vibrations, the coefficients x_{sk} are, in general, all approximately equal to zero if the s-th atom is outside the locally vibrating region and have values of the order of 1 if the s-th atom is in the locally vibrating region. When the impurity is electronically excited, shifts of the potential minima of the order of the Cartesian shifts and changes of the frequencies of vibration of the order $\Delta(\sqrt{D})/\sqrt{D}$ will occur.

Consider now a molecular impurity in a crystal with its natural frequency of vibration ω_m in resonance with some frequency in an allowed band. One of the following circumstances may arise:

a. The forces binding the impurity to the crystal are weak
b. The symmetry of the molecular vibration of frequency ω_m does not match the symmetry of the crystal vibration
c. The density of lattice phonon modes at $\omega = \omega_m$ is very low.

In any of the above circumstances the coupling of the intramolecular vibration to the lattice vibrations will be small and the transfer of energy from the molecular impurity to the crystal will be slow; that is, if τ is the time it takes the molecular impurity to transfer energy to the lattice and $T = 2\pi/\omega_m$, it will be $\tau \gg T$. In such a situation we have what we call with Rebane[1] a *quasilocalized* or *pseudolocalized vibration*.

It is possible to extend the concept of pseudolocalized vibration to the case in which the impurity is represented simply by an atom or ion. An atom substituting at a particular site the natural component of a crystal has in general mass and interatomic force constants different from those of the replaced atom. Two cases are of interest:

a. The force constants of the impurity are larger and the mass smaller than those of the replaced atom. In this case the impurity atom will tend to participate more in the high-frequency vibrations; that is, the amplitude of its vibrations will be larger at high frequency than those of the replaced atom. Since $\sum_k x_{sk}^2 = 1$ in any case, this enhancement will take place at the expense of the low-frequency vibrations, which will decrease in amplitude.
b. The force constants of the impurity are smaller and the mass is larger than those of the replaced atom. In this case the amplitudes of low-frequency vibrations will be enhanced at the expense of the amplitude of the frequency vibrations.

In effect, because of, and quantitatively depending on, the physical differences between substituting and substituted atoms, a tendency will develop to set in a dependence of the coefficients x_{ik} ($s = i$ is the index of impurity or of immediately surrounding atoms) on k, so that x_{ik} may indeed present a peak in correspondence to certain k, even if in the perfect crystal no such peak occurs in the phonon spectrum. This tendency can produce a localized vibration if the perturbation introduced by the impurity is strong enough; otherwise a pseudolocalized vibration will set in. In a sense a pseudolocalized vibration is a localized vibration that did not make it.

The following observations can now be made:

1. A localized vibration is a normal mode vibration; a pseudolocalized vibration is not. Rather, a pseudolocalized vibration is a wave

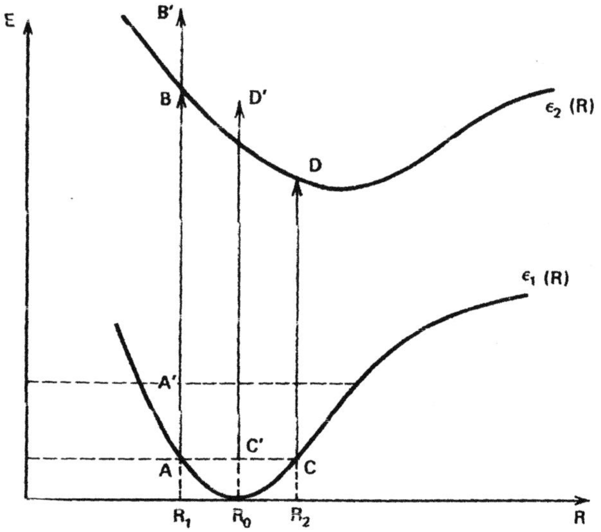

Fig. 10.2. Diagram illustrating the Franck-Condon principle.

packet with frequencies confined in a small region of the phonon spectrum.

2. A localized vibration can only relax anharmonically. A pseudolocalized vibration relaxes harmonically and anharmonically; this means that relatively speaking localized vibrations have longer lifetimes and are spectrally more sharply defined.

10.6. The Franck-Condon Principle

Let us consider for simplicity a diatomic molecule. The potential curve diagram for such a system is given in Fig. 10.2; in it R is the distance between the two nuclei. The Franck-Condon principle can be stated in three different versions.

a. The Classical Franck-Condon Principle

By using purely classical arguments the Franck-Condon principle is stated as follows:

> "During an electronic transition the electronic state changes so fast that (1) the nuclei do not move and (2) the nuclei do not change their momenta."

Condition (1) above means that during an electronic transition the internuclear distance R must remain constant; condition (2) implies that the kinetic energy must also remain constant, and this means that if the molecule's initial state is given by the point A′ the molecule after the electronic transition will be found at B′ with AA′ = BB′.

A third condition may be added considering that the harmonic oscillator which represents the vibrational motion of the molecule spends most of its time at its turning points at which the kinetic energy is zero. This condition (3) states that at the instant at which the electronic transition takes place the oscillator is found at these turning points. Therefore according to this condition the following transitions are allowed:

$$\mathrm{AB}, \quad \mathrm{CD}$$

and the following transitions are forbidden:

$$\mathrm{A'B'}, \quad \mathrm{C'D'}$$

b. The Semiclassical Franck-Condon Principle

The semiclassical Franck-Condon principle consists of three conditions:

> (1) R = const, (2) p = const, and (3) an electronic transition can take place for any value of R with a probability W(R), as given by quantum mechanics.

If the system is in particular vibrational quantum state, say the i-th vibrational state, then

$$\mathrm{W(R)} = |\phi_i(\mathrm{R})|^2. \tag{10.6.1}$$

If the system is in thermal equilibrium at a temperature T,

$$\mathrm{W_T(R)} = \frac{\sum_i e^{-E_i/kT} |\phi_i(\mathrm{R})|^2}{\sum_i e^{-E_i/kT}}. \tag{10.6.2}$$

For most diatomic molecules at ordinary temperature $\hbar\omega$ is so large (of the order 0.1 eV) that

$$\hbar\omega \gg kT, \tag{10.6.3}$$

(room temperature $kT \simeq 0.025\,\mathrm{eV}$). Note that $1.5°\mathrm{K} \simeq 1.25 \times 10^{-4}\,\mathrm{eV}$.

It may be useful to recall the following formulas that apply to a harmonic oscillator ($\beta = 1/kT$). The energy of the i-th state is given by

$$E_i = \left(n_i + \frac{1}{2}\right)\hbar\omega \quad n_i = 0, 1, 2, 3, \ldots, \tag{10.6.4}$$

and the partition sum by

$$\zeta_v = \sum_{i=0}^{\infty} e^{-\beta E_i} = \sum_{i=0}^{\infty} e^{-\beta\hbar\omega(n_i+1/2)} = \frac{e^{-(1/2)\beta\hbar\omega}}{1 - e^{-\beta\hbar\omega}}. \tag{10.6.5}$$

If $\hbar\omega = 0.1\,\mathrm{eV}$ and the system is at room temperature the number of molecules that reside in the ground state is

$$\begin{aligned}\frac{Ne^{-(1/2)\beta\hbar\omega}}{\zeta_v} &= N(1 - e^{-\beta\hbar\omega}) = N(1 - e^{-0.1/0.025}) \\ &= N(1 - e^{-4}) = N(1 - 0.018) = 0.982\,N, \end{aligned} \tag{10.6.6}$$

where N is the total number of molecules. Under these conditions the relevant probability $W(R)$ may well be given by (10.6.1) with $i = 0$.

The above considerations can be generalized by treating a molecule with many vibrational degrees of freedom and, of course, by treating a crystal in which the relevant coordinates are the normal coordinates. In these more complex cases the relevant probability is given by

$$W_T = \prod_k W_{kT}(q_k), \tag{10.6.7}$$

where W_{kT} is of the type (10.6.2) with $\omega = \omega_k$ and $R = q_k$; in the limit of low temperatures or high ω, W_{kT} is simply of the type (10.6.1).

Considering now again the diagram in Fig. 10.2 we note that the semiclassical Franck-Condon principle allows the transitions AB, CD, and C′D′, which have probabilities

$$|\phi_i(R_1)|^2, \quad |\phi_i(R_2)|^2, \quad |\phi_i(R_o)|^2,$$

respectively; in particular if $i = 0$ the probability will have its maximum in correspondence to R_o and transitions such as C′D′ have the greatest importance.

c. The Quantum-Mechanical Franck-Condon Principle

The transition probability for a radiative transition in the dipole approximation is proportional to the square of the matrix element:

$$M_{fi} = \langle \underset{\sim}{\pi}_\alpha^\lambda \cdot \underset{\sim}{M} \rangle_{fi} = \langle \underset{\sim}{\pi}_\alpha^\lambda \cdot \sum_\ell q_\ell \underset{\sim}{r}_\ell \rangle_{fi}, \tag{10.6.8}$$

where q_ℓ and $\underset{\sim}{r}_\ell$ are the charge and the position of the ℓ-th charged particle, respectively. The initial state of the system is given in the adiabatic approximation (see 1.1.17) by

$$\Psi_i(\underset{\sim}{r}, \underset{\sim}{R}) = \Psi_k(\underset{\sim}{r}, \underset{\sim}{R})\phi_{k\ell}(\underset{\sim}{R}), \tag{10.6.9}$$

and the final state by

$$\Psi_f(\underset{\sim}{r}, \underset{\sim}{R}) = \Psi_m(\underset{\sim}{r}, \underset{\sim}{R})\phi_{mn}(\underset{\sim}{R}). \tag{10.6.10}$$

The electric dipole operators can be written as

$$\underset{\sim}{M} = -e\sum_i \underset{\sim}{r}_i + e\sum_\alpha Z_\alpha \underset{\sim}{R}_\alpha, \tag{10.6.11}$$

where $\underset{\sim}{r}_i$ is the coordinate of the i-th electron, $\underset{\sim}{R}_\alpha$ is the coordinate of the α-th nucleus, and Z_α is the charge of the α-th nucleus. Then,

$$\begin{aligned} \underset{\sim}{\pi} \cdot \underset{\sim}{M} &= -e\sum_i \underset{\sim}{r}_i \cdot \underset{\sim}{\pi} + e\sum_\alpha Z_\alpha \underset{\sim}{R}_\alpha \cdot \underset{\sim}{\pi} \\ &= D_e(\underset{\sim}{r}) + D_n(\underset{\sim}{R}). \end{aligned} \tag{10.6.12}$$

The relevant matrix element is now given by

$$\begin{aligned} M_{fi} = M_{mn;k\ell} &= \iint d^3\underset{\sim}{r}d^3\underset{\sim}{R}\psi_f(\underset{\sim}{r}, \underset{\sim}{R})^*[D_e(\underset{\sim}{r}) + D_n(\underset{\sim}{R})\psi_i(\underset{\sim}{r}, \underset{\sim}{R}) \\ &= \iint d^3\underset{\sim}{r}d^3\underset{\sim}{R}\psi_m(\underset{\sim}{r}, \underset{\sim}{R})^*\phi_{mn}(\underset{\sim}{R})^*[D_e(\underset{\sim}{r}) + D_n(\underset{\sim}{R})]\psi_k(\underset{\sim}{r}, \underset{\sim}{R})\phi_{k\ell}(\underset{\sim}{R}) \\ &= \int d^3\underset{\sim}{R}\phi_{mn}(\underset{\sim}{R})^* \left[\int d^3\underset{\sim}{r}\psi_m(\underset{\sim}{r}, \underset{\sim}{R})D_e(\underset{\sim}{R})\psi_k(\underset{\sim}{r}, \underset{\sim}{R})\right] \phi_{k\ell}(\underset{\sim}{R}) \end{aligned}$$

$$+\int d^3\underset{\sim}{R}\phi_{mn}(\underset{\sim}{R})^* D_n(\underset{\sim}{R})\phi_{k\ell}(\underset{\sim}{R})\left[\int d^3 r\psi_m(\underset{\sim}{r},\underset{\sim}{R})^*\psi_k(\underset{\sim}{r},\underset{\sim}{R})\right]$$

$$=\int d^3\underset{\sim}{R}\phi_{mn}(\underset{\sim}{R})^* D_{mk}(\underset{\sim}{R})\phi_{k\ell}(\underset{\sim}{R})+\int d^3\underset{\sim}{R}\phi_{mn}(\underset{\sim}{R})^* D_n(\underset{\sim}{R})\phi_{k\ell}(\underset{\sim}{R})\delta_{mk}, \tag{10.6.13}$$

where

$$D_{mk}(\underset{\sim}{R})=\int d^3\underset{\sim}{r}\psi_m(\underset{\sim}{r},\underset{\sim}{R})^* D_e(\underset{\sim}{r})\psi_k(\underset{\sim}{r},\underset{\sim}{R}). \tag{10.6.14}$$

The following observations can now be made:

1. The matrix element M_{fi} consists of two terms. The second term contributes to the matrix element only when $m = k$, that is, when the electronic state does not change in the transition and gives rise to infrared absorption by the vibrations.
2. The first term contains the dipole moment of the electrons and corresponds to transitions between different electronic states.
3. If $m = k$ the first term becomes

$$\int d^3\underset{\sim}{R}\phi_{mn}(\underset{\sim}{R})^* D_{mm}(\underset{\sim}{R})\phi_{m\ell}(\underset{\sim}{R}), \tag{10.6.15}$$

where

$$D_{mm}(\underset{\sim}{R})=\int d^3\underset{\sim}{r}|\psi_m(\underset{\sim}{r},\underset{\sim}{R})|^2 D_e(\underset{\sim}{r}). \tag{10.6.16}$$

Since $D_e(\underset{\sim}{r})$ is an odd operator, $D_{mm}(\underset{\sim}{R})$ is different from zero only if the molecule lacks inversion symmetry. If this is the case the first term in M_{fi} also contributes to the infrared transition.

If we assume that the "electronic" matrix element $D_{mk}(\underset{\sim}{R})$ does not depend strongly on the nuclear coordinates, we may expand it about the equilibrium positions of the nuclei as follows:

$$D_{mk}(\underset{\sim}{R})=D^{o}_{mk}+\sum_{s\nu\alpha} D^{1,s\nu\alpha}_{mk}u_{s\nu\alpha}+\sum_{\substack{s\nu\alpha\\ s'\nu'\alpha'}} D^{2;s\nu\alpha;s'\nu'\alpha'}_{mk}u_{s\nu\alpha}u_{s'\nu'\alpha'}+\cdots, \tag{10.6.17}$$

where $u_{s\nu\alpha}$ is the displacement in the α direction of the ν-th atom nucleus in the s-th unit cell.

The Franck-Condon principle states that it is legitimate to replace the matrix element $D_{mk}(\underset{\sim}{R})$ by the first term in the expression D^o_{mk} that does not depend on the nuclear coordinates.

In this approximation if $m \neq k$, the square of the relevant matrix element is given by

$$|M_{fi}|^2 = |D^o_{mk}|^2 \left| \int \phi_{mn}(\underset{\sim}{R})^* \phi_{k\ell}(\underset{\sim}{R}) d^3\underset{\sim}{R} \right|^2 . \qquad (10.6.18)$$

The transition probability is then proportional to the square of the overlap integral of the vibrational eigenfunctions of the initial and final states. It is this overlap that controls (apart from D^o_{mk}) the strength of the transition probabilities. It is to be noted here that, as we said in Section 1.2, $\phi_{mn}(\underset{\sim}{R})$ and $\phi_{k\ell}(\underset{\sim}{R})$ are solutions of different Schröedinger equations; they actually belong to different sets of orthonormal eigenfunctions and the subscripts m and k do not play the role of good quantum numbers. Therefore if $m \neq k$ the overlap integral in general may not go to zero even if $n \neq \ell$.

The diagram in Fig. 10.3 illustrates the role played by the overlap integral in determining the strength of a transition; it is evident from this diagram that the "vertical" transition AA′ is the one of highest probability and that the "sloped" transitions AB′ and AC′ have much lower probabilities; in particular in the former case there is practically no overlap, while in the latter case the overlap is small because of the oscillating nature of the upper vibrational eigenfunction.

The following considerations can be made:

1. The quantum-mechanical Franck-Condon principle replaces the requirements $R = \text{const}$, $p = \text{const}$ of its semiclassical formulation with the notion that transitions for which such conditions are not fulfilled are very unlikely.
2. The Franck-Condon principle in its quantum-mechanical formulation is not a selection rule in the conventional sense. It does not say that certain transitions cannot occur but, rather, that they are highly improbable. Selection rules are generally derived from the matrix element of an operator taken between two eigenfunctions of the *same* orthonormal set, contrary to the present situation where the relevant entity is an overlap integral of two wavefunctions belonging to different orthonormal sets.

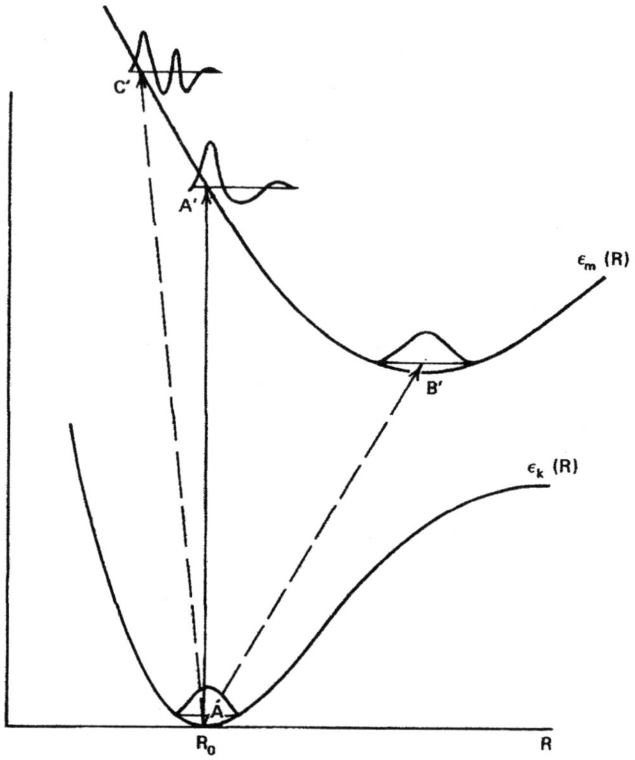

Fig. 10.3. Diagram illustrating the quantum-mechanical Franck-Condon principle.

3. If the potential curves of the two electronic states are the same

$$\varepsilon_k(\underset{\sim}{R}) = \varepsilon_m(\underset{\sim}{R}) + \text{const}, \tag{10.6.19}$$

then the two functions ϕ_{mn} and $\phi_{k\ell}$ belong to the same set and

$$\int \phi_{mn}(\underset{\sim}{R})^* \phi_{k\ell}(\underset{\sim}{R}) d^3\underset{\sim}{R} = \delta_{n\ell}. \tag{10.6.20}$$

In this case the Franck-Condon principle can be expressed as the selection rule: "If the adiabatic potentials of two electronic states are the same, no transition can take place in which the molecule (crystal) changes its vibrational state." This selection rule is independent from the nature (electric dipole, magnetic dipole, or electric quadrupole) of the transition.

10.7. Absorption and Emission in Crystals

We have already derived the following formula (see 9.5.40) for the transition probability per unit time in the electric dipole approximation for one-photon processes ($\underset{\sim}{k} = \underset{\sim}{k}_\alpha$, polarization $\underset{\sim}{\pi}_\alpha^\lambda$):

$$P_\alpha^\lambda d\Omega_\alpha = \frac{4\pi^2 I(\omega_\alpha;\lambda)}{\hbar^2 c}|M_{fi}|^2 d\Omega_\alpha$$
$$= \left\{\frac{4\pi^2 I(\omega_\alpha;\lambda)}{\hbar^2 c} + \frac{\omega_\alpha^3}{\hbar c^3}\right\}|M_{fi}|^2 d\Omega_\alpha, \tag{10.7.1}$$

where

$$|M_{fi}|^2 = |\langle \underset{\sim}{\pi}_\alpha^\lambda \cdot \underset{\sim}{M}\rangle|^2 = \left|\left\langle \underset{\sim}{\pi}_\alpha^\lambda \cdot \sum_\ell q_\ell \underset{\sim}{r}_\ell \right\rangle\right|^2. \tag{10.7.2}$$

If these processes take place in a crystal some additional considerations are in order.

The probability for spontaneous emission, from relation (10.7.1), is given by

$$P_\alpha^\lambda(sp) = \frac{\omega_\alpha^3}{hc^3}|M_{fi}|^2. \tag{10.7.3}$$

If this transition takes place in an atom imbedded in a crystal the above formula has to be modified to take into account the following two facts:

1. For electric dipole transitions the transition probability is proportional to the square of the matrix element of the electric dipole moment and therefore to the square of the electric field at the atom site. The expression (10.7.3) must be multiplied by the factor $(E_{eff}/E)^2$ where E_{eff} is the electric field at the atom and E is the electric field in vacuum. The ratio of these quantities can be expressed as

$$\frac{E_{eff}}{E} = \frac{E_{eff}}{E_c}\frac{E_c}{E}, \tag{10.7.4}$$

where E_c is the average value of the field in the crystal. E and E_c must correspond to the same photon density,

$$\frac{E^2}{4\pi} = \frac{\varepsilon E_c^2}{4\pi}, \tag{10.7.5}$$

where ε is the dielectric constant. Therefore

$$\frac{E_{eff}}{E} = \frac{E_{eff}}{E_c}\frac{E_c}{E} = \frac{1}{\sqrt{\varepsilon}}\frac{E_{eff}}{E_c}, \tag{10.7.6}$$

and

$$\left(\frac{E_{eff}}{E}\right)^2 = \frac{1}{\varepsilon}\left(\frac{E_{eff}}{E_c}\right)^2. \tag{10.7.7}$$

2. The transition probability is proportional to the density of final states [see (9.5.25)]:

$$g(\omega_\alpha) = \frac{V\omega_\alpha^2}{8\pi^3 c^3} d\Omega_\alpha, \tag{10.7.8}$$

and since in the crystal the velocity of light is c/n, n being the index of refraction, an extra factor of n^3 has to be included in the expression for $P_\alpha^\lambda(sp)$.

The expression for $P_\alpha^\lambda(sp)$ becomes

$$P_\alpha^\lambda(sp) = \frac{n^3}{\varepsilon}\left(\frac{E_{eff}}{E_c}\right)^2 \frac{\omega_\alpha^3}{hc^3}|M_{fi}|^2. \tag{10.7.9}$$

The radiated power is given by

$$P_\alpha^\lambda(sp) d\Omega_\alpha \hbar\omega_\alpha = \frac{n^3}{\varepsilon}\left(\frac{E_{eff}}{E_c}\right)^2 \frac{\omega_\alpha^4}{2\pi c^3}|M_{fi}|^2 d\Omega_\alpha. \tag{10.7.10}$$

If several pairs of initial and final states are connected by transitions corresponding to the same frequency, wave vector, and polarization, the power radiated per unit energy range, is given by

$$\begin{aligned}&\frac{n^3}{\varepsilon}\left(\frac{E_{eff}}{E_c}\right)^2 \frac{\omega_\alpha^4}{2\pi c^3}\sum_i n_i \sum_f |M_{fi}|^2 \delta(E_f - E_i + \hbar\omega_\alpha) d\Omega_\alpha \\ &= A_+ \sum_i n_i \sum_f |M_{fi}|^2 \delta(E_f - E_i + \hbar\omega_\alpha) d\Omega_\alpha,\end{aligned} \tag{10.7.11}$$

where n_i number of atoms occupying the state i and

$$A_+ = \frac{n^3}{\varepsilon}\left(\frac{E_{eff}}{E}\right)^2 \frac{\omega_\alpha^4}{2\pi c^3}. \tag{10.7.12}$$

Let us consider now the expression for the absorption and emission transition probability,

$$P^{\lambda}_{\alpha}(\mathrm{abs};\mathrm{emi}) = \frac{4\pi^2 I(\omega_\alpha;\lambda)}{\hbar^2 c}|M_{fi}|^2. \tag{10.7.13}$$

As above this expression has to be corrected for transition probabilities in the crystal as follows:

$$P^{\lambda}_{\alpha}(\mathrm{abs};\mathrm{emi}) = \frac{n^3}{\varepsilon}\left(\frac{E_{\mathrm{eff}}}{E}\right)^2 \frac{4\pi^2 I(\omega_\alpha;\lambda)}{\hbar^2 c}|M_{fi}|^2. \tag{10.7.14}$$

Correspondingly, the power absorbed will be

$$P^{\lambda}_{\alpha} d\Omega_\alpha h\omega_\alpha = \frac{n^3}{\varepsilon}\left(\frac{E_{\mathrm{eff}}}{E}\right)^2 \frac{4\pi^2 \omega_\alpha I(\omega_\alpha;\lambda)}{\hbar c}|M_{fi}|^2 d\Omega_\alpha, \tag{10.7.15}$$

and if several pairs of levels may be connected by transitions of the same frequency ω_α, wave vector, and polarization, the power absorbed per unit energy range is given by

$$\begin{aligned} &\frac{n^3}{\varepsilon}\left(\frac{E_{\mathrm{eff}}}{E}\right)^2 \frac{4\pi^2 \omega_\alpha I(\omega_\alpha;\lambda)}{\hbar c}\sum_i n_i \sum_f |M_{fi}|^2 \delta(E_f - E_i - \hbar\omega_\alpha) d\Omega_\alpha \\ &= A_- \sum_i n_i \sum_f |M_{fi}|^2 \delta(E_f - E_i - \hbar\omega_\alpha) d\Omega_\alpha, \end{aligned} \tag{10.7.16}$$

where

$$A_- = \frac{n^3}{\varepsilon}\left(\frac{E_{\mathrm{eff}}}{E}\right)^2 \frac{4\pi^2 \omega_\alpha I(\omega_\alpha;\lambda)}{\hbar c}. \tag{10.7.17}$$

We can write (10.7.11) and (10.7.16) concisely by using the expression

$$A_\pm \sum_i n_i \sum_f |M_{fi}|^2 \delta(E_f - E_i \pm \hbar\omega_\alpha) d\Omega_\alpha, \tag{10.7.18}$$

where the plus and minus signs stand for spontaneous emission and absorption, respectively, and A_+ and A_- are given by (10.7.12) and (10.7.17), respectively.

10.8. Purely Electronic (Zero-Phonon) Transitions

We now present a simple treatment of the interaction of radiation with an impurity center in a crystal that is based on the following assumptions:

1. The validity of the adiabatic approximation for treating the impurity eigenfunctions.
2. The validity of the quantum-mechanical Franck-Condon principle for treating radiative transitions.
3. The validity of the harmonic approximation for treating the lattice vibrations.

The adiabatic potential curves of the impurity in the crystal depend on all the normal coordinates. A diagram similar to the one given for a diatomic molecule in Fig. 10.2 may illustrate the dependence of the electronic energy eigenvalue of the impurity center on the value of a normal coordinate. In dealing with the interaction of radiation with an impurity center we use the following expression for the emitted and absorbed radiative power per unit energy range:

$$A_{\pm}\sum_{i} n_i \sum_{f} |M_{fi}|^2 \delta(E_f - E_i \pm \hbar\omega_\alpha) d\Omega_\alpha. \tag{10.8.1}$$

The quantities A_+ and A_- have much weaker frequency dependence than the factor $|M_{fi}|^2$ and for the present purpose will be considered constant.

Consider the diagram in Fig. 10.4 indicating two adiabatic potential curves, one corresponding to the electronic ground state and the other to an electronic excited state. In general the equilibrium position and the curvatures (i.e., interatomic force constants) will differ. *Purely electronic transitions* are those corresponding to the energy gaps

$$\begin{aligned}
&E_{10} - E_{00}\\
&E_{11} - E_{01} = (E_{10} + \hbar\omega^1) - (E_{00} + \hbar\omega^o)\\
&\qquad\qquad = E_{10} - E_{00} + \hbar(\omega^1 - \omega^o)\\
&E_{12} - E_{02} = (E_{10} + 2\hbar\omega^1) - (E_{00} + 2\hbar\omega^o)\\
&\qquad\qquad = E_{10} - E_{00} + 2\hbar(\omega^1 - \omega^o)
\end{aligned}$$

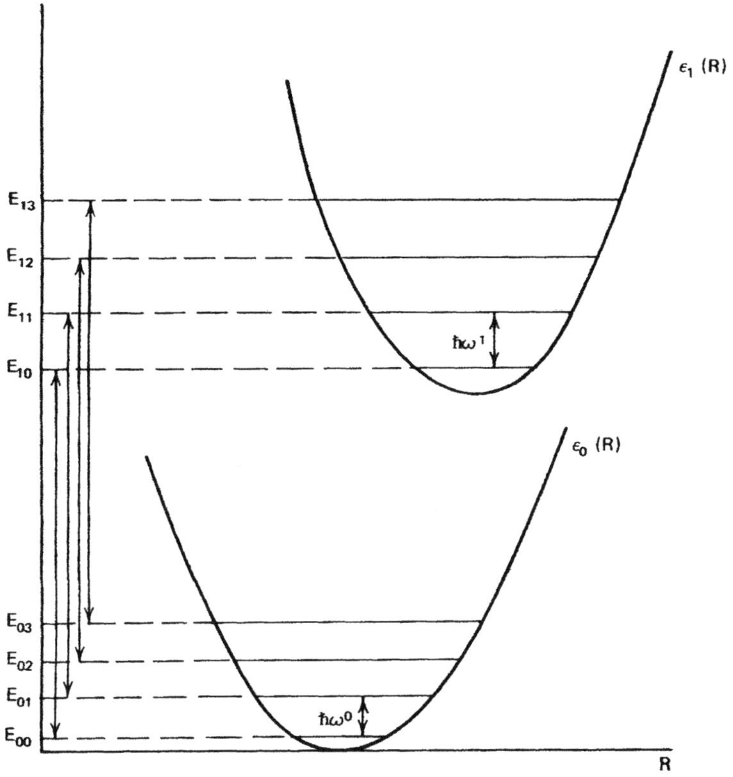

Fig. 10.4. Purely electronic transitions in a diatomic molecule.

$$\begin{aligned} E_{13} - E_{03} &= (E_{10} + 3\hbar\omega^1) - (E_{00} + 3\hbar\omega^o) \\ &= E_{10} - E_{00} + 3\hbar(\omega^1 - \omega^o). \end{aligned} \tag{10.8.2}$$

...

Considering the 0–0 transition, the vibrational energy of the crystal changes by

$$\sum_{\underset{\sim}{k}\lambda} \frac{1}{2} \hbar\omega_{k\lambda} - \sum_{k'\lambda'} \frac{1}{2} \hbar\omega'_{\underset{\sim}{k}'\lambda'}. \tag{10.8.3}$$

Similarly, following the 1–1 transition, the vibrational energy changes by

$$\sum_{k\lambda} \frac{3}{2} \hbar\omega^o_{k\lambda} - \sum_{k'\lambda'} \frac{3}{2} \hbar\omega'_{k'\lambda'}, \tag{10.8.4}$$

and so on. It should be noted that, in general, because of the change in the interatomic force constants going from the ground electronic state to an excited electronic state, the system of normal coordinates in the two states will be different; this is the reason for using a different sum for the two terms in (10.8.3) and (10.8.4).

If $\omega^0_{\tilde{k}\lambda} = \omega^1_{\tilde{k}'\lambda'}$, then the vibrational energy of the crystal does not change and the purely electronic transition can be defined as that in which the vibrational energy of the crystal does not change. In this case all the zero-phonon transitions give a single line of frequency,

$$\omega = \frac{E_{10} - E_{00}}{\hbar} = \frac{E_{11} - E_{01}}{\hbar} = \frac{E_{12} - E_{02}}{\hbar} = \cdots, \qquad (10.8.5)$$

a line that is actually the superposition of the 0–0, 1–1, 2–2,... transition.

For the purpose of studying such a line let us make the following additional assumptions:

a. In correspondence to each normal coordinate it is $\omega^0 = \omega^1$.
b. The electronic excitation of the impurity introduces very little change in the normal coordinates so that the ground and the excited electronic states can be represented in the same diagram as in Fig. 10.5.
c. No localized vibrations are present.

With these assumptions the zero-phonon line is formed by transitions in which the vibrational state of the crystal does *not* change.

In the Franck-Condon approximation the emitted or absorbed power per unit solid angle corresponding to a zero-phonon line connecting the electronic states m and k of a diatomic molecule is given by

$$A \sum_i n_i |M_{fi}|^2, \qquad (10.8.6)$$

where

$$|M_{fi}|^2 = |D^0_{mk}|^2 \left| \int \phi_{mn_i}(R)^* \phi_{kn_i}(R) dR \right|^2. \qquad (10.8.7)$$

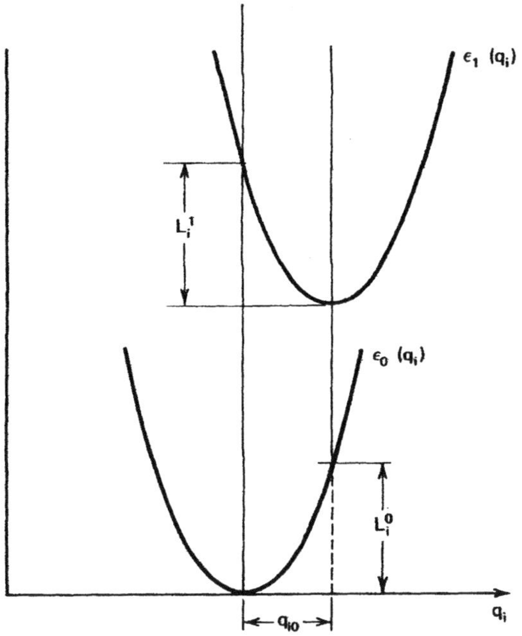

Fig. 10.5. Adiabatic potential curves for a band vibration.

Therefore the emitted or absorbed power per unit solid angle per molecule corresponding to a zero-phonon line is given by

$$P(T) = A \sum_i \overline{n_i} |M_{fi}|^2$$

$$= A|D|^2 \frac{\sum_i e^{-(n_i+1/2)\hbar\omega_i/kT} |\int \phi_{mn_i}(R)^* \phi_{kn_i}(R) dR|^2}{\sum_i e^{-(n_i+1/2)\hbar\omega_i/kT}}, \tag{10.8.8}$$

where D stands for D^0_{mk}.

If a crystal has J atoms per unit cell and N unit cells, the vibrational state of the crystal is defined by 3NJ numbers that give the excitation state of each oscillator; the final state for a no-phonon transition defines also the final vibrational state. In this case the emitted or absorbed power per unit solid angle per impurity is given, dropping the unnecessary asterisk, by

$$A|D|^2 \frac{\sum_{[n_i]} \exp\left[-\sum_{\{n_i\}} (n_i + 1/2)\hbar\omega_i/kT\right] \prod_{\{n_i\}} \times |\int \phi_{mn_i}(q_i)\phi_{kn_i}(q_i)dq_i|^2}{\sum_{[n_i]} \exp\left[-\sum_{\{n_i\}} (n_i + 1/2)\hbar\omega_i/kT\right]}, \tag{10.8.9}$$

where $\{n_i\}$ indicates an operation to be performed over all the 3NJ oscillators of a set and $[n_i]$ indicates an operation to be performed over all possible sets. The thermal average in (10.8.9) can be performed for each oscillator independently. We have then

$$A \sum_i \overline{n_i}|M_{fi}|^2 = A|D|^2 \prod_{i=1}^{3NJ} \left\langle \left| \int \phi_{mn_i}\phi_{kn_i}dq_i \right|^2 \right\rangle_T . \tag{10.8.10}$$

Referring now to Fig. 10.5 in order to apply formula (10.8.9) to the case of a non-phonon line in absorption we set $m = 1$ and $k = 0$ in (10.8.9). It is

$$\phi_{1n_i}(q_i) = \phi_{0n_i}(q_i - q_{io}), \tag{10.8.11}$$

where q_{io} is the shift in the equilibrium position. Expanding the function $\phi_{1n_i}(q_i)$ in a Taylor series about the point q_i we obtain

$$\begin{aligned} \phi_{1n_i}(q_i) &= \phi_{0n_i}(q_i - q_{io}) \\ &= \phi_{0n_i}(q_i) - \frac{d\phi_{0n_i}}{dq_i}q_{io} + \frac{1}{2}\frac{d^2\phi_{0n_i}}{dq_i^2}q_{io}^2 + \cdots . \end{aligned} \tag{10.8.12}$$

The overlap integral can then be expanded as follows:

$$\begin{aligned} \int \phi_{1n_i}\phi_{0n_i}dq_i &= \int \phi_{0n_i}\phi_{0n_i}dq_i - q_{io}\int \phi_{0n_i}\frac{d\phi_{0n_i}}{dq_i}dq_i \\ &\quad + \frac{1}{2}q_{io}^2 \int \phi_{0n_i}\frac{d^2\phi_{0n_i}}{dq_i^2}dq_i + \cdots \\ &= 1 + \frac{1}{2}q_{io}^2 \int \phi_{0n_i}\frac{d^2\phi_{0n_i}}{dq_i^2}dq_i + \cdots , \end{aligned} \tag{10.8.13}$$

where the second term in the expansion has been set equal to zero because of the symmetry properties of the harmonic oscillator eigenfunctions. The kinetic energy operator of the i-th oscillator is given by

$$T = \frac{P_i^2}{2m_i} = -\frac{\hbar^2}{2m_i}\frac{d^2}{dq_i^2}, \tag{10.8.14}$$

where the mass m_i has been introduced in order to make the dimension of the normal coordinates homogeneous with a length. Comparing (10.8.13) and (10.8.14) we may write, using the virial theorem,

$$\begin{aligned}\frac{1}{2}\int \phi_{0n_i}\frac{d^2\phi_{0n_i}}{dq_i^2}dq_i &= -\frac{m_i}{\hbar^2}\int \phi_{0n_i} T\phi_{0n_i} dq_i \\ &= -\frac{m_i}{\hbar^2}T_i = -\frac{m_i}{2\hbar^2}E_i,\end{aligned} \tag{10.8.15}$$

where T_i is the kinetic energy of the i-th oscillator and E_i is the total energy of the i-th oscillator. Therefore

$$\int \phi_{1n_i}\phi_{0n_i} dq_i = 1 - \frac{q_{io}^2 m_i}{2\hbar^2}E_i + \cdots, \tag{10.8.16}$$

and

$$\left|\int \phi_{1n_i}\phi_{0n_i} dq_i\right|^2 \simeq 1 - \frac{q_{io}^2 m_i}{\hbar^2}E_i, \tag{10.8.17}$$

where

$$E_i = \left(n_i + \frac{1}{2}\right)\hbar\omega_i. \tag{10.8.18}$$

It is then

$$\left\langle\left|\int \phi_{1n_i}\phi_{0n_i} dq_i\right|^2\right\rangle_T = 1 - \frac{q_{io}^2 m_i}{\hbar^2}E_i(T), \tag{10.8.19}$$

where

$$\begin{aligned}E_i(T) = \hbar\omega_i\left(\bar{n}_i + \frac{1}{2}\right) &= \hbar\omega_i\left(\frac{1}{e^{\hbar\omega_i/kT} - 1} + \frac{1}{2}\right) \\ &= \frac{\hbar\omega_i}{2}\coth\frac{\hbar\omega_i}{2kT}.\end{aligned} \tag{10.8.20}$$

Therefore

$$A\sum_i \bar{n}_i|M_{fi}|^2 = A|D|^2\prod_{i=1}^{3NJ}\left[1 - \frac{q_{io}^2 m_i}{\hbar^2}E_i(T)\right]. \tag{10.8.21}$$

But

$$\ln \prod_{i=1}^{3NJ}\left[1-\frac{q_{io}^2 m_i}{\hbar^2}E_i(T)\right] = \sum_{i=1}^{3NJ}\ln\left[1-\frac{q_{io}^2 m_i}{\hbar^2}E_i(T)\right]$$
$$= -\sum_{i=1}^{3NJ}\frac{q_{io}^2 m_i}{\hbar^2}E_i(T), \qquad (10.8.22)$$

and

$$\prod_{i=1}^{3NJ}\left[1-\frac{q_{io}^2 m_i}{\hbar^2}E_i(T)\right] \simeq \exp\left[-\sum_{i=1}^{3NJ}\frac{q_{io}^2 m_i}{\hbar^2}E_i(T)\right]. \qquad (10.8.23)$$

We may then write

$$A\sum_i \bar{n}_i|M_{fi}|^2 = A|D|^2 \exp\left[-\sum_{i=1}^{3NJ}\frac{q_{io}^2 m_i}{\hbar^2}E_i(T)\right]$$
$$= A|D|^2 \exp\left[-\sum_{i=1}^{3NJ}\frac{q_{io}^2 m_i}{\hbar^2}\hbar\omega_i\left(\bar{n}_i+\frac{1}{2}\right)\right]$$
$$= A|D|^2 \exp\left[-\sum_{i=1}^{3NJ}\frac{q_{io}^2 m_i}{\hbar^2}kT_i\right], \qquad (10.8.24)$$

where T_i is the *effective temperature* for the i-th oscillator, defined by

$$kT_i = E_i(T) = \hbar\omega_i\left(\bar{n}_i+\frac{1}{2}\right) = \frac{\hbar\omega_i}{2}\coth\frac{\hbar\omega_i}{2kT}. \qquad (10.8.25)$$

At very low temperatures ($kT \ll \hbar\omega_i$), $T_i \to \frac{\hbar\omega_i}{2k}$;
at very high temperatures ($kT \gg \hbar\omega_i$), $T_i \to T$.

Considering now Fig. 10.5 we may introduce the notation

$$L_i^o = \frac{1}{2}m_i\omega_i^2 q_{io}^2 = L_i^1 = L_i, \qquad (10.8.26)$$

since $\omega_i^o = \omega_i^1$. The quantity $2L_i$ is the difference in energy between an absorption and an emission transition at very low temperatures, and is called the *Stokes shift* between absorption and emission. In the case where $\omega_i^o \neq \omega_i^l$ it is

$$L_i^o = \frac{1}{2} m_i (\omega_i^o)^2 q_{io}^2, \qquad L_i^l = \frac{1}{2} m_i (\omega_i^l)^2 q_{io}^2, \tag{10.8.27}$$

and the Stokes shift is given by $L_i^o + L_i^l$. The quantity

$$\frac{1}{2}(L_i^o + L_i^l), \tag{10.8.28}$$

is called the *Stokes energy loss* and represents the average energy lost to vibrations of the q_i mode when the system goes through an absorption or emission transition at very low temperatures. In the present case ($\omega^o = \omega^l$) we can rewrite (10.8.24) as follows, taking (10.8.26) into account:

$$\begin{aligned} P(T) &= A \sum_i \bar{n}_i |M_{fi}|^2 = A|D|^2 \exp\left[-\sum_{i=1}^{3NJ} \frac{q_{io}^2 m_i}{\hbar^2} E_i(T) \right] \\ &= A|D|^2 \exp\left[-\sum_{i=1}^{3NJ} \frac{2L_i}{\hbar\omega_i} \left(\bar{n}_i + \frac{1}{2} \right) \right] \\ &= A|D|^2 \exp\left[-\sum_{i=1}^{3NJ} \frac{2L_i}{\hbar\omega_i} \frac{kT_i}{\hbar\omega_i} \right]. \end{aligned} \tag{10.8.29}$$

Let us examine now the behavior of this function for two limiting cases:

1. At very low temperatures, such that $kT \ll \hbar\omega_{min}$ where ω_{min} is the minimum (acoustic) frequency of all modes interacting with the impurity $kT_i \rightarrow \hbar\omega_i/2$ and

$$P \rightarrow A|D|^2 \exp\left(-\sum_{i=1}^{3NJ} \frac{L_i}{\hbar\omega_i} \right) = A|D|^2 \exp\left(-\sum_{i=1}^{3NJ} \ell_i \right), \tag{10.8.30}$$

where $\ell_i = L_i/\hbar\omega_i$ is the Stokes loss expressed in number of phonons. The above formula shows that the intensity of the zero-phonon line is the greater at low temperature, the smaller are the Stokes losses.

2. At high temperatures such that $kT \gg \hbar\omega_{max}$, $kT_i \to kT$ and

$$P \to A|D|^2 \exp\left(-kT \sum_{i=1}^{3NJ} \frac{2L_i}{(\hbar\omega_i)^2}\right). \tag{10.8.31}$$

The above formula shows that at high temperature the intensity of the zero-phonon line decreases exponentially with temperature.

The following observations can be made:

1. At low temperatures a zero-phonon line has intensity independent of temperature; it is large for small Stokes losses. Two contrasting cases are the one in alkali halide crystal systems, like KCl doped with Tl for which ℓ_i is of the order of 10 (see Ref. 1, p. 49) and the one of laser crystals, ionic solids doped with rare earth ions, in which ℓ_i is very small.
2. At high temperatures a zero-phonon line has intensity that depends both on temperature and on Stokes losses. For zero Stokes losses a zero-phonon line may be expected to have an intensity independent of temperature.

10.9. Characteristics of the Zero-Phonon Lines

A spectral line can be represented by a function $I(\varepsilon)$ where $I(\varepsilon)d\varepsilon$ is the energy of the line in the energy interval $(\varepsilon, \varepsilon + d\varepsilon)$. The moment of ℓth order of such a line is given by

$$M_\ell = \int_{-\infty}^{+\infty} \varepsilon^\ell I(\varepsilon) d\varepsilon. \tag{10.9.1}$$

The zero moment is given by

$$M_0 = \int_{-\infty}^{+\infty} I(\varepsilon) d\varepsilon, \tag{10.9.2}$$

and expresses the integrated intensity of the line. The first and second moments are given by

$$M_1 = \int_{-\infty}^{\infty} \varepsilon I(\varepsilon) d\varepsilon, \qquad (10.9.3)$$

$$M_2 = \int_{-\infty}^{\infty} \varepsilon^2 I(\varepsilon) d\varepsilon. \qquad (10.9.4)$$

On the other hand

$$\bar{\varepsilon} = \frac{\int_{\infty}^{\infty} \varepsilon I(\varepsilon) d\varepsilon}{\int_{-\infty}^{\infty} I(\varepsilon) d\varepsilon} = \frac{M_1}{M_0}, \qquad (10.9.5)$$

and

$$\overline{\varepsilon^2} = \frac{\int_{-\infty}^{\infty} \varepsilon^2 I(\varepsilon) d\varepsilon}{\int_{-\infty}^{\infty} I(\varepsilon) d\varepsilon} = \frac{M_2}{M_0}. \qquad (10.9.6)$$

The dispersion is given by

$$\overline{(\varepsilon - \varepsilon)^2} = \overline{\varepsilon^2} - (\bar{\varepsilon})^2 = \frac{M_2}{M_0} - \left[\frac{M_1}{M_0}\right]^2. \qquad (10.9.7)$$

For an absorption or emission transition the ℓ-th moment is given by

$$\sum_i n_i \sum_f W_{if} (E_f - E_i)^{\ell}, \qquad (10.9.8)$$

where n_i is the occupation number of the i-th initial state, W_{if} is the transition probability per unit time, E_i is the energy of the initial state, and E_f is the energy of the final state. In the most general case

$$E_i = \sum_k \hbar\omega_k^i \left(n_k^i + \frac{1}{2}\right), \qquad (10.9.9)$$

$$E_f = \sum_{k'} \hbar\omega_{k'}^f \left(n_{k'}^f + \frac{1}{2}\right) + \varepsilon_o, \qquad (10.9.10)$$

where ε_o corresponds to the purely electronic excitation. The ℓ-th moment is given by

$$M_\ell = \sum_i n_i \sum_f W_{if} (E_f - E_i)^{\ell}$$

$$= \sum_i n_i \sum_f \left[\varepsilon_o + \sum_{k'} \hbar\omega_{k'}^f \left(n_{k'}^f + \frac{1}{2} \right) \right.$$

$$\left. - \sum_k \hbar\omega_k^i \left(n_k^i + \frac{1}{2} \right) \right]^\ell W_{if}. \tag{10.9.11}$$

For zero-phonon lines $n_{k'}^f = n_k^i$; assuming $\omega_{k'}^f = \omega_k^i$,

$$M_\ell = \varepsilon_o^\ell \sum_i n_i W_{if} = \varepsilon_o^\ell M_0, \tag{10.9.12}$$

where M_0 is the integrated intensity of the zero-phonon line. From the above formula we derive

$$\bar{\varepsilon} = \frac{M_1}{M_0} = \frac{\varepsilon_o P(T)}{P(T)} = \varepsilon_o, \tag{10.9.13}$$

and

$$\overline{(\varepsilon - \varepsilon)^2} = \frac{M_2}{M_0} - \left[\frac{M_2}{M_0} \right]^2$$

$$= \frac{\varepsilon_o^2 M_0}{M_0} - \left[\frac{\varepsilon_o M_0}{M_0} \right]^2 = 0. \tag{10.9.14}$$

The conclusions to be derived from the two formulas above are that a zero-phonon line in the present simple model is centered at the energy corresponding to the electronic excitation and has zero width.

10.10. Phonon-Assisted Transitions

An absorption or emission zero-phonon line is accompanied by other electronic transitions in which the vibrational state of the system undergoes a change; these transitions are called *vibrational-electronic* or *vibronic* or *phonon assisted.* The zero phonon transition and the accompanying vibronic transitions produce a spectral "band".

In order to calculate the integrated intensity of such a band we have to use the formula

$$A \sum_i n_i \sum_f |M_{fi}|^2, \tag{10.10.1}$$

where i is the initial state, f is the final state, and M_{fi} is given by (10.6.13). With the notation

$$\begin{aligned} &\mathrm{i:}\quad 0(\text{electronic}),\quad n_i(\text{vibrational})\\ &\mathrm{f:}\quad 1(\text{electronic}),\quad n_f(\text{vibrational}) \end{aligned}$$

the integrated intensity of the band is given by

$$M_0(T) = A\sum_i n_i \sum_f \left| \int \phi_{1n_f}(Q)^* D_{12}(Q)\phi_{0n_i}(Q)dQ \right|^2, \quad (10.10.2)$$

where

$$D_{12}(Q) = \int d^3\underset{\sim}{r}\,\psi_1(\underset{\sim}{r}, Q)^* D_e(\underset{\sim}{r})\psi_o(\underset{\sim}{r}, Q), \qquad (10.10.3)$$

and Q stands for all the coordinates describing the normal vibrations.

We now have

$$\begin{aligned} M_0(T) &= A\sum_i n_i \sum_f \left| \int \phi_{1n_f}(Q)^* D_{12}(Q)\phi_{0n_i}(Q)dQ \right|^2 \\ &= A\sum_i n_i \sum_f \int \phi_{1n_f}(Q)^* D_{12}(Q)\phi_{0n_i}(Q)dQ \\ &\quad \times \int \phi_{1n_f}(Q')D_{12}(Q')^*\phi_{0n_f}(Q')^* dQ' \\ &= A\sum_i n_i \iint dQ\, dQ' \left[\sum_f \phi_{1n_f}(Q)^* \phi_{1n_f}(Q') \right] \\ &\quad \times D_{12}(Q)D_{12}(Q')^* \phi_{0n_i}(Q)\phi_{0n_i}(Q')^* \\ &= A\sum_i n_i \iint dQ\, dQ'\delta(Q - Q')D_{12}(Q)D_{12}(Q')^* \\ &\quad \times \phi_{0n_i}(Q)\phi_{0n_i}(Q')^* \\ &= A\sum_i n_i \int dQ|D_{12}(Q)|^2|\phi_{0n_i}(Q)|^2. \end{aligned} \qquad (10.10.4)$$

The above formula has been derived with the assumption of the validity of the adiabatic approximation. If the latter approximation is considered valid, then

$$\begin{aligned} M_o(T) &= A \sum_i n_i |D^o_{12}|^2 \int dQ |\phi_{0n_i}(Q)^2| \\ &= A|D^o_{12}|^2 \sum_i n_i = A|D^o_{12}|^2, \end{aligned} \tag{10.10.5}$$

where n_i has been taken as the probability for an impurity to be in the vibrational state n_i.

The above result indicates that the overall integrated intensity of the *entire* vibronic band, consisting of the zero-phonon line and the accompanying vibronic lines, is independent of temperature within the limits of the present model, which implies:

a. The coefficient A is a constant, independent of frequency.
b. The adiabatic approximation is valid.
c. The Franck–Condon approximation is valid.

The fact that the entire vibronic band has constant intensity implies that the energy lost by the zero-phonon line according to (10.8.31) or, more precisely, according to (10.8.29), is gained by the purely vibronic part of the band.

Let us examine now the vibronic transitions; in particular, let us consider those transitions in which the vibrational state of one normal mode changes by 1, that is the vibronic transitions in which one phonon is created or annihilated. As for the case of zero-phonon transitions we make the following assumptions:

a. the validity of the adiabatic approximation,
b. the validity of the Franck-Condon approximation,
c. the validity of the harmonic approximation,
d. no dependence of frequencies on impurity excitation, and
e. absence of localized vibrations.

Under these assumptions if the phonon corresponds to the k-th mode, the intensity of a vibronic line where one phonon is being created is given, according to (10.8.10), by

$$
\begin{aligned}
&A\sum_i n_i \sum_f |M_{fi}|^2 \\
&= A|D|^2 \left[\prod_{i=1}^{3N}{}' \left\langle \left| \int \phi_{ln_i}(q_i)\phi_{0n_i}(q_i)dq_i \right|^2 \right\rangle_T \right] \\
&\times \left\langle \left| \int \phi_{l,n_{k+1}}(q_k)\phi_{0n_k}(q_k)dq_k \right|^2 \right\rangle_T, \qquad (10.10.6)
\end{aligned}
$$

where the product $\prod'$ is exclusive of the k-th normal mode. The overlap integral of interest is now the one related to the k-th mode,

$$
\int \phi_{l,n_{k+1}}(q_k)\phi_{0n_k}(q_k)dq_k. \qquad (10.10.7)
$$

It may be written

$$
\begin{aligned}
\phi_{l,n_{k+1}}(q_k) &= \phi_{0,n_{k+1}}(q_k - q_{ko}) \\
&= \phi_{0,n_k+1}(q_k) - q_{ko}\frac{d\phi_{0,n_{k+1}}}{dq_k} + \frac{1}{2} + \frac{d^2\phi_{0,n_{k+1}}}{dq_k^2} + \cdots,
\end{aligned}
$$

and

$$
\begin{aligned}
\int \phi_{l,n_k+1}\phi_{0n_k}dq_k &= \int \phi_{0,n_k+1}\phi_{0n_k}dq_k - q_{ko}\int \frac{d\phi_{0,n_k+1}}{dq_k}\phi_{0n_k}dq_k \\
&\quad + \frac{1}{2}q_{k0}^2 \int \frac{d^2\phi_{0,n_k+1}}{dq_k}\phi_{0n_k}dq_k + \cdots \\
&= -q_{ko}\int \frac{d\phi_{0,n_k+1}}{dq_k}\phi_{0n_k}dq_k. \qquad (10.10.8)
\end{aligned}
$$

For harmonic oscillator eigenfunctions the following relations can be written:

$$
\frac{d\phi_{n+1}(q)}{dq} = q\frac{m\omega}{\hbar}\phi_{n+1}(q) - \sqrt{2(n+2)}\phi_{n+2}(q) - \sqrt{\frac{m\omega}{\hbar}}, \qquad (10.10.9)
$$

$$
\phi_n(q) = \frac{1}{q}\sqrt{\frac{\hbar}{m\omega}}\sqrt{\frac{n+1}{2}}\phi_{n+1}(q) + \frac{1}{q}\sqrt{\frac{\hbar}{m\omega}}\sqrt{\frac{n}{2}}\phi_{n-1}(q). \qquad (10.10.10)
$$

Therefore

$$\int \phi_{1,n_k+1}\phi_{0n_k}dq_k$$

$$= -q_{ko}\int \frac{d\phi_{0,n_k+1}}{dq_k}\phi_{on_k}dq_k$$

$$= -q_{ko}\int q_k\frac{m_k\omega_k}{\hbar}\frac{1}{q_k}\sqrt{\frac{\hbar}{m_k\omega_k}}\sqrt{\frac{n_{k+1}}{2}}\phi_{0,n_k+1}\phi_{0,n_k+1}dq_k$$

$$= -q_{ko}\sqrt{\frac{m_k\omega_k(n_k+1)}{2\hbar}}, \tag{10.10.11}$$

and

$$\left\langle\left|\int \phi_{1,n_k+1}(q_k)\phi_{0n_k}(q_k)dq_k\right|^2\right\rangle_T = q_{io}^2\frac{m_k\omega_k(\bar{n}_k+1)}{2\hbar}. \tag{10.10.12}$$

Therefore the intensity of a vibronic line corresponding to the creation or annihilation of a phonon of frequency ω_k is given by

$$A\sum_i n_i\sum_f |M_{fi}|^2$$

$$A|D|^2\left[\prod_{i=1}^{3N}{}'\left\langle\left|\int \phi_{ln_i}\phi_{0n_i}dq_i\right|^2\right\rangle_T\right]q_{ko}^2\frac{m_k\omega_k\left(\bar{n}_k+\frac{1}{2}\pm\frac{1}{2}\right)}{2\hbar}$$

$$A|D|^2\exp\left[-\prod_{i=1}^{3N}{}'\frac{q_{io}^2m_i}{\hbar^2}E_i(T)\right]q_{ko}^2\frac{m_k\omega_k\left(\bar{n}_k+\frac{1}{2}\pm\frac{1}{2}\right)}{2\hbar}$$

$$A|D|^2\exp\left[-\prod_{i=1}^{3N}{}'\frac{2L_i}{\hbar\omega_i}\left(\bar{n}_i+\frac{1}{2}\right)\right]q_{ko}^2\frac{m_k\omega_k\left(\bar{n}_k+\frac{1}{2}\pm\frac{1}{2}\right)}{2\hbar}, \tag{10.10.13}$$

where use has been made of (10.8.29) and where the upper (lower) sign corresponds to the creation (annihilation) of a phonon of frequency ω_k.

We can make the following observations:

1. The probability for one-phonon vibronic transitions increases with $\bar{n}_k$, that is, with temperature.

2. In absorption the frequencies of the zero-phonon line and of a vibronic line are related as follows:

   zero-phonon: ω_o
   vibronic with creation of a phonon: $\omega_o + \omega_k$
   (Stokes line)
   vibronic with annihilation of a phonon: $\omega_o - \omega_k$
   (anti-Stokes line).

   At low temperature the anti-Stokes lines disappear.
3. In emission the frequencies of the zero-phonon line of a vibronic line are related as follows:

   zero-phonon: ω_o
   vibronic with creation of a phonon: $\omega_o - \omega_k$
   (Stokes line)
   vibronic with annihilation of a phonon: $\omega_o + \omega_k$
   (anti-Stokes line).

   At low temperature the anti-Stokes lines disappear. These results are represented in Fig. 10.6.
4. The probability for one-phonon vibronic transitions is directly proportional to q_{ko}^2, quantity that goes as $\sim 1/N$. It should be remembered, however, that the phonon spectrum of a solid, is, because of the large value of N, to all effects a continuous spectrum. The vibronic transition probability is then proportional to the function $q_o^2(\omega)g(\omega)$; this means that many modes of vibration concentrated near $\omega = \omega_k$ can contribute to create a vibronic peak.

Turning now to vibronic transitions which involve two phonons we may distinguish the following cases:

1. Creation of two phonons of frequencies ω_k and $\omega_{k'}$ (in particular, it may be $\omega_k = \omega_{k'}$).
2. Annihilation of two phonons of frequencies ω_k and ω_k (in particular it may be $\omega_k = \omega_{k'}$).

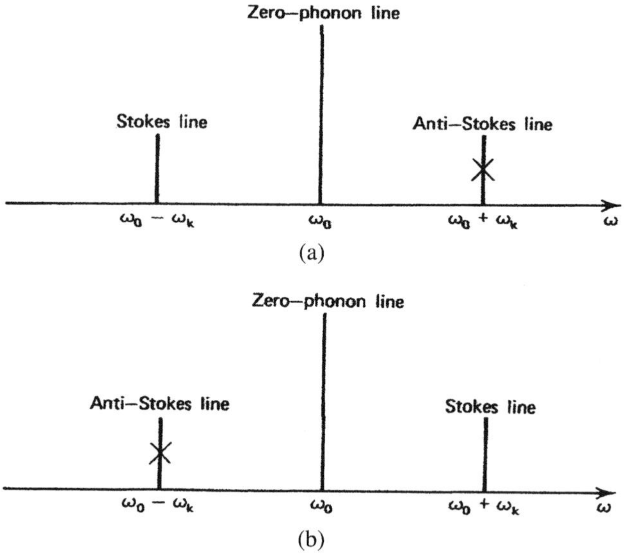

Fig. 10.6. One-phonon vibronic line in (a) emission and (b) absorption.

3. Creation of a phonon of frequency ω_k and annihilation of a phonon of frequency $\omega_{k'}$.

The position of a line brought about by one of the above processes in emission is the mirror image of the position of the same line in absorption with respect to the zero-phonon line. The situation is represented in Fig. 10.7.

It is easy to derive an expression for the intensity of a two-phonon vibronic line. Considering (10.10.13) it is simple to derive for this intensity the formula

$$A|D|^2 \exp\left(-\sum_{i=1}^{3N}{}'' \frac{2L_i}{\hbar\omega_i}\left(\bar{n}_i+\frac{1}{2}\right)\right) q_{ko}^2 q_{k'o}^2 m_k m_{k'} \times \left(\frac{\bar{n}_k+\frac{1}{2}\pm\frac{1}{2}}{2\hbar}\right)\left(\frac{\overline{n_{k'}}+\frac{1}{2}\pm\frac{1}{2}}{2\hbar}\right)\omega_k\omega_{k'}, \qquad (10.10.14)$$

where the upper sign corresponds to creation and the lower sign to annihilation of phonons. For a two-phonon vibronic line an intensity proportional to $1/N^2$ is then to be expected. On the other hand, in certain frequency intervals $\Delta\omega$, the number of two-phonon vibronic

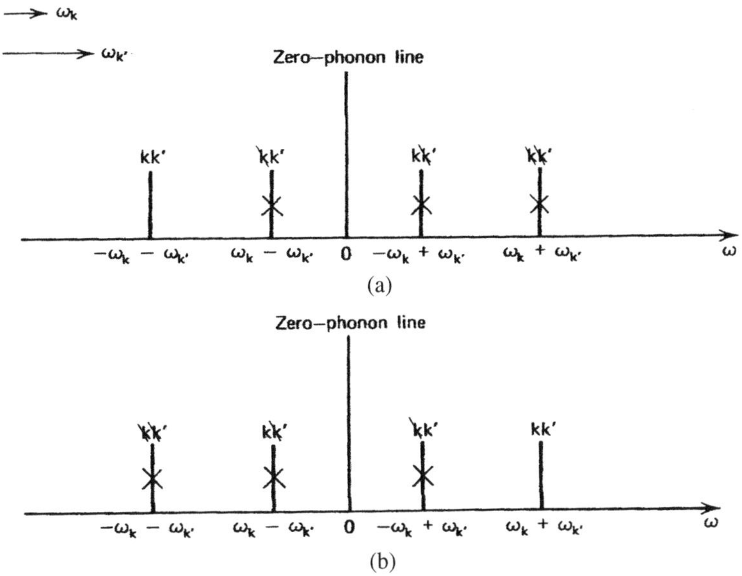

Fig. 10.7. Two-phonon vibronic lines in (a) emission and (b) absorption. Lines with x sign disappear at low temperatures. Mode indices with a slash indicate phonons annihilated, mode indices without a slash indicate phonons created.

transitions, is proportional to N^2; the result of this is that two-phonon vibronic transitions may give a detectable contribution to the vibronic part of the spectrum in both absorption and emission.

We may generalize these considerations to include n-phonon vibronic transitions. The intensity of an individual vibronic transition of this type is proportional to q_0^{2n}, and therefore to $1/N^n$; on the other hand, the number of transitions of this type with phonon frequency in an interval $\Delta\omega$ is proportional to N^n, producing the possibility of a detectable contribution of n-phonon vibronic transitions to the vibronic spectrum.

It should be noted that:

1. The temperature dependence of a vibronic transition increases with the number of phonons; that is, as the temperature increases, we may expect an increase of the multi phonon background of the vibronic spectrum.
2. The one-phonon vibronic spectrum will to some extent resemble the phonon spectral distribution of the crystal. Multi-phonon

transitions from their very nature will tend to smooth out the peaks so that an almost continuous background may result.

3. The intensity of a multiphonon transition is proportional to

$$q_{10}^2 q_{20}^2 \cdots q_{n0}^2,$$

where $1, 2, \ldots, n$ are the phonons involved in the transition. This product is proportional to N^{-n}, a factor that may be in some way compensated by the fact that the number of possible n-phonon vibronic transitions in a frequency interval $\Delta\omega$ is proportional to N^n. However, each q_0 can be expressed as

$$q_0 \approx u_0/\sqrt{N},$$

where u_0 is a measure of the interaction between the impurity and the lattice vibrations. If this interaction is very small, the product of n factors u_0^2 will be very small, ensuring the prominence of the one-phonon vibronic spectrum over the multiphonon contributions. Quantitatively the quantities u_0 are small when the displacements of the atoms of the impurity due to the electronic excitation of the impurity are small in comparison with the range of the zero point displacements of these atoms.

For the purpose of illustrating the above points the entire fluorescence spectrum of the impurity V^{2+} in MgO is given in Fig. 10.8, including the zero-phonon line located at ~8700 Å and the vibronic sidebands.[4]

10.11. Radiative Transitions in the Presence of Localized Vibrations

We make the following assumptions:

a. One mode of localized vibration of wave vector $\underset{\sim}{K}$ and frequency Ω is present in the crystal.
b. $\hbar\Omega \gg kT$, so that the local vibration oscillator normally resides in its ground state. The range of $\hbar\Omega$ is in usual cases 0.01 to 0.1 eV, corresponding to about 100 to 1000 cm^{-1}, so that condition (b) is satisfied for $T \lesssim 77°K$.

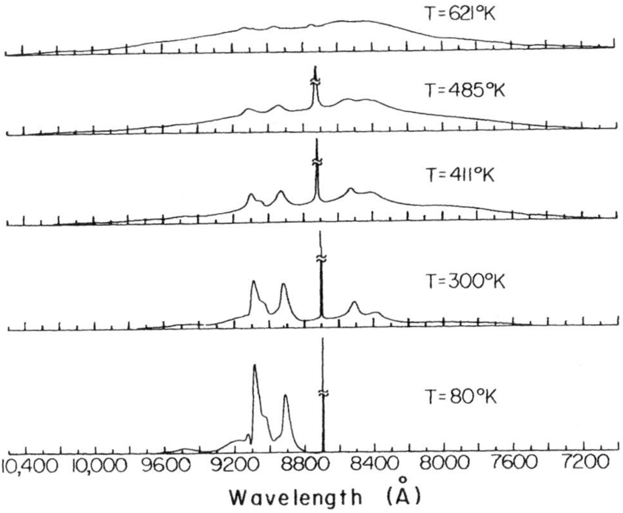

Fig. 10.8. Fluorescence spectrum of MgO:V^{2+} at various temperatures, showing the R (zero-phonon) line at ~8700 Å and the vibronic sidebands.

c. The electronic excitation of the impurity produces a shift in the equilibrium position of the adiabatic potential for the localized normal coordinate Q_K that is "large" and independent of N.
d. The frequency Ω is the same for the ground and excited electronic state of the impurity. This condition is evidently necessary in order to perform proper calculations whose value may be merely indicative. Under these conditions, the intensity of a vibronic line in which, say, n_K "localized" phonons are produced is given by

$$
\begin{aligned}
A\sum_i n_i \sum_f |M_{fi}|^2 \\
&= A|D|^2 \left[\prod_{i=1}^{3N}{}' \left\langle \left| \int \phi_{\ln_i}\phi_{0n_i}dq_i \right|^2 \right\rangle_T \right] \left\langle \left| \int \phi_{\ln_k}\phi_{00}dQ_k \right|^2 \right\rangle_T \\
&= A|D|^2 \exp\left(-\sum_{i=1}^{3N}{}' \frac{2L_i}{\hbar\omega_i}\left(n_i + \frac{1}{2}\right)\right) \left| \int \phi_{\ln_k}\phi_{00}dQ_k \right|^2,
\end{aligned}
\tag{10.11.1}
$$

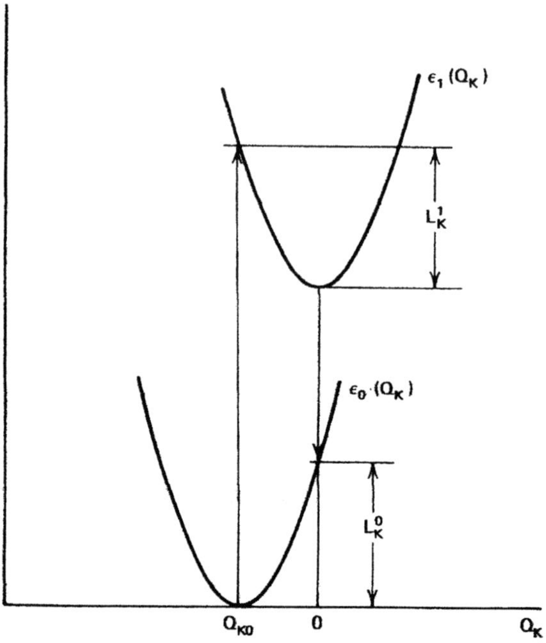

Fig. 10.9. Adiabatic potential curves for a localized vibration.

where, because of condition (b), no thermal averaging is necessary for the localized vibration oscillator. We have now to perform the task of evaluating the overlap integral in (10.11.1); because of the large shift it is no longer possible to use a Taylor expansion. It is, from (10.4.27) and Fig. 10.9,

$$\begin{aligned}\phi_{00}(Q) &= \phi_0(Q + Q_o)\\ &= \frac{1}{\sqrt{\bar{Q}}}\frac{1}{\sqrt[4]{\pi}}\exp\left[-\frac{1}{2}(\xi+\xi_o)^2\right]\\ &= \frac{1}{\sqrt{\bar{Q}}}\frac{1}{\sqrt[4]{\pi}}\exp\left[-\frac{\xi^2}{2}-\frac{\xi_o^2}{2}-\xi\xi_o\right] \qquad (10.11.2)\end{aligned}$$

$$\phi_{1n}(Q) = \frac{1}{\sqrt{\bar{Q}}}\frac{(-1)^n}{\sqrt{2^n n!\sqrt{\pi}}}e^{\xi^2/2}\frac{d^n}{d\xi^n}e^{-\xi^2}, \qquad (10.11.3)$$

where

$$\xi = \frac{Q}{\bar{Q}} = \frac{Q}{\sqrt{\hbar/m\Omega}} \quad \text{and} \quad \xi_o = \frac{Q_o}{\bar{Q}}. \qquad (10.11.4)$$

Then

$$\int_{-\infty}^{+\infty} \phi_{ln}(Q)\phi_{00}(Q)dQ$$

$$= \frac{1}{\bar{Q}} \frac{(-1)^n}{\sqrt{2^n n! \pi}} e^{-\xi_o^2/2} \int_{-\infty}^{+\infty} e^{-\xi\xi_o} \frac{d^n}{d\xi^n} e^{-\xi^2} dQ$$

$$= \frac{(-1)^n}{\sqrt{2^n n! \pi}} e^{-\xi_o^2/2} \int_{-\infty}^{+\infty} e^{-\xi\xi_o} \frac{d^n}{d\xi^n} e^{-\xi^2} d\xi. \qquad (10.11.5)$$

By successive integrations by parts we obtain

$$\int_{-\infty}^{+\infty} e^{-\xi\xi_o} \frac{d^n}{d\xi^n} e^{-\xi^2} d\xi$$

$$= \int_{-\infty}^{+\infty} e^{-\xi\xi_o} \frac{d}{d\xi} \left(\frac{d^{n-1}}{d\xi^{n-1}} e^{-\xi^2} \right) d\xi$$

$$= \left[e^{-\xi\xi_o} \frac{d^{n-1}}{d\xi^{n-1}} e^{-\xi^2} \right]_{-\infty}^{+\infty} + \xi_o \int_{-\infty}^{+\infty} \left(\frac{d^{n-1}}{d\xi^{n-1}} e^{-\xi^2} \right) e^{-\xi\xi_o} d\xi$$

$$= \xi_o \int_{-\infty}^{+\infty} \left(\frac{d^{n-1}}{d\xi^{n-1}} e^{-\xi^2} \right) e^{-\xi\xi_o} d\xi = \xi_o^n \int_{-\infty}^{+\infty} e^{-\xi^2 - \xi\xi_o} d\xi. \qquad (10.11.6)$$

We evaluate this integral by completing the square in the exponent:

$$\int_{-\infty}^{+\infty} e^{-\xi^2 - \xi\xi_o} d\xi = e^{\xi_o^2/4} \int_{-\infty}^{+\infty} e^{-(\xi_o/2 + \xi)^2} d\xi. \qquad (10.11.7)$$

We can set

$$\frac{\xi_o}{2} + \xi = y.$$

Then

$$d\xi = dy,$$

and

$$\int_{-\infty}^{+\infty} e^{-\xi^2 - \xi\xi_o} d\xi = e^{\xi_o^2/4} \int_{-\infty}^{+\infty} e^{-y^2} dy = e^{\xi_o^2/4} \sqrt{\pi}. \qquad (10.11.8)$$

Therefore

$$\int_{-\infty}^{+\infty} \phi_{ln}(Q)\phi_{00}(Q)dQ = \frac{(-1)^n}{\sqrt{2^n n!\pi}} e^{-\xi_o^2/2}\xi_o^n e^{\xi_o^2/4}\sqrt{\pi}$$

$$= \frac{(-\xi_o)^n}{\sqrt{2^n n!}} e^{-\xi_o^2/4}, \qquad (10.11.9)$$

and

$$\left|\int_{-\infty}^{+\infty} \phi_{ln}(Q)\phi_{00}(Q)dQ\right|^2 = e^{-\xi_o^2/2}\left[\frac{\xi_o^2}{2}\right]^n \frac{1}{n!}. \qquad (10.11.10)$$

But

$$\frac{\xi_o^2}{2} = \frac{Q_o^2}{2(\bar{Q})^2} = \frac{Q_o^2 m\Omega}{2\hbar} = \frac{m\Omega^2 Q_o^2}{2\hbar\Omega}. \qquad (10.11.11)$$

Therefore the quantity $\xi_o^2/2$ indicates the Stokes losses in number of phonons, which we shall designate by the symbol ℓ_K. The overlap integral is then given by

$$\left|\int \phi_{ln}(Q_k)\phi_{00}(Q_k)dQ_k\right|^2 = e^{-\ell_k}\frac{\ell_k^n}{n!}. \qquad (10.11.12)$$

The intensity of a vibronic line in which n localized phonons are produced is then obtained by replacing (10.11.12) in (10.11.1):

$$A\sum_i n_i \sum_f |M_{fi}|^2 = A|D|^2 \exp\left(-{\sum_{i=1}^{3N}}' \frac{2L_i}{\hbar\omega_i}\left(\bar{n}_i + \frac{1}{2}\right)\right) e^{-\ell_k}\frac{\ell_k^n}{n!}. \qquad (10.11.13)$$

In order to visualize the effect of a localized vibration, the function in (10.11.12) is represented in Fig. 10.10 for different cases of Stokes losses ℓ. We note here that similar patterns correspond to those observed in the electric spectra of diatomic molecules, apart from the effects caused by the rotation of the molecule, which produces additional fine structure to each vibronic line.

A vibronic transition could include the creation or annihilation of one or more band phonons, in which case proper factors such as

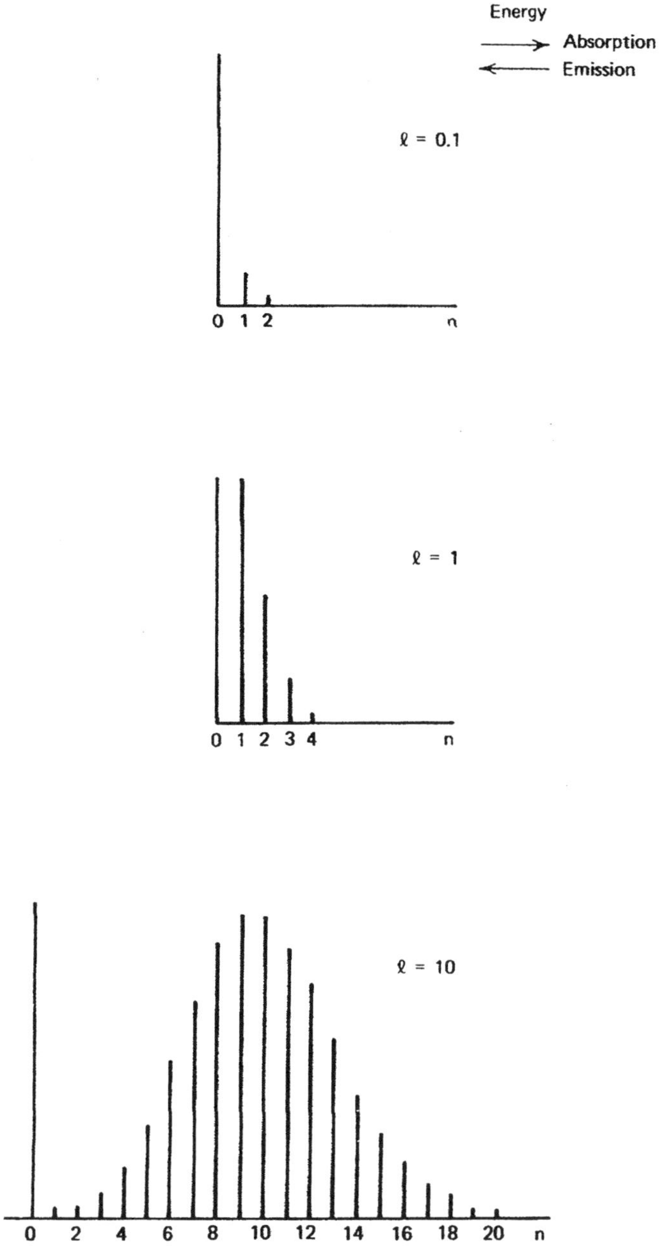

Fig. 10.10. Typical vibronic transitions with creation of localized phonons.[1]

(10.10.12) should be introduced in the expression (10.11.13) and the sum in (10.11.13) should be exclusive of the normal modes that relate to the band phonons involved in the process.

With regard to the thermal dependence of the absorption or emission spectra we have already ascertained the following facts:

1. The intensity of the zero-phonon line in the absence of localized vibrations is given, according to (10.8.29), by

$$A|D|^2 \exp\left(-\sum_{i=1}^{3N} \ell_i(2\bar{n}_i + 1)\right), \tag{10.11.14}$$

   and decreases with temperature.
2. The overall integrated intensity of the *entire* vibronic band including a zero-phonon line and the accompanying vibronic lines is independent of temperature in the Franck-Condon approximation. This result is expressed by the formula (10.10.5).

Let us now consider the effects of the presence of a localized vibration:

1. If the temperature is such that $\hbar\Omega \gg kT$ (Ω is the frequency of the localized oscillation), the localized oscillator in the initial state of a transition can be considered to be always in its ground state. In this case the intensity of the a n-localized and zero-band phonon line will be given by (10.11.13) and the temperature dependence of the intensity (but not the absolute intensity) of the zero-band phonon line will be the same as that of a zero-phonon line.
2. If $\hbar\Omega \gg kT$, the most general vibronic transition will have a transition probability given by the product of three factors:

   a. the zero-phonon part,
   b. the part related to creation or annihilation of band phonons,
   c. the part related to the localized vibration.

   Of these three parts only (a) and (b) are thermally dependent.
3. If Ω is much greater than all the band frequencies, the presence of a localized vibration will produce replicas of the entire vibronic (band) spectrum at frequencies differing by the zero-phonon line

position by

$$0, \pm\Omega, \pm 2\Omega, \ldots, \pm n\Omega, \ldots,$$

where the upper (lower) sign corresponds to absorption (emission). Each replica will be affected by the factor

$$\eta_{0n} = e^{-\ell}\frac{\ell^n}{n!}. \tag{10.11.15}$$

If $I_B(\omega, T)$ is the basic shape of the band vibronic spectrum, the entire spectrum can be described by the function

$$\begin{aligned} I(\omega, T) = \eta_{00} I_B(\omega, T) + \eta_{01} I_B(\omega \mp \Omega, T) \\ + \eta_{02} I_B(\omega \mp 2\Omega, T) + \cdots, \end{aligned} \tag{10.11.16}$$

where the origin of the frequencies is taken at the zero-phonon line and the upper (lower) sign corresponds to absorption (emission).

A typical spectrum for the case of a localized vibration with frequency Ω and $\ell = 1$ is given in Fig. 10.11.

4. The general result (10.11.16) is valid even in the presence of anharmonic interactions, *provided* these interactions are limited to band modes and no anharmonic interaction between band modes and the localized mode is present.
5. The "similarity law" expressed by (10.11.16) implies that for each replica of the basic (band) vibronic shape the integrated intensity is independent of temperature.
6. In the case where the condition $\hbar\Omega \gg kT$ is not fulfilled replicas of the basic (band) vibronic shape appear in the anti-Stokes region of the spectrum. Also, the quantity $\eta_{o\lambda}$ has to be replaced by a quantity which takes into account the thermal averaging (and which is thermally dependent).
7. The general result remains valid that the *total* intensity of the *entire* vibronic spectrum is constant within the limits of (a) independence of A from frequency, (b) validity of the adiabatic approximation, and (c) validity of the Franck-Condon approximation.
8. It is simple to generalize the expression (10.11.16) to the case of L localized vibrations. If for each localized vibration $\hbar\Omega \gg kT$,

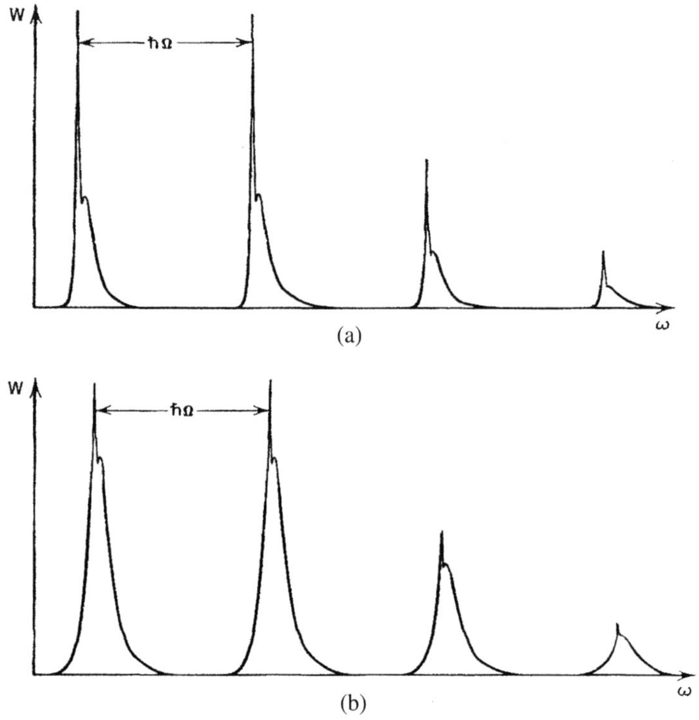

Fig. 10.11. Vibronic spectrum in the presence of a localized vibration of frequency Ω and Stokes loss $\ell = 1$ at (a) 5°K and (b) 22.5°K.[1]

then

$$I(\omega, T) = \sum_{[\lambda_i]} I_B(\omega \mp \lambda_i \Omega_i) \prod_{\{\lambda_i\}} \eta_{0\lambda_i}, \tag{10.11.17}$$

where $\{\lambda_i\}$ indicates a set of L numbers corresponding to the L localized oscillators and $[\lambda_i]$ indicates an operation to be performed over all possible sets. In (10.11.17) it is

$$\prod_{\{\lambda_i\}} \eta_{0\lambda_i} = \left(e^{-\ell_1} \frac{\ell_1^{\lambda_1}}{\lambda_1!}\right) \left(e^{-\ell_2} \frac{\ell_2^{\lambda_2}}{\lambda_2!}\right) \cdots \left(e^{-\ell_L} \frac{\ell_L^{\lambda_L}}{\lambda_L!}\right). \tag{10.11.18}$$

It is important to remind ourselves at this point the basic shape of the vibronic spectrum can have a structure of its own. The peaks that arise in the basic shape of the (band) vibronic spectrum can always be traced back to peaks in the one-phonon vibronic spectrum.

We cannot have a peak in correspondence to, say, a two-phonon vibronic involving two phonons of frequencies ω_1 and ω_2 if there is no peak in the one-phonon vibronic spectrum at either ω_1 or ω_2. It is then important to look in detail to the causes of the presence of peaks in the one-phonon spectrum, as derived in (10.10.13). At low temperature we may observe one-phonon vibronic transitions in the Stokes region proportional to

$$\frac{q_{ko}^2 m_k \omega_k^2}{2\hbar\omega_k} = \ell_k. \qquad (10.11.19)$$

Because of the close spacing of the frequencies of vibrations we can consider the dimensionless Stokes loss ℓ_k as a continuous function $\ell(\omega)$. In this case $\ell(\omega)$ represents the dimensionless Stokes losses for each oscillator in the frequency interval $(\omega, \omega + d\omega)$. If $g(\omega)$ is the density of phonon states in the interval $(\omega, \omega+d\omega)$, then one-phonon vibronic transitions result proportional to $g(\omega)\ell(\omega)$.

It is now evident that peaks in the one-phonon vibronic spectrum can result in two circumstances:

1. A peak may occur in the phonon distribution function $g(\omega)$. Peaks of this type may correspond to some critical points in the dispersion relations, generally occurring at points of high symmetry in the Brillouin zone. If the crystal contains impurities additional peaks, in the form of δ functions, may arise because of the presence of localized vibrations; it should be noted that $\ell(\omega)$ is also "large" in correspondence to a localized vibration.
2. A peak may occur in the Stokes loss function $\ell(\omega)$. This phenomenon is related to the existence of pseudolocalized vibrations and is due to a selective interaction of the impurity with band vibrations. Pseudolocalized vibrations give, in general, rise to peaks in the one-phonon vibronic spectrum that cannot be related to any peak in the phonon distribution of the perfect crystal or to any localized vibration.

One additional point of comment may be necessary in regard to the cases (1) and (2) above. Rigorously speaking, not all the peaks in the function $g(\omega)\ell(\omega)$ may be seen in the one-phonon spectrum. It may happen that certain modes of vibration may not take part in the vibronic spectrum because they are not allowed by the selection rules determined by the symmetry at the impurity site.

When pseudolocalized vibrations are sharply defined, their effect is comparable to that of localized vibrations and the formulas derived earlier in this section for localized vibrations can be used to a good approximation also for pseudolocalized vibrations.

We may conclude this section by listing the factors that enter the most general type of transition:

1. Zero-phonon:

$$\mathrm{A}|\mathrm{D}|^2 \exp\left(-\sum_{\mathrm{i}=1}^{3\mathrm{N}}{}' 2\ell_{\mathrm{i}}\left(\bar{\mathrm{n}}_{\mathrm{i}}+\frac{1}{2}\right)\right).$$

2. One-phonon (band):

$$\ell_{\mathrm{k}}\left(\bar{\mathrm{n}}_{\mathrm{k}}+\frac{1}{2}\pm\frac{1}{2}\right).$$

3. n-phonon (localized or pseudolocalized):

$$\mathrm{e}^{-\ell_{\mathrm{k}}}(\ell_{\mathrm{k}}^{\mathrm{n}}/\mathrm{n}!).$$

10.12. Classification of Vibronic Spectra

It may be impossible to consider all the possibilities that may arise with vibronic spectra of solids. However, some typical cases may be listed:

a. Absence of localized and pseudolocalized vibrations. The vibronic spectrum at low temperatures is essentially the one-phonon spectrum. This is a case frequently observed in laser crystals where a transition metal ion or a rare earth is present as a substitutional impurity.[4]
b. "Absence" of band vibrations but presence of localized vibrations. Band vibrations are not really absent, but because of the temperature the band spectrum may be so completely smeared out as to lose its structure; on the other hand, the temperature may be such that localized oscillators are in their ground state. The observed spectrum is in this case represented by a series of well-separated

"lines." An example of this situation is presented by the emission spectrum of KBr: 0_2^- at 77°K.[5,6]

c. Presence of both localized and band vibrations. The spectrum appears as a series of replicas of the basic band vibronic spectrum. Example: spectrum of KBr: 0_2^- at 4.2°K.
d. Any intermediate case may arise because of the frequencies of the localized vibrations, possible overlaps, and strength of the interaction of the impurity with both localized and band vibrations. Certain systems present, even at 4.2°K, a structureless, bell-shaped vibronic spectrum. Example: ultraviolet emission band of KCl:Tl.[7]

The entire treatment up to this point has been based on the following assumptions:

1. validity of the adiabatic approximation,
2. validity of the harmonic approximation,
3. validity of the Franck-Condon approximation, and
4. changes in the minima, but no changes in the shapes of the adiabatic potential.

The concepts developed here concerning zero-phonon and phonon-assisted transitions represent the fundamental physics underlying the operation of single wavelength and wavelength tunable solid state lasers. The application of these concepts to laser systems is discussed in detail in Chapter 13.

References

1. K. K. Rebane, *Impurity Spectra of Solids* (translated from Russian), Plenum, New York, 1970.
2. C. Kittel, *Introduction to Solid State Physics*, 2nd ed., Wiley, New York, 1953, p. 491.
3. A. Karo and J. Hardy, *Phys. Rev.* 129, 2024 (1963).
4. B. Di Bartolo and R. Peccei, *Phys. Rev.* 137, A1770 (1965).
5. J. J. Rolfe, *J. Chem. Phys.* 40, 1664 (1964).
6. K. K. Rebane, L. A. Rebane, and O. I. Sil'd, *Proceedings of the International Conference on Luminescence*, Budapest, 1966.
7. D. A. Patterson and C. C. Klick, *Phys. Rev.* 105, 401 (1957).

Chapter 11

Optical Spectra of Impurities in Solids II

This chapter continues the discussion started in Chapter 10 concerning the optical spectroscopic properties of impurity ions in solids. Some of the simplifying assumptions made previously that were useful in understanding the general properties of the spectra are lifted to show how more detailed properties of the spectra can be explained. This allows for the treatment of radiationless transitions within the same electronic state and between electronic states.

An effective Hamiltonian for a solid state laser material is derived in Sec. 11.4. Using this model, the properties of radiative, vibronic, and radiationless transitions of impurity ions in solids are discussed. The effects of temperature on the optical spectra are also described, especially the broadening and shifting of sharp spectral lines. The results described in this chapter provide the foundation for understanding the dynamics of laser operation described in Chapter 13.

11.1. Summary of Previous Results

The entire treatment of the optical spectra of impurities in solids, up to this point has been based on the following assumptions:

1. Validity of the harmonic approximation for treating the lattice vibrations.
2. Validity of the adiabatic approximation.
3. Validity of the Franck–Condon principle.

4. Change in the position of the minima, but no change in the curvature of the adiabatic potentials in correspondence to an electronic excitation of the impurity. No change in the normal coordinate system.

The most important results of this treatment are recounted:

1. Because of the absence of anharmonicity the lattice vibrations are described in terms of normal coordinates. Normal modes of vibrations can be delocalized (band modes) or localized. As long as the broadening $\Delta\omega_k$ of a mode of frequency ω_k brought about by anharmonicity is such that $\Delta\omega_k \ll \omega_k$, the description of lattice vibrations in terms of normal modes is a valid one.
2. The stationary states of the impurity in the solid are described (adiabatic approximation) by eigenfunctions of the type

$$\psi_n(\underset{\sim}{r}, \underset{\sim}{R}) = \psi_k(\underset{\sim}{r}, \underset{\sim}{R})\phi_{kl}(\underset{\sim}{R}), \tag{11.1.1}$$

where $\underset{\sim}{r}$ stands for the electron coordinates and $\underset{\sim}{R}$ stands for the nuclear coordinates. In (11.1.1) the "electronic" eigenfunctions $\psi_k(\underset{\sim}{r}, \underset{\sim}{R})$ are solutions of the Schrödinger equation

$$-\frac{\hbar^2}{2m}\sum_{i=1}^{n}\nabla_i^2\psi_k(\underset{\sim}{r}, \underset{\sim}{R}) + V(\underset{\sim}{r}, \underset{\sim}{R})\psi_k(\underset{\sim}{r}, \underset{\sim}{R}) = \varepsilon_k(\underset{\sim}{R})\psi_k(\underset{\sim}{r}, \underset{\sim}{R}), \tag{11.1.2}$$

and satisfy orthonormality and closure relations:

$$\int \psi_k(\underset{\sim}{r}, \underset{\sim}{R})^*\psi_{k'}(\underset{\sim}{r}, \underset{\sim}{R})d^3\underset{\sim}{r} = \delta_{kk'}, \tag{11.1.3}$$

$$\sum_k \psi_k(\underset{\sim}{r}, \underset{\sim}{R})^*\psi_k(\underset{\sim}{r}', \underset{\sim}{R}) = \delta(\underset{\sim}{r} - \underset{\sim}{r}'). \tag{11.1.4}$$

Also in (11.1.1) the "vibrational" eigenfunctions are solutions of the Schrödinger equation

$$-\frac{\hbar^2}{2}\sum_{\alpha=1}^{N}\frac{1}{M_\alpha}\nabla_\alpha^2\phi_{kl}(\underset{\sim}{R}) + \varepsilon_k(\underset{\sim}{R})\phi_{kl}(\underset{\sim}{R}) = E_{kl}\phi_{kl}(\underset{\sim}{R}). \tag{11.1.5}$$

It is noted that k does not play the role of a quantum number for ϕ; this means that $\phi_{kl}(\underset{\sim}{R})$ and $\phi_{k'l'}(\underset{\sim}{R})$, in general, are not mutually orthogonal even if $k \neq k'$. However, if

$$\varepsilon_k(\underset{\sim}{R}) = \varepsilon_{k'}(\underset{\sim}{R}) + \text{const}, \tag{11.1.6}$$

that is, if the adiabatic potentials are parallel, it is in effect

$$\phi_{kl}(\underset{\sim}{R}) = \phi_{k'l}(\underset{\sim}{R}). \tag{11.1.7}$$

3. Given a couple of electronic states $\psi_k(\underset{\sim}{r}, \underset{\sim}{R})$ and $\psi_m(\underset{\sim}{r}, \underset{\sim}{R})$ the existence of a transition between them presupposes that the following matrix element is different from zero:

$$D_{mk}(\underset{\sim}{R}) = \int d^3\underset{\sim}{r}\,\psi_m(\underset{\sim}{r}, \underset{\sim}{R})^* D_e(\underset{\sim}{r})\psi_k(\underset{\sim}{r}, \underset{\sim}{R}), \tag{11.1.8}$$

where $\psi_k(\underset{\sim}{r}, \underset{\sim}{R})$ is the eigenfunction of the initial electronic state, $\psi_m(\underset{\sim}{r}, \underset{\sim}{R})$ is the eigenfunction of the final electronic state, and $D_e(\underset{\sim}{r})$ is the electric dipole operator. In addition, when expanding the matrix element D_{mk} about the equilibrium nuclear configuration, only the first term D^o_{mk} in the expansion is retained (Franck-Condon principle); for this reason the existence of an electronic transition implies that $D^o_{mk} \neq 0$.
4. The most general electronic transition gives rise to a spectral band that consists of the zero-phonon line and the accompanying vibronic side bands. The zero-phonon line has zero width and corresponds to a purely electronic change of state with no change in the vibrational state; this line occupies the same position when observed in emission and in absorption. The phonon-assisted part of the electronic transition gives rise to the vibronic bands which at very low temperature occur at higher (lower) energy than the zero-phonon line in absorption (emission).
5. The intensity of a zero-phonon line is given by

$$A|D|^2 \exp\left[-\sum_i \ell_i(2\bar{n}_i + 1)\right]. \tag{11.1.9}$$

At low temperatures ($kT \ll \hbar\omega_{min}$) the intensity becomes independent of temperature,

$$A|D|^2 \exp\left[-\sum_i \ell_i\right], \tag{11.1.10}$$

but at high temperature ($kT \gg \hbar\omega_{max}$) its intensity decreases exponentially with increasing temperature,

$$A|D|^2 \exp\left[-kT \sum_i \frac{2\ell_i}{\hbar\omega_i}\right]. \tag{11.1.11}$$

We note here the dependence of the zero-phonon line intensity on the Stokes losses; if all $\ell_i = 0$, then the intensity of the zero-phonon line is independent of temperature and equal to $A|D|^2$.

6. The relevant factor for the one band-phonon vibronic factor is given by

$$l_k\left(\overline{n_k} + \frac{1}{2} \pm \frac{1}{2}\right), \tag{11.1.12}$$

with the upper (lower) sign corresponding to creation (annihilation) of one phonon. Peaks in the one-phonon vibronic spectrum are either due to a peak in the phonon spectrum or a peak in the Stokes loss $\ell(\omega)$. It is to be noted here that in the present scheme of things no vibronic transition can take place in absence of Stokes losses.

7. The entire spectral band corresponding to the zero-phonon line and the accompanying vibronic sidebands has an integrated intensity independent of temperature and given by

$$M_o(T) = A|D|^2, \tag{11.1.13}$$

if the Franck-Condon principle holds.

8. At temperatures such that the localized vibrations are unexcited, the relevant vibronic factor corresponding to the creation of n localized phonons is given by

$$e^{-\ell_k}\frac{\ell_K^n}{n!}. \tag{11.1.14}$$

9. The localized vibrations produce replicas of the zero-phonon line at distance $\pm n\Omega$ (Ω is the frequency of localized vibration, n is an integer number) from the zero-phonon line with the plus (minus) sign corresponding to absorption (emission). If $P(T)$ is the intensity of the zero-phonon line the replicas will have

intensities given by

$$\frac{\ell_K^n}{n!}P(T). \tag{11.1.15}$$

10. The localized vibrations produce also replicas of the entire vibronic band (zero-phonon lines and sidebands). In a temperature region in which the localized vibrations are unexcited, the integrated intensity of each replica does not change with temperature.
11. Because the adiabatic potentials have in the present model the same curvature, independently of the electronic state in consideration, the shape of an electronic spectral band in emission should be at all temperatures the mirror image of the band in absorption with respect to the zero-phonon line.

11.2. Deviations from the Franck-Condon Approximation

We now study the possible effects on the optical spectra of impurities in solids due to the breakdown of the Franck-Condon approximation. This approximation consists simply of neglecting the possible dependence of the (oscillator) strength of an electronic transition on the displacement of the nuclei from their equilibrium positions.

In order to single out the effects of the "breakdown" of the Franck-Condon approximation we continue to assume that both the harmonic approximation and the adiabatic approximations are valid and, moreover, that the frequencies of vibration and normal coordinate are not affected by the electronic state of the impurity.

With reference to Section 10.6 we shall call the initial state of the system

$$\psi_i(\underset{\sim}{r}, \underset{\sim}{R}) = \psi_k(\underset{\sim}{r}, \underset{\sim}{R})\phi_{kl}(\underset{\sim}{R}), \tag{11.2.1}$$

and the final state of the system

$$\psi_f(\underset{\sim}{r}, \underset{\sim}{R}) = \psi_m(\underset{\sim}{r}, \underset{\sim}{R})\phi_{mn}(\underset{\sim}{R}). \tag{11.2.2}$$

If $k \neq m$, the transition between the two states has a probability proportional to the square of the absolute value of the following matrix element, as derived in (10.6.13):

$$M_{fi} = M_{mn;kl} = \int d^3\underset{\sim}{R}\phi_{mn}(\underset{\sim}{R})^* D_{mk}(\underset{\sim}{R})\phi_{kl}(\underset{\sim}{R}), \tag{11.2.3}$$

where

$$D_{mk}(\underset{\sim}{R}) = \int d^3\underset{\sim}{r}\psi_m(\underset{\sim}{r}, \underset{\sim}{R})^* D_e(\underset{\sim}{r})\psi_k(\underset{\sim}{r}, \underset{\sim}{R}). \tag{11.2.4}$$

The Franck-Condon approximation consists simply of neglecting the dependence of D_{mk} on the positions of the nuclei and in setting it equal to the value that it takes in correspondence to the *equilibrium* nuclear configuration. In order to take into account the possible effects of the deviations from such a simple assumption, we could expand the "electronic" matrix element $D_{mk}(\underset{\sim}{R})$ in terms of the displacements of the nuclei as in (10.6.17); we will instead expand it in terms of the normal coordinates of the lattice vibrations. Dropping the subscripts indicating the electronic states, we can write

$$D(q_1 q_2 \cdots q_{3N}) = D^o + \sum_{r=1}^{3N} D_r^1 q_r + \sum_{r=1}^{3N}\sum_{t=1}^{3N} D_{rt}^2 q_r q_t + \cdots. \tag{11.2.5}$$

The integral in (11.2.3) can be evaluated by noting that the vibrational eigenfunctions $\phi_{kl}(\underset{\sim}{R})$ and $\phi_{mn}(\underset{\sim}{R})$ can be expressed as products of 3N harmonic oscillator eigenfunctions

$$\phi_{kl} = \prod_{s=1}^{3N} \phi_{ki_s}(q_s), \tag{11.2.6}$$

$$\phi_{mn} = \prod_{s'=1}^{3N} \phi_{mf_{s'}}(q_{s'}), \tag{11.2.7}$$

where i_s and $f_{s'}$ indicate the excitation number of the corresponding oscillators.

We have, then, dropping for the moment the subscripts k and m in the harmonic oscillator eigenfunctions,

$$M_{fi} = M_{mn;kl} = \iint \cdots \int \left\{ \prod_{s'=1}^{3N} \phi_{f_{s'}}(q_{s'}) \right.$$

$$\times \left[D^o + \sum_{r=1}^{3N} D_r^1 q_r + \sum_{r=1}^{3N}\sum_{t=1}^{3N} D_{rt}^2 q_r q_t + \cdots\right] \prod_{s=1}^{3N} \phi_{i_s}(q_s)\Bigg\}$$

$$\times\, dq_1 dq_2 \cdots dq_{3N}$$

$$= D^o \iint \cdots \int \left[\prod_{s'=1}^{3N} \phi_{f_{s'}}(q_{s'})\right]\left[\prod_{s=1}^{3N} \phi_{i_s}(q_s)\right] dq_1 dq_2 \cdots dq_{3N}$$

$$+ \sum_{r=1}^{3N} D_r^1 \iint \cdots \int \left\{\left[\prod_{s'=1}^{3N} \phi_{f_{s'}}(q_{s'})\right] q_r \left[\prod_{s=1}^{3N} \phi_{i_s}(q_s)\right]\right\}$$

$$\times\, dq_1 dq_2 \cdots dq_{3N}$$

$$+ \sum_{r=1}^{3N}\sum_{t=1}^{3N} D_{rt}^2 \iint \cdots \int \left\{\left[\prod_{s'=1}^{3N} \phi_{f_{s'}}(q_{s'})\right] q_r q_t \left[\prod_{s=1}^{3N} \phi_{i_s}(q_s)\right]\right\}$$

$$\times\, dq_1 dq_2 \cdots dq_{3N}$$

$$= D^o \prod_{s=1}^{3N} \int \phi_{f_s}(q_s)\phi_{i_s}(q_s)dq_s + \sum_{r=1}^{3N} D_r^1 \int \phi_{f_r}(q_r) q_r \phi_{i_r}(q_r) dq_r$$

$$\times \prod_{\substack{s=1\\ s\neq r}}^{3N} \int \phi_{f_s}(q_s)\phi_{i_s}(q_s)dq_s + \sum_{r=1}^{3N} D_{rr}^2 \int \phi_{f_r}(q_r) q_r^2 \phi_{i_r}(q_r) dq_r$$

$$\times \prod_{\substack{s=1\\ s\neq r}}^{3N} \int \phi_{f_s}(q_s)\phi_{i_s}(q_s)dq_s + \sum_{r=1}^{3N} \underset{r\neq t}{} \sum_{t=1}^{3N} D_{rt}^2 \int \phi_{f_r}(q_2) q_r \phi_{i_r}(q_r) dq_r$$

$$\times \int \phi_{f_t}(q_t) q_t \phi_{i_f}(q_t) dq_t \prod_{\substack{s=1\\ s\neq r\\ s\neq t}}^{3N} \int \phi_{f_s}(q_s)\phi_{i_s}(q_s)dq_s$$

$$= D^o \prod_{s=1}^{3N} \langle mf_s|ki_s\rangle + \sum_{r=1}^{3N} [D_r^1 \langle mf_r|q_r|ki_r\rangle + D_{rr}^2 \langle mf_r|q_r^2|ki_r\rangle]$$

$$\times \prod_{\substack{s=1\\ s\neq r}}^{3N} \langle mf_s|ki_s\rangle + \sum_{r=1}^{3N} \underset{r\neq t}{} \sum_{t=1}^{3N} D_{rt}^2 \langle mf_r|q_r|ki_r\rangle \langle mf_t|q_t|ki_t\rangle$$

$$\times \prod_{\substack{s=1\\ s\neq r,t}}^{3N} \langle mf_s|ki_s\rangle, \qquad (11.2.8)$$

where

$$\langle mf_s|ki_s\rangle = \int \phi_{mf_s}(q_s)\phi_{ki_s}(q_s)dq_s, \qquad (11.2.9)$$

$$\langle mf_s|q_s|ki_s\rangle = \int \phi_{mf_s}(q_s)q_s\phi_{ki_s}(q_s)dq_s, \qquad (11.2.10)$$

$$\langle mf_s|q_s^2|ki_s\rangle = \int \phi_{mf_s}(q_s)q_s^2\phi_{ki_s}(q_s)dq_s. \qquad (11.2.11)$$

We notice immediately that if all $D^1_r(r = 1, 2, \ldots, 3N)$ and all $D^2_{rt}(r, t = 1, 2, \ldots, 3N)$ are set equal to zero, we fall back into the Franck-Condon case. In many cases we may expect the term in D^o to be the predominant one; on the other hand, a case that may be of interest here is the one in which D^o is zero. We want to elaborate on this point: When is $D^o = 0$?

We may recount that, reintroducing the subscripts that identify the electronic states, it is

$$D^o_{mk} = \int d^3\underset{\sim}{r}\psi_m(\underset{\sim}{r}, \underset{\sim}{R}_o)^* D_e(\underset{\sim}{r})\psi_k(\underset{\sim}{r}, \underset{\sim}{R}_o), \qquad (11.2.12)$$

where $\underset{\sim}{R}_o$ stands for the equilibrium nuclear coordinates. We may point out that $\psi_m(\underset{\sim}{r}, \underset{\sim}{R}_o)$ and $\psi_k(\underset{\sim}{r}, \underset{\sim}{R}_o)$ are eigenfunctions of the Hamiltonian

$$H_e = -\frac{\hbar^2}{2m}\sum_{i=1}^{3N}\nabla_i^2 + V(\underset{\sim}{r}, \underset{\sim}{R}_o). \qquad (11.2.13)$$

This Hamiltonian is equal to H_e given in (1.4.7) when $\underset{\sim}{R}$ is set equal to $\underset{\sim}{R}_o$. As indicated in Section 1.4, the Hamiltonian H_e is invariant under all those operations on the electronic coordinates which, when used on the nuclear coordinate parameters, send identical nuclei into one another. This property applies at each instant in time, regardless of what the nuclear configuration is, and it certainly applies for an equilibrium nuclear configuration.

With reference to Section 2.11 on the connection between quantum mechanics and group theory, we can state that the eigenfunctions that belong to the same eigenvalue of H_e must form a basis for an irreducible representation that leaves H_e invariant. (The dimension of this irreducible representation is equal to the degree of degeneracy of the energy eigenvalue.) What this means is that the electron charge distribution must be compatible with the symmetry of the "nuclear skeleton" of the system. Great use is made of this point in designing the possible electron charge distributions of molecules whose symmetry is known: molecular orbitals are calculated for the stable equilibrium configuration corresponding to the ground electronic state of the molecule.

Let us see how these principles can apply to a simple case: let us assume a molecule is in its ground electronic state and that in correspondence to its equilibrium configuration $\underset{\sim}{R}_o$ it presents inversion symmetry. Under these conditions the Hamiltonian H_e is invariant under the inversion operation and all the electronic eigenfunctions $\psi(\underset{\sim}{r}, \underset{\sim}{R}_o)$ are either even or odd. Electric dipole transitions will then be allowed ($D^o_{mk} \neq 0$) between the ground state and another electronic state if these two states have different parity and will be forbidden ($D^o_{mk} = 0$) if these two states have equal parity. However, the fact that $D_{mk}(\underset{\sim}{R})$ is zero in correspondence to the equilibrium configuration $\underset{\sim}{R} = \underset{\sim}{R}_o$ does not necessarily imply that for $\underset{\sim}{R} \neq \underset{\sim}{R}_o$ it will also vanish: it may happen that for a configuration $\underset{\sim}{R}$ the molecule loses its inversion symmetry, in which case the eigenfunctions will be states of mixed parity and $D_{mk}(\underset{\sim}{R})$ may be different from zero even if $D^o_{mk} = D_{mk}(\underset{\sim}{R}_o) = 0$.

Another interesting question can be raised: If $D^o_{mk} = 0$ (k is the initial ground state, m is the final state), does this necessarily imply that $D^o_{km} = 0$? Or, in other words, if an absorption transition is forbidden according to the Franck-Condon principle, can we infer that the inversion (emission) transition is also forbidden? The answer is no for the most general case: it may happen that the equilibrium nuclear configuration when the molecule is in an excited electronic state may have symmetry properties different from those of the molecule in its ground state ($\underset{\sim}{R}_o$ is a function of the electronic state). This is certainly not the case for a diatomic molecule that preserves its symmetry properties regardless of its electronic excitation.

We can now go back to the expression for the matrix element M_{fi} in (11.2.8) and consider the integrals (11.2.9), (11.2.10), and (11.2.11) that appear in this expression. The integral (11.2.9) is simply the overlap integral of the two eigenfunctions ϕ_{mf_s} and ϕ_{ki_s}; this integral has already been evaluated for different cases:

1. f = i, q is the band vibration normal coordinate; as given in (10.8.16),

$$\langle mi|ki\rangle = 1 - \frac{q_o^2 m\omega}{2\hbar}\left(i + \frac{1}{2}\right) = 1 - \frac{1}{2}\frac{q_o^2}{\bar{q}^2}\left(i + \frac{1}{2}\right), \tag{11.2.14}$$

   where $\bar{q} = \sqrt{\hbar/m\omega}$. If $q_o = 0$, this integral is equal to 1.

2. $f = i \pm 1$, q is the band vibration coordinate; as given in (10.10.11),

$$\begin{aligned} \langle m, i \pm 1|ki\rangle &= -q_o\sqrt{\frac{m\omega\left(i + \frac{1}{2} \pm \frac{1}{2}\right)}{2\hbar}} \\ &= -\frac{q_o}{q}\sqrt{\frac{i + \frac{1}{2} \pm \frac{1}{2}}{2}}. \end{aligned} \tag{11.2.15}$$

   The overlap integral in this case is zero if $q_o = 0$.

3. $f = i \pm 2$, q is the band vibration normal coordinate. We have not previously evaluated the overlap integral for this case; it is left to the reader to show that

$$\begin{aligned} \langle m, i \pm 2|ki\rangle &= \frac{1}{2}q_o^2\frac{m\omega}{2\hbar}\sqrt{\left(i + \frac{1}{2} \pm \frac{1}{2}\right)\left(i + \frac{1}{2} \pm \frac{3}{2}\right)} \\ &= \frac{1}{4}\frac{q_o^2}{\bar{q}^2}\sqrt{\left(i + \frac{1}{2} \pm \frac{1}{2}\right)\left(i + \frac{1}{2} \pm \frac{3}{2}\right)}. \end{aligned} \tag{11.2.16}$$

   This overlap integral vanishes if $q_o = 0$.

4. f = n, i = 0, q is the localized vibration normal coordinate; as given in (10.11.19),

$$\langle mn|ko\rangle = \left(-\frac{q_o}{\bar{q}}\right)^n \frac{1}{\sqrt{2^n n!}}\exp\left[-\frac{1}{4}\left(\frac{q_o}{\bar{q}}\right)^2\right]. \tag{11.2.17}$$

   This overlap integral is zero if $q_o = 0$, $n \neq 0$ and is equal to 1 if $q_o = 0$, $n = 0$.

We consider now the integral (11.2.10), which, according to Ref. 1, is given by

$$\langle \mathrm{mf}|\mathrm{q}|\mathrm{ki}\rangle = -\frac{\bar{\mathrm{q}}^2}{\mathrm{q_o}}\left(|\mathrm{f}-\mathrm{i}| - \frac{\mathrm{q_o^2}}{2\bar{\mathrm{q}}^2}\right)\langle \mathrm{mf}|\mathrm{ki}\rangle. \qquad (11.2.18)$$

We have for the different cases:

1. f = i, q is the band vibration normal coordinate:

$$\langle \mathrm{mi}|\mathrm{q}|\mathrm{ki}\rangle = \frac{\mathrm{q_o}}{2}\left[1 - \frac{1}{2}\frac{\mathrm{q_o^2}}{\bar{\mathrm{q}}^2}\left(\mathrm{i}+\frac{1}{2}\right)\right]. \qquad (11.2.19)$$

If $\mathrm{q_o} = 0$, this integral vanishes.

2. f = i ± 1, q is the band vibration normal coordinate:

$$\begin{aligned}\langle \mathrm{m,i}\pm 1|\mathrm{q}|\mathrm{ki}\rangle &= -\frac{\bar{\mathrm{q}}^2}{\mathrm{q_o}}\left(1 - \frac{\mathrm{q_o^2}}{2\bar{\mathrm{q}}^2}\right)\langle \mathrm{m,i}\pm 1|\mathrm{ki}\rangle \\ &= -\frac{\bar{\mathrm{q}}^2}{\mathrm{q_o}}\left(1 - \frac{\mathrm{q_o^2}}{2\bar{\mathrm{q}}^2}\right)\left[-\frac{\mathrm{q_o}}{\bar{\mathrm{q}}}\sqrt{\frac{\mathrm{i}+\frac{1}{2}\pm\frac{1}{2}}{2}}\right] \\ &= \bar{\mathrm{q}}\left(1 - \frac{\mathrm{q_o^2}}{2\bar{\mathrm{q}}^2}\right)\sqrt{\frac{\mathrm{i}+\frac{1}{2}\pm\frac{1}{2}}{2}}. \end{aligned} \qquad (11.2.20)$$

If $\mathrm{q_o} = 0$

$$\langle \mathrm{m,i}\pm 1|\mathrm{q}|\mathrm{ki}\rangle = \bar{\mathrm{q}}\sqrt{\frac{\mathrm{i}+\frac{1}{2}\pm\frac{1}{2}}{2}}. \qquad (11.2.21)$$

3. f = i ± 2, q is the band vibration normal coordinate,

$$\begin{aligned}&\langle \mathrm{m,i}\pm 2|\mathrm{q}|\mathrm{ki}\rangle \\ &\quad= -\frac{\bar{\mathrm{q}}^2}{\mathrm{q_o}}\left(2 - \frac{\mathrm{q_o^2}}{2\bar{\mathrm{q}}^2}\right)\langle \mathrm{m,i}\pm 2|\mathrm{ki}\rangle \\ &\quad= -\frac{\bar{\mathrm{q}}^2}{\mathrm{q_o}}\left(2 - \frac{\mathrm{q_o^2}}{2\bar{\mathrm{q}}^2}\right)\frac{1}{4}\frac{\mathrm{q_o^2}}{\bar{\mathrm{q}}^2}\sqrt{\left(\mathrm{i}+\frac{1}{2}\pm\frac{1}{2}\right)\left(\mathrm{i}+\frac{1}{2}\pm\frac{3}{2}\right)} \\ &\quad= -\frac{\mathrm{q_o}}{2}\left(2 - \frac{\mathrm{q_o^2}}{2\bar{\mathrm{q}}^2}\right)\frac{\sqrt{\left(\mathrm{i}+\frac{1}{2}\pm\frac{1}{2}\right)\left(\mathrm{i}+\frac{1}{2}\pm\frac{3}{2}\right)}}{2}. \end{aligned} \qquad (11.2.22)$$

If $\mathrm{q_o} = 0$, this integral vanishes.

4. $\mathrm{f} = \mathrm{n}$, $\mathrm{i} = 0$, q is the localized vibration normal coordinate,

$$\begin{aligned}\langle \mathrm{mn}|\mathrm{q}|\mathrm{ko}\rangle &= -\frac{\bar{\mathrm{q}}^2}{\mathrm{q_o}}\left(\mathrm{n} - \frac{\mathrm{q_o^2}}{2\bar{\mathrm{q}}^2}\right)\langle \mathrm{mn}|\mathrm{ko}\rangle \\ &= -\frac{\bar{\mathrm{q}}^2}{\mathrm{q_o}}\left(\mathrm{n} - \frac{\mathrm{q_o^2}}{2\bar{\mathrm{q}}^2}\right)\left(-\frac{\mathrm{q_o}}{\bar{\mathrm{q}}}\right)^{\mathrm{n}}\frac{1}{\sqrt{2^{\mathrm{n}}\mathrm{n}!}}\exp\left[-\frac{1}{4}\left(\frac{\mathrm{q_o}}{\mathrm{q}}\right)^2\right].\end{aligned} \tag{11.2.23}$$

For $\mathrm{n} = 0$

$$\langle \mathrm{mo}|\mathrm{q}|\mathrm{ko}\rangle = \frac{\mathrm{q_o}}{2}\exp\left[-\frac{1}{4}\left(\frac{\mathrm{q_o}}{\bar{\mathrm{q}}}\right)^2\right], \tag{11.2.24}$$

and the integral vanishes for $\mathrm{q_o} = 0$. For $\mathrm{n} = 1$:

$$\begin{aligned}\langle \mathrm{ml}|\mathrm{q}|\mathrm{kl}\rangle &= -\frac{\bar{\mathrm{q}}^2}{\mathrm{q_o}}\left(1 - \frac{\mathrm{q_o^2}}{2\bar{\mathrm{q}}^2}\right)\left(-\frac{\mathrm{q_o}}{\bar{\mathrm{q}}}\right)\frac{1}{\sqrt{2}}\exp\left[-\frac{1}{4}\left(\frac{\mathrm{q_o}}{\bar{\mathrm{q}}}\right)^2\right] \\ &= \frac{\bar{\mathrm{q}}}{\sqrt{2}}\left(1 - \frac{\mathrm{q_o^2}}{2\bar{\mathrm{q}}^2}\right)\exp\left[-\frac{1}{4}\left(\frac{\mathrm{q_o}}{\bar{\mathrm{q}}}\right)^2\right],\end{aligned} \tag{11.2.25}$$

and for $\mathrm{q_o} = 0$ this integral reduces to

$$\langle \mathrm{ml}|\mathrm{q}|\mathrm{ko}\rangle = \frac{\bar{\mathrm{q}}}{\sqrt{2}}. \tag{11.2.26}$$

For $\mathrm{n} = 2$:

$$\begin{aligned}\langle \mathrm{m}^2|\mathrm{q}|\mathrm{ko}\rangle &= -\frac{\bar{\mathrm{q}}^2}{\mathrm{q_o}}\left(2 - \frac{\mathrm{q_o^2}}{2\bar{\mathrm{q}}^2}\right)\frac{\mathrm{q_o^2}}{\bar{\mathrm{q}}^2}\frac{1}{2\sqrt{2}}\exp\left[-\frac{1}{4}\left(\frac{\mathrm{q_o}}{\bar{\mathrm{q}}}\right)^2\right] \\ &= -\frac{\mathrm{q_o}}{2}\left(2 - \frac{\mathrm{q_o^2}}{2\bar{\mathrm{q}}^2}\right)\frac{1}{\sqrt{2}}\exp\left[-\frac{1}{4}\left(\frac{\mathrm{q_o}}{\bar{\mathrm{q}}}\right)^2\right],\end{aligned} \tag{11.2.27}$$

and vanishes if $\mathrm{q_o} = 0$.

We consider now the integral (11.2.11), which, according to Ref. 1, is given by

$$\begin{aligned}\langle \mathrm{mf}|\mathrm{q}^2|\mathrm{ki}\rangle = {} & \frac{\bar{\mathrm{q}}^4}{\mathrm{q_o^2}}\left(|\mathrm{f}-\mathrm{i}| - \frac{\mathrm{q_o^2}}{2\bar{\mathrm{q}}^2}\right)\left(|\mathrm{f}-\mathrm{i}| + 1 - \frac{\mathrm{q_o^2}}{2\bar{\mathrm{q}}^2}\right)\langle \mathrm{mf}|\mathrm{ki}\rangle \\ & + \frac{\bar{\mathrm{q}}^3}{\mathrm{q_o}}\sqrt{2(\mathrm{i}+1)}\langle \mathrm{mf}|\mathrm{k}, \mathrm{i}+1\rangle.\end{aligned} \tag{11.2.28}$$

We do not calculate this integral for the case $q_o \neq 0$; rather we perform the calculation for the simpler case in which $q_o = 0$:

1. $f = i$, q is the band vibration normal coordinate,

$$\begin{aligned} \langle mi|q^2|ki\rangle &= \langle i|q^2|i\rangle \\ &= \langle i|q|i-1\rangle\langle i-1|q|i\rangle + \langle i|q|i+1\rangle\langle i+1|q|i\rangle \\ &= |\langle i|q|i-1\rangle|^2 + |\langle i|q|i+1\rangle|^2 \\ &= \frac{i}{2}\bar{q}^2 + \frac{i+1}{2}\bar{q}^2 = \frac{2i+1}{2}\bar{q}^2. \end{aligned} \tag{11.2.29}$$

2. $f = i \pm 1$, q is the band vibration normal coordinate,

$$\langle m, i \pm 1|q^2|ki\rangle = \langle i \pm 1|q^2|ki\rangle = 0. \tag{11.2.30}$$

3. $f = i + 2$, q is the band vibration normal coordinate,

$$\begin{aligned} \langle m, i+2|q^2|ki\rangle &= \langle i+2|q^2|i\rangle \\ &= \langle i+2|q|i+1\rangle\langle i+1|q|i\rangle \\ &= \sqrt{\frac{i+2}{2}}\bar{q}\sqrt{\frac{i+1}{2}}\bar{q} = \bar{q}^2\frac{\sqrt{(i+1)(i+2)}}{2}. \end{aligned} \tag{11.2.31}$$

$f = i - 2$, q is the band vibration normal coordinate,

$$\begin{aligned} \langle m, i-2|q^2|ki\rangle &= \langle i-2|q^2|i\rangle \\ &= \langle i-2|q|i-1\rangle\langle i-1|q|i\rangle \\ &= \sqrt{\frac{i-1}{2}}\bar{q}\sqrt{\frac{i}{2}}\bar{q} = \bar{q}^2\frac{\sqrt{i(i-1)}}{2}. \end{aligned} \tag{11.2.32}$$

Concisely, we can write

$$\langle m, i \pm 2|q^2|ki\rangle = \bar{q}^2\frac{\sqrt{(i \pm 1)(i+1 \pm 1)}}{2}. \tag{11.2.33}$$

4. $f = n$, $i = 0$, q is the localized vibration

$$\langle m0|q^2|k0\rangle = \langle 0|q^2|0\rangle = \frac{\bar{q}^2}{2}, \tag{11.2.34}$$

$$\langle m1|q^2|k0\rangle = \langle 0|q^2|1\rangle = 0, \tag{11.2.35}$$

$$\langle m2|q^2|k0\rangle = \frac{\bar{q}^2}{\sqrt{2}}. \tag{11.2.36}$$

For all other values of n, this integral vanishes. The results of all these calculations are presented in Tables 11.1 to 11.3.

It is important at this point to write down the expression for the matrix element M_{fi} in (11.2.8) for the simple case in which all q_o are zero. This matrix element can now be written, taking into account the results reported in Tables 11.1 to 11.3, as follows:

$$\begin{aligned}
M_{fi} &= M_{mn;kl} \\
&= D^o \prod_{s=1}^{3N} \langle f_s|i_s\rangle + \sum_{r=1}^{3N} [D_r^1 \langle f_r|q_r|i_r\rangle + D_{rr}^2 \langle f_r|q_r^2|i_r\rangle] \prod_{\substack{s=1\\ s\neq r}}^{3N} \langle f_s|i_s\rangle \\
&\quad + \sum_{r=1}^{3N} \underset{r\neq t}{} \sum_{t=1}^{3N} D_{rt}^2 \langle f_r|q_r|i_r\rangle \langle f_t|q_t|i_t\rangle \prod_{\substack{s=1\\ s\neq r,t}}^{3N} \langle f_s|i_s\rangle \\
&= D^o \prod_{s=1}^{3N} \delta_{f_s i_s} + \sum_{r=1}^{3N} \left[D_r^1 \bar{q}_r \sqrt{\frac{i+\frac{1}{2}\pm\frac{1}{2}}{2}} \delta_{f_r, i_r \pm 1} + D_{rr}^2 \bar{q}_{rr}^2 \right. \\
&\quad \left. \times \left(\frac{2i_r+1}{2} \delta_{f_r i_r} + \frac{\sqrt{(i_r \pm 1)(i_r + 1 \pm 1)}}{2} \delta_{f_r, i_r \pm 2} \right) \right] \prod_{\substack{s=1\\ s\neq r}}^{3N} \delta_{f_s i_s} \\
&\quad + \sum_{r=1}^{3N} \underset{r\neq t}{} \sum_{t=1}^{3N} \left[D_{rt}^2 \bar{q}_r \bar{q}_t \frac{\sqrt{\left(i_r+\frac{1}{2}\pm\frac{1}{2}\right)\left(i_t+\frac{1}{2}\pm\frac{1}{2}\right)}}{2} \delta_{f_r, i_r \pm 1} \right] \\
&\quad \times \prod_{\substack{s=1\\ s\neq r,t}}^{3N} \delta_{f_s i_s}.
\end{aligned} \tag{11.2.37}$$

This result can be the basis of several important considerations:

1. *Zero-phonon Line.* In this case the initial state is given by

$$|k; i_1, i_2, \ldots, i_{3N}\rangle = |k\rangle|i_1\rangle|i_2\rangle \ldots |i_{3N}\rangle, \tag{11.2.38}$$

Table 11.1. Values of overlap integral $\langle \mathrm{mf}|\mathrm{ki}\rangle$.

	$q_o \neq 0$	$q_o = 0$
f = i (band)	$1 - \frac{1}{2}\frac{q_o^2}{\bar{q}^2}\left(i + \frac{1}{2}\right)$	1
f = i ± 1 (band)	$-\frac{q_o}{\bar{q}}\sqrt{\frac{i + \frac{1}{2} \pm \frac{1}{2}}{2}}$	0
f = i ± 2 (band)	$\frac{1}{4}\frac{q_o^2}{\bar{q}^2}\sqrt{\left(i + \frac{1}{2} \pm \frac{1}{2}\right)\left(i + \frac{1}{2} \pm \frac{3}{2}\right)}$	0
f = n, i = 0 (localized)	$\left(-\frac{q_o^2}{\bar{q}}\right)^n \frac{1}{\sqrt{2^n n!}} \exp\left[-\frac{1}{4}\left(\frac{q_o}{\bar{q}}\right)^2\right]$	0 if n ≠ 0, 1 if n = 0

Table 11.2. Values of matrix element $\langle \mathrm{mf}|q|\mathrm{ki}\rangle$.

	$q_o \neq 0$	$q_o = 0$
f = i (band)	$\frac{q_o}{2}\left[1 - \frac{1}{2}\frac{q_o^2}{\bar{q}^2}\left(i + \frac{1}{2}\right)\right]$	0
f = i ± 1 (band)	$\bar{q}\left(1 - \frac{q_o^2}{\bar{q}^2}\right)\sqrt{\frac{i + \frac{1}{2} \pm \frac{1}{2}}{2}}$	$\bar{q}\sqrt{\frac{i + \frac{1}{2} \pm \frac{1}{2}}{2}}$
f = i ± 2 (band)	$-\frac{q_o}{2}\left(2 - \frac{q_o^2}{2\bar{q}^2}\right)\frac{\sqrt{\left(i + \frac{1}{2} \pm \frac{1}{2}\right)\left(i + \frac{1}{2} \pm \frac{3}{2}\right)}}{2}$	0
f = 0, i = 0 (localized)	$\frac{q_o}{2}\exp\left[-\frac{1}{4}\left(\frac{q_o}{\bar{q}}\right)^2\right]$	0
f = 1, i = 0 (localized)	$\frac{\bar{q}}{\sqrt{2}}\left(1 - \frac{q_o^2}{2\bar{q}^2}\right)\exp\left[-\frac{1}{4}\left(\frac{q_o}{\bar{q}}\right)^2\right]$	$\bar{q}\frac{1}{\sqrt{2}}$
f = 2, i = 0 (localized)	$-\frac{q_o}{2}\left(2 - \frac{q_o^2}{2\bar{q}^2}\right)\frac{1}{\sqrt{2}}\exp\left[-\frac{1}{4}\left(\frac{q_o}{\bar{q}}\right)^2\right]$	0

Table 11.3. Values of the matrix element $\langle \mathrm{mf}|\mathrm{q}^2|\mathrm{ki}\rangle$.

	$\mathrm{q_o} = 0$
$\mathrm{f} = \mathrm{i}$ (band)	$\bar{\mathrm{q}}^2 \frac{2\mathrm{i}+1}{2}$
$\mathrm{f} = \mathrm{i} \pm 1$ (band)	0
$\mathrm{f} = \mathrm{i} + 2$ (band)	$\bar{\mathrm{q}}^2 \frac{\sqrt{(2+1)(\mathrm{i}+2)}}{2}$
$\mathrm{f} = \mathrm{i} - 2$ (band)	$\bar{\mathrm{q}}^2 \frac{\sqrt{\mathrm{i}(\mathrm{i}-1)}}{2}$
$\mathrm{f} = 0$, $\mathrm{i} = 0$ (localized)	$\frac{\bar{\mathrm{q}}^2}{2}$
$\mathrm{f} = 1$, $\mathrm{i} = 0$ (localized)	0
$\mathrm{f} = 2$, $\mathrm{i} = 0$ (localized)	$\bar{\mathrm{q}}^2 \frac{\sqrt{2}}{2}$

and the final state by

$$|\mathrm{m}; \mathrm{i}_1, \mathrm{i}_2, \ldots, \mathrm{i}_{3N}\rangle = |\mathrm{m}\rangle|\mathrm{i}_1\rangle|\mathrm{i}_2\rangle \ldots |\mathrm{i}_{3N}\rangle. \tag{11.2.39}$$

The matrix element reduces to

$$\mathrm{M_{fi}} = \mathrm{M_{ml;kl}} = \mathrm{D^o} + \sum_{\mathrm{r}} \mathrm{D^2_{rr}} \bar{\mathrm{q}}_{\mathrm{r}}^2 \frac{2\mathrm{i_r}+1}{2}. \tag{11.2.40}$$

The first term is simply the electronic matrix element in correspondence with the equilibrium nuclear configuration; the second term corresponds, from a perturbation theory point of view, to a process whereby the system goes from an electronic state k to an intermediate state j and simultaneously emits (or absorbs) a phonon of wave vector $\underset{\sim}{\mathrm{k}}_{\mathrm{r}}$ and then moves from the intermediate state j to the electronic state m simultaneously absorbing (or emitting) a phonon with the same wave vector.

2. *One-phonon Vibronic.* Let us assume that a phonon of wave vector $\underset{\sim}{k}_r$ is created. The initial state is given by

$$|k; i_1, i_2, \ldots, i_r, \ldots, i_{3N}\rangle = |k\rangle|i_1\rangle|i_2\rangle \ldots |i_r\rangle \ldots |i_{3N}\rangle, \quad (11.2.41)$$

and the final state by

$$\begin{aligned} &|m; i_1, i_2, \ldots, i_{r-1}, i_r + 1, i_{r+1}, \ldots, i_{3N}\rangle \\ &\quad = |m\rangle|i_1\rangle|i_2\rangle \ldots |i_{r-1}\rangle|i_r + 1\rangle|i_{r+1}\rangle \ldots |i_{3N}\rangle. \end{aligned} \quad (11.2.42)$$

The matrix element reduces to

$$M_{fi} = M_{mn;kl} = D_r^1 \bar{q}_r \sqrt{\frac{i_r + 1}{2}}. \quad (11.2.43)$$

We note here that, under the present conditions ($q_{oi} = 0$, that is, all Stokes losses negligible) the one-phonon vibronic transition can take place only if $D_r^1 \neq 0$. Clearly, if one phonon is annihilated the matrix element becomes

$$M_{fi} = M_{mn;kl} = D_r^1 \bar{q}_r \sqrt{\frac{i_r}{2}}. \quad (11.2.44)$$

3. *Two-phonon Vibronic Transitions.* If the two phonons have wave vectors $\underset{\sim}{k}_r$ and $\underset{\sim}{k}_t$ and are both created, the initial state is given by

$$|k\rangle|i_1\rangle|i_2\rangle \ldots |i_r\rangle|i_t\rangle \ldots |i_{3N}\rangle, \quad (11.2.45)$$

and the final state by

$$|m\rangle|i_1\rangle|i_2\rangle \ldots |i_r + 1\rangle|i_t + 1\rangle \ldots |i_{3N}\rangle. \quad (11.2.46)$$

The matrix element is now given by

$$\begin{aligned} M_{fi} = M_{mn;kl} &= 2D_{rt}^2 \bar{q}_r \bar{q}_t \frac{\sqrt{(i_r + 1)(i_t + 1)}}{2} \\ &= D_{rt}^2 \bar{q}_r \bar{q}_t \sqrt{(i_r + 1)(i_t + 1)}. \end{aligned} \quad (11.2.47)$$

On the other hand, if the two phonons have equal $\underset{\sim}{k}$ vector, say $\underset{\sim}{k}_r$, then

$$M_{fi} = M_{mn;kl} = D^2_{rr}\bar{q}^2_r \frac{\sqrt{(i_r + 1)(i_r + 2)}}{2}. \tag{11.2.48}$$

4. *Multiphonon Vibronic Transitions.* If a vibronic transition is accompanied by the creation (annihilation) of several band phonons, and if the Stokes losses are zero, the transitions can take place only considering higher terms in the expansion of D^o_{mk}.
5. *Transitions Involving Localized Phonons.* It is to be noted that the considerations (2) to (4) above apply also to the case in which the phonons involved in the transition are of localized type. However, one must be aware of the fact that the condition of absence of Stokes losses is less easily realizable in regard to localized vibrations.

11.3. Deviations from the Adiabatic Approximation. Radiationless Transitions

In the adiabatic approximation, stationary states of a crystal or molecule are represented by eigenfunctions of the type

$$\psi_i(\underset{\sim}{r}, \underset{\sim}{R}) = \psi_k(\underset{\sim}{r}, \underset{\sim}{R})\phi_{kl}(\underset{\sim}{R}). \tag{11.3.1}$$

It is clear, however, that these states are not stationary in a strict sense and that it is possible for the system to move from one such state to another of approximately equal energy.

In different words, the quantum states of the type (11.3.1) are eigenstates of the approximate Hamiltonian given in (1.1.18)

$$H = H_e + H_v, \tag{11.3.2}$$

where

$$H_e = -\frac{\hbar^2}{2m}\sum_{i=1}^{n} \nabla_i^2 + V(\underset{\sim}{r}, \underset{\sim}{R}), \tag{11.3.3}$$

$$H_v = -\frac{\hbar^2}{2}\sum_{\alpha=1}^{N} \frac{\nabla_\alpha^2}{M_\alpha} + \varepsilon_k(\underset{\sim}{R}). \tag{11.3.4}$$

The more "exact" Hamiltonian is given by H + A where A is the "nonadiabaticity operator" defined by

$$A\psi_k(\underset{\sim}{r}, \underset{\sim}{R})\phi_{kl}(\underset{\sim}{R}) = -\sum_{\alpha=1}^{N} \frac{\hbar^2}{M_\alpha} \underset{\sim}{\nabla}_\alpha \phi_{kl}(\underset{\sim}{R}) \cdot \underset{\sim}{\nabla}_\alpha \psi_k(\underset{\sim}{r}, \underset{\sim}{R})$$

$$-\sum_{\alpha=1}^{N} \frac{\hbar^2}{2M_\alpha} \phi_{kl}(\underset{\sim}{R}) \nabla_\alpha^2 \psi_k(\underset{\sim}{r}, \underset{\sim}{R}). \qquad (11.3.5)$$

Since the function $A\psi_k\phi_{kl}$ can he expanded in terms of the complete set (11.3.1), it is clear that if it is

$$\langle \psi_{k'}\phi_{k'l'}|A|\psi_k\phi_{kl}\rangle \neq 0, \qquad (11.3.6)$$

then the system can undergo a change of state, provided the energy of the bra state and the energy of the ket state are approximately equal.

Having established the mechanism by which the system can undergo a change of "quasistationary" state, we examine the sequence of events following the absorption of a photon that brings the system from the ground to an excited electronic state. We use Fig. 11.1 to describe the events: in this figure two adiabatic

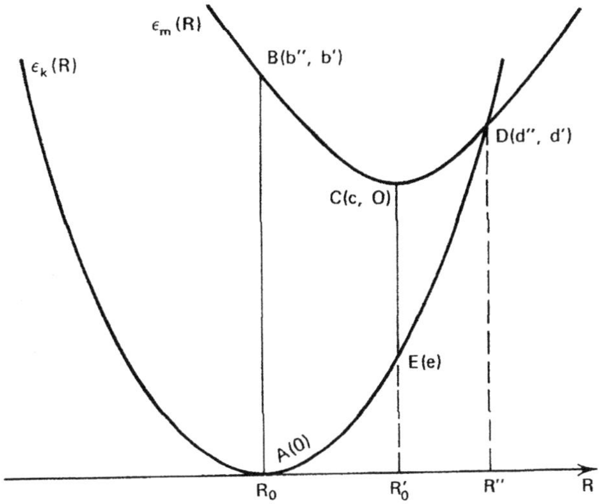

Fig. 11.1. Diagram illustrating the possible modes of decay of an electronically excited molecule.

potentials for a diatomic molecule (or normal coordinate) are presented. Note that each point in the diagram has two numbers in parentheses: the first number, say b″, indicates the vibrational level in the lower electronic state, the second number, say b′, indicates the vibrational level in the excited electronic state. The considerations below apply to diatomic molecules as well as to impurities in solids: we just remind ourselves that thermal equilibrium in a gas of molecules is achieved by the intervention of collisions, whereas in a solid the anharmonic forces play a similar role. With reference to Fig. 11.1[2],

1. The primary act is the absorption of a photon that brings the system from A to B.
2. Once at B, the system is in the quasistationary state described by

$$\psi_m(\underset{\sim}{r}, \underset{\sim}{R}_o)\phi_{mb'}(\underset{\sim}{R}_o). \tag{11.3.7}$$

   At this point the system can, in principle, do one of the following things:

   a. It may go back to A, by emitting a photon equal to the one previously absorbed. This is an unlikely possibility considering the instability to thermal relaxation of the system when at B. If $R_o = R_o'$, this possibility is, however, strongly enhanced.
   b. It can go to the state

$$\psi_k(\underset{\sim}{r}, \underset{\sim}{R}_o)\phi_{kb''}(\underset{\sim}{R}_o),$$

   from which it can then quickly decay to A via thermal relaxation. This possibility also is unlikely considering the poor overlap of $\phi_{kb''}(\underset{\sim}{R}_o)$ and $\phi_{mb'}(\underset{\sim}{R}_o)$.
   c. The system can quickly decay by thermal relaxation to C, remaining in the same electronic state. Considering the arrangement in the figure where point D is *below* point C, this possibility may also be unlikely for the following reason: as the system on its way to C has reached the level of D, the overlap between $\phi_{md'}$ and $\phi_{kd''}$ will be greatly enhanced and will enhance the probability for the system to change its electronic state.

3. Considering the most likely route the system is now at D in a state described by

$$\psi_k(\underset{\sim}{r}, \underset{\sim}{R}'')\phi_{kd''}(\underset{\sim}{R}'').$$

   At this point the system will quickly decay to A by thermal relaxation. When this sequence of steps has been followed the energy of the absorbed photon has been entirely transformed into (thermal) energy of vibration.

One could ask the question: What if the crossing point D is *above* B? This would clearly enhance the possibility for the system to decay quickly to C and then reemit a photon of lower energy. However, once the system is in C, especially if the lifetime of the electronic state is long enough, the system could still jump by thermal activation to the vibrational level d′ and then follow its way to A by radiationless relaxation.

What we have here is a range of possibilities whose possible occurrence depend to a large extent on the values of such parameters as lifetimes, relaxation times, and so on.

11.4. A Simple Model for Laser Crystals: An Effective Hamiltonian

The model of optical spectra of impurities in solids that makes use of adiabatic potentials explains to a large extent the essential features of these spectra. The case in which all Stokes losses are zero or very small is a particular case that falls within the bounds of this model. In such a case, however, a simpler approach may be desirable, with the use of a model that would allow more speedy calculations of transition probabilities and simple application of group-theoretical tools.

We shall consider the application of such a model to "laser crystals", that is, ionic solids doped with magnetic ion impurities. Several groups of atoms in their ionic states have one or more of their shells only partly filled with electrons; this results in some magnetic and optical properties of these ions when they are present in a crystal. The electrons of the unfilled shell provide a net magnetic moment and account for the magnetic properties of these ions. The same electrons set the energy levels of the ion and transitions among these levels are responsible for the absorption and fluorescence spectra. The magnetic ions can be divided into five categories:

1. *Transition Metal Ions of the First Series (Iron Group).* The atoms of this series have electronic configurations of the type

$$1s^2 2s^2 2p^6 3s^2 3p^6 3d^{\mathrm{n}} 4s^{\mathrm{m}} = (\mathrm{Ar\ core})^{18} 3d^{\mathrm{n}} 4s^{\mathrm{m}},$$

   where $\mathrm{m} = 1, 2$ and $\mathrm{n} = 1, 2, \ldots, 10$. Their atomic numbers range from $\mathrm{Z} = 21$ (Sc) to $\mathrm{Z} = 30$ (Zn). The divalent and trivalent ions of this series present configurations of the type $(\mathrm{Ar\ core})^{18}$ $3\mathrm{d}^{\mathrm{n}}(\mathrm{n} = 1, 2, \ldots, 10)$ in which the unfilled 3d shell is the outermost shell. When in a crystal the orbital motion of the 3d electrons are strongly perturbed by the Coulomb field of the surrounding ions and energies of the order of $10{,}000\,\mathrm{cm}^{-1}$ ($\equiv 15{,}000°\mathrm{K} \equiv 1.25\,\mathrm{eV} \equiv 2 \times 10^{-6}$ erg) are associated with the coupling of the orbitals with this field.
2. *Transition Metal Ions of the Second Series (Palladium Group).* The atoms of this series have electronic configurations of the type

$$(\mathrm{Kr\ core})^{35} 4\mathrm{d}^{\mathrm{n}} 5\mathrm{s}^{\mathrm{m}},$$

   where $\mathrm{m} = 1, 2$ and $\mathrm{n} = 1, 2, \ldots, 10$. Their atomic numbers range from $\mathrm{Z} = 39$ (Y) to $\mathrm{Z} = 48$ (Cd) . The divalent ions of this series present configurations of the type $(\mathrm{Kr\ core})^{36}$ $4\mathrm{d}^{\mathrm{n}}$, where the unfilled 4d shell is the outermost shell.
3. *Transition Metal Ions of the Third Series (Platinum Group).* The atoms of this series have electronic configurations of the type

$$(\mathrm{Pd\ core})^{46}\ 4\mathrm{f}^{14}\ 5\mathrm{s}^2\ 5\mathrm{p}^6\ 5\mathrm{d}^{\mathrm{n}}\ 6\mathrm{s}^{\mathrm{m}},$$

   where $\mathrm{m} = 1, 2$ and $\mathrm{n} = 2, 3, 4, 5, 6, 9$, and 10. Their atomic numbers range from $\mathrm{Z} = 72$ (Hf) to $\mathrm{Z} = 80$ (Hg). The divalent ions present configurations of the type $(\mathrm{Pd\ core})^{46}$ $4\mathrm{f}^{14} 5\mathrm{s}^2 5\mathrm{p}^6 5\mathrm{d}^{\mathrm{n}}$, where the unfilled 4d shell is the outermost shell.
4. *Rare Earth Ions (Group of the Lanthanides).* The atoms of this group have electronic configurations of the type

$$(\mathrm{Pd\ core})^{46}\ 4\mathrm{f}^{\mathrm{n}} 5\mathrm{s}^2 5\mathrm{p}^6 5\mathrm{d}^{\mathrm{m}} 6\mathrm{s}^2,$$

where $m = 1, 2$ and $n = 2, 3, \ldots, 13$. Their atomic numbers range from $Z = 58$ (Ce) to $Z = 70$ (Yb). The trivalent ions present configurations of the type $(\text{Pd core})^{46}\ 4f^n 5s^2 5p^6$ and the unfilled 4f shell is an inner shell screened by the 5s and tp shells. (The 4f electrons have orbital radii of 0.6 to 0.35 from Ce to Lu; for the latter this is less than half the ionic radius.) When in a crystal the orbital motion of the 4f electrons are not much affected by the Coulomb field of the surrounding ions, as the interaction energy is in most cases much less than $1000\,\text{cm}^{-1}$; on the other hand, the spin-orbit energy is greater than for the 3d ions (600 to $3000\,\text{cm}^{-1}$ from Ce to Yb).

5. *Actinide Ions.* The atoms of this group have electronic configurations of the type

$$(\text{Pt core})^{78}\ 5f^n 6d^m 7s^2,$$

where $m = 1, 2$ and $n = 0, 2, 3, \ldots$ and their atomic numbers range from $Z = 90$ (Th) up. The trivalent ions present configurations of the type $(\text{Pt core})^{78}\ 5f^n 7s^2$.

In a magnetic ionic impurity the electrons occupy orbits that are highly localized about the ion. Each electron, however, feels the influence of the electrons belonging to the other ions (a repulsion) and of the nuclei belonging to the other ions (an attraction). This influence is taken into account by considering that the electrons of the impurity ion are subjected to the action of a *crystalline field.* The crystalline field is completely external to the ion and has a definite symmetry set by the configuration of the ions coordinating the magnetic ion. In the crystalline field approximation the charge of the ligand ions does not penetrate into the region occupied by the magnetic ion and the perturbing potential satisfies Laplace's equation

$$\nabla^2 V = 0. \tag{11.4.1}$$

The solutions of (11.4.1) are of the type $r^l Y_l^m(\theta, \phi)$ where Y is a spherical harmonic. The general solution is

$$V(r, \theta, \phi) = \sum_{l,m} A_{lm} r^l Y_l^m(\theta, \phi). \tag{11.4.2}$$

The Hamiltonian of an ion in a crystal can now be written

$$H = \sum_{i=1}^{n} \left[-\frac{\hbar^2}{2m}\nabla_i^2 - \frac{e^2 Z}{r_i} + \frac{1}{2}\sum_{\substack{j=1 \\ j\neq i}}^{n} \frac{e^2}{|\underset{\sim}{r}_i - \underset{\sim}{r}_j|} \right] + \text{spin-orbit interaction} + \sum_{i=1}^{n} eV(\underset{\sim}{r}_i). \qquad (11.4.3)$$

The crystalline field perturbation $\sum_i eV(\underset{\sim}{r}_i)$ destroys the spherical symmetry of the environment; such operators as L^2 and L_z no longer commute with the Hamiltonian. S^2 and S_z commute with the crystal field operator, which does not depend on spin coordinates; they can still be considered constants of the motion if spin-orbit interaction is disregarded. Parity, which in a free ion is always a good quantum number, remains so only if the ion sits at a point that is a center of inversion.

The energy levels of the magnetic ion can, in principle, be found by simply calculating the eigenvalues of the Hamiltonian (11.4.3). In practice what is known may just be the symmetry of the perturbing term; by proper use of group theory it is possible to predict the splitting of the energy levels and the transformation properties of the eigenfunctions representing the electronic states. By comparing the spectral data with the theoretical prediction it is then possible to derive what group theory cannot furnish, the *strength* of the crystalline perturbation. For the treatment of the energy levels of an ion in a crystal the reader is referred to more specialized articles and books such as those cited as Refs. 3 and 4.

Within the framework of this approach we now investigate how various transitions between electronic states such as zero-phonon, phonon-assisted, and radiationless, transitions are treated. The specific application to solid state lasers is discussed in Chapter 13.

11.5. Radiative, Vibronic, and Radiationless Transitions of Magnetic Impurities

The crystalline field hypothesis leads to the conceptually simple Hamiltonian (11.4.3), which can in principle serve to establish the energy levels and eigenfunctions of a magnetic impurity in a crystal. Transitions between these levels can be brought about by the interaction of radiation with the impurity ion in a way similar to that for free atoms; obviously, selection rules will be determined for the various types of radiation (electric dipole, magnetic dipole) by the symmetry of the environment of the impurity ion. The problem, even if more complicated, is, as we said, similar conceptually to that for free atoms. The transitions produced by the ion-radiation interaction in the present scheme of things are what we have called purely radiative or zero-phonon transitions. But we well know that a magnetic impurity can also undergo (vibronic) transitions in which the absorption or emission of a photon is accompanied by the absorption or emission of one or more phonons. Also, a magnetic impurity in an excited energy state can undergo a (radiationless) transition where the entire electronic excited energy is transformed into vibrational energy. Clearly we are operating with conditions in which deviations from both the Franck-Condon principle and the adiabatic approximation must be allowed. However, now we can improve on our Hamiltonian to take phonon effects into account.

From a physical point of view the crystalline field hypothesis lends itself to an extension that includes the effects of lattice vibrations. The crystalline field at a magnetic ion site is set by the ligand ions and these ions are participating in the thermal motion of the crystal's constituents; as a result of this motion the electrons of the magnetic ion are seeing a fluctuating potential field, rather than a field independent of time as introduced in (11.4.2). The effective potential-energy term to be introduced in the Hamiltonian (11.4.3) has to take these fluctuations into account; that is, it is a function of the instantaneous positions of the optically active electrons of the magnetic ion and of the other ions in the crystal.

Before expressing this functional dependence let us assume that the crystal has J ions per unit cell and N unit cells; let us also assume that there are n optically active electrons in the impurity ion and that the impurity ion is located at the point 11 (unit cell 1, ion 1 in the unit cell), we can now write

$$\mathrm{V} = \mathrm{V}(\ldots, \underset{\sim}{\mathrm{U}}_{s\nu} + \underset{\sim}{\mathrm{R}}_{s\nu}, \ldots, \underset{\sim}{\mathrm{r}}_{t}, \ldots), \tag{11.5.1}$$

where $\underset{\sim}{R}_{s\nu}$ is the equilibrium position of the ν-th atom in the s-th unit cell, $\underset{\sim}{U}_{s\nu}$ is the displacement of the ν-th atom of the s-th unit cell from its equilibrium position, $\underset{\sim}{r}_t$ is the instantaneous position of the t-th electron of the magnetic ion $= \underset{\sim}{r}'_t + \underset{\sim}{U}_{11}$, and where

$$s = 1, 2, \ldots, N, \quad \nu = 2, 3, \ldots, N.$$

The interaction energy V can be Taylor expanded about the equilibrium positions of the ions of the crystal:

$$\begin{aligned} V = V_o &+ \sum_{s\nu\alpha}{}' \left.\frac{\partial V}{U_{s\nu\alpha}}\right|_o U_{s\nu\alpha} + \sum_{t\alpha} \left.\frac{\partial V}{\partial r_{t\alpha}}\right|_o U_{11\alpha} \\ &+ \frac{1}{2}\sum_{s\nu\alpha}{}' \sum_{s'\nu'\beta}{}' \left.\frac{\partial^2 V}{\partial U_{s\nu\alpha}\partial U_{s'\nu'\beta}}\right|_o U_{s\nu\alpha}U_{s'\nu'\beta} \\ &+ \frac{1}{2}\sum_{t\alpha}\sum_{t'\beta} \left.\frac{\partial^2 V}{\partial r_{t\alpha}\partial r_{t'\beta}}\right|_o U_{11\alpha}U_{11\beta} \\ &+ \sum_{s\nu\alpha}{}'\sum_{t\beta} \left.\frac{\partial^2 V}{\partial U_{s\nu\alpha}\partial r_{t\beta}}\right|_o U_{s\nu\alpha}U_{11\beta} + \cdots, \end{aligned} \tag{11.5.2}$$

where the sum $\sum'$ are exclusive of $s\nu = 11$. Each term in the expansion above has coefficients that depend on the electron coordinates $\underset{\sim}{r}'_t$ of the impurity ion at equilibrium. The term V_o is the independent static perturbation that appears as $\sum_i eV(r_i^o)$ in (11.4.3) and produces the splitting of the free ion levels.

The dynamic part of V, caused by the vibrations of the ions, is

$$H_{ev} = V - V_o = \left[\sum_{s\nu} \underset{\sim}{U}_{s\nu} \cdot \underset{\sim}{\nabla}_{s\nu} + \frac{1}{2}\left(\sum_{s\nu} \underset{\sim}{U}_{s\nu} \cdot \underset{\sim}{\nabla}_{s\nu}\right)^2 + \cdots\right] V, \tag{11.5.3}$$

where $\underset{\sim}{\nabla}_{11}$ means differentiation with respect to the coordinates of the n electrons. If we assume that the presence of the impurity does not affect the phonon spectrum of the crystal we may expect the displacements of the ions in terms of the normal coordinates as we did in (6.5.77):

$$U_{s\nu\alpha} = \sum_{\underset{\sim}{k}j} V_{\nu\alpha}(\underset{\sim}{k}, j) e^{i\underset{\sim}{k}\cdot\underset{\sim}{R}_s} q^j_{\underset{\sim}{k}}. \tag{11.5.4}$$

Using this expression in (11.5.3) we obtain

$$\begin{aligned}
H_{ev} &= \sum_{s\nu\alpha} U_{s\nu\alpha}\nabla_{s\nu\alpha} V\Big|_o + \frac{1}{2}\left(\sum_{s\nu\alpha} U_{s\nu\alpha}\nabla_{s\nu\alpha}\right)\left(\sum_{s'\nu'\beta} U_{s'\nu'\beta}\nabla_{s'\nu'\beta}\right) V\Big|_o \\
&= \sum_{\underset{\sim}{k}j}\left[\sum_{s\nu\alpha} V_{\nu\alpha}(\underset{\sim}{k}, j) e^{i\underset{\sim}{k}\cdot\underset{\sim}{R}_s}\nabla_{s\nu\alpha} V\Big|_o\right] q^j_{\underset{\sim}{k}} + \frac{1}{2}\sum_{\underset{\sim}{k}j}\sum_{\underset{\sim}{k}'j'} \\
&\quad \times \left[\sum_{s\nu\alpha}\sum_{s'\nu'\beta} V_{\nu\alpha}(\underset{\sim}{k}, j) V_{\nu'\beta}(\underset{\sim}{k}', j') + e^{i\underset{\sim}{k}\cdot\underset{\sim}{R}_s} e^{i\underset{\sim}{k}'\cdot\underset{\sim}{R}_{s'}} \nabla_{s\nu\alpha}\nabla_{s'\nu'\beta} V\Big|_o\right] \\
&\quad \times q^j_{\underset{\sim}{k}} q^{j'}_{\underset{\sim}{k}'} + \cdots \\
&= \sum_{\underset{\sim}{k}j} V^j_{\underset{\sim}{k}} q^j_{\underset{\sim}{k}} + \sum_{\underset{\sim}{k}j}\sum_{\underset{\sim}{k}'j'} V^{ij'}_{\underset{\sim}{k}\underset{\sim}{k}'} q^j_{\underset{\sim}{k}} q^{j'}_{\underset{\sim}{k}'} + \cdots.
\end{aligned} \tag{11.5.5}$$

The coefficients $V^j_{\underset{\sim}{k}}$ and $V^{jj'}_{\underset{\sim}{k}\underset{\sim}{k}}$ are functions only of the electronic coordinates of the impurity ion at equilibrium; the normal coordinates $q^j_{\underset{\sim}{k}}$ are functions only of the displacements of the ions.

Having established the form of the interaction between the impurity ion and the lattice vibrations, we can examine now the problem of the vibronic transitions. We shall consider the system that consists of the magnetic ion, the lattice vibrations, and the radiative field. The Hamiltonian of the system is given by

$$H = H_{ion} + H_{rf} + H_{vib} + H_{er} + H_{ev}, \tag{11.5.6}$$

where

$$\begin{aligned}
H_{ion} &= \sum_{i=1}^{n}\left[-\frac{\hbar^2}{2m}\nabla_i^2 - \frac{e^2 Z}{r_i} + \frac{1}{2}\sum_{\substack{j=1\\ j\neq i}}^{n}\frac{e^2}{|\underset{\sim}{r}_i - \underset{\sim}{r}_j|}\right] \\
&\quad + \text{spin-orbit interaction } + \sum_{i=1}^{n} eV(\underset{\sim}{r}_i),
\end{aligned} \tag{11.5.7}$$

$$H_{rf} = \sum_{\underset{\sim}{\alpha}} \sum_{\lambda} \hbar\omega_{\underset{\sim}{\alpha}} \left(a_{\underset{\sim}{\alpha}}^{\lambda+} a_{\underset{\sim}{\alpha}}^{\lambda} + \frac{1}{2} \right), \tag{11.5.8}$$

$$H_{vib} = \sum_{\underset{\sim}{k}} \sum_{j} \hbar\omega_{\underset{\sim}{k}j} \left(b_{\underset{\sim}{k}}^{j+} b_{\underset{\sim}{k}}^{j} + \frac{1}{2} \right), \tag{11.5.9}$$

$$H_{er} = \sum_{i=1}^{n} \frac{e}{mc} \underset{\sim}{A}(\underset{\sim}{r}_i) \cdot \underset{\sim}{p}_i + \sum_{i=1}^{n} \frac{e^2}{2mc^2} [\underset{\sim}{A}(\underset{\sim}{r}_i)]^2, \tag{11.5.10}$$

$$H_{ev} = \sum_{\underset{\sim}{k}j} V_{\underset{\sim}{k}}^{j} q_{\underset{\sim}{k}}^{j} + \sum_{\underset{\sim}{k}j} \sum_{\underset{\sim}{k}'j'} V_{\underset{\sim}{k}\underset{\sim}{k}'}^{jj'} q_{\underset{\sim}{k}}^{j} q_{\underset{\sim}{k}'}^{j'} + \cdots . \tag{11.5.11}$$

H_{ion} represents the Hamiltonian of the free ion plus the static term $V_o = \sum_{i=1}^{n} eV(\underset{\sim}{r}_i)$; H_{rf} represents the Hamiltonian of the radiative field with $a_{\underset{\sim}{\alpha}}^{\lambda}$ and $a_{\underset{\sim}{\alpha}}^{\lambda+}$ photon annihilation and creation operators, respectively; H_{vib} represents the Hamiltonian of the lattice vibrations with $b_{\underset{\sim}{k}}^{j}$ and $b_{\underset{\sim}{k}}^{j+}$ phonon annihilation and creation operators, respectively; H_{er} is the interaction Hamiltonian between the ion and the radiative field, and H_{ev} is the interaction between the ion and the lattice vibrations. We assume that the radiative field interacts solely with the impurity ion, that is, that the radiative field has no effect on the rest of the crystal.

The "unperturbed" Hamiltonian is given by

$$H_o = H_{ion} + H_{em} + H_v. \tag{11.5.8}$$

A state of this system can be identified by a set of quantum numbers that has to include:

a. quantum numbers necessary to identify the electronic state of the impurity;
b. an infinite set of quantum numbers identifying the various degrees of excitations of the radiative field oscillators, and
c. a set of 3NJ quantum numbers identifying the various degrees of excitation of the vibrational oscillators.

The most general state function of the unperturbed system is given by

$$\psi = |\psi^{el}; n_1 n_2 \ldots n_{\underset{\sim}{k}}^{j} \ldots n_{3NJ}; n_1 n_2 \ldots n_{\underset{\sim}{\alpha}}^{\lambda} \ldots\rangle$$
$$= |\psi^{el}\rangle |n_1\rangle |n_2\rangle \ldots |n_{\underset{\sim}{k}}^{j}\rangle \ldots |n_{3NJ}\rangle |n_1\rangle |n_2\rangle \ldots |n_{\underset{\sim}{\alpha}}^{\lambda}\rangle, \quad (11.5.12)$$

where ψ^{el} is the electronic wavefunction of the impurity ion (eigenfunction of H_{ion}).

The interaction between the radiative field and the impurity ion is brought about by the Hamiltonian H_{er} in which, according to (9.2.8), $\underset{\sim}{A}$ is given by

$$\underset{\sim}{A}(\underset{\sim}{r}) = \sum_{\underset{\sim}{\alpha}\lambda} \sqrt{\frac{hc^2}{\omega_{\underset{\sim}{\alpha}} V}} \underset{\sim}{\pi}_{\underset{\sim}{\alpha}}^{\lambda} (e^{i\underset{\sim}{\alpha}\cdot\underset{\sim}{r}} a_{\underset{\sim}{\alpha}}^{\lambda} + e^{-i\underset{\sim}{\alpha}\cdot\underset{\sim}{r}} a_{\underset{\sim}{\alpha}}^{\lambda+}). \quad (11.5.13)$$

For visible light, the variation of $\underset{\sim}{A}$ over the dimensions of the impurity can be neglected, i.e., the exponentials in (11.5.13) can be set equal to 1. Also, since the velocities of the electrons are much greater than the velocity of the impurity's nucleus, we can set $p_{t'} \approx p_t$.

The interaction between the lattice vibrations and the impurity ion is brought about by the Hamiltonian H_{ev} in which the normal coordinates can be expressed in terms of annihilation and creation phonon operators as in (6.5.72)

$$q_{\underset{\sim}{k}}^{j} = \left(\frac{\hbar}{2M\omega_{\underset{\sim}{k}j}}\right)^{\frac{1}{2}} (a_{\underset{\sim}{k}}^{j} + a_{\underset{\sim}{k}}^{j+}). \quad (11.5.14)$$

The use of H_{er} alone when calculating transitions between states (11.5.12) results in purely radiative electronic transitions; the use of H_{ev} alone when calculating transitions between states (11.5.12) results in radiationless transitions. The use of both H_{er} and H_{ev} results in vibronic transitions; this can be easily seen by calculating the matrix element involved in the simultaneous creation or annihilation of a photon of frequency $\omega_{\underset{\sim}{\alpha}}$ and polarization $\underset{\sim}{\pi}_{\underset{\sim}{\alpha}}^{\lambda}$ and creation or annihilation of a phonon of frequency $\omega_{\underset{\sim}{k}j}$:

$$\sum_{p} \frac{\langle \psi_f^{el}; n_{\underset{\sim}{k}}^{j} + 1; n_{\underset{\sim}{\alpha}}^{\lambda} + 1 | H_{er} + H_{ev} | \psi_p \rangle \langle \psi_p | H_{er} + H_{ev} | \psi_i^{el}; n_{\underset{\sim}{k}}^{j}; n_{\underset{\sim}{\alpha}}^{\lambda} \rangle}{E_i - E_p}$$

$$= \sum_{p} \left[\frac{\langle \psi_f^{el}; n_{\underset{\sim}{k}}^j + 1; n_{\underset{\sim}{\alpha}}^\lambda + 1 | H_{er} | \psi_p^{el}; n_{\underset{\sim}{k}}^j + 1; n_{\underset{\sim}{\alpha}}^\lambda \rangle \times \langle \psi_p^{el}; n_{\underset{\sim}{k}}^j + 1; n_{\underset{\sim}{\alpha}}^\lambda | H_{ev} | \psi_i^{el}; n_{\underset{\sim}{k}}^j; n_{\underset{\sim}{\alpha}}^\lambda \rangle}{E_i^{el} - (E_p^{el} + \hbar\omega_{\underset{\sim}{k}j})} \right.$$

$$\left. + \frac{\langle \psi_f^{el}; n_{\underset{\sim}{k}}^j + 1; n_{\underset{\sim}{\alpha}}^\lambda + 1 | H_{ev} | \psi_p^{el}; n_{\underset{\sim}{k}}^j; n_{\underset{\sim}{\alpha}}^\lambda + 1 \rangle \times \langle \psi_p^{el}; n_{\underset{\sim}{k}}^j; n_{\underset{\sim}{\alpha}}^\lambda + 1 | H_{er} | \psi_i^{el}; n_{\underset{\sim}{k}}^j; n_{\underset{\sim}{\alpha}}^\lambda \rangle}{E_i^{el} - (E_p^{el} + \hbar\omega_{\underset{\sim}{\alpha}})} \right]$$

$$= \sum_{p} \left[\frac{\langle \psi_f^{el}; n_{\underset{\sim}{\alpha}}^\lambda + 1 | H_{er} | \psi_p^{el}; n_{\underset{\sim}{\alpha}}^\lambda \rangle \langle \psi_p^{el} | V_{\underset{\sim}{k}}^j | \psi_i^{el} \rangle}{E_i^{el} - (E_p^{el} + \hbar\omega_{\underset{\sim}{k}j})} \right.$$

$$\left. + \frac{\langle \psi_f^{el} | V_{\underset{\sim}{k}}^j | \psi_p^{el} \rangle \langle \psi_p^{el}; n_{\underset{\sim}{\alpha}}^\lambda + 1 | H_{er} | \psi_i^{el}; n_{\underset{\sim}{\alpha}}^\lambda \rangle}{E_i^{el} - (E_p^{el} + \hbar\omega_{\underset{\sim}{\alpha}})} \right]$$

$$\times \left(\frac{\hbar}{2M\omega_{\underset{\sim}{k}j}} \right)^{\frac{1}{2}} (n_{\underset{\sim}{k}}^j + 1)^{\frac{1}{2}}$$

$$= \frac{e}{m} \left(\frac{\hbar}{2M\omega_{\underset{\sim}{k}j}} \right)^{\frac{1}{2}} \left(\frac{h}{\omega_{\underset{\sim}{\alpha}} V} \right)^{\frac{1}{2}} (n_{\underset{\sim}{k}}^j + 1)^{\frac{1}{2}} (n_{\underset{\sim}{\alpha}}^\lambda + 1)^{\frac{1}{2}}$$

$$\times \sum_{p} \left[\frac{\left\langle \psi_f^{el} \left| \sum_t e^{-i\underset{\sim}{\alpha} \cdot \underset{\sim}{r}'_t} \underset{\sim}{p}'_t \cdot \underset{\sim}{\pi}_{\underset{\sim}{\alpha}}^\lambda \right| \psi_p^{el} \right\rangle \langle \psi_p^{el} | V_{\underset{\sim}{k}}^j | \psi_i^{el} \rangle}{E_i^{el} - (E_p^{el} + \hbar\omega_{\underset{\sim}{k}j})} \right.$$

$$\left. + \frac{\langle \psi_f^{el} | V_{\underset{\sim}{k}}^j | \psi_p^{el} \rangle \left\langle \psi_p^{el} \left| \sum_t e^{-i\underset{\sim}{\alpha} \cdot \underset{\sim}{r}'_t} \underset{\sim}{p}'_t \cdot \underset{\sim}{\pi}_{\underset{\sim}{\alpha}}^\lambda \right| \psi_i^{el} \right\rangle}{E_i^{el} - (E_p^{el} + \hbar\omega_{\underset{\sim}{\alpha}})} \right]. \qquad (11.5.15)$$

Briefly the matrix element for the process "creation of a photon and creation of a phonon (emission in the low-energy band)" is given by

$$\frac{e}{m}\left(\frac{\hbar}{2M\omega_{\underset{\sim}{k}j}}\right)^{\frac{1}{2}}\left(\frac{h}{\omega_{\underset{\sim}{\alpha}}V}\right)^{\frac{1}{2}}\overline{(n^j_{\underset{\sim}{k}}+1)^{\frac{1}{2}}(n^\lambda_{\underset{\sim}{\alpha}}+1)^{\frac{1}{2}}}$$

$$\times\sum_p\left[\frac{\left\langle\psi^{el}_f\middle|\sum_t e^{-i\underset{\sim}{\alpha}\cdot\underset{\sim}{r}'_t}\underset{\sim}{p}'_t\cdot\underset{\sim}{\pi}^\lambda_{\underset{\sim}{\alpha}}\middle|\psi^{el}_p\right\rangle\langle\psi^{el}_p|V^j_{\underset{\sim}{k}}|\psi^{el}_i\rangle}{E^{el}_i-(E^{el}_p+\hbar\omega_{\underset{\sim}{k}j})}\right.$$

$$\left.+\frac{\langle\psi^{el}_f|V^j_{\underset{\sim}{k}}|\psi^{el}_p\rangle\left\langle\psi^{el}_p\middle|\sum_t e^{-i\underset{\sim}{\alpha}\cdot\underset{\sim}{r}'_t}\underset{\sim}{p}'_t\cdot\underset{\sim}{\pi}^\lambda_{\underset{\sim}{\alpha}}\middle|\psi^{el}_i\right\rangle}{E^{el}_i-(E^{el}_p+\hbar\omega_\alpha)}\right]. \qquad (11.5.16)$$

The matrix element of the process "creation of a photon and absorption of a phonon (emission in the high-energy band)" is given by

$$\frac{e}{m}\left(\frac{\hbar}{2M\omega_{\underset{\sim}{k}j}}\right)^{\frac{1}{2}}\left(\frac{h}{\omega_{\underset{\sim}{\alpha}}V}\right)^{\frac{1}{2}}\overline{(n^j_{\underset{\sim}{k}})^{\frac{1}{2}}(n^\lambda_{\underset{\sim}{\alpha}}+1)^{\frac{1}{2}}}$$

$$\times\sum_p\left[\frac{\left\langle\psi^{el}_f\middle|\sum_t e^{-i\underset{\sim}{\alpha}\cdot\underset{\sim}{r}'_t}\underset{\sim}{p}'_t\cdot\underset{\sim}{\pi}^\lambda_{\underset{\sim}{\alpha}}\middle|\psi^{el}_p\right\rangle\langle\psi^{el}_p|V^j_{\underset{\sim}{k}}|\psi^{el}_i\rangle}{E^{el}_i-(E^{el}_p-\hbar\omega_{\underset{\sim}{k}j})}\right.$$

$$\left.+\frac{\langle\psi^{el}_f|V^j_{\underset{\sim}{k}}|\psi^{el}_p\rangle\left\langle\psi^{el}_p\middle|\sum_t e^{-i\underset{\sim}{\alpha}\cdot\underset{\sim}{r}'_t}\underset{\sim}{p}'_t\cdot\underset{\sim}{\pi}^\lambda_{\underset{\sim}{\alpha}}\middle|\psi^{el}_i\right\rangle}{E^{el}_i-(E^{el}_p+\hbar\omega_{\underset{\sim}{\alpha}})}\right]. \qquad (11.5.17)$$

The matrix element of the process "absorption of a photon and creation of a phonon (absorption in the high-energy band)" is given by

$$\frac{e}{m}\left(\frac{\hbar}{2M\omega_{\underset{\sim}{k}j}}\right)^{\frac{1}{2}}\left(\frac{h}{\omega_{\underset{\sim}{\alpha}}V}\right)^{\frac{1}{2}}(\overline{n^{j}_{\underset{\sim}{k}}}+1)^{\frac{1}{2}}(n^{\lambda}_{\underset{\sim}{\alpha}})^{\frac{1}{2}}$$

$$\times\sum_{p}\left[\frac{\left\langle\psi^{el}_{f}\left|\sum_{t}e^{i\underset{\sim}{\alpha}\cdot\underset{\sim}{r}'_{t}}\underset{\sim}{p}'_{t}\cdot\underset{\sim}{\pi}^{\lambda}_{\underset{\sim}{\alpha}}\right|\psi^{el}_{p}\right\rangle\langle\psi^{el}_{p}|V^{j}_{\underset{\sim}{k}}|\psi^{el}_{i}\rangle}{E^{el}_{i}-(E^{el}_{p}+\hbar\omega_{\underset{\sim}{k}j})}\right.$$

$$\left.+\frac{\langle\psi^{el}_{f}|V^{j}_{\underset{\sim}{k}}|\psi^{el}_{p}\rangle\left\langle\psi^{el}_{p}\left|\sum_{t}e^{i\underset{\sim}{\alpha}\cdot\underset{\sim}{r}'_{t}}\underset{\sim}{p}'_{t}\cdot\underset{\sim}{\pi}^{\lambda}_{\underset{\sim}{\alpha}}\right|\psi^{el}_{i}\right\rangle}{E^{el}_{i}-(E^{el}_{p}-\hbar\omega_{\underset{\sim}{\alpha}})}\right]. \qquad (11.5.18)$$

The matrix element of the process "absorption of a photon and absorption of a phonon (absorption in the low-energy band)" is given by

$$\frac{e}{m}\left(\frac{\hbar}{2M\omega_{\underset{\sim}{k}j}}\right)^{\frac{1}{2}}\left(\frac{h}{\omega_{\underset{\sim}{\alpha}}V}\right)^{\frac{1}{2}}(\overline{n^{j}_{\underset{\sim}{k}}})^{\frac{1}{2}}(n^{\lambda}_{\underset{\sim}{\alpha}})^{\frac{1}{2}}$$

$$\times\sum_{p}\left[\frac{\left\langle\psi^{el}_{f}\left|\sum_{t}e^{i\underset{\sim}{\alpha}\cdot\underset{\sim}{r}'_{t}}\underset{\sim}{p}'_{t}\cdot\underset{\sim}{\pi}^{\lambda}_{\underset{\sim}{\alpha}}\right|\psi^{el}_{p}\right\rangle\langle\psi^{el}_{p}|V^{j}_{\underset{\sim}{k}}|\psi^{el}_{i}\rangle}{E^{el}_{i}-(E^{el}_{p}-\hbar\omega_{\underset{\sim}{k}j})}\right.$$

$$\left.+\frac{\langle\psi^{el}_{f}|V^{j}_{\underset{\sim}{k}}|\psi^{el}_{p}\rangle\left\langle\psi^{el}_{p}\left|\sum_{t}e^{i\underset{\sim}{\alpha}\cdot\underset{\sim}{r}'_{t}}\underset{\sim}{p}'_{t}\cdot\underset{\sim}{\pi}^{\lambda}_{\underset{\sim}{\alpha}}\right|\psi^{el}_{i}\right\rangle}{E^{el}_{i}-(E^{el}_{p}-\hbar\omega_{\underset{\sim}{\alpha}})}\right]. \qquad (11.5.19)$$

In the formulas above the photon has frequency $\omega_{\underset{\sim}{\alpha}}$, wave vector $\underset{\sim}{\alpha}$ and polarization $\underset{\sim}{\pi}^{\lambda}_{\underset{\sim}{\alpha}}$; the phonon has frequency $\omega_{\underset{\sim}{k}j}$, wave vector $\underset{\sim}{k}$, and belongs to the j-th branch. Also thermal averages have been introduced for the phonon occupation numbers; these thermal averages are responsible for the temperature dependence of the matrix element. It is clear that processes that involve the absorption of a phonon are not possible at very low temperatures; this means that the high-energy vibronic sidebands in emission and the low-energy vibronic sidebands in absorption disappear at low temperature.

The intensity of the radiative emission in the low-energy band in correspondence to wave vector $\underset{\sim}{\alpha}$ and polarization λ has the proportionality

$$I(\underset{\sim}{\alpha}, \lambda) \propto \sum_{\underset{\sim}{k}j} \int_{\Delta E_i^{el}} dE_i^{el} \int_{\Delta E_f^{el}} dE_f^{el} \rho(E_i^{el}) \rho(E_f^{el})$$
$$\times |M(\underset{\sim}{k}j; \underset{\sim}{\alpha}, \lambda)|^2 \rho(\omega_{\underset{\sim}{\alpha}}) \delta(E_i^{el} - E_f^{el} - \hbar\omega_{\underset{\sim}{k}j} - \hbar\omega_{\underset{\sim}{\alpha}}), \quad (11.5.20)$$

where $\rho(\omega_{\underset{\sim}{\alpha}})$ is the density of photon states. $\rho(\omega_{\underset{\sim}{\alpha}})$, as seen in (9.5.20), has a $\omega_{\underset{\sim}{\alpha}}^2$ dependence but can be considered constant over the range of the frequencies of the vibronic transitions. If the width of the purely electronic transition $i \to f$ is narrow in comparison to the widths of the peaks in the vibronic structure, then the electronic levels i and f are also narrow and $\rho(E_i^{el})$ and $\rho(E_f^{el})$ may be approximated with δ functions. This leads to the proportionality

$$I(\underset{\sim}{\alpha}, \lambda) \propto \sum_{\underset{\sim}{k}j} |M(\underset{\sim}{k}j; \underset{\sim}{\alpha}, \lambda)|^2 \, \delta(E_i^{el} - E_f^{el} - \hbar\omega_{\underset{\sim}{k}j} - \hbar\omega_{\underset{\sim}{\alpha}}). \quad (11.5.21)$$

The presence of the δ function indicates that the only phonon modes that may participate in the vibronic process are those with frequency

$$\Omega = \frac{1}{\hbar}(E_i^{el} - E_f^{el}) - \omega_{\underset{\sim}{\alpha}}. \quad (11.5.22)$$

In addition the selection rules for the matrix element $M(\underset{\sim}{k}j; \underset{\sim}{\alpha}\lambda)$ may further limit the modes allowed by the energy conservation.

Measuring the frequency of the vibronic emission from the frequency of the purely electronic transition, and also neglecting the individuality of modes in the matrix element we can write[5]

$$I \propto \int |M(\omega; \underset{\sim}{\alpha}\lambda)|^2 \, g(\omega)\delta(\omega - \Omega) d\omega = |M(\Omega; \underset{\sim}{\alpha}\lambda)|^2 g(\Omega). \quad (11.5.23)$$

The emission intensity is proportional to the product of two functions of the phonon frequency Ω, the density of phonon states, and the square of the matrix element. If the latter depends smoothly on Ω, maxima in the vibronic emission correspond to maxima in the phonon frequency distribution. In effect, however, the selection rules of the matrix element $M(\Omega; \underset{\sim}{\alpha}, \lambda)$ may limit the contribution of some modes. The effect of the selection rules will be particularly apparent if they do not allow the participation in the vibronic process by modes at a critical point of the Brillouin zone that corresponds to a peak in the phonon density of states.[5]

11.6. Selection Rules for Vibronic Transitions

Vibronic transitions that involve only one phonon depend essentially on a matrix element of the type

$$\langle \psi_f^{el} | e^{\pm i \underset{\sim}{\alpha} \cdot \underset{\sim}{r}} \underset{\sim}{p} \cdot \underset{\sim}{\pi} | \psi_p^{el} \rangle \langle \psi_p^{el} | V_{\underset{\sim}{k}}^{j} | \psi_i \rangle. \tag{11.6.1}$$

Selection rules can be derived from the following group-theoretical considerations:

1. The presence of the impurity ion destroys the translational invariance of the crystal.
2. The wavefunctions ψ_i^{el}, ψ_j^{el}, and ψ_f^{el} of the impurity ion transform according to irreducible representations of the point group S with origin at the ion site.
3. The normal coordinates in the perfect crystal form the basis for irreducible representations of the space group G of the crystal. In the presence of the impurity, the normal coordinates must form a basis for irreducible representations of the point group S. In general, the degeneracies of the perfect crystal modes will be partly lifted. The transformation properties of the normal coordinates in the presence of the impurity are found by reducing the space group representations of the normal coordinates to the irreducible representations of the point group S (see Ref. 6, p. 398). Such a reduction has been carried out by London for some space groups[7]; Wall[5] has also performed similar reductions.
4. The interaction Hamiltonian H_{ev} between the impurity ion and the lattice vibrations is invariant under all the operations of the point group S (i.e., transforms as the identity representation of S). Each term in the expansion

$$H_{ev} = \sum_{\underset{\sim}{k}j} V_{\underset{\sim}{k}}^{j} q_{\underset{\sim}{k}}^{j} + \sum_{\underset{\sim}{k}j} \sum_{\underset{\sim}{k}'j'} V_{\underset{\sim}{k}\underset{\sim}{k}'}^{jj'} q_{\underset{\sim}{k}}^{j} q_{\underset{\sim}{k}'}^{j'} + \cdots, \tag{11.6.2}$$

is also invariant under all the operations of S. But we know that the normal coordinate $q^j_{\underset{\sim}{k}}$ transforms according to an irreducible representation of S, say $\Gamma(q^j_{\underset{\sim}{k}})$; therefore the coefficient $V^j_{\underset{\sim}{k}}$ must transform according to the representation $\Gamma(q^j_{\underset{\sim}{k}})^*$. Similarly the coefficient $V^{jj'}_{\underset{\sim}{k}\underset{\sim}{k}'}$ transforms according to $\Gamma(q^j_{\underset{\sim}{k}})^* \times \Gamma(q^{j'}_{\underset{\sim}{k}'})^*$.

5. The factor $e^{i\underset{\sim}{\alpha}\cdot\underset{\sim}{r}}$ in the matrix element (11.6.1) may be expanded in series; the product $e^{i\underset{\sim}{\alpha}\cdot\underset{\sim}{r}}\underset{\sim}{p}$ will then result in the sum of electric dipole, and electric quadrupole operators, and so on. The electric dipole operator transforms as a polar vector, the magnetic dipole operator as an axial vector. Each component of an operator transforms according to an irreducible representation of S.

One can, in principle, ascertain which component of an operator is associated with a transition by conducting polarization studies.

The matrix element for a vibronic transition is characterized by (1) the initial electronic state of the impurity [representation $\Gamma(\psi^{el}_i)$], (2) the final electronic state of the impurity [representation $\Gamma(\psi^{el}_f)$], (3) a particular radiation operator [representation Γ(Rad op)], and (4) a particular mode of vibration [representation $\Gamma(q^j_{\underset{\sim}{k}})$]. On the basis of group theory a vibronic transition is allowed if

$$\Gamma(\psi^{el}_f) \in [\Gamma(\text{Rad op}) \times \Gamma(q^j_{\underset{\sim}{k}}) \times \Gamma(\psi^{el}_i)]. \qquad (11.6.3)$$

11.7. Effect of Temperature on the Position and Shape of a Purely Electronic Line

The effects of phonons on purely electronic (zero-phonon) lines of impurities in solids can be classified as follows:

1. The appearance of vibronic sidebands,
2. the change in the line position with temperature, and
3. the increase of the width of the line with increasing temperature.

We have already examined the first effect in the previous section; we shall now consider the other two thermal effects.

11.7.1. *Thermal Line Shift*

The thermal shift of a purely electronic line is the algebraic sum of the shifts of the two levels involved in the transition. We may then consider an ion in a level i and look for the mechanism that may shift the energy of this state. We may assume the energy of a level to be a thermodynamic quantity depending on the two independent variables T, the temperature, and V, the volume, of the crystal[8]:

$$\mathrm{E} = \mathrm{E}(\mathrm{T}, \mathrm{V}). \tag{11.7.1}$$

Therefore

$$\mathrm{dE} = \left(\frac{\partial \mathrm{E}}{\partial \mathrm{T}}\right)_{\mathrm{V}} \mathrm{dT} + \left(\frac{\partial \mathrm{E}}{\partial \mathrm{V}}\right)_{\mathrm{T}} \mathrm{dV}, \tag{11.7.2}$$

and

$$\left(\frac{\partial \mathrm{E}}{\partial \mathrm{T}}\right)_{\mathrm{P}} = \left(\frac{\partial \mathrm{E}}{\partial \mathrm{T}}\right)_{\mathrm{V}} + \left(\frac{\partial \mathrm{E}}{\partial \mathrm{V}}\right)_{\mathrm{T}} \left(\frac{\partial \mathrm{V}}{\partial \mathrm{T}}\right)_{\mathrm{P}}. \tag{11.7.3}$$

This means that the shift may be due partly to the thermal expansion of the crystal (that will produce a decrease in the strength of the crystalline field) and partly to an intrinsic temperature dependence. The latter effect can be estimated to some degree by knowing how the level shifts with hydrostatic pressure and by using comprensibility and thermal expansion data. The estimated shift cannot account for all the observed thermal shift and sometimes it is even of different sign than the observed shift.[8] An additional mechanism is evidently at work.

The ion-vibration interaction Hamiltonian [see (11.5.11)] has form that is similar in some ways to that of the ion-radiation interaction Hamiltonian [see (11.5.10)]; both Hamiltonians have terms linear in the field and terms quadratic in the field. In the radiative case the interaction Hamiltonian is not diagonal in the first order in the field $\underset{\sim}{\mathrm{A}}$, but in second order accounts for the electromagnetic (Lamb) shift (see Ref. 6, p. 339). In a similar way the ion-vibration interaction Hamiltonian H_{ev} is not diagonal in the first order in $\mathrm{q}_{\underset{\sim}{\mathrm{k}}}$, but in second order may be diagonal and may contribute to the energy of the systems. The result is a temperature-dependent contribution to the energy of the system,

$$\delta E_i = \sum_j \frac{\langle i|H'|j\rangle\langle j|H'|i\rangle}{E_i - E_j} + \langle i|H''|i\rangle, \qquad (11.7.4)$$

where H' and H'' are the parts of H_{ev} linear and quadratic in the normal coordinates, respectively, and where $|i\rangle$ and $|j\rangle$ represent states of the (ion + vibrations) system. The details of the calculation of δE_i will not be given here, but they can be found in Ref. 6, p. 372; the result of the calculation* is

$$\delta E = \alpha \left(\frac{T}{T_D}\right)^4 \int_o^{T_D/T} \frac{x^3}{e^x - 1} dx, \qquad (11.7.5)$$

where T_D is the Debye temperature of the crystal and α is an adjustable parameter, independent of temperature. A table of is also given in Ref. 6.

In Section 6.4 we derived for the total heat of the crystal the expression [see (6.4.40)]

$$E(T) = 9N\,kT \left(\frac{T}{T_D}\right)^3 \int_o^{T_D/T} \frac{x^3}{e^x - 1} dx. \qquad (11.7.6)$$

Therefore one would expect the same thermal dependence for the thermal line shift and for the total heat of the crystal; this fact has been confirmed experimentally.[9]

11.7.2. *Thermal Broadening of Sharp Lines*

The shapes and widths of purely electronic lines of impurities in solid present the following general pattern of behavior:

1. At low temperatures ($T < 77°K$) the width of the line appear to be independent of temperature; in the same temperature range the line presents a profile (called the *Voigt profile*) that is the convolution of a Lorentzian and a Gaussian frequency distribution (see Ref. 6, p. 362).

*In this calculation a simplified form for V_{ev} is assumed; in particular it is assumed that each $v^j_{\underset{\sim}{k}}$ is proportional to $\omega_{\underset{\sim}{k}}$ and that the phonon spectrum has a Debye frequency distribution.

2. At higher temperatures the width of the line grows rapidly; for temperatures sufficiently high (≈300°K),[10] the profile of the line is practically Lorentzian.

These general features can be explained by considering that at low temperatures the line is actually the superposition of many spectral lines at slightly different center frequencies. These different lines correspond to impurities, which, being at different locations in the crystal, sit in slightly different environments because of the presence of internal strains, other impurities, and point defects.[11] At low temperature the homogeneous broadening of each of these lines is very small and the profile of the compound spectral line reflects the spatial distribution of the inhomogeneities in the crystal. If this distribution, as is often the case, is random, the shape of the compound line should be Gaussian; the small but finite homogeneous (Lorentzian) broadening of the individual lines causes the Gaussian curve to go over to a Voigt profile. The Voigt profile is given by a normalized function

$$f(\nu) = \frac{2\ln 2}{\pi\sqrt{\pi}}\frac{\Delta\nu_L}{\Delta\nu_G^2}\int_{-\infty}^{+\infty}\frac{e^{-y^2}}{a^2+(b-y)^2}dy, \qquad (11.7.7)$$

where

$$b = \frac{2(\nu-\nu_o)\sqrt{\ln 2}}{\Delta\nu_G}, \qquad a = \frac{\Delta\nu_L}{\Delta\nu_G}\sqrt{\ln 2},$$

and ν represents frequency and ν_o is the center frequency. For $\Delta\nu_G = 0$, $f(\nu)$ reduces to

$$f_L(\nu) = \frac{\Delta\nu_L}{2\pi}\frac{1}{(\nu-\nu_o)^2+\left(\frac{\Delta\nu_L}{2}\right)^2}, \qquad (11.7.8)$$

which is a normalized Lorentzian frequency distribution; for $\Delta\nu_L = 0$, $f(\nu)$ reduces to

$$f_G(\nu) = \frac{2\sqrt{\ln 2}}{\Delta\nu_G\sqrt{\pi}}\exp\left[-\frac{2(\nu-\nu_o)}{\Delta\nu_G}\sqrt{\ln 2}\right]^2, \qquad (11.7.9)$$

which is a normalized Gaussian frequency distribution. Therefore $\Delta\nu_L$ and $\Delta\nu_G$ are parameters related to the Lorentzian and Gaussian components of the line, respectively. Tables for the Voigt profile for various values of $\Delta\nu_L$ and $\Delta\nu_G$ have been prepared by Posener.[12]

The parameter $\Delta\nu_G$ is in a way the measure of the inhomogeneities in the crystal; it is not temperature dependent and may be different for two chemically identical crystals doped with the same impurity, but grown differently. On the other hand the parameter $\Delta\nu_L$ is expected to increase with temperature. This explains the general behavior of the spectral line:

1. At low temperature $\Delta\nu_L$ is small but finite, and $\Delta\nu_G$ is certainly different from zero; the result is a Voigt line, which is the closer in shape to a Gaussian line, the smaller is $\Delta\nu_L$ in comparison to $\Delta\nu_G$.
2. At higher temperatures $\Delta\nu_L$ increases so as to become much greater than $\Delta\nu_G$; the result is practically a Lorentzian line.

We have still to calculate the thermal dependence of $\Delta\nu_L$. For this purpose it is proper to consider the thermal broadening of a single electric level of an impurity as the thermal width is the sum of the thermal widths of the two levels involved in the transition. It may be expected that of the two levels, the excited level is broadened by decay processes radiative and radiationless that tend to remove the ion from this level; the inverse of the sum of the transition rates of all these decay processes is the lifetime (τ) of the excited level. Simply on the basis of the Heisenberg principle the broadening of an excited level should be

$$\Delta E\,(\mathrm{cm}^{-1}) \simeq \frac{5.3 \times 10^{-12}}{\tau\,(\mathrm{sec})}. \qquad (11.7.10)$$

Consider now what is perhaps the best-known purely electronic line: the R_1 (laser) line of ruby ($Al_2O_3 : Cr^{3+}$). This line corresponds to a radiative transition to the ground state from a metastable state that has a lifetime of ~3 msec at room temperature. The corresponding lifetime broadening is only about $2 \times 10^{-9}\,\mathrm{cm}^{-1}$, whereas in effect a linewidth of about $10\,\mathrm{cm}^{-1}$ is observed. Similar observations can in general be made for impurity zero-phonon lines; this shows that (1) lifetime broadening cannot account for the width of the line and (2) the processes responsible for the broadening of the excited level must be such that they do *not* remove the ion from this level.

On the basis of the above considerations, one can reflect on the fact that the variations of the crystalline field caused by the lattice vibrations are in effect adiabatic (or slow), since the frequency of an activated lattice mode is in general much smaller than the frequency of an optical transition. For this reason a process that allows the Raman scattering of phonons by the ion in the excited state may take place without removing the ion from this state and without affecting the lifetime of the state.

Such processes, which consist of the absorption of a phonon and the emission of another phonon, can be treated by considering again the Hamiltonian H_{ev} of (11.5.11), which represents the interaction between the ion and the lattice vibrations. The "system" in this case, as in the previous case of the thermal shift, consists of ion plus vibrations. In the case of the thermal shift the diagonal matrix element of H_{ev} was the relevant matrix element; in the present case we have to consider the off-diagonal matrix elements

$$\langle f|H_{ev}|i\rangle = \langle \psi_i^{el}; n_{\underset{\sim}{k}} - 1; n_{\underset{\sim}{k}'} + 1|H_{ev}|\psi_i^{el}; n_{\underset{\sim}{k}}; n_{\underset{\sim}{k}'}\rangle. \quad (11.7.11)$$

The total probability W per unit time of all Raman processes can be obtained by summing over all $\underset{\sim}{k}$ and $\underset{\sim}{k}'$; the details of this calculation are not given here but may be found in Ref. 4, p. 391. The resultant width, given by $\sim$hW, is

$$\Delta E(T) = \bar{\alpha}\left(\frac{T}{T_D}\right)^7 \int_o^{T_D/T} \frac{x^6 e^x}{(e^x - 1)^2} dx, \quad (11.7.12)$$

where T_D is the Debye temperature of the crystal and $\bar{\alpha}$ is an intrinsically positive, adjustable parameter. A table of $\Delta E(T)/\bar{\alpha}$ is also given in Ref. 6.

In general, the magnitude of the thermal width and of the thermal shift of purely electronic lines are comparable[8]; the experimental evidence points to the fact that the coupling coefficients $|\alpha|$ and $\bar{\alpha}$ are for rare earth ions $\lesssim 100\,cm^{-1}$, whereas for sharp lines of transition metal ions they are $\gtrsim 300\,cm^{-1}$.[13] The thermal broadening and shifting of spectral lines is important for solid state lasers as discussed in Chapter 13.

References

1. R. A. Preem, *Trudy Inst. Fix. Astron. Akad. Nauk. Eston.* SSR No. 16, 57 (1961).
2. D. L. Dexter, C. C. Klick, and G. A. Russell, *Phys. Rev.* 100, 603 (1955).
3. D. S. McClure, "Electronic Spectra of Molecules and Ions in Crystals. Part II. Spectra of Ions in Crystals," in *Solid State Physics*, Vol. 9, edited by F. Seitz and D. Turnbull, Academic, New York, 1959, p. 399.
4. Y. Tanabe and S. Sugano, *J. Phys. Soc. Japan* 9, 753 (1954), *ibid.* 9, 766 (1954); *ibid.* 11, 864 (1956).
5. W. A. Wall, "Selection Rules for Vibronic Transitions by Ion Impurities in Solids," Doctoral Dissertation, Boston College, 1973, unpublished.
6. B. Di Bartolo, *Optical Interactions in Solids*, Wiley, New York, 1968.
7. R. Loudon, *Proc. Phys. Soc. (London)* 84, 379 (1964).
8. D. B. Fitchen, "Zero Phonon Transitions," in *Physics of Color Centers*, edited by W. Beall Fowler, Academic, New York, 1964, p. 293.
9. G. F. Imbusch, W. M. Yen, A. L. Schawlow, D. E. McCumber, and M. D. Sturge, *Phys. Rev.* 133, A1029 (1964).
10. B. Di Bartolo and R. Peccei, *Phys. Rev.* 137, A1770 (1965).
11. K. K. Rebane, *Impurity Spectra of Solids*, Plenum, New York, 1970, p. 150.
12. D. W. Posener, *Austral. J. Phys.* 12, 184 (1959).
13. J. T. Karpick and B. Di Bartolo, *Nuovo Cimento* 7B, 62 (1972).

Chapter 12

Interaction of Light with Lattice Vibrations: Infrared Absorption and Inelastic Light Scattering

As well as interacting with the electronic states of materials, electromagnetic radiation can also interact directly with the vibrational states. The interaction can occur in either of two ways. The first is through absorption of light. Since this is a resonant process, the frequency of the electromagnetic radiation must have photons with energies equal to the energy of the transition between vibrational states. This will occur in the infrared region of the electromagnetic spectrum. From a quantum mechanical perspective, infrared absorption processes involve the destruction of a photon with the concurrent creation of a phonon. Infrared absorption spectra provide information about the phonons in the material. The second type of interaction involves the inelastic scattering of light. For this case, the electromagnetic radiation can be in the visible region of the spectrum. The outgoing light beam (scattered light) will have a component with a frequency different from that of the incoming light beam. The difference in the frequencies of these beams will be equal to the frequency of one of the phonon modes of the material. The process of transferring energy between the incoming light energy and the vibrational energy of the material is known as *Raman scattering* for optical phonons and *Brillouin scattering* for acoustic phonons. Raman and Brillouin spectroscopy provide information about the phonon modes of the material.

This chapter develops the formalisms use to describe infrared absorption, Raman spectroscopy and Brillouin spectroscopy. Because of the differences in selection rules and conservation conditions between these processes, they provide complementary information about the vibrational states of the material. The data obtained in by these spectroscopic techniques are helpful in understanding the vibrational properties of materials described in Part II of this book.

12.1. General Characteristics of Infrared Absorption by Crystals

Infrared spectroscopy provides a technique for directly studying the vibrational characteristics of solids. In this case the phonons themselves are the object of the investigation and the resonant transitions of interest are those taking place between the vibrational levels of the electronic ground state of the solid. The energy of these transitions lies in the infrared region of the spectrum which ranges in wavelength from approximately $0.75\,\mu$ to 1 mm. The process of infrared absorption by lattice vibrations consists in the absorption of a photon accompanied by the production of one or more phonons.

We have already seen in the previous two chapters how the electrons of an atom in a solid may interact with the electromagnetic radiation to bring about the "optical spectra." Infrared absorption may be simply thought of as due to the interaction of the radiative field with the motion of the nuclei, that is, lattice vibrations; since the nuclear motions take place at lower frequencies than the electronic motion, one can expect spectra at longer wavelengths than the optical spectra. In considering the infrared spectra of solids one should, however, be aware of the fact that when the problem is stated in more exact terms the electrons have also to be considered; in order to elucidate this point we consider in the next section the infrared spectra of a molecular system.

12.2. Infrared Transitions in a Molecular System

Consider a molecular system in which two quantum states are defined, in the adiabatic approximation, by

$$\psi_{\rm i}(\underset{\sim}{r}, \underset{\sim}{R}) = \psi_{\rm k}(\underset{\sim}{r}, \underset{\sim}{R})\phi_{\rm kl}(\underset{\sim}{R}) \tag{12.2.1}$$

$$\psi_{\rm f}(\underset{\sim}{r}, \underset{\sim}{R}) = \psi_{\rm m}(\underset{\sim}{r}, \underset{\sim}{R})\phi_{\rm mn}(\underset{\sim}{R}) \tag{12.2.2}$$

The transition probability for a radiative transition between these two states in the dipole approximation is proportional to the square of the matrix element:

$$M_{fi} = \langle \underset{\sim}{\pi}_\alpha^\lambda \cdot \underset{\sim}{M} \rangle_{fi}. \tag{12.2.3}$$

The electric dipole operator $\underset{\sim}{M}$ is given by

$$\underset{\sim}{M} = -e \sum_i \underset{\sim}{r}_i + e \sum_\alpha Z_\alpha \underset{\sim}{R}_\alpha, \tag{12.2.4}$$

where $\underset{\sim}{r}_i$ is the coordinate of the i-th electron, $\underset{\sim}{R}_\alpha$ is the coordinate of the α-th nucleus, and Z_α is the charge of the α-th nucleus. From (12.2.4)

$$\begin{aligned} \underset{\sim}{\pi} \cdot \underset{\sim}{M} &= -e \sum_i \underset{\sim}{r}_i \cdot \underset{\sim}{\pi} + e \sum_\alpha Z_\alpha \underset{\sim}{R}_\alpha \cdot \underset{\sim}{\pi} \\ &= D_e(\underset{\sim}{r}) + D_n(\underset{\sim}{R}), \end{aligned} \tag{12.2.5}$$

and, in accordance with (10.6.13),

$$\begin{aligned} M_{fi} &= M_{mn;kl} \\ &= \int d^3\underset{\sim}{R}\phi_{mn}(\underset{\sim}{R})^* D_{mk}(\underset{\sim}{R})\phi_{kl}(\underset{\sim}{R}) \\ &\quad + \int d^3\underset{\sim}{R}\phi_{mn}(\underset{\sim}{R})^* D_n(\underset{\sim}{R})\phi_{kl}(\underset{\sim}{R})\delta_{mk}, \end{aligned} \tag{12.2.6}$$

where

$$D_{mk}(\underset{\sim}{R}) = d^3\underset{\sim}{r}\psi_m(\underset{\sim}{r}, \underset{\sim}{R})^* D_e(\underset{\sim}{r})\psi_k(\underset{\sim}{r}, \underset{\sim}{R}). \tag{12.2.7}$$

If we assume now that the two states $\psi_i(\underset{\sim}{r}, \underset{\sim}{R})$ and $\psi_f(\underset{\sim}{r}, \underset{\sim}{R})$ correspond to the same electronic state ($m = k$) we can write

$$\begin{aligned} M_{fi} &= M_{kn;kl} \\ &= \int d^3\underset{\sim}{R}\phi_{kn}(\underset{\sim}{R})^* D_{kk}(\underset{\sim}{R})\phi_{kl}(\underset{\sim}{R}) \\ &\quad + \int d^3\underset{\sim}{R}\phi_{kn}(\underset{\sim}{R}) D_n(\underset{\sim}{R})\phi_{kl}(\underset{\sim}{R}) \end{aligned}$$

$$= \int d^3\underset{\sim}{R}\left[\int d^3\underset{\sim}{r}\psi_k(\underset{\sim}{r},\underset{\sim}{R})^* D_e(\underset{\sim}{r})\psi_k(\underset{\sim}{r},\underset{\sim}{R})\right]\phi_{kn}(\underset{\sim}{R})^*\phi_{kl}(\underset{\sim}{R})$$
$$+\int d^3\underset{\sim}{R}\phi_{kn}(\underset{\sim}{R})^* D_n(\underset{\sim}{R})\phi_{kl}(\underset{\sim}{R}). \qquad (12.2.8)$$

In this matrix element the first term contributes only if the system lacks a center of symmetry [this is because $D_e(\underset{\sim}{r})$ is an odd operator]; the second term is present only because there is no change in the electronic state for the transitions now being considered.

One could think at this point that, since a crystal is nothing but a "big" molecule, we could extend the present treatment to the case in which similar transitions take place while the atoms remain in their (ground) electronic state. A difficulty, however, arises: the wavelength of the light is *not* in this case much greater than the "size of the system." The system now consists of the vibrational oscillators that spread throughout the crystal. A consequence of this fact is that the usual electric dipole approximation on which the above treatment is based cannot be applied and the problem of the interaction of radiation with the "system" has to be considered anew.

12.3. Momentum and Energy Conservation in Infrared Absorption

In the process that consists of the absorption of a photon and the production of a phonon it is necessary for momentum (and, therefore, wave vector) to be conserved. The conservation of momentum (which will be discussed in detail in the next section) is expressed as

$$\underset{\sim}{\alpha} = \underset{\sim}{k} + \underset{\sim}{K}, \qquad (12.3.1)$$

where $\underset{\sim}{\alpha}$ is the photon wave vector, $\underset{\sim}{k}$ is the phonon wave vector, and $\underset{\sim}{K}$ is the primitive vector of the reciprocal lattice.

The conservation of energy, on the other hand, requires the phonon angular frequency to be

$$\omega = c|\underset{\sim}{\alpha}|, \qquad (12.3.2)$$

where c is the velocity of light.

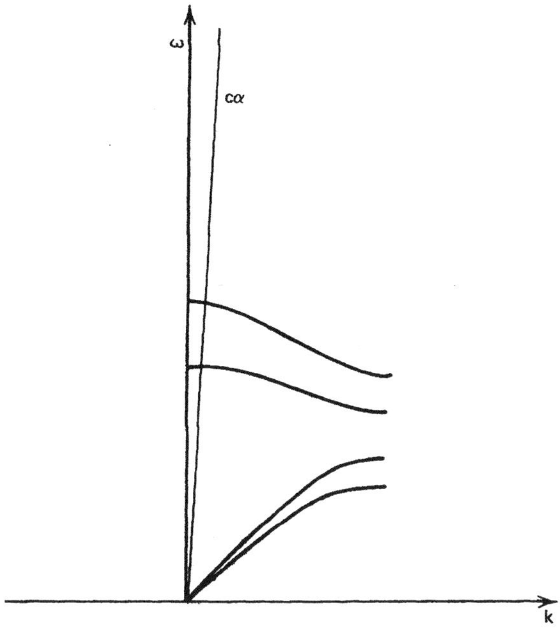

Fig. 12.1. Diagram illustrating conservation laws.

Conditions (12.3.1) and (12.3.2) are illustrated in Fig. 12.1, in which the intersections of the line $c\alpha(\alpha = |\underset{\sim}{\alpha}|)$ with different $\omega(\underset{\sim}{k})$ lines gives the allowed values of α. Since $c \sim 10^5$ times the velocity of sound, the line representing cα is very steep and therefore only very small values of α are allowed. It is also evident that $\underset{\sim}{K} = 0$ in (12.3.1) and that the acoustical branches cannot contribute to the absorption process.

Considering that for $\underset{\sim}{k} \approx 0$, ω in correspondence to the optical branches is of the order 10^{13} to 10^{14} sec^{-1}, the conservation law requires the absorbed photon wave vector to be about $10^3\,\mathrm{cm}^{-1}$, while a phonon wave vector at the boundary of the Brillouin zone has a value of about $10^8\,\mathrm{cm}^{-1}$.

In order to clarify the mechanism by which electromagnetic radiation interacts with "optical" phonons, it is worthwhile to consider the simple case of a linear crystal whose unit cell consists of two ions of opposite charges and unequal masses. A similar system was considered in Section 6.4 and Fig. 6.4. In correspondence to $\underset{\sim}{k} \approx 0$ the optical mode of vibration presents a pattern of motion (see Fig. 6.6b) in which the center of mass of the unit cell is stationary and the two differently charged ions move with respect to each other creating a fluctuating dipole moment of frequency $\omega(\underset{\sim}{k} = 0)$ with which the radiative field can interact.

Under the action of the radiative field the equations of motion already considered in (6.4.15) become

$$\begin{aligned} M\ddot{u}_{n1} &= \beta u_{n2} + \beta u_{n-1,2} - 2\beta u_{n1} + eE_o e^{-i\omega t} \\ m\ddot{u}_{n2} &= \beta u_{n+1,1} + \beta u_{n1} - 2\beta u_{n2} + eE_o e^{-i\omega t}, \end{aligned} \tag{12.3.3}$$

where we have assumed that the ions have unit charge. In order to decouple these equations we look for solutions of the type given in (6.4.7):

$$u_{s\nu} = V_\nu e^{-i\omega t + ikR_s}, \tag{12.3.4}$$

where V_ν is polarization vector and R_s indicates the position of the sth unit cell. These equations are greatly simplified if we restrict ourselves to the case $k = 0$ of interest here. Taking this condition into account, equations (12.3.3) become

$$\begin{aligned} -\omega^2 M V_1 &= 2\beta(V_2 - V_1) + eE_o \\ -\omega^2 m V_2 &= -2\beta(V_2 - V_1) - eE_o. \end{aligned} \tag{12.3.5}$$

From these equations we derive

$$V_1 = -\frac{m}{M} V_2 \tag{12.3.6}$$

in agreement with (6.4.30). Also

$$\begin{aligned} V_1 &= \frac{eE_o}{M}\left[\frac{1}{\omega_o^2 - \omega^2}\right] \\ V_2 &= -\frac{eE_o}{m}\left[\frac{1}{\omega_o^2 - \omega^2}\right], \end{aligned} \tag{12.3.7}$$

where

$$\omega_0^2 = 2\beta\left[\frac{1}{M} + \frac{1}{m}\right]. \tag{12.3.8}$$

Comparing this result with Fig. 6.5 one can see that ω_o corresponds to the $k = 0$ limit of the optical branch. According to (12.3.7) the process of infrared absorption will present a maximum for $\omega = \omega_o$.

The motions of ions of opposite charges exemplified above with the ions moving towards each other within the unit cell but the center of mass remaining stationary bring about the characteristic infrared absorption of ionic crystals. The order of magnitude of the wavelength of such absorption can be easily calculated for a crystal like NaCl, for which Kittel[1] gives $\beta \approx 1.5 \times 10^4$ dynes/cm as derived from elastic stiffness data; from (12.3.8),

$$\omega_o^2 = 2 \times 1.5 \times 10^4 \times \left[\frac{1}{35.5} + \frac{1}{23}\right] 6 \times 10^{23} \simeq 13 \times 10^{26},$$

and

$$\omega_o \approx 3.6 \times 10^{13}\,\mathrm{sec}^{-1},$$

or

$$\lambda_o = \frac{2\pi c}{\omega_o} \approx 50\,\mu$$

in reasonable agreement with the experimental value $\lambda_o = 61\,\mu$. This absorption in ionic crystals is called *Reststrahl absorption.*

12.4. Quantum Theory of Infrared Absorption

Let us consider a perfect crystal with J atoms in the unit cell and N unit cells. At one particular instant in time the electron coordinates will be given by $\underset{\sim}{r}_i$ $(i = 1, 2, \ldots, n)$ and the nuclear coordinates by

$$\underset{\sim}{X}_{s\nu} = \underset{\sim}{R}_{s\nu} + \underset{\sim}{u}_{s\nu}, \tag{12.4.1}$$

where $\underset{\sim}{R}_{s\nu}$ is the equilibrium position of the ν-th atom of the s-th unit cell, $\underset{\sim}{u}_{s\nu}$ is the displacement from the equilibrium position of the ν-th atom of the s-th unit cell, and

$$s = 1, 2, \ldots, N, \quad \nu = 1, 2, \ldots, J.$$

The total number of atoms in the crystal is NJ.

The wave function of the crystal, in the adiabatic approximation, is given by

$$\psi(\underset{\sim}{r}, \underset{\sim}{X}) = \psi(\underset{\sim}{r}, \underset{\sim}{X})\phi(\underset{\sim}{X}), \tag{12.4.2}$$

where $\underset{\sim}{r}$ stands for the coordinates of the electrons and $\underset{\sim}{X}$ stands for the coordinates of the nuclei. The electronic coordinates now include *all* the electrons in the crystals. In order to treat properly the electronic wave functions it is necessary to specify that we shall be dealing with two types of crystals[2]:

a. ionic crystals (in these crystals the electrons are attached to the various ions and form closed shells), and
b. homopolar crystals (in which the electrons belong to filled valence bands).

In both cases the determinantal wave function can always be expressed as determinant of localized orbitals; also, since we will be dealing with one-body operators, the wave function can be expressed as a Hartree product of localized orbitals.[2]

We now examine the problem of infrared absorption by considering the system that consists of (crystal plus radiative field). The Hamiltonian of the system can be written

$$H = H_{cryst} + H_{em} + H', \tag{12.4.3}$$

where

$$\begin{aligned} H_{cryst} &= H_e + H_v \\ &= \left[-\frac{\hbar^2}{2m}\sum_{i=1}^{n}\nabla_i^2 + V(\underset{\sim}{r}, \underset{\sim}{X})\right] + \left[-\frac{\hbar^2}{2}\sum_{s\nu}^{NJ}\frac{\nabla_{s\nu}^2}{M_\nu} + \varepsilon_k(\underset{\sim}{X})\right], \end{aligned} \tag{12.4.4}$$

$$H_{em} = \sum_{\underset{\sim}{\alpha}}\sum_{\lambda}\hbar\omega_{\underset{\sim}{\alpha}}\left[a_{\underset{\sim}{\alpha}}^{\lambda+}a_{\underset{\sim}{\alpha}}^{\lambda} + \frac{1}{2}\right], \tag{12.4.5}$$

$$H' = \sum_{i=1}^{n}\frac{e}{mc}\underset{\sim}{A}(\underset{\sim}{r}_i)\cdot\underset{\sim}{p}_i - \sum_{s\nu}^{NJ}\frac{z_\nu e}{M_\nu c}\underset{\sim}{A}(\underset{\sim}{X}_{s\nu})\cdot\underset{\sim}{P}_{s\nu}. \tag{12.4.6}$$

Considering the process of "absorption of one photon" of wave vector $\underset{\sim}{\alpha}$ and polarization λ, the initial and final wave functions of the system are given by

$$\psi_i = \underbrace{|\psi_i(\underset{\sim}{r}, \underset{\sim}{X}) > |\phi_i(\underset{\sim}{X}) >}_{\text{crystal}} \quad \underbrace{|n_1 > |n_2 > \cdots |1^{\lambda}_{\underset{\sim}{\alpha}} > \cdots}_{\text{radiative field}}, \tag{12.4.7}$$

$$\psi_f = |\psi_i(\underset{\sim}{r}, \underset{\sim}{X}) > |\phi_f(\underset{\sim}{X}) > |n_1 > |n_2 > \cdots |0^{\lambda}_{\underset{\sim}{\alpha}} > \cdots, \tag{12.4.8}$$

where, for simplicity, we have assumed that only one photon of wave vector $\underset{\sim}{\alpha}$ and polarization $\pi^{\lambda}_{\underset{\sim}{\alpha}}$ is originally present in the field.

The vector field $\underset{\sim}{A}$ can be expressed, according to (9.2.8), by

$$\underset{\sim}{A}(\underset{\sim}{r}) = \sum_{\underset{\sim}{\alpha}} \sum_{\lambda} \left[\frac{hc^2}{\omega_{\underset{\sim}{\alpha}} V}\right]^{1/2} \underset{\sim}{\pi}^{\lambda}_{\underset{\sim}{\alpha}} [e^{i\underset{\sim}{\alpha}\cdot\underset{\sim}{r}} a^{\lambda}_{\underset{\sim}{\alpha}} + e^{-i\underset{\sim}{\alpha}\cdot\underset{\sim}{r}} a^{\lambda+}_{\underset{\sim}{\alpha}}]. \tag{12.4.9}$$

The interaction Hamiltonian H′ can now be expressed as

$$H' = \sum_{\underset{\sim}{\alpha}} \sum_{\lambda} \left[\frac{hc^2}{\omega_{\underset{\sim}{\alpha}} V}\right]^{1/2} \underset{\sim}{\pi}^{\lambda}_{\underset{\sim}{\alpha}} \cdot \left[\sum_{i=1}^{n} \frac{e}{mc} e^{i\underset{\sim}{\alpha}\cdot\underset{\sim}{r}_i} - \sum_{s\nu}^{NJ} \frac{Z_{\nu} e}{M_{\nu} c} e^{i\underset{\sim}{\alpha}\cdot\underset{\sim}{X}_{s\nu}} \underset{\sim}{P}_{s\nu}\right] a^{\lambda}_{\underset{\sim}{\alpha}}, \tag{12.4.10}$$

where we have neglected terms with a creation photon operator. Changing the notation for the electronic coordinates, by identifying each electron with three subscripts, two of which give the position of the related nucleus, and taking (9.5.7) into account,

$$\begin{aligned} H' = {} & \sum_{\underset{\sim}{\alpha}} \sum_{\lambda} \left[\frac{hc^2}{\omega_{\underset{\sim}{\alpha}} V}\right]^{1/2} \underset{\sim}{\pi}^{\lambda}_{\underset{\sim}{\alpha}} \cdot \sum_{s\nu} \left[\sum_{t} \frac{e}{mc} e^{i\underset{\sim}{\alpha}\cdot\underset{\sim}{r}_{s\nu t}} i m \omega_{\underset{\sim}{\alpha}} \underset{\sim}{r}_{s\nu t} \right. \\ & \left. - \frac{Z_{\nu} e}{M_{\nu} c} e^{i\underset{\sim}{\alpha}\cdot\underset{\sim}{X}_{s\nu}} i M_{\nu} \omega_{\underset{\sim}{\alpha}} \underset{\sim}{X}_{s\nu}\right] a^{\lambda}_{\underset{\sim}{\alpha}} \\ \approx {} & \sum_{\underset{\sim}{\alpha}} \sum_{\lambda} \left[\frac{h\omega_{\underset{\sim}{\alpha}}}{V}\right]^{1/2} i \underset{\sim}{\pi}^{\lambda}_{\underset{\sim}{\alpha}} \cdot \sum_{s\nu} \left[e^{i\underset{\sim}{\alpha}\cdot\underset{\sim}{R}_{s\nu}} \left(\sum_{t} e\underset{\sim}{r}_{s\nu t} - Z_{\nu} e \underset{\sim}{X}_{s\nu}\right)\right] a^{\lambda}_{\underset{\sim}{\alpha}}. \end{aligned} \tag{12.4.11}$$

In the last passage we have made use of the approximation

$$e^{i\underset{\sim}{\alpha}\cdot\underset{\sim}{r}_{s\nu t}} \approx e^{i\underset{\sim}{\alpha}\cdot\underset{\sim}{X}_{s\nu}} \approx e^{i\underset{\sim}{\alpha}\cdot\underset{\sim}{R}_{s\nu}} \tag{12.4.12}$$

The matrix element of the interaction Hamiltonian taken between the initial and the final state of the system is given by

$$\begin{aligned}\langle f|H'|i\rangle &= \langle kn; kl\rangle \\ &= i\left[\frac{h\omega_{\underset{\sim}{\alpha}}}{V}\right]^{1/2} \underset{\sim}{\pi}_{\underset{\sim}{\alpha}}^{\lambda}\cdot\left[\langle\phi_f(\underset{\sim}{X})|\langle\psi_i(\underset{\sim}{r},\underset{\sim}{x})|\sum_{s\nu} e^{i\underset{\sim}{\alpha}\cdot\underset{\sim}{R}_{s\nu}}\right.\\ &\quad\times\left.\sum_t e\underset{\sim}{r}_{s\nu t} - Z_\nu e X_{s\nu}|\psi_i(\underset{\sim}{r},\underset{\sim}{X})\rangle\phi_i(\underset{\sim}{X})\rangle\right]\\ &= i\left[\frac{h\omega_{\underset{\sim}{\alpha}}}{V}\right]^{1/2} \underset{\sim}{\pi}_{\underset{\sim}{\alpha}}^{\lambda}\cdot\langle\phi_f(\underset{\sim}{X})|M_{ii}(\underset{\sim}{\alpha},\underset{\sim}{X})|\phi_i(\underset{\sim}{X})\rangle,\end{aligned} \tag{12.4.13}$$

where

$$\begin{aligned}\underset{\sim}{M}_{ii}(\underset{\sim}{\alpha},\underset{\sim}{X}) &= \sum_{s\nu} e^{i\underset{\sim}{\alpha}\cdot\underset{\sim}{R}_{s\nu}}\langle\psi_i(\underset{\sim}{r},\underset{\sim}{X})\left|\sum_t e\underset{\sim}{r}_{s\nu t} - Z_\nu e\underset{\sim}{X}_{s\nu}\right|\psi_i(\underset{\sim}{r},\underset{\sim}{X})\rangle\\ &= \sum_{s\nu} e^{i\underset{\sim}{\alpha}\cdot\underset{\sim}{R}_{s\nu}}\int \psi_i(\underset{\sim}{r},\underset{\sim}{X})^*\left[\sum_t e\underset{\sim}{r}_{s\nu t} - Z_\nu e\underset{\sim}{X}_{s\nu}\right]\psi_i(\underset{\sim}{r},\underset{\sim}{X})d^3\underset{\sim}{r}\\ &= \sum_{s\nu} e^{i\underset{\sim}{\alpha}\cdot\underset{\sim}{R}_{s\nu}}\left[-eZ_\nu\underset{\sim}{X}_{s\nu} + \int\psi_i(\underset{\sim}{r},\underset{\sim}{X})^*\sum_t e\underset{\sim}{r}_{s\nu t}\psi_i(\underset{\sim}{r},\underset{\sim}{X})d^3\underset{\sim}{r}\right].\end{aligned} \tag{12.4.14}$$

It is interesting to compare (12.4.13) with the similar formula (12.2.8) that applies to molecular system; we note that

1. Both formulas contain two terms, one corresponding to the interaction of the radiative field with the electrons and the other corresponding to the interaction of the radiative field with the nuclei.

2. The only difference between the two formulas is in the factor $e^{i\underset{\sim}{\alpha}\cdot\underset{\sim}{R}_{s\nu}}$. It is clear that the two formulas become equal if this factor is set equal to 1.

We proceed with the present treatment by making the following change of variables:

$$\begin{aligned} \underset{\sim}{\rho}_{s\nu t} &= \underset{\sim}{r}_{s\nu t} - \underset{\sim}{X}_{s\nu} \\ \psi_i(\underset{\sim}{r}, \underset{\sim}{X}) &= \psi_i(\underset{\sim}{\rho} + \underset{\sim}{X}, \underset{\sim}{X}) = \chi_i(\underset{\sim}{\rho}, \underset{\sim}{X}). \end{aligned} \tag{12.4.15}$$

Then

$$\begin{aligned} \underset{\sim}{M}_{ii}(\underset{\sim}{\alpha}, \underset{\sim}{X}) = \sum_{s\nu} e^{i\underset{\sim}{\alpha}\cdot\underset{\sim}{R}_{s\nu}} \Bigg\{ &-eZ_\nu \underset{\sim}{X}_{s\nu} + e \int \chi_i(\underset{\sim}{\rho}, \underset{\sim}{X})^* \\ &\times \sum_t [\underset{\sim}{\rho}_{s\nu t} + \underset{\sim}{X}_{s\nu}] \chi_i(\underset{\sim}{\rho}, \underset{\sim}{X}) d^3\underset{\sim}{\rho} \Bigg\}. \end{aligned} \tag{12.4.16}$$

We note that the wave function $\chi(\underset{\sim}{\rho}, \underset{\sim}{X})$ may depend for certain electrons only on ρ and not on $\underset{\sim}{X}$; this is certainly the case for the *core* electrons. For these electrons, when summing over t', we obtain

$$\sum_{t'} e \int \chi_i(\underset{\sim}{\rho})^* (\underset{\sim}{\rho}_{s\nu t'} + \underset{\sim}{X}_{s\nu}) \chi_i(\underset{\sim}{\rho}) d^3\underset{\sim}{\rho} = \sum_{t'} e\underset{\sim}{X}_{s\nu} \int |\chi_i(\underset{\sim}{\rho})|^2 d^3\underset{\sim}{\rho} = Z'_\nu e\underset{\sim}{X}_{s\nu},$$

where Z'_ν is the number of electrons attached to the ν-th nucleus. The expression for $\underset{\sim}{M}_{ii}(\underset{\sim}{\alpha}, \underset{\sim}{X})$ can now be written

$$\underset{\sim}{M}_{ii}(\underset{\sim}{\alpha}, \underset{\sim}{X}) = \sum_{s\nu} e^{i\underset{\sim}{\alpha}\cdot\underset{\sim}{R}_{s\nu}} \left[-ez_\nu^{ion} \underset{\sim}{X}_{s\nu} \sum_t \int \chi_i(\underset{\sim}{\rho}, \underset{\sim}{X})^* e\underset{\sim}{\rho}_{s\nu t} \chi_i(\underset{\sim}{\rho}, \underset{\sim}{X}) d^3\underset{\sim}{\rho} \right], \tag{12.4.17}$$

where $Z_\nu^{ion} = Z_\nu - Z'_\nu$ and where the $\sum_t$ is extended only to the *deformable electrons* (for which χ depends on both $\underset{\sim}{\rho}$ and $\underset{\sim}{X}$). The value of $\underset{\sim}{M}_{ii}$ in correspondence to $\underset{\sim}{X} = \underset{\sim}{R}$ (equilibrium configuration) is generally zero; if not, it represents a static moment that has no effect on the transitions.

We may expand the integral that appears in (12.4.17) about the equilibrium nuclear configuration:

$$\underset{\sim}{\mu}(s, \nu, t; \underset{\sim}{X}) = \int \chi(\underset{\sim}{\rho}, \underset{\sim}{X})^* e\underset{\sim}{\rho}_{s\nu t}\chi(\underset{\sim}{\rho}, \underset{\sim}{X})d^3\underset{\sim}{\rho}$$
$$= \underset{\sim}{\mu}^{(o)}(s\nu t) + \sum_{s'\nu'} \underset{=}{\mu}^{(1)}(s\nu t; s'\nu') \cdot \underset{\sim}{u}_{s'\nu'}$$
$$+ \sum_{s'\nu'}\sum_{s''\nu''} \underset{\equiv}{\mu}^{(2)}(s\nu t; s'\nu'; s''\nu'') : \underset{\sim}{u}_{s'\nu'}\underset{\sim}{u}_{s''\nu''} + \cdots, \tag{12.4.18}$$

where

$$\underset{\sim}{\mu}^{(o)}(s\nu t) = \underset{\sim}{\mu}(s\nu t)|_{u=0} \tag{12.4.19}$$

$$\underset{=}{\mu}^{(1)}(s\nu t; s'\nu') = \left.\frac{\partial\underset{\sim}{\mu}(s\nu t; \underset{\sim}{X})}{\partial\underset{\sim}{u}_{s'\nu'}}\right|_{u=0} \tag{12.4.20}$$

$$\underset{\equiv}{\mu}^{(2)}(s\nu t; s'\nu'; s''\nu'') = \left.\frac{1}{2}\frac{\partial^2\underset{\sim}{\mu}(s\nu t; \underset{\sim}{X})}{\partial\underset{\sim}{u}_{s'\nu'}\partial\underset{\sim}{u}_{s''\nu''}}\right|_{u=0}. \tag{12.4.21}$$

$$\cdots$$

Using the above expansion in (12.4.17) and retaining only the terms that, depending on the displacement of the atoms, can give rise to transitions, we have

$$\underset{\sim}{M}_{ii}(\underset{\sim}{\alpha}, \underset{\sim}{X}) = \sum_{s\nu} e^{i\underset{\sim}{\alpha}\cdot\underset{\sim}{R}_{s\nu}}\left[-ez_\nu^{ion}\underset{\sim}{u}_{s\nu} + \sum_{ts'\nu'}\underset{=}{\mu}^{(1)}(s\nu t; s'\nu')\cdot\underset{\sim}{u}_{s'\nu'}\right.$$
$$\left.+\sum_{ts'\nu'}\sum_{s''\nu''}\underset{\equiv}{\mu}^{(2)}(s\nu t; s'\nu'; s''\nu'') : \underset{\sim}{u}_{s'\nu'}\underset{\sim}{u}_{s''\nu''}\right]$$
$$= \sum_{s\nu} e^{i\underset{\sim}{\alpha}\cdot\underset{\sim}{R}_{s\nu}}(-eZ_\nu^{ion})\underset{\sim}{u}_{s\nu} + \sum_{s\nu t} e^{i\underset{\sim}{\alpha}\cdot\underset{\sim}{R}_{s\nu}}\sum_{s'\nu'}\underset{=}{\mu}^{(1)}(s\nu t; s'\nu')\cdot\underset{\sim}{u}_{s'\nu'}$$
$$+\sum_{s\nu t} e^{i\underset{\sim}{\alpha}\cdot\underset{\sim}{R}_{s\nu}}\sum_{s'\nu'}\sum_{s''\nu''}\underset{\equiv}{\mu}^{(2)}(s\nu t; s'\nu'; s''\nu'') : \underset{\sim}{u}_{s'\nu'}\underset{\sim}{u}_{s''\nu''}$$

$$= \sum_{s\nu} e^{i\underset{\sim}{q}\cdot\underset{\sim}{R}_{s\nu}}(-eZ_\nu^{ion})\underset{\sim}{u}_{s\nu} + \sum_{s'\nu't'} e^{i\underset{\sim}{q}\cdot\underset{\sim}{R}_{s'\nu'}} \sum_{s\nu} \underline{\underline{\mu}}^{(1)}(s'\nu't';s\nu)\cdot\underset{\sim}{u}_{s\nu}$$

$$+ \sum_{s'\nu't'} e^{i\underset{\sim}{q}\cdot\underset{\sim}{R}_{s'\nu'}} \sum_{s\nu}\sum_{s''\nu''} \underline{\underline{\underline{\mu}}}^{(2)}(s'\nu't';s\nu;s''\nu'') : \underset{\sim}{u}_{s\nu}\underset{\sim}{u}_{s''\nu''}$$

$$= \sum_{s\nu} e^{i\underset{\sim}{q}\cdot\underset{\sim}{R}_{s\nu}} \left[eZ_\nu^{ion} + \sum_{s'\nu't'} \underline{\underline{\mu}}^{(1)}(s'\nu't';s\nu) e^{i\underset{\sim}{q}\cdot(\underset{\sim}{R}_{s'\nu'}-\underset{\sim}{R}_{s\nu})}\right]\cdot\underset{\sim}{u}_{s\nu}$$

$$+ \sum_{s\nu}\sum_{s''\nu''} e^{i\underset{\sim}{q}\cdot\underset{\sim}{R}_{s\nu}} \left[\sum_{s'\nu't'} \underline{\underline{\underline{\mu}}}^{(2)}(s'\nu't;s\nu;s''\nu'') \right.$$

$$\left. \times\ e^{i\underset{\sim}{q}\cdot(\underset{\sim}{R}_{s'\nu'}-\underset{\sim}{R}_{s\nu})} \right] : \underset{\sim}{u}_{s\nu}\underset{\sim}{u}_{s''\nu}$$

$$= \sum_{s\nu} e^{i\underset{\sim}{q}\cdot\underset{\sim}{R}_{s\nu}} \left\{ \left[-eZ_\nu^{ion} + \sum_{s'\nu't} \underline{\underline{\mu}}^{(1)}(s'\nu't';s\nu) \right.\right.$$

$$\left. \times\ e^{i\underset{\sim}{q}\cdot(\underset{\sim}{R}_{s'\nu'}-\underset{\sim}{R}_{s\nu})} \right] \cdot \underset{\sim}{u}_{s\nu}$$

$$\left. + \sum_{s''\nu''} \left[\sum_{s'\nu't} \underline{\underline{\underline{\mu}}}^{(2)}(s'\nu't;s\nu's''\nu'') e^{i\underset{\sim}{q}\cdot(\underset{\sim}{R}_{s'\nu'}-\underset{\sim}{R}_{s\nu})} \right] : \underset{\sim}{u}_{s\nu}\underset{\sim}{u}_{s''\nu''} \right\}$$

$$= \sum_{s\nu} e^{i\underset{\sim}{q}\cdot\underset{\sim}{R}_{s\nu}} \left\{ \underline{\underline{m}}(s\nu)\cdot\underset{\sim}{u}_{s\nu} + \sum_{s''\nu''} \underline{\underline{\underline{m}}}(s\nu;s''\nu'') : \underset{\sim}{u}_{s\nu}\underset{\sim}{u}_{s''\nu''} \right\}, \tag{12.4.22}$$

where

$$\underline{\underline{m}}(s\nu) = -eZ_\nu^{ion} + \sum_{s'\nu't} \underline{\underline{\mu}}^{(1)}(s'\nu't;s\nu) e^{i\underset{\sim}{q}\cdot(\underset{\sim}{R}_{s'\nu'}-\underset{\sim}{R}_{s\nu})}, \tag{12.4.23}$$

$$\underline{\underline{\underline{m}}}(s\nu;s''\nu'') = \sum_{s'\nu't} \underline{\underline{\underline{\mu}}}^{(2)}(s'\nu't;s\nu;s''\nu'') e^{i\underset{\sim}{q}\cdot(\underset{\sim}{R}_{s'\nu'}-\underset{\sim}{R}_{s\nu})}. \tag{12.4.24}$$

The quantity $\underline{\underline{m}}(s\nu)$ can be interpreted as the effective charge of the $s\nu$-th atom in correspondence to the equilibrium nuclear configuration; the quantity $\underline{\underline{\underline{m}}}(s\nu;s''\nu'')\cdot\underset{\sim}{u}_{s''\nu''}$ can be interpreted as the charge induced at the sν-th site by the displacement of the s″ν″-th atom from its equilibrium position.

Due to the translational invariance of the crystal $\underline{\underline{m}}(s\nu)$ should be independent of the position of the unit cell s and $\underline{\underline{\underline{m}}}(s\nu; s''\nu'')$ should depend only on the relative distance of the two unit cells:

$$\underline{\underline{m}}(s\nu) = \underline{\underline{m}}(\nu), \tag{12.4.25}$$

$$\underline{\underline{\underline{m}}}(s\nu; s''\nu'') = \underline{\underline{\underline{m}}}(s - s'', \nu; o, \nu''). \tag{12.4.26}$$

We may expect that in (12.4.23) and (12.4.24) the sums will converge rapidly because only neighboring atoms will be affecting each other; since the wavelength of the infrared radiation is much larger than the dimension of the unit cell, we may replace $\underline{\underline{m}}(s\nu)$ and $\underline{\underline{\underline{m}}}(s\nu; s''\nu'')$ by their approximate values:

$$\begin{aligned}\underline{\underline{m}}(s\nu) &= \underline{\underline{m}}(\nu) \approx \underline{\underline{m}}^{(o)}(\nu)\\ &= -eZ_\nu^{\text{ion}} + \sum_{s'\nu't} \mu^{(1)}(s'\nu't; s\nu),\end{aligned} \tag{12.4.27}$$

$$\begin{aligned}\underline{\underline{\underline{m}}}(s\nu; s''\nu'') &= \underline{\underline{\underline{m}}}(s - s'', \nu; o, \nu'') \approx \underline{\underline{\underline{m}}}^{(o)}(s - s'', \nu; o, \nu'')\\ &= \sum_{s'\nu't} \mu^{(2)}(s'\nu't; s\nu; s''\nu'').\end{aligned} \tag{12.4.28}$$

$\underset{\sim}{M}_{ii}(\underset{\sim}{\alpha}, \underset{\sim}{X})$ then becomes

$$\begin{aligned}\underset{\sim}{M}_{ii}(\underset{\sim}{\alpha}, \underset{\sim}{X}) &= \sum_{s\nu} e^{i\underset{\sim}{\alpha}\cdot\underset{\sim}{R}_{s\nu}} \Bigg\{ \underline{\underline{m}}^{(o)}(\nu) \cdot \underset{\sim}{u}_{s\nu}\\ &\quad + \sum_{s''\nu''} \underline{\underline{\underline{m}}}^{(o)}(s - s'', \nu; o, \nu'') : \underset{\sim}{u}_{s\nu}\underset{\sim}{u}_{s''\nu''} \Bigg\}\\ &= \underset{\sim}{M}^{(1)}(\underset{\sim}{\alpha}, \underset{\sim}{X}) + \underset{\sim}{M}^{(2)}(\underset{\sim}{\alpha}, \underset{\sim}{X}),\end{aligned} \tag{12.4.29}$$

where

$$\underset{\sim}{M}^{(1)}(\underset{\sim}{\alpha}, \underset{\sim}{X}) = \sum_{s\nu} e^{i\underset{\sim}{\alpha}\cdot\underset{\sim}{R}_{s\nu}} \underline{\underline{m}}^{(o)} \cdot \underset{\sim}{u}_{s\nu}, \tag{12.4.30}$$

$$\underset{\sim}{M}^{(2)}(\underset{\sim}{\alpha}, \underset{\sim}{X}) = \sum_{s\nu}\sum_{s''\nu''} e^{i\underset{\sim}{\alpha}\cdot\underset{\sim}{R}_{s\nu}} \underline{\underline{\underline{m}}}^{(o)}(s - s'', \nu; o, \nu'') : \underset{\sim}{u}_{s\nu}\underset{\sim}{u}_{s''\nu''}. \tag{12.4.31}$$

Some more invariance requirements can be derived considering the expression of $\underset{\sim}{M}_{ii}$ in correspondence to very long wavelengths ($\underset{\sim}{\alpha} \simeq 0$):

$$\underset{\sim}{M}_{ii}(\underset{\sim}{X}) = \sum_{s\nu} \underline{\underline{m}}^{(o)}(\nu) \cdot \underset{\sim}{u}_{s\nu} + \sum_{s\nu} \sum_{s''\nu''} \underline{\underline{\underline{m}}}^{(o)}(s - s'', \nu; o, \nu'') : \underset{\sim}{u}_{s\nu} \underset{\sim}{u}_{s''\nu''}. \tag{12.4.32}$$

This (static) moment is invariant under an arbitrary translation $\underset{\sim}{c}$:

$$\begin{aligned}
\underset{\sim}{M}_{ii}(\underset{\sim}{X} + \underset{\sim}{c}) &= \sum_{s\nu} \underline{\underline{m}}^{(o)}(\nu) \cdot (\underset{\sim}{u}_{s\nu} + \underset{\sim}{c}) \\
&\quad + \sum_{s\nu} \sum_{s''\nu''} \underline{\underline{\underline{m}}}^{(o)}(s - s'', \nu; o, \nu'') : (\underset{\sim}{u}_{s\nu} + \underset{\sim}{c})(\underset{\sim}{u}_{s''\nu''} + \underset{\sim}{c}) \\
&= \sum_{s\nu} \underline{\underline{m}}^{(o)}(\nu) \cdot \underset{\sim}{u}_{s\nu} + \left[\sum_{s\nu} \underline{\underline{m}}^{(o)}(\nu) \right] \cdot \underset{\sim}{c} \\
&\quad + \sum_{s\nu} \sum_{s''\nu''} \underline{\underline{\underline{m}}}^{(o)}(s - s'', \nu; o, \nu'') : \underset{\sim}{u}_{s\nu} \underset{\sim}{u}_{s''\nu''} \\
&\quad + \sum_{s\nu} \left[\sum_{s''\nu''} \underline{\underline{\underline{m}}}^{(o)}(s - s'', \nu; o, \nu'') \right] : \underset{\sim}{u}_{s\nu} \underset{\sim}{c} \\
&\quad + \sum_{s''\nu''} \left[\sum_{s\nu} \underline{\underline{\underline{m}}}^{(o)}(s - s'', \nu; o, \nu'') \right] : \underset{\sim}{u}_{s''\nu''} \underset{\sim}{c} \\
&\quad + \sum_{s\nu} \left[\sum_{s''\nu''} \underline{\underline{\underline{m}}}^{(o)}(s - s'', \nu; o, \nu'') \right] : \underset{\sim}{c} \underset{\sim}{c} = \underset{\sim}{M}_{ii}(\underset{\sim}{X}).
\end{aligned} \tag{12.4.33}$$

From this it follows that

$$\sum_{\nu} \underline{\underline{m}}^{(o)}(\nu) = 0, \tag{12.4.34}$$

$$\sum_{s\nu} \underline{\underline{\underline{m}}}^{(o)}(s - s'', \nu; o, \nu'') = \sum_{s''\nu''} \underline{\underline{\underline{m}}}^{(o)}(s - s'', \nu; o, \nu''). \tag{12.4.35}$$

The result (12.4.34) can be simply interpreted in the sense that the sum of the effective charges of all the atoms in the unit cell is zero.

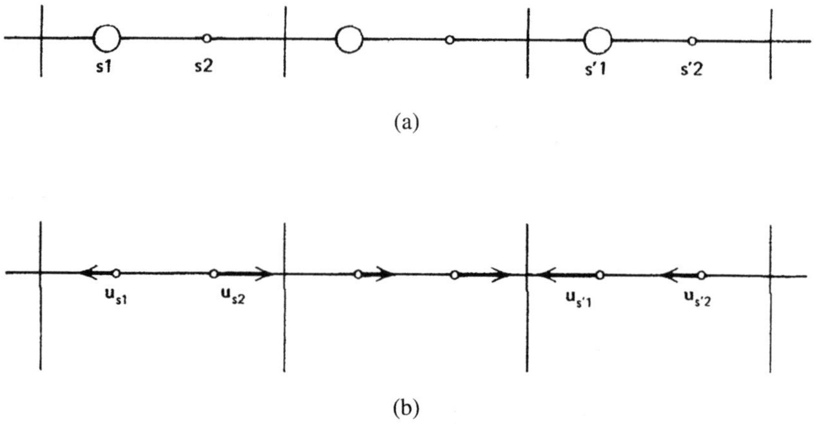

Fig. 12.2. (a) Linear ionic crystal; (b) linear homopolar crystal.

It is worthwhile to pursue the present tract by using the property of the static moment, $\underset{\sim}{M}_{ii}(\underset{\sim}{X})$. We use the following simple argument based on the article by Lax and Burstein[2] to show that the linear part of the static moment $\underset{\sim}{M}(\underset{\sim}{X})$ vanishes for homopolar crystals. We shall refer to Fig. 12.2, in which for simplicity we represent (a) an ionic linear crystal and (b) a homopolar linear crystal, with two atoms per unit cell. It is clear from the figure that crystal (b) has an inversion symmetry with respect to the center of mass of the unit cell, a symmetry that is lacking in crystal (a). For the homopolar crystal the linear part of $\underset{\sim}{M}(\underset{\sim}{X})$ is given by

$$\begin{aligned}\underset{\sim}{M}^{(1)}(\underset{\sim}{X}) &= \sum_{s} \underline{\underline{m}}^{(o)}(1)\underset{\sim}{u}_{s1} + \sum_{s} \underline{\underline{m}}^{(o)}(2)\underset{\sim}{u}_{s2} \\ &= \sum_{s} \underline{\underline{m}}^{(o)}(1)\underset{\sim}{u}_{s1} - \sum_{s} \underline{\underline{m}}^{(o)}(1)\underset{\sim}{u}_{s2} \\ &= \sum_{s} \underline{\underline{m}}^{(o)}(1)[\underset{\sim}{u}_{s1} - \underset{\sim}{u}_{s2}], \qquad (12.4.36)\end{aligned}$$

where use has been made of (12.4.34). The effect of the operation inversion about the center of mass of a particular unit cell is to change, say, $\underset{\sim}{u}_{s1}$ into some $-\underset{\sim}{u}_{s'2}$ and $\underset{\sim}{u}_{s2}$ into $-\underset{\sim}{u}_{s'1}$. Because of this the static moment will remain unchanged; on the other hand, we expect it to change sign on account of the inversion. It is then zero and, moreover, since it must remain so for arbitrary displacements, it is clear that for the case of homopolar crystals the effective charges of $\underline{\underline{m}}^{(o)}(\nu)$ must all be zero.

12.5. Reststrahl (One-Phonon) Absorption

We remind ourselves at this point that the relevant matrix element of the interaction Hamiltonian that can bring about an infrared absorption transition is given, as in (12.4.13), apart from constants, by

$$\pi^{\lambda}_{\underset{\sim}{\alpha}} \cdot \langle \phi_f(\underset{\sim}{X})|\underset{\sim}{M}_{ii}(\underset{\sim}{\alpha}, \underset{\sim}{X})|\phi_i(\underset{\sim}{X})\rangle, \tag{12.5.1}$$

where the operator $\underset{\sim}{M}_{ii}$ is expressed by (12.4.29). We shall examine in this section the effect of the term in the expansion of $\underset{\sim}{M}_{ii}$ that is linear in the nuclear displacement:

$$\underset{\sim}{M}^{(1)}(\underset{\sim}{\alpha}, \underset{\sim}{X}) = \sum_{s\nu} e^{i\underset{\sim}{\alpha}\cdot\underset{\sim}{R}_{s\nu}} \underline{\underline{m}}^{(o)}(\nu) \cdot \underset{\sim}{u}_{s\nu}, \tag{12.5.2}$$

where

$$\underline{\underline{m}}^{(o)}(\nu) = -eZ^{\nu}_{ion} + \sum_{s'\nu't} \mu^{(1)}(s'\nu't; s\nu). \tag{12.5.3}$$

In order to let $\underset{\sim}{M}^{(1)}(\underset{\sim}{\alpha}, \underset{\sim}{X})$ operate on the vibrational eigenfunction $\phi_i(\underset{\sim}{X})$ we express the displacement operation $\underset{\sim}{u}_{s\nu}$ in terms of creation and annihilation phonon operators as in (5.5.7):

$$\underset{\sim}{u}_{s\nu} = \sum_{\underset{\sim}{k}j} \left(\frac{\hbar}{2M\omega_{\underset{\sim}{k}j}}\right)^{1/2} [\underset{\sim}{V}_{\nu}(\underset{\sim}{k}, j) e^{i\underset{\sim}{k}\cdot\underset{\sim}{R}_s} b^{j}_{\underset{\sim}{k}} + \underset{\sim}{V}_{\nu}(\underset{\sim}{k}, j)^* e^{-i\underset{\sim}{k}\cdot\underset{\sim}{R}_s} b^{j+}_{\underset{\sim}{k}}]. \tag{12.5.4}$$

Using now this expression in (12.5.2) and retaining only the terms containing a creation phonon operator, we obtain

$$\underset{\sim}{M}^{(1)}(\underset{\sim}{\alpha}, \underset{\sim}{X}) = \sum_{s\nu} e^{i\underset{\sim}{\alpha}\cdot\underset{\sim}{R}_s} \left\{ \underline{\underline{m}}^{(o)}(\nu) \cdot \sum_{\underset{\sim}{k}j} \left(\frac{\hbar}{2M\omega_{\underset{\sim}{k}j}}\right)^{1/2} \times [\underset{\sim}{V}_{\nu}(\underset{\sim}{k}, j) e^{-i\underset{\sim}{k}\cdot\underset{\sim}{R}_s} b^{j+}_{\underset{\sim}{k}}] \right\}$$

$$\approx \sum_{\nu} \sum_{\underset{\sim}{k}j} \left(\frac{\hbar}{2M\omega_{\underset{\sim}{k}j}} \right)^{1/2} \underline{\underline{m}}^{(o)}(\nu) \cdot \underset{\sim}{V}_{\nu}(\underset{\sim}{k}, j)^* \sum_{s} e^{i(\underset{\sim}{\alpha}-\underset{\sim}{k})\cdot \underset{\sim}{R}_s} b^{j+}_{\underset{\sim}{k}}$$

$$= \sum_{\nu} \sum_{\underset{\sim}{k}j} \left(\frac{\hbar}{2M\omega_{\underset{\sim}{k}j}} \right)^{1/2} \underline{\underline{m}}^{(o)}(\nu) \cdot \underset{\sim}{V}_{\nu}(\underset{\sim}{k}, j)^* N\delta_{\underset{\sim}{\alpha}\underset{\sim}{k}} b^{j+}_{\underset{\sim}{k}}$$

$$= \sum_{j} \left(\frac{\hbar N^2}{2M\omega_{\underset{\sim}{\alpha}j}} \right)^{1/2} b^{j+}_{\underset{\sim}{\alpha}} \sum_{\nu} \underline{\underline{m}}^{(o)}(\nu) \cdot \underset{\sim}{V}_{\nu}(\underset{\sim}{k}, j)^*, \qquad (12.5.5)$$

where use has been made of the formula

$$\sum_{s} e^{i(\underset{\sim}{\alpha}-\underset{\sim}{k})\cdot \underset{\sim}{R}_s} = N\delta_{\underset{\sim}{\alpha}\underset{\sim}{k}}, \qquad (12.5.6)$$

already reported in Chapter 6 as (6.5.35), for the case in which both $\underset{\sim}{\alpha}$ and $\underset{\sim}{k}$ are restricted, as it is the case here, to the first Brillouin zone. The transition probability according to (12.5.1) is proportional to the square of the expression

$$\underset{\sim}{\pi}^{\lambda}_{\underset{\sim}{\alpha}} \cdot \left\langle \phi_f(\underset{\sim}{X}) \sum_{j} \left(\frac{\hbar N^2}{2M\omega_{\underset{\sim}{\alpha}j}} \right)^{1/2} b^{j+}_{\underset{\sim}{\alpha}} \sum_{\nu} \underline{\underline{m}}^{(o)}(\nu) \cdot \underset{\sim}{V}_{\nu}(\underset{\sim}{\alpha}, j)^* | \phi_i(\underset{\sim}{X}) \right\rangle$$

$$= \sum_{j} \left(\frac{\hbar N^2}{2M\omega_{\underset{\sim}{\alpha}j}} \right)^{1/2} \langle \phi_f(\underset{\sim}{X}) | b^{j+}_{\underset{\sim}{\alpha}} | \phi_i(\underset{\sim}{X}) \rangle \sum_{\nu} \underline{\underline{m}}^{(o)}(\nu) \cdot \underset{\sim}{V}_{\nu}(\underset{\sim}{\alpha}, j)^* \cdot \underset{\sim}{\pi}^{\lambda}_{\underset{\sim}{\alpha}}$$

$$= \sum_{j} \left(\frac{\hbar N^2}{2M\omega_{\underset{\sim}{\alpha}j}} \right)^{1/2} \langle \phi_f(\underset{\sim}{X}) | b^{j+}_{\underset{\sim}{\alpha}} | \phi_i(\underset{\sim}{X}) \rangle$$

$$\times \sum_{\nu} \left[-eZ^{\text{ion}}_{\nu} + \sum_{s'\nu' t} \mu^{(1)}(s'\nu' t; s\nu) \right] \underset{\sim}{V}_{\nu}(\underset{\sim}{\alpha}, j)^* \cdot \underset{\sim}{\pi}^{j}_{\underset{\sim}{\alpha}}. \qquad (12.5.7)$$

Various observations can be made on this result:

1. The term $\underset{\sim}{M}^{(1)}$, linear in the displacements of the atoms, gives rise to one-phonon absorption transitions. The matrix element for a

transition of this type is proportional to

$$\langle n^j_{\underset{\sim}{\alpha}} + 1 | b^{j+}_{\underset{\sim}{\alpha}} | n^j_{\underset{\sim}{\alpha}} \rangle = \sqrt{n^j_{\underset{\sim}{\alpha}} + 1}, \tag{12.5.8}$$

and the transition probability is proportional to $(n^j_{\underset{\sim}{\alpha}} + 1)$. For a crystal in thermal equilibrium at temperature T, the transition probability results proportional to $\overline{n^j_{\underset{\sim}{\alpha}}} + 1$; at very low temperatures $n^j_{\underset{\sim}{\alpha}} \approx 0$, and the infrared absorption is practically temperature independent.

2. The conservation of momentum is valid here; that is, the wave vector of the created phonon must be equal to the wave vector of the absorbed photon. Note that we restricted the wave vectors $\underset{\sim}{\alpha}$ and $\underset{\sim}{k}$ in (12.5.6) to the first Brillouin zone because of the energy conservation condition.
3. In the expression (12.5.7), $\underline{\underline{m}}^{(o)}$ is generally a dyad but, if the ionic term is predominant, it is practically a scalar. In this case only those modes that have polarization vectors with components in the direction of the light polarization can interact with the radiative field. For some systems it is possible to define directions of the wave vectors that allow the separation of the vibrational modes into transverse and longitudinal; in this case it is clear that only transverse (optical) modes can participate in the absorption process.
4. Since the effective charge $\underline{\underline{m}}^{(o)}(\nu)$ is zero for homopolar crystals, no Reststrahl absorption can be expected for such crystals.
5. For ionic crystals the one-phonon infrared spectra consists of a set of absorption lines that appear in correspondence to the frequencies of "transverse" optical modes for $\underset{\sim}{\alpha} \simeq 0$.

12.6. Two-Phonon Absorption

We examine in this section the effect of the term in the expansion of $\underset{\sim}{M}_{ii}$ that is quadratic in the nuclear displacement.

$$\underset{\sim}{M}^{(2)}(\underset{\sim}{\alpha}, \underset{\sim}{X}) = \sum_{s\nu} \sum_{s'\nu'} e^{i\underset{\sim}{\alpha}\cdot\underset{\sim}{R}_{s\nu}} \underline{\underline{\underline{m}}}^{(o)}(s - s', \nu; o, \nu'') : \underset{\sim}{u}_{s\nu} \underset{\sim}{u}_{s'\nu'}, \tag{12.6.1}$$

where

$$\underset{\equiv}{m}^{(o)}(s - s', \nu; o, \nu') = \sum_{s''\nu''} \underset{\equiv}{\mu}^{(2)}(s''\nu''t; s\nu, s'\nu'). \qquad (12.6.2)$$

Using the expression (12.5.4) for the displacement vector in (12.6.1) we obtain the following pages of equations. Here $\underset{\sim}{R}_\sigma = \underset{\sim}{R}_s - \underset{\sim}{R}_{s'}$ and we have neglected terms in $b^j_{\underset{\sim}{k}}\, b^{j'}_{\underset{\sim}{k}'}$, corresponding to the annihilation of two phonons, a process that cannot take place since it would not conserve energy.

The mechanism for two-phonon absorption can be thought of in the following terms[2]: a vibrational mode induces charges on the atoms and a second mode makes these charges vibrate, producing an electric dipole moment that interacts with the radiative field. It may be expected that the modes that will produce the charge deformation will be those at short wavelength (high $\underset{\sim}{k}$), since they correspond to asymmetric displacements of the atoms within the unit cell. Thus

$$\underset{\sim}{M}^{(2)}(\underset{\sim}{\alpha}, \underset{\sim}{X}) = \sum_{s\nu}\sum_{s'\nu'} e^{i\underset{\sim}{\alpha}\cdot\underset{\sim}{R}_{s\nu}} \underset{\equiv}{m}^{(o)}(s - s', \nu; o, \nu')$$

$$: \sum_{\underset{\sim}{k}j} \left(\frac{\hbar}{2M\omega_{\underset{\sim}{k}j}}\right)^{1/2} \Big[\underset{\sim}{V}_\nu(\underset{\sim}{k}, j) e^{i\underset{\sim}{k}\cdot\underset{\sim}{R}_s} b^j_{\underset{\sim}{k}}$$

$$+ \underset{\sim}{V}_\nu(\underset{\sim}{k}, j)^* e^{-i\underset{\sim}{k}\cdot\underset{\sim}{R}_s} b^{j+}_{\underset{\sim}{k}}\Big] \sum_{\underset{\sim}{k}'j'} \left(\frac{\hbar}{2M\omega_{\underset{\sim}{k}'j'}}\right)^{1/2}$$

$$\times \Big[\underset{\sim}{V}_{\nu'}(\underset{\sim}{k}', j') e^{i\underset{\sim}{k}'\cdot\underset{\sim}{R}_{s'}} b^{j'}_{\underset{\sim}{k}'} + \underset{\sim}{V}_{\nu'}(\underset{\sim}{k}', j')^* e^{-i\underset{\sim}{k}'\cdot\underset{\sim}{R}_{s'}} b^{j'+}_{\underset{\sim}{k}'}\Big]$$

$$= \sum_{\underset{\sim}{k}j}\sum_{\underset{\sim}{k}'j'} \frac{\hbar}{2M} \frac{1}{\sqrt{\omega_{\underset{\sim}{k}j}\omega_{\underset{\sim}{k}'j'}}} \sum_{s\nu}\sum_{s'\nu'} e^{i\underset{\sim}{\alpha}\cdot\underset{\sim}{R}_{s\nu}} \underset{\equiv}{m}^{(o)}(s - s', \nu; o, \nu')$$

$$: \Big[\underset{\sim}{V}_\nu(\underset{\sim}{k}, j)\underset{\sim}{V}_{\nu'}(\underset{\sim}{k}', j') e^{i\underset{\sim}{k}\cdot\underset{\sim}{R}_s + i\underset{\sim}{k}'\cdot\underset{\sim}{R}_{s'}} b^j_{\underset{\sim}{k}} b^{j'}_{\underset{\sim}{k}'}$$

$$+ \underset{\sim}{V}_\nu(\underset{\sim}{k}, j)\underset{\sim}{V}_{\nu'}(\underset{\sim}{k}', j')^* e^{i\underset{\sim}{k}\cdot\underset{\sim}{R}_s - i\underset{\sim}{k}'\cdot\underset{\sim}{R}_s} b^j_{\underset{\sim}{k}} b^{j'+}_{\underset{\sim}{k}'}$$

$$+ V_\nu(\underset{\sim}{k}, j)^* \underset{\sim}{V}_{\nu'}(\underset{\sim}{k}', j') e^{-i\underset{\sim}{k}\cdot\underset{\sim}{R}_s + i\underset{\sim}{k}'\cdot\underset{\sim}{R}_s} b^{j+}_{\underset{\sim}{k}} b^{j'}_{\underset{\sim}{k}'}$$

$$+ V_\nu(\underset{\sim}{k}, j)^* V_{\nu'}(\underset{\sim}{k}', j')^* e^{-i\underset{\sim}{k}\cdot\underset{\sim}{R}_s - i\underset{\sim}{k}'\cdot\underset{\sim}{R}_{s'}} b^{j+}_{\underset{\sim}{k}} b^{j'+}_{\underset{\sim}{k}'}\Big]$$

$$
= \sum_{\underset{\sim}{k}j} \sum_{\underset{\sim}{k}'j'} \frac{\hbar}{2M} \frac{1}{\sqrt{\omega_{\underset{\sim}{k}j}\omega_{\underset{\sim}{k}'j'}}} \sum_{\nu} \sum_{\sigma\nu'} \underset{\equiv}{m}^{(o)}(\sigma, \nu; o, \nu')
$$

$$
: \left\{ \underset{\sim}{V}_{\nu}(\underset{\sim}{k}, j) \underset{\sim}{V}_{\nu'}(\underset{\sim}{k}', j')^{*} \left[\sum_{s} e^{i(\underset{\sim}{\alpha}+\underset{\sim}{k}-\underset{\sim}{k}')\cdot \underset{\sim}{R}_s} \right] e^{i\underset{\sim}{k}'\cdot \underset{\sim}{R}_\sigma} b^{j}_{\underset{\sim}{k}} b^{j'+}_{\underset{\sim}{k}'} \right.
$$

$$
+ \underset{\sim}{V}_{\nu}(\underset{\sim}{k}, j)^{*} \underset{\sim}{V}_{\nu'}(\underset{\sim}{k}', j') \left[\sum_{s} e^{i(\underset{\sim}{\alpha}-\underset{\sim}{k}+\underset{\sim}{k}')\cdot \underset{\sim}{R}_s} \right] e^{-i\underset{\sim}{k}'\cdot \underset{\sim}{R}_\sigma} b^{j+}_{\underset{\sim}{k}} b^{j'}_{\underset{\sim}{k}'}
$$

$$
\left. + \underset{\sim}{V}_{\nu}(\underset{\sim}{k}, j)^{*} \underset{\sim}{V}_{\nu'}(\underset{\sim}{k}', j')^{*} \left[\sum_{s} e^{i(\underset{\sim}{\alpha}-\underset{\sim}{k}-\underset{\sim}{k}')\cdot \underset{\sim}{R}_s} \right] e^{i\underset{\sim}{k}'\cdot \underset{\sim}{R}_\sigma} b^{j+}_{\underset{\sim}{k}} b^{j'+}_{\underset{\sim}{k}'} \right\}
$$

$$
= \sum_{\underset{\sim}{k}j} \sum_{\underset{\sim}{k}'j'} \frac{\hbar}{2M} \frac{1}{\sqrt{\omega_{\underset{\sim}{k}j}\omega_{\underset{\sim}{k}'j'}}} \sum_{\sigma} \sum_{\nu\nu'} \underset{\equiv}{m}^{(o)}(\sigma, \nu; o, \nu')
$$

$$
: \left\{ \underset{\sim}{V}_{\nu}(\underset{\sim}{k}, j) \underset{\sim}{V}_{\nu'}(\underset{\sim}{k}', j')^{*} \delta_{\underset{\sim}{\alpha},(-\underset{\sim}{k}+\underset{\sim}{k}'+\underset{\sim}{K})} e^{i\underset{\sim}{k}'\cdot \underset{\sim}{R}_\sigma} b^{j}_{\underset{\sim}{k}} b^{j'+}_{\underset{\sim}{k}'} \right.
$$

$$
+ \underset{\sim}{V}_{\nu}(\underset{\sim}{k}, j)^{*} \underset{\sim}{V}_{\nu'}(\underset{\sim}{k}', j') \delta_{\underset{\sim}{\alpha},(\underset{\sim}{k}-\underset{\sim}{k}'+\underset{\sim}{K})} e^{-i\underset{\sim}{k}'\cdot \underset{\sim}{R}_\sigma} b^{j}_{\underset{\sim}{k}} b^{j'}_{\underset{\sim}{k}'}
$$

$$
\left. + \underset{\sim}{V}_{\nu}(\underset{\sim}{k}, j)^{*} \underset{\sim}{V}_{\nu'}(\underset{\sim}{k}', j')^{*} \delta_{\underset{\sim}{\alpha},(\underset{\sim}{k}+\underset{\sim}{k}'+\underset{\sim}{K})} e^{i\underset{\sim}{k}'\cdot \underset{\sim}{R}_\sigma} b^{j+}_{\underset{\sim}{k}} b^{j+}_{\underset{\sim}{k}'} \right\}
$$

$$
= \sum_{\underset{\sim}{k}} \sum_{jj'} \frac{\hbar}{2M\sqrt{\omega_{\underset{\sim}{k}j}}} \sum_{\sigma} \sum_{\nu\nu'} \underset{\equiv}{m}^{(o)}(\sigma, \nu; o, \nu')
$$

$$
: \left\{ \frac{1}{\sqrt{\omega_{\underset{\sim}{\alpha}+\underset{\sim}{k},j'}}} \underset{\sim}{V}_{\nu}(\underset{\sim}{k}, j) \underset{\sim}{V}_{\nu'}(\underset{\sim}{\alpha}+\underset{\sim}{k}, j')^{*} e^{i(\underset{\sim}{\alpha}+\underset{\sim}{k})\cdot \underset{\sim}{R}_\sigma} b^{j}_{\underset{\sim}{k}} b^{j'+}_{\underset{\sim}{\alpha}+\underset{\sim}{k}-\underset{\sim}{K}} \right.
$$

$$
+ \frac{1}{\sqrt{\omega_{-\underset{\sim}{\alpha}+\underset{\sim}{k},j'}}} \underset{\sim}{V}_{\nu}(\underset{\sim}{k}, j)^{*} \underset{\sim}{V}_{\nu'}(-\underset{\sim}{\alpha}+\underset{\sim}{k}, j') e^{-i(-\underset{\sim}{\alpha}+\underset{\sim}{k})\cdot \underset{\sim}{R}_\sigma} b^{j+}_{\underset{\sim}{k}} b^{j'}_{-\underset{\sim}{\alpha}+\underset{\sim}{k}+\underset{\sim}{K}}
$$

$$
\left. + \frac{1}{\sqrt{\omega_{\underset{\sim}{\alpha}-\underset{\sim}{k},j'}}} \underset{\sim}{V}_{\nu}(\underset{\sim}{k}, j)^{*} \underset{\sim}{V}_{\nu'}(\underset{\sim}{\alpha}-\underset{\sim}{k}, j')^{*} e^{i(\underset{\sim}{\alpha}-\underset{\sim}{k})\cdot \underset{\sim}{R}_\sigma} b^{j+}_{\underset{\sim}{k}} b^{j'+}_{\underset{\sim}{\alpha}-\underset{\sim}{k}-\underset{\sim}{K}} \right\}, \tag{12.6.3}
$$

where $\underset{\sim}{K}$ is the primitive vector of the reciprocal lattice.

Let us consider the consequences of the result (12.6.3). Three processes are possible:

1. A process in which a phonon of wave vector $\underset{\sim}{k}$ is destroyed and a phonon of wave vector $\underset{\sim}{k}'$ is created. The transition probability for such a process is proportional to

$$\overline{n^{j}_{\underset{\sim}{k}}}\ (\overline{n^{j'}_{\underset{\sim}{k}'}} + 1), \tag{12.6.4}$$

and tends to zero as $T \to 0$. Notice here that no restriction is placed on j, j', and $\underset{\sim}{k}$, provided energy is conserved. For very low temperatures, $\overline{n^{j}_{\underset{\sim}{k}}} \approx 0$ if j designates an optical branch; therefore processes in which an optical phonon is absorbed and an optical or acoustical phonon created are highly improbable. On the other hand, $\overline{n^{j}_{\underset{\sim}{k}}}$ may be different from zero if j designates an acoustical branch; in this case the process in which an acoustical phonon is destroyed and an optical or acoustical phonon created may be possible; if in such a case T is such that $\overline{n^{j}_{\underset{\sim}{k}}} \approx$ classical limit $\approx kT/\hbar\omega$, the absorption associated with this *difference band* is proportional to the temperature.

2. A process in which a phonon of wave vector $\underset{\sim}{k}'$ is destroyed and a phonon of wave vector $\underset{\sim}{k}$ is created. The transition probability for such a process is proportional to

$$(\overline{n^{j}_{\underset{\sim}{k}}} + 1)\ \overline{n^{j'}_{\underset{\sim}{k}}}. \tag{12.6.5}$$

The considerations made above apply also to this *difference band.*

3. A process in which two phonons of wave vectors $\underset{\sim}{k}$ and $\underset{\sim}{k}'$ are created. The transition probability for such a process is proportional to

$$(\overline{n^{j}_{\underset{\sim}{k}}} + 1)\ \overline{n^{j'}_{\underset{\sim}{k}'}} + 1), \tag{12.6.6}$$

and is different from zero even if $T = 0$. In the limit of very low temperature this *summation band* is independent of temperature.

Any mode of vibration can participate in two-phonon processes; one would expect from the combination of two modes a continuous absorption spectrum. Actually, the absorption spectrum may present peaks that reflect a structure in the combined density of states of the phonons participating in the process.[3,4]

In the above considerations the harmonic approximation has been considered valid. The anharmonic forces result in the presence of higher-order terms in the expansion (6.5.3) of the potential energy of the crystal. The third-order harmonic term can also bring about two-phonon processes; in such processes a typical matrix element has the form

$$\langle n^{j'}_{\underset{\sim}{k}'} \pm 1, n^{j''}_{\underset{\sim}{k}''} \pm 1, n^{j}_{\underset{\sim}{k}} | V^{(3)} | n^{j'}_{\underset{\sim}{k}'}, n^{j''}_{\underset{\sim}{k}''}, n^{j}_{\underset{\sim}{k}+1} \rangle \langle n^{j}_{\underset{\sim}{k}} + 1 | \underset{\sim}{\pi} \cdot \underset{\sim}{M}^{(1)} | n^{j}_{\underset{\sim}{k}} \rangle. \tag{12.6.7}$$

In physical terms this means that the radiative field interacts with a $\underset{\sim}{k} = \underset{\sim}{\alpha}$ "transverse" optical mode that is coupled by the anharmonic term V^3 to two other modes. As a result of this interaction two phonons can be created or one phonon created and one destroyed. It can be shown[5] that the wave vector is conserved in an anharmonic interaction; therefore the wave vector is conserved from the initial to the intermediate (virtual) state and from the intermediate to the final state. Since the energy need not be conserved in the intermediate state, any transverse optical phonon at $\underset{\sim}{k} = \underset{\sim}{\alpha}$ can be a virtual phonon in this process. It is to be noted that the anharmonic two-phonon process cannot occur in homopolar crystals, because in this crystal the operator $\underset{\sim}{M}^1$ is identically zero; however, it is believed that for polar crystals the anharmonic process predominates over the two-phonon process described previously.[3]

12.7. Selection Rules for Infrared Absorption

Selection rules for infrared absorption can be derived by considering the matrix element (12.4.13). The relevant quantity is

$$\langle \phi_f(\underset{\sim}{X}) | \underset{\sim}{M}_{ii} | \phi_i(\underset{\sim}{X}) \rangle. \tag{12.7.1}$$

We shall assume that the static dipole of (12.4.32) may be used in (12.8.1), that is, that the wave vector $\underset{\sim}{\alpha} \approx 0$. The static dipole $\underset{\sim}{M}_{ii}(\underset{\sim}{X})$ transforms as a polar vector; also, in the expansion (12.4.32) all the terms that are different from zero transform in the same way. The expansion of $\underset{\sim}{M}_{ii}(\underset{\sim}{X})$ can be rewritten in terms of normal coordinates; in this case as in (12.6.32) the coefficients of the expansion are quantities calculated in correspondence to the equilibrium nuclear configuration of the crystal and are therefore invariant under the operations of the space group G of the crystal. Therefore in the expansion the coefficients in the first-order terms will be different from zero only for those normal coordinates $q^j_{\underset{\sim}{k}}(\underset{\sim}{k} \approx 0)$ that transform under the operations of the space group $\tilde{G}$ according to the irreducible representations of the crystallographic point group G_o that have as basis functions x, y, and z. For infrared single-phonon absorption only the normal modes that correspond to these representations can be activated.

At low enough temperatures, all the phonon oscillators are in their ground state ($n^j_{\underset{\sim}{k}} \approx 0$) and the initial vibrational state $\phi_i(\underset{\sim}{X})$ transforms as the identity representation. In this case a single-phonon infrared absorption will occur if a normal coordinate for $\underset{\sim}{k} = 0$ and one or more of the electric dipole coordinates (x, y, z) belong to the same irreducible representation of the crystallographic point group G.

It is of particular interest to consider the case of a crystal of simple structure, such as diatomic cubic crystals. In such crystals the $\underset{\sim}{k} = 0$ optical modes transform as x, y, and z are expected to be triply degenerate; this is indeed the case for homopolar crystals, such as diamond and germanium.[6] In ionic crystals at $\underset{\sim}{k} \approx 0$ the optical modes split into two components of different frequency with the longitudinal optical (LO) branch higher in frequency than the two degenerate transversal optical (TO) branches. This is because the ions, when vibrating in the LO modes, feel an additional restoring force due to the charge polarization. In this case, since practically only TO modes can participate in the absorption process, the frequency of the LO branch has to be derived by other means such as the Lyddane, Sachs, and Teller relation for diatomic cubic crystals,[7]

$$\omega^2_{LO}(\underset{\sim}{k} \sim O) = \frac{\varepsilon_s}{\varepsilon_\infty}\omega^2_{TO}(\underset{\sim}{k} \sim O), \tag{12.7.2}$$

where ε_s is the static dielectric constant and ε_∞ is the high-frequency dielectric constant. This result has been generalized by Cochran and

Cowley to cubic crystals with more complicated unit cells[8] and by Cochran to certain noncubic crystals.[9]

In two-phonon absorption processes, the two modes participating in the process may have, in general, wave vectors different from zero, and, as such, they do not transform according to the point group G_o representations, but, rather, according to representations of the space group G. The product of two normal coordinates $q^j_{\underset{\sim}{k}} q^{j'}_{\underset{\sim}{k}'}$ transforms as the dircct product of the two related space group representations. If this direct product representation, when reduced into irreducible representations of the space group, contains one of the irreducible representations of G_o that have x, y, or z as basis functions, then the corresponding term in the expansion of $\underset{\sim}{M}(\underset{\sim}{X})$ may have its coefficient different from zero and the two modes $(\underset{\sim}{k}, j)$ and $(\underset{\sim}{k}', j')$ may participate in the absorption process. Such selection rules, as shown by Cornwell,[10] require the conservation of $\underset{\sim}{k}$ vector within a primitive $\underset{\sim}{K}$ vector of the reciprocal lattice.

12.8. The Effect of Impurities on Infrared Absorption Spectra

A (one-phonon) infrared absorption transition to an allowed optical branch produces the excitation of a single normal mode with wave vector $\underset{\sim}{k}$. The translational invariance of the crystal imposes the selection rule that the wave vector of the phonon created must be equal to the wave vector of the photon absorbed.

The introduction of impurity ions into the crystal destroys the translational invariance and may result in the following:

1. The activation of phonons already present in the perfect crystal and correspondingly an absorption spectrum that resembles, apart from a frequency-dependent weighting function, the phonon densities of states. The resemblance to the actual phonon spectrum will be close if the presence of the impurity does not perturb the vibrations of the crystal.
2. The activation of localized or pseudolocalized modes of vibration.

The normal modes of vibration of a perfect crystal transform according to the irreducible representations of the space group; when the symmetry is reduced because of the presence of an impurity, the new symmetry group includes only those operations of the space group that leave the impurity invariant. Correspondingly, the space group representations are reduced in terms of the irreducible representations of the impurity site symmetry.

Impurity-induced one-phonon infrared absorption has been found effective in studying the phonon spectrum of the pure crystal and the interatomic forces around the impurity.[11,12]

12.9. Infrared Absorption in Homopolar Crystals

The above discussion of reststrahlen (allowed one-phonon infrared absorption) transitions applies specifically to ionic crystals where the vibrations of ions of opposite electrical charge in their unit cell creates an oscillating dipole moment. The application of the same rigid-ion model to covalently bonded crystals made up of atoms with identical charges would imply that no reststrahlen transitions can occur in these types of materials. However there are some covalent elemental materials that do exhibit reststrahlen transitions. To understand the observed infrared absorption spectra of these materials, it is necessary to expand the treatment of lattice dynamics to include dynamic charge redistribution around the atoms in a unit cell.[13] If an optical phonon mode at the center of the Brillouin zone induces an electric dipole moment due to this charge redistribution, a reststrahlen transition can occur.

Knowing the crystal structure and symmetry of the material, group theory can be used to determine if dynamic charge redistribution will result in an allowed reststrahlen transition or not. This is accomplished using the basic concepts developed in Chapters 2, 3 and 4 along with the treatment of lattice vibrations discussed in Chapter 5. The discussion here follows closely the argument developed in Ref. 13.

As described in Chapter 3, a crystal structure is composed of one of 14 Bravais lattices upon which is superimposed a basis of one or more atoms per unit cell. This structure determines the symmetry properties of the crystal. There are a limited number of ways that the basis atoms can be arraigned on the Bravais lattice to maintain the symmetry of the structure. If there are s atoms in a primitive unit cell, the atomic displacements are represented by a 3s vector made up of Cartesian coordinates attached to each atom. An operation of the symmetry group of the crystal acts on this vector through a 3s × 3s transformation matrix. The set of transformation matrices for all the symmetry operations of the crystallographic point group act as the vibrational representation of the group with the atomic displacement vectors as the basis for the representation. As shown in Chapter 5, the transformation matrices are reducible representations that can be expressed in box-diagonal form with the boxes on the diagonal being 3 × 3 matrices representing the effect of the symmetry operation on the Cartesian coordinates of a specific atom. Thus they are three-dimensional polar vector representations of the symmetry operator. These submatrices are irreducible representations of the group and are non-zero only for atoms that remain invariant under the symmetry operation of the group.

From the preceding discussion and Eq. (2.6.1), the character of a transformation matrix in the vibrational representation is the sum over all of the irreducible representations of the group of the product of the character of an irreducible representation times the number of times that irreducible representation is contained in the reduction of the vibrational representation

$$\chi_V(R) = \sum_j c_j \chi_j^V(R). \tag{12.9.1}$$

The reduction of the reducible vibrational representation in terms of irreducible representations can be accomplished as described in Chapter 2. The number of vibrational modes of the material equals the sum of the number of irreducible representations contained in the reduction of the vibrational representation. According to Eq. (2.6.3), each irreducible representation will appear in the reduction c_j times as given by

$$c_j = (1/g) \sum_R \chi_V(R) \chi_j^V(R)^*. \tag{12.9.2}$$

The three dimensional polar vector representation of the group Γ_P in general may also be a reducible representation that can be reduced in terms of its irreducible representations Γ_j^P. Since the polar vector submatrix appears along the diagonal of the vibrational matrix operator for each symmetry operation R for each atom the remains invariant under R, the characters of Γ_P anc Γ_V are related by

$$\chi_V(R) = s(R)\chi_P(R), \tag{12.9.3}$$

where s(R) represents the number of atoms remaining invariant under operation R. One of the irreducible representations of the polar vector represents acoustical vibrations and must be subtracted from the total number of vibrational modes to obtain the number of optical modes thai might be infrared active

$$n_v(ir) = \sum_{\Gamma_j^P} \left[c(\Gamma_j^P, \Gamma_V) - c(\Gamma_j^P, \Gamma_P)\right]. \tag{12.9.4}$$

Here $c(\Gamma_i^P, \Gamma_V)$ is the number of times the jth irreducible representation of the polar vector appears in the reduction of the vibrational representation and $c(\Gamma_j^P, \Gamma_P)$ is the number of times the same irreducible representation appears in the reduction of the reducible polar vector representation.

Using Eq. (12.9.2) for each of the terms on the right hand side of Eq. (12.9.4) along with the relationship between the characters given in Eq. (12.9.3) leads to the expression for the number of distinct infrared frequencies $n_v(ir)$ is given by

$$n_v(ir) = (1/g) \sum_{R,\Gamma_j^P} [s(R) - 1]\, \chi_P(R)\chi_j^P(R)^*, \tag{12.9.5}$$

where g is the order of the factor group G of the crystal, R denotes a specific symmetry operation of the group, s(R) is the number of atoms in the unit cell remaining invariant under operation R, Γ_j^P is an irreducible representation contained in the representation of the polar vector in G, and χ_P(R) is the character of the R operation in the reducible representation of the polar vector while χ_j^P(R) is the character of the same operation in the irreducible representation of the polar vector.

Equation (12.9.5) can be used to analyze various crystal structures to determine the number of infrared active frequencies.[13] In this section the specific interest is in crystals consisting of one type of atom. The following paragraphs treat crystals structures with different numbers of atoms per unit cell.

First, for structures with one atom per unit cell s = 1. The atom in the unit cell will be invariant for all R transformations so s(R) = 1. Substituting this into Eqs. (12.9.5) shows that $n_v(ir) = 0$ so this type of crystal will have no infrared active optical phonon modes.

For structures with two identical atoms per unit cell s = 2. All crystal symmetries of this type have inversion as one of their group operators with the center of symmetry located halfway between the two primitive cell atoms. Thus the crystallographic point group G of order g can be factored into two groups each of order g/2 given by $G_0X\{l, \bar{1}\}$ where 1 is the identity operator and $\bar{1}$ is the inversion operator and the set of R_0's represents the symmetry elements contained in G_0. The sum in Eq. (12.9.5) can then be separated into two parts

$$n_v(ir) = (1/g) \sum_{R_0 \Gamma_j^P} [s(R_0) + s(R_0\bar{1}) - 2]\, \chi_P(R_0)\chi_j^P(R_0)^*, \quad (12.9.6)$$

where $\chi_P(R_0\bar{1}) = -\chi_P(R_0)$ and $\chi_j^P(R_0\bar{1}) = -\chi_j^P(R_0)$. Under the inversion operation the two atoms interchange positions so $s(\bar{1}) = 0$. If one of the other symmetry operations R_0 leaves both atoms invariant, $s(R_0) = 2$ but $s(R_0\bar{1}) = 0$. On the other hand, if $s(R_0) = 0$ then $s(R_0\bar{1}) = 0$. In any case the quantity in brackets in Eq. (12.9.6) is identically equal to zero so there are no reststrahlen transitions for crystals of this type.

Finally, for materials with three or more identical atoms per unit cell $s \geq 3$ and the symmetry properties of each of the 32 crystal classes must be considered. First consider the 27 higher symmetry classes excluding the 5 monoclinic and triclinic classes. For these crystal structures the sum in Eq. (12.9.5) can be factored into separate sums over R and Γ_j^P

$$n_v(ir) = (1/g) \sum_R \left\{ [s(R) - 1]\, \chi_P(R) \sum_{\Gamma_j^P} \chi_j^P(R)^* \right\}. \quad (12.9.7)$$

For the 27 crystal classes being considered each irreducible representation Γ_j^P appears just once in the reduction of the polar vector representation Γ_P. Thus the second sum in Eq. (12.9.7) becomes

$$\chi_P(R)^* = \sum_{\Gamma_j^P} \chi_j^P(R)^*. \tag{12.9.8}$$

Substituting this expression into Eq. (12.9.7) gives a simplified expression for the number of infrared active frequencies that occur in these classes of crystals

$$\mathrm{n_v(ir)} = (1/\mathrm{g}) \sum_R [s(R) - 1] \, |\chi_P(R)|^2. \tag{12.9.9}$$

If a symmetry operation R does not leave any of the atoms invariant, s(R) = 0 and the terms in the sum in Eq. (12.9.9) are negative. The total negative contribution is given by

$$\mathrm{n_v(neg)} = -(1/\mathrm{g}) \sum_{R \subset s(R)=0} |\chi_P(R)|^2, \tag{12.9.10}$$

where the sum is over all symmetry operations for which s(R) = 0. The minimum negative contribution is given by restricting the sum to all operations R except the identity operation R = 1. For R being the identity operator s(1) = s and the character for $\chi_P(1)$ equals the dimension of the representation which is 3 for a vector representation. This term in the sum over R always makes a positive contribution of 9(s − 1). This is a lower limit on the positive contribution to $\mathrm{gn_v(ir)}$. Combining the lower limits of the positive contribution with that of the negative contribution gives a lower limit for the number of possible infrared active modes of

$$\mathrm{g}n_v(\mathrm{ir}) \geq 9(s-1) - \sum_{R \neq 1} |\chi_P(R)|^2 = 9s - \sum_R |\chi_P(R)|^2. \tag{12.9.11}$$

From the orthogonality conditions for irreducible representations discussed in Chapter 2 and the fact that for each of these 27 crystal classes each irreducible representation Γ_j^P occurs only once in the reduction of the polar vector representation of the group (see the character tables given in Ref. 4), the sum in Eq. (12.9.11) is equal to the product of the order of the group g and the number of Γ_j^P's given by n_p. The inequality in Eq. (12.9.11) then becomes

$$gn_\nu(ir) \geq 9s - gn_p. \tag{12.9.12}$$

For all except four of the 27 classes of crystals being considered here[4] the product $gn_p \leq 24$. Thus for crystals with 3 identical atoms per unit cell (or greater) the number of infrared vibrational frequencies will always be greater than zero so reststrahlen transitions may appear in their infrared absorption spectra.

For the other four crystallographic point groups O_h, D_{4h}, D_{6h}, and C_{6h} the product gn_p is much larger and Eq. (12.9.12) is not useful in its present form for determining the number of infrared active modes. Thus the lower limit condition used in deriving Eq. (12.9.11) must be strengthened. In deriving this inequality the lower limit represented by the equals sign was derived using the assumption that the identity was the only operation that leaves at least one of the atoms in the unit cell invariant. In other words, no atom is located on an element of symmetry. This is generally not true so for these four classes of crystals the nature of each symmetry operation must be considered. Equation (12.9.11) can be rewritten as

$$gn_\nu(ir) \geq 9(s-1) - \sum_{R \subset s(R)=0} |\chi_P(R)|^2 = 18 - \sum_{R \subset s(R)=0} |\chi_P(R)|^2, \tag{12.9.13}$$

where the assumption of three atoms per unit cell (s = 3) has been made. For crystals with s = 3, the symmetry of the lattice structure dictates that only the operations C_3 and S_6 will leave no atom unchanged and therefore contribute to the sum over s(R) = 0 elements in Eq. (12.9.13).

First consider a crystal of the cubic class with O_h point group symmetry. The order of the group is g = 48 and a polar vector transforms as one irreducible representation,[4] T_{1u}. The character table for O_h, shows that $\chi_{T_{1u}}(C_3) = 0$ and $\chi_{T_{1u}}(S_6) = 0$. Thus the sum in Eq. (12.9.13) is equal to zero and $gn_\nu(ir) \geq 18$. Therefore $n_\nu(ir)$ is always greater than zero and crystals of this class with three atoms per unit cell may exhibit reststrahlen transitions.

Next consider a crystal of the tetragonal class with D_{4h} point group symmetry. The order of the group is g = 16 and a polar vector transforms as the two irreducible representations[14] A_{2u} and E_u. The character table for D_{4h} shows that there are no C_3 or S_6 symmetry elements and thus the sum over s(R) = 0 elements is equal to zero. Therefore $gn_\upsilon(ir) \geq 18$, which shows that $n_\upsilon(ir)$ is always greater than zero so crystals of this class with three atoms per unit cell may exhibit reststrahlen transitions.

Finally consider tetragonal crystals with three identical atoms per unit cell. Two classes must be examined. The first is the crystal class with point group D_{6h}. The character table[14] shows that the order of this group is g = 24 and a polar vector transforms as the irreducible representations A_{2u} and E_{1u}. The characters of the s(R) = 0 elements in these representations are $\chi_{A_{2u}}(C_3) = 1$, $\chi_{A_{2u}}(S_6) = -1$, $\chi_{E_{1u}}(C_3) = 1$ and $\chi_{E_{1u}}(S_6) = -1$. Thus the sum in Eq. (12.9.10) equals eight and $gn_\upsilon(ir) \geq 10$. This shows that $n_\upsilon(ir)$ is always greater than zero and crystals of this class with three atoms per unit cell may exhibit reststrahlen transitions. Exactly the same results are obtained from this type of analysis made for the other tetragonal class of crystals with C_{6h} symmetry. In this case the order of the group is g = 12 and a polar vector transforms as the A_u, E'_{1u} and E''_{1u} irreducible representations.

The final step in this analysis is to consider the triclinic C_1 and C_i and the monoclinic C_2, C_s, and C_{2h} crystal symmetries. For three of these symmetries, C_1, C_2, and C_s, the irreducible representations according to which the polar vector transforms includes all the irreducible representations of the group.[4] Then Eq. (12.9.5) can be modified to account for this and to include a factor of $\chi_j^P(1) = \chi_j(1)$,

$$n(ir) = (1/g)\sum_R \left\{ [s(R) - 1]\, \chi_P(R) \sum_{\Gamma_j} [\chi_j(R)^* \chi_j(1)] \right\}. \quad (12.9.14)$$

Since this last factor is just the dimension of the jth irreducible representation of the polar vector, its inclusion takes into account any degeneracy by converting the number of infrared active frequencies $n_\upsilon(ir)$ to the number if infrared active vibrations n(ir). The second sum in Eq. (12.9.14) can be evaluated through the use of orthogonality conditions for irreducible representations. This results in a delta function that can then be used to evaluate the first sum

$$\begin{aligned} \mathrm{n(ir)} &= (1/\mathrm{g}) \sum_{R} [s(R) - 1]\, \chi_P(R)\delta_{R,1} \\ &= [s(1) - 1]\, \chi_P(1) = 3\mathrm{s} - 3. \end{aligned} \tag{12.9.15}$$

In the final expression the facts that the identity operator leaves all the ions in the unit cell invariant and that the dimension of the representation of a polar vector is 3 have been used. Equation (12.9.15) shows that for homopolar crystals with 3 atoms per unit cell, reststrahlen transitions are symmetry allowed.

For crystals with symmetries C_i the order of the group is $g = 2$ with the two operations being the identity and inversion. There are two irreducible representations[4], A_g and A_u. All three components of the polar vector transform as the second of these. Thus Eq. (12.9.7) becomes

$$\begin{aligned} \mathrm{n}_\upsilon(\mathrm{ir}) &= (1/2) \sum_{R=1,\bar{1}} [s(R) - 1]\, \chi_{A_u(x,y,z)}(R) \sum_{A_u} \chi^{A_u}(R)^* \\ &= (1/2)\{[s(1) - 1]\,3 + [s(\bar{1}) - 1]\,3\} = (3/2)\,[s(1) + s(\bar{1}) - 2] \\ &= (3/2)\,[3 + \mathrm{s}(\bar{1})]\,. \end{aligned} \tag{12.9.16}$$

In the last step the assumption of three atoms per unit cell has been made. Since the number of atoms remaining unchanged under inversion will be zero or one, $\mathrm{n}_\upsilon(\mathrm{ir})$ will always be a positive number so reststrahlen transitions may be observed.

The final symmetry group to consider is C_{2h}. The order of the group is $g = 4$ and the polar vector transforms as the A_u (z-components) and B_u (x and y components) irreducible representations. Again using Eq. (12.9.7) yields the same result shown in Eq. (12.9.16) so reststrahlen transitions are symmetry allowed for crystals with 3 atoms per unit cell.

The discussions in this section show how symmetry and group theory considerations can be used to predict the possible presence of one-phonon infrared absorption transitions in the infrared spectra of homopolar crystals. The results show that no reststrahlen transitions will occur in the infrared spectra of crystals of this type with only one or two atoms per unit cell. However, the results of this analysis show that it is possible to have reststrahlen transitions in the infrared spectra of crystals with three identical atoms per unit cell for all crystal symmetry classes. Similar arguments are used in ref. 3 to show that reststrahlen transitions are also symmetry allowed for crystals with greater that three atoms of the same type per unit cell. Thus the condition stated in the previous section for observing reststrahlen transition must be expanded to include both ionic crystals and covalent crystals with three or more atoms per unit cell. However, being allowed by symmetry conditions does not insure that a crystal will exhibit reststrahlen transitions. The size of the dipole moment induced by atomic vibrations of atoms of the same type is determined by the deformability of the charge cloud surrounding each atomic nucleus. If the atom is fairly ridged, the dipole induced by vibrational deformability will be so small that the spectral line strength of a reststrahlen transition may be too small to observe.

12.10. General Characteristics of Raman Scattering from Crystals

Light scattering from crystals is another technique that directly yields information on the vibrational characteristics of solids. As is the case with infrared absorption, light scattering involves the interaction between an external electromagnetic field and the atoms of the solid. The basic difference between the two mechanisms is that in infrared absorption the energy of the incident radiation is in resonance with the energy of the lattice vibrations, while in light scattering the energy of the incident radiation is greater than that of the lattice vibration states but less than that of the electronic excited states. Thus in the former mechanism the incident radiation interacts directly with the fluctuations of the local dipole moments produced by the lattice vibrations, whereas in the latter mechanism the frequency of the fluctuating radiation field is too great for the vibrating atoms to follow. However, the electrons on these atoms can respond to the incident radiation field, and this gives rise to an "induced" oscillating electric dipole moment at each atom. These oscillating dipoles can absorb energy from the radiation field and can reemit radiation of the same frequency. This radiation is detected as scattered light and the phenomenon is known as *Rayleigh scattering.* The oscillations of these induced dipoles can be modulated by the vibrations of the lattice. This is the same as saying that phonons are created or destroyed during a light-scattering process, and thus scattered light can appear at frequencies higher (anti-Stokes lines) and lower (Stokes lines) than the Rayleigh line by amounts equal to the characteristic phonon frequencies. The modulation of light by optical phonons is known as *Raman scattering* while the modulation by acoustical phonons gives rise to *Brillouin scattering.* Higher-order processes may involve a combination of optical and acoustical phonons and produce *high-order scattering.*

Usually, in light-scattering experiments, the frequency distribution of the light scattered is measured at a fixed angle. This information along with the conservation of energy and momentum can be used to determine the phonon frequencies. Every phonon is not active in light scattering, just as every phonon is not active in infrared absorption. Since different physical mechanisms are involved in these two processes, different phonons may be active in producing infrared and light-scattering spectra. Thus these techniques are supplementary in obtaining information on the normal vibrational modes of a crystal.

Light scattering also differs from the other two scattering mechanisms discussed previously, X-ray scattering (Chapter 7) and neutron scattering (Chapter 8). The main difference between X-ray and light scattering is that the energy of an X-ray photon is on the order of 10^4 eV whereas that of a light photon is between about 0.1 and 4 eV. The energy transferred from phonons to photons with a typical order of magnitude of 10 meV is so small that it is difficult to detect when compared to the X-ray energy, whereas it can be measured accurately in light scattering. The situation is essentially reversed for momentum transfer, which can be precisely determined in X-ray scattering but is difficult to measure in light scattering. However, with laser sources it is now possible to have a range of scattering angles from about 1° to about 179°, which corresponds to a difference of the wave vectors of the incident and scattered light of 1.7×10^3 to $2 \times 10^5\,\text{cm}^{-1}$.

Compared to neutron scattering, the energies of the phonons that are active can be measured more accurately by light scattering. However, all of the phonons are active in first-order neutron scattering and the momentum transfer can be measured more accurately by this technique than by light scattering.

Similar to the case of infrared absorption, the wave vector $\underset{\sim}{\alpha}$ of the incident light is much smaller than the dimensions of the Brillouin zone, so that we may set $\underset{\sim}{\alpha} \approx 0$. In first-order scatter-processes, energy and momentum conservation imply

$$\begin{aligned} \omega_{\text{i}} &= \omega_{\text{s}} \pm \omega(\underset{\sim}{k}), \\ \underset{\sim}{\alpha}_{\text{i}} &= \underset{\sim}{\alpha}_{\text{s}} \pm \underset{\sim}{k}, \end{aligned} \tag{12.10.1}$$

where $\underset{\sim}{\alpha}_{\text{i}}$ is the wave vector of incident photon, $\underset{\sim}{\alpha}_{\text{s}}$ is the wave vector of scattered photon, ω_{i} is the frequency of incident photon, ω_{s} is the frequency of scattered photon, $\omega(\underset{\sim}{k})$ is the frequency of phonon, $\underset{\sim}{k}$ is the wave vector of phonon, and the plus and minus signs refer to the process of phonon creation and annihilation, respectively. Since the scattering experiments are generally carried out with photon frequencies where there is no dispersion in the index of refraction, to a good approximation

$$|\underset{\sim}{\alpha}_{\text{i}}| = |\underset{\sim}{\alpha}_{\text{s}}|, \tag{12.10.2}$$

and

$$|\underset{\sim}{k}| = 2|\underset{\sim}{\alpha}_{\text{i}}| \sin\theta = \frac{2\omega_{\text{i}}\eta(\omega_{\text{i}})\sin\theta}{c}, \tag{12.10.3}$$

where 2θ is the angle between the incident and the scattered photons, and where $\eta(\omega_{\text{i}})$ is the index of refraction at $\omega = \omega_{\text{i}}$.

The frequencies of the optical phonon branches near $\underset{\sim}{k} \approx 0$ are quite independent of the wave vectors; for this reason the shifts in frequency for Raman scattering are rather insensitive to changes in θ. On the other hand the frequencies of the acoustical branches are, near $\underset{\sim}{k} \approx 0$, proportional to $|\underset{\sim}{k}|$; therefore, for Brillouin scattering the frequency shifts depend on θ.

12.11. Theory of Raman Scattering

As usual we consider a perfect crystal with J atoms per unit cell and N unit cells, where the electron coordinates are given by $\underset{\sim}{r}_i (i = 1, 2, n)$ and the nuclear coordinates by

$$\underset{\sim}{X}_{s\nu} = \underset{\sim}{R}_{s\nu} + \underset{\sim}{u}_{s\nu}, \tag{12.11.1}$$

where $\underset{\sim}{R}_{s\nu}$ is the equilibrium position of the ν-th atom of the s-th unit cell and $\underset{\sim}{u}_{s\nu}$ is the displacement from the equilibrium position of the ν-th atom of the s-th unit cell. The wave function of the crystal is given, in the adiabatic approximation, by

$$\Psi(\underset{\sim}{r}, \underset{\sim}{X}) = \psi(\underset{\sim}{r}, \underset{\sim}{X})\phi(\underset{\sim}{X}). \tag{12.11.2}$$

We shall now examine the problem of light scattering by crystals by using a semiclassical approach in which the radiation field is treated classically and the lattice vibrations quantum mechanically.[3] For a treatment in which both the radiation field and the lattice vibrations are quantized the reader is referred to Refs. 13 and 14. In the present treatment the interaction between the radiative field and the lattice vibrations is brought about by the interaction of the time-dependent electric field and the electric moment of the crystal; the perturbation is represented by the Hamiltonian

$$H' = -\underset{\sim}{M} \cdot \underset{\sim}{\mathscr{E}}(t), \tag{12.11.3}$$

where the electric field is given by

$$\underset{\sim}{\mathscr{E}}(t) = \underset{\sim}{\mathscr{E}}^{+} e^{i\omega t} + \underset{\sim}{\mathscr{E}}^{-} e^{-i\omega t}, \tag{12.11.4}$$

where $(\underset{\sim}{\mathscr{E}}^{+})^{*} = \underset{\sim}{\mathscr{E}}^{-}$ represents the spatially constant electric field ($\underset{\sim}{\alpha} = o$ approximation) and where the electric dipole moment operator is given, according to (12.4.17), by

$$\underset{\sim}{M}(\underset{\sim}{X}, \underset{\sim}{r}) = \sum_{s\nu} e z_{\nu}^{\text{ion}} \underset{\sim}{x}_{s\nu} - \sum_{s\nu j} e \underset{\sim}{r}_{s\nu j}. \tag{12.11.5}$$

The problem is now one of time-dependent perturbation theory. The time-dependent imperturbed wave functions of the system

($\equiv$ crystal) are given by

$$\psi_i(t) = e^{-i(E_i/\hbar)t}\psi_i(o), \tag{12.11.6}$$

where the $\psi_i(o)$ are orthonormal solutions of the eigenvalue equation

$$H_o\psi_i(o) = E_i\psi_i(o). \tag{12.11.7}$$

In the adiabatic approximation the wavefunctions $\psi_i(o)$ have the form given in (12.11.2).

The perturbed wave functions of the system are solutions of the time-dependent Schrödinger equation

$$\begin{aligned} H\psi_I(t) &= (H_o + H')\psi_I(t) \\ &= (H_o - \underset{\sim}{M} \cdot \underset{\sim}{\mathscr{E}}^+ e^{i\omega t} - \underset{\sim}{M} \cdot \underset{\sim}{\mathscr{E}}^- e^{-i\omega t})\psi_I(t) \\ &= i\hbar\frac{\partial\psi_I}{\partial t}. \end{aligned} \tag{12.11.8}$$

As we have done above, we designate the unperturbed and the perturbed wavefunction with small and capital letters, respectively. We can consider the following trial form for $\psi_I(t)$:

$$\psi_I(t) = \psi_i(o)e^{-i(E_i/\hbar)t} + \left[\psi_i^+ e^{i\omega t} + \psi_i^- e^{-i\omega t}\right] e^{-i(E_i/\hbar)t}, \tag{12.11.9}$$

where ψ_i^+ and ψ_i^- are of the first-order with respect to the perturbation, as will be shown below. Replacing this expression for $\psi_I(t)$ in (12.11.8) and neglecting terms of the second order with respect to the perturbation,

$$\begin{aligned} &H_o\psi_i(o)e^{-i(E_i/\hbar)t} + H_o\left[\psi_i^+ e^{i\omega t} + \psi_i^- e^{-i\omega t}\right] e^{-i(E_i/\hbar)t} \\ &\quad - \left[\underset{\sim}{M} \cdot \underset{\sim}{\mathscr{E}}^+ e^{i\omega t} + \underset{\sim}{M} \cdot \underset{\sim}{\mathscr{E}}^- e^{-i\omega t}\right] \psi_i(o)e^{-i(E_i/\hbar)t} \\ &= E_i\psi_i(o)e^{-i(E_i/\hbar)t} + (E_i - \hbar\omega)\psi_i^+ e^{i\omega t}e^{-i(E_i/\hbar)t} \\ &\quad + (E_i + \hbar\omega)\psi_i^- e^{-i\omega t}e^{i(E_i/\hbar)t}, \end{aligned} \tag{12.11.10}$$

or

$$\begin{aligned} &H_o\psi_i^+ e^{i\omega t} + H_o\psi_i^- e^{-i\omega t} - \underset{\sim}{M} \cdot \underset{\sim}{\mathscr{E}}^+ e^{i\omega t}\psi_i(o) - \underset{\sim}{M} \cdot \underset{\sim}{\mathscr{E}}^- e^{-i\omega t}\psi_i(o) \\ &\quad = (E_i - \hbar\omega)\psi_i^+ e^{i\omega t} + (E_i + \hbar\omega)\psi_i^- e^{-i\omega t}. \end{aligned} \tag{12.11.11}$$

The above relation gives the following two relations:

$$(H_o - E_i \pm \hbar\omega)\psi_i^{\pm} = \underset{\sim}{M} \cdot \mathscr{E}^{\pm}\psi_i(o). \qquad (12.11.12)$$

The functions $\psi^{\pm}$ can be expanded in terms of the unperturbed wavefunctions,

$$\psi_i^{\pm} = \sum_{\ell} c_{\ell}^{\pm}\psi_{\ell}(o). \qquad (12.11.13)$$

Using (12.11.13) in (12.11.12) we obtain

$$(H_o - E_i \pm \hbar\omega)\sum_{\ell} c_{\ell}^{\pm}\psi_{\ell}(o) = \underset{\sim}{M} \cdot \underset{\sim}{\mathscr{E}}^{\pm}\psi_i(o),$$

or

$$\sum_{\ell}(E_{\ell} - E_i \pm \hbar\omega)c_{\ell}^{\pm}\psi_{\ell}(o) = \underset{\sim}{M} \cdot \underset{\sim}{\mathscr{E}}^{\pm}\psi_i(o). \qquad (12.11.14)$$

Multiplying by $\psi_j(o)^*$ and integrating over all the particle coordinates we find

$$c_j^{\pm} = \frac{1}{\hbar}\frac{\langle j|\underset{\sim}{M}|i\rangle \cdot \underset{\sim}{\mathscr{E}}^{\pm}}{\omega_{ji} \pm \omega}, \qquad (12.11.15)$$

where $|i> = \psi_i(o)$ and $\omega_{ji} = (E_j - E_i)\hbar$. The wavefunction $\psi_I(t)$ can now be expressed as

$$\begin{aligned}\psi_I(t) &= \psi_i(o)e^{-i(E_i/\hbar)t} \\ &\quad + \frac{1}{\hbar}\sum_j\sum_{\beta}\frac{\langle j|M_{\beta}|i\rangle}{\omega_{ji} + \omega}\mathscr{E}_{\beta}^{+}e^{i(\omega - E_i/\hbar)t}\psi_j(o) \\ &\quad + \frac{1}{\hbar}\sum_j\sum_{\beta}\frac{\langle j|M_{\beta}|i\rangle}{\omega_{ji} - \omega}\mathscr{E}_{\beta}^{-}e^{-i(\omega + E_i/\hbar)t}\psi_i(o). \qquad (12.11.16)\end{aligned}$$

On the other hand, in correspondence to another perturbed state,

$$\begin{aligned}\psi_F(t)^* &= \psi_f(o)^* e^{i(E_f/\hbar)t} \\ &\quad + \frac{1}{\hbar}\sum_j\sum_{\beta}\frac{\langle f|M_{\beta}|j\rangle\underset{\sim}{\mathscr{E}}_{\beta}^{-}}{\omega_{if} + \omega}e^{-i(\omega - E_f/\hbar)t}\psi_j(o)^* \\ &\quad + \frac{1}{\hbar}\sum_j\frac{\langle f|M_{\beta}|j\rangle\underset{\sim}{\mathscr{E}}_{\beta}^{+}}{\omega_{if} - \omega}e^{i(\omega + E_f/\hbar)t}\psi_j(o)^*. \qquad (12.11.17)\end{aligned}$$

The induced transition electric moment between states ψ_I and ψ_F is given by

$$\int (\psi_F^* \underset{\sim}{M} \psi_I + \psi_I^* \underset{\sim}{M} \psi_F) d\tau. \tag{12.11.18}$$

The α component of this moment is given, neglecting terms of the second order with respect to the perturbation by

$$\begin{aligned}
&\int (\psi_F^* M_\alpha \psi_I + \psi_I^* M_\alpha \psi_F) d\tau \\
&= \langle f|M_\alpha|i\rangle e^{-i\omega_{if}t} + \langle i|M_\alpha|f\rangle e^{i\omega_{if}t} \\
&\quad + \frac{1}{\hbar} \sum_j \sum_\beta \left[\frac{\langle f|M_\beta|j\rangle\langle j|M_\alpha|i\rangle}{\omega_{if} + \omega} \mathscr{E}_\beta^- e^{-i(\omega+\omega_{if})t} \right. \\
&\quad + \frac{\langle i|M_\alpha|j\rangle\langle j|M_\beta|f\rangle}{\omega_{if} + \omega} \mathscr{E}_\beta^+ e^{i(\omega+\omega_{if})t} + \frac{\langle f|M_\beta|j\rangle\langle j|M_\alpha|i\rangle}{\omega_{if} - \omega} \mathscr{E}_\beta^+ e^{i(\omega-\omega_{if})t} \\
&\quad + \frac{\langle i|M_\alpha|j\rangle\langle j|M_\beta|f\rangle}{\omega_{if} - \omega} \mathscr{E}_\beta^- e^{-i(\omega-\omega_{if})t} + \frac{\langle f|M_\alpha|j\rangle\langle j|M_\beta|i\rangle}{\omega_{ji} + \omega} \mathscr{E}_\beta^+ e^{-(\omega-\omega_{if})t} \\
&\quad + \frac{\langle i|M_\beta|j\rangle\langle j|M_\alpha|f\rangle}{\omega_{ji} + \omega} \mathscr{E}_\beta^- e^{-i(\omega-\omega_{if})t} + \frac{\langle f|M_\alpha|j\rangle\langle j|M_\beta|i\rangle}{\omega_{ji} - \omega} \mathscr{E}_\beta^- e^{-i(\omega+\omega_{if})t} \\
&\quad \left. + \frac{\langle i|M_\beta|j\rangle\langle j|M_\alpha|f\rangle}{\omega_{ji} - \omega} \mathscr{E}_\beta^+ e^{i(\omega+\omega_{if})t} \right].
\end{aligned} \tag{12.11.19}$$

The ten terms above account for the following:

a. The first two terms, which do not depend on the electric field, are related to the spontaneous emission from the state ψ_f to the state ψ_i, assuming that the former is higher in energy.
b. The terms with exponentials $e^{\pm i(\omega+\omega_{if})t}$ are related to the scattering process that takes place with the system undergoing the $i \rightarrow f$ transition.

c. The terms with exponentials $e^{\pm i(\omega-\omega_{if})t}$ are related to the scattering process that takes place with the system undergoing the $f \to i$ transition.

12.12. Transition Polarizability

The electric moment that is responsible for the Raman scattering that takes place when the system goes from the state i to the state f can be written as follows:

$$
\begin{aligned}
m_\alpha(t) &= \frac{1}{\hbar}\sum_j \sum_\beta \left\{ \left[\frac{\langle i|M_\alpha|j\rangle\langle j|M_\beta|f\rangle}{\omega_{if}+\omega} \right.\right. \\
&\quad \left. + \frac{\langle i|M_\beta|j\rangle\langle j|M_\alpha|f\rangle}{\omega_{ji}-\omega} \right] \underset{\sim}{\mathscr{E}}{}_\beta^{+} e^{i(\omega+\omega_{if})t} \\
&\quad \left. + \left[\frac{\langle f|M_\beta|j\rangle\langle j|M_\alpha|i\rangle}{\omega_{if}+\omega} + \frac{\langle f|M_\alpha|j\rangle\langle j|M_\beta|i\rangle}{\omega_{ji}-\omega} \right] \underset{\sim}{\mathscr{E}}{}_\beta^{-} e^{-i(\omega+\omega_{if})t} \right\} \\
&= \sum_\beta P_{\alpha\beta}^{if}(\omega) \underset{\sim}{\mathscr{E}}{}_\beta^{+} e^{i(\omega+\omega_{if})t} + \sum_\beta [P_{\alpha\beta}^{if}(\omega)]^* \underset{\sim}{\mathscr{E}}{}_\beta^{-} e^{-i(\omega+\omega_{if})t},
\end{aligned}
\tag{12.12.1}
$$

where

$$
P_{\alpha\beta}^{if}(\omega) = \frac{1}{\hbar}\sum_j \left[\frac{\langle i|M_\alpha|j\rangle\langle j|M_\beta|f\rangle}{\omega_{if}+\omega} + \frac{\langle i|M_\beta|j\rangle\langle j|M_\alpha|f\rangle}{\omega_{ji}-\omega} \right]. \tag{12.12.2}
$$

The tensor defined above is called the *transition polarizability* from state i to state f.

If $i = f$, (12.12.2), becomes

$$
P_{\alpha\beta}^{ii}(\omega) = \frac{1}{\hbar}\sum_j \left[\frac{\langle i|M_\alpha|j\rangle\langle j|M_\beta|i\rangle}{\omega_{ji}+\omega} + \frac{\langle i|M_\beta|j\rangle\langle j|M_\alpha|i\rangle}{\omega_{ji}-\omega} \right], \tag{12.12.3}
$$

which is the *polarizability* of the system in state i. By putting $\omega = 0$ in (12.12.3) we obtain the *static polarizability.*

For Raman scattering by lattice vibrations the initial and final states of the crystal differ only in their vibrational part; we have then

$$\psi_i(\underset{\sim}{r}, \underset{\sim}{X}) = \psi_o(\underset{\sim}{r}, \underset{\sim}{X})\phi_{on}(\underset{\sim}{x}), \tag{12.12.4}$$

$$\psi_f(\underset{\sim}{r}, \underset{\sim}{X}) = \psi_o(\underset{\sim}{r}, \underset{\sim}{X})\phi_{on'}(\underset{\sim}{x}), \tag{12.12.5}$$

where n and n′ designate two sets of phonon occupation numbers; the energies of the two state ψ_i and ψ_f are given by

$$E_i = E(o, n) = \varepsilon_o(\underset{\sim}{R}) + \varepsilon_{on}, \tag{12.12.6}$$

$$E_f = E(o, n') = \varepsilon_o(\underset{\sim}{R}) + \varepsilon_{on'}, \tag{12.12.7}$$

respectively. In the two relations above $\underset{\sim}{R}$ designates the equilibrium nuclear configuration and $\varepsilon_o(\underset{\sim}{R})$ and ε_{on} represent the electronic and vibrational energies, respectively.

The intermediate states that appear in (12.12.2) are given by

$$\Psi_j(\underset{\sim}{r}, \underset{\sim}{X}) = \Psi_j(\underset{\sim}{r}, \underset{\sim}{X})\phi_{jn''}(\underset{\sim}{X}), \tag{12.12.8}$$

with energies given by

$$E_j = E(jn'') = \varepsilon_j(\underset{\sim}{R}) + \varepsilon_{jn''}. \tag{12.12.9}$$

With these provisions the transition polarizability (12.12.2) becomes:

$$\begin{aligned}
P_{\alpha\beta}^{nn'} &= \sum_{jn''}\left[\frac{\langle on|M_\alpha|jn''\rangle\langle jn''|M_\beta|on'\rangle}{E(jn'') - E(on') + \hbar\omega} + \frac{\langle on|M_\beta|jn''\rangle\langle jn''|M_\alpha|on'\rangle}{E(jn'') - E(on) + \hbar\omega}\right] \\
&\approx \sum_{jn''}\left[\frac{\langle on|M_\alpha|on''\rangle\langle on''|M_\beta|on'\rangle}{\varepsilon_{on''} - \varepsilon_{on'} + \hbar\omega} + \frac{\langle on|M_\beta|on''\rangle\langle on''|M_\alpha|on'\rangle}{\varepsilon_{on} - \varepsilon_{on} - \hbar\omega}\right] \\
&\quad + \sum_{j\neq 0}\left[\frac{\sum_{n''}\langle on|M_\alpha|jn''\rangle\langle jn''|M_\beta|on'\rangle}{\varepsilon_j - \varepsilon_o + \hbar\omega}\right. \\
&\quad \left. + \frac{\sum_{n''}\langle on|M_\alpha|jn''\rangle\langle jn''|M_\alpha|on'\rangle}{\varepsilon_j - \varepsilon_o - \hbar\omega}\right],
\end{aligned} \tag{12.12.10}$$

where we have made the approximations

$$E(jn'') - E(on') = (\varepsilon_j + \varepsilon_{jn''}) - (\varepsilon_o + \varepsilon_{on'}) \approx \varepsilon_j - \varepsilon_o, \quad (12.12.11)$$

$$E(jn'') - E(on) = (\varepsilon_j + \varepsilon_{jn''}) - (\varepsilon_o + \varepsilon_{on}) \approx \varepsilon_j - \varepsilon_{o'}. \quad (12.12.12)$$

We want now to consider the two sums in (12.12.10) separately. First let us take the first sum over n″; a matrix element that appears in it can be written as follows:

$$\begin{aligned}
&\langle on|M_\alpha(\underset{\sim}{r}, \underset{\sim}{X})|on''\rangle \\
&= \iint \psi_o(\underset{\sim}{r}, \underset{\sim}{X})^* \phi_{on}(\underset{\sim}{X})^* M_\alpha(\underset{\sim}{r}, \underset{\sim}{X}) \psi_o(\underset{\sim}{r}, \underset{\sim}{X}) \phi_{on''}(\underset{\sim}{X}) d^3\underset{\sim}{r} d^3\underset{\sim}{X} \\
&= \int \phi_{on}(\underset{\sim}{X})^* \left[\int \psi_o(\underset{\sim}{r}, \underset{\sim}{X})^* M_\alpha(\underset{\sim}{r}, \underset{\sim}{X}) \psi_o(\underset{\sim}{r}, \underset{\sim}{X}) d^3\underset{\sim}{r}\right] \phi_{on''}(\underset{\sim}{X}) d^3\underset{\sim}{X} \\
&= \int \phi_{on}(\underset{\sim}{X})^* M^\alpha_{oo}(\underset{\sim}{X}) \phi_{on''}(\underset{\sim}{X}) d^3\underset{\sim}{X} = \langle n|M^\alpha_{oo}(\underset{\sim}{X})|n''\rangle, \quad (12.12.13)
\end{aligned}$$

where

$$M^\alpha_{oo}(\underset{\sim}{X}) = \int \psi_o(\underset{\sim}{r}, \underset{\sim}{X})^* M_\alpha(\underset{\sim}{r}, \underset{\sim}{X}) \psi_o(\underset{\sim}{r}, \underset{\sim}{X}) d^3\underset{\sim}{X}. \quad (12.12.14)$$

The first sum in (12.11.10) can now be written as

$$\sum_{n''} \left[\frac{\langle n|M^\alpha_{oo}(\underset{\sim}{X})|n''\rangle\langle n''|M^\beta_{oo}(\underset{\sim}{X})|n'\rangle}{\varepsilon_{on''} - \varepsilon_{on'} + \hbar\omega} + \frac{\langle n|M^\beta_{oo}(\underset{\sim}{X})|n''\rangle\langle n''|M^\alpha_{oo}(\underset{\sim}{X})|n'\rangle}{\varepsilon_{on''} - \varepsilon_{on} - \hbar\omega}\right]. \quad (12.12.15)$$

This part of the transition polarizability is called the *ionic part of the polarizability* because it does not correspond to any electronic transitions.

In the second sum in (12.12.10) we have

$$\begin{aligned}
&\sum_{n''} \langle on|M_\alpha|jn''\rangle\langle jn''|M_\beta|on'\rangle \\
&\quad = \sum_{n''} \left[\int \psi_o(\underset{\sim}{r}, \underset{\sim}{X})^* \phi_{on}(\underset{\sim}{X})^* M_\alpha(\underset{\sim}{r}, \underset{\sim}{X}) \psi_j(\underset{\sim}{r}, \underset{\sim}{X}) \phi_{jn''}(\underset{\sim}{X}) d^3\underset{\sim}{r} d^3\underset{\sim}{X}\right]
\end{aligned}$$

$$\times \left[\int \psi_j(\underset{\sim}{r}', \underset{\sim}{X}')^* \phi_{jn''}(\underset{\sim}{X}')^* M_\beta(\underset{\sim}{r}', \underset{\sim}{X}') \psi_o(\underset{\sim}{r}', \underset{\sim}{X}') \phi_{on'}(\underset{\sim}{X}') d^3\underset{\sim}{r}' d^3\underset{\sim}{X}' \right]$$

$$= \int d^3\underset{\sim}{r} \int d^3\underset{\sim}{r}' \int d^3\underset{\sim}{X} \int d^3\underset{\sim}{X}' \phi_{on}(\underset{\sim}{X})^* \left[\psi_o(\underset{\sim}{r}, \underset{\sim}{X})^* M_\alpha(\underset{\sim}{r}, \underset{\sim}{X}) \psi_j(\underset{\sim}{r}, \underset{\sim}{X}) \right]$$

$$\times \left[\sum_n \phi_{jn''}(\underset{\sim}{X}) \phi_{jn''}(\underset{\sim}{X}')^* \right] \left[\psi_j(\underset{\sim}{r}', \underset{\sim}{X}')^* M_\beta(\underset{\sim}{r}', \underset{\sim}{X}') \psi_o(\underset{\sim}{r}', \underset{\sim}{X}') \right] \phi_{on'}(\underset{\sim}{X}')$$

$$= \int \phi_{on}(\underset{\sim}{X}) \left[\int \psi_o(\underset{\sim}{r}, \underset{\sim}{X})^* M_\alpha(\underset{\sim}{r}, \underset{\sim}{X}) \psi_j(\underset{\sim}{r}, \underset{\sim}{X}) d^3\underset{\sim}{r} \right]$$

$$\times \left[\int \psi_j(\underset{\sim}{r}', \underset{\sim}{X})^* M_\beta(\underset{\sim}{r}', \underset{\sim}{X}) \psi_o(\underset{\sim}{r}', \underset{\sim}{X}) d^3\underset{\sim}{r}' \right] \phi_{on'}(\underset{\sim}{X}) d^3\underset{\sim}{X}$$

$$= \langle n | M^\alpha_{oj}(\underset{\sim}{X}) M^\beta_{jo} | n' \rangle, \tag{12.12.16}$$

where we have taken advantage of the relation

$$\sum_{n''} \phi_{jn''}(\underset{\sim}{X}')^* \phi_{jn''}(\underset{\sim}{X}) = \delta(\underset{\sim}{X} - \underset{\sim}{X}'), \tag{12.12.17}$$

and where

$$M^\alpha_{ij}(\underset{\sim}{X}) = \int \psi_i(\underset{\sim}{r}, \underset{\sim}{X})^* M_\alpha(\underset{\sim}{r}, \underset{\sim}{X}) \psi_j(\underset{\sim}{r}, \underset{\sim}{X}) d^3\underset{\sim}{r}. \tag{12.12.18}$$

The second sum in (12.11.10) can now be written as

$$\sum_{j \neq 0} \left[\frac{\langle n | M^\alpha_{oj}(\underset{\sim}{X}) M^\beta_{jo}(\underset{\sim}{X}) | n' \rangle}{\varepsilon_j - \varepsilon_o + \hbar\omega} + \frac{\langle n | M^\beta_{oj}(\underset{\sim}{X}) M^\alpha_{jo}(\underset{\sim}{X}) | n' \rangle}{\varepsilon_j - \varepsilon_o - \hbar\omega} \right]$$

$$= \langle n | \sum_{j \neq 0} \left[\frac{M^\alpha_{oj}(\underset{\sim}{X}) M^\beta_{jo}(\underset{\sim}{X})}{\varepsilon_j - \varepsilon_o + \hbar\omega} + \frac{M^\beta_{oj}(\underset{\sim}{X}) M^\alpha_{jo}(\underset{\sim}{X})}{\varepsilon_j - \varepsilon_o - \hbar\omega} \right] | n' \rangle$$

$$= \langle n | P_{\alpha\beta}(\omega, \underset{\sim}{X}) | n' \rangle, \tag{12.12.19}$$

where

$$P_{\alpha\beta}(\omega,\underset{\sim}{X}) = \sum_{j\neq 0}\left[\frac{M^{\alpha}_{oj}(\underset{\sim}{X})M^{\beta}_{jo}(\underset{\sim}{X})}{\varepsilon_j-\varepsilon_o+\hbar\omega}+\frac{M^{\beta}_{oj}(\underset{\sim}{X})M^{\alpha}_{jo}(\underset{\sim}{X})}{\varepsilon_j-\varepsilon_o-\hbar\omega}\right]. \tag{12.12.20}$$

This part of the polarizability is called the *electronic part of the polarizability.*

Taking now (12.11.15) and (12.11.19) into account we can express (12.11.10) as

$$\begin{aligned} P^{nn'}_{\alpha\beta} = \sum_{n''}\Bigg[&\frac{\langle n|M^{\alpha}_{oo}(\underset{\sim}{X})|n''\rangle\langle n''|M^{\beta}_{oo}(\underset{\sim}{X})|n'\rangle}{\varepsilon_{on''}-\varepsilon_{on'}+\hbar\omega}\\ &+\frac{\langle n|M^{\beta}_{oo}(\underset{\sim}{X})|n''\rangle\langle n''|M^{\alpha}_{oo}(\underset{\sim}{X})|n'\rangle}{\varepsilon_{on''}-\varepsilon_{on'}-\hbar\omega}\Bigg]\\ &+\sum_{j\neq 0}\left[\frac{\langle n|M^{\alpha}_{oj}(\underset{\sim}{X})M^{\beta}_{jo}(\underset{\sim}{X})|n'\rangle}{\varepsilon_j-\varepsilon_o+\hbar\omega}+\frac{\langle n|M^{\beta}_{oj}(\underset{\sim}{X})M^{\alpha}_{jo}(\underset{\sim}{X})|n'\rangle}{\varepsilon_j-\varepsilon_o-\hbar\omega}\right]. \end{aligned} \tag{12.12.21}$$

The static polarizability can be obtained by setting $\omega = o$ and $n' = n$ is the formula above; it is known[3] that for ionic crystals the ionic and the electronic portions contribute roughly equal amount to the static polarizability. In Raman scattering experiments, in general, it is always

$$\hbar\omega \gg |E(on'') - E(on)|, \tag{12.12.22}$$

and the electronic part of the polarizability will predominate over the ionic part. Therefore we neglect the ionic contribution to the transition polarizability and simply write

$$P^{nn'}_{\alpha\beta}(\omega) = \langle n|P_{\alpha\beta}(\omega,\underset{\sim}{X})|n'\rangle. \tag{12.12.23}$$

12.13. Energy Scattered in Raman Scattering Experiments

Following the considerations of the previous section the electric moment associated with a transition $n \to n'$ can be written

$$\begin{aligned} m_\alpha(t) &= \left\{ \sum_\beta P_{\alpha\beta}^{nn'}(\omega)\mathscr{E}_\beta^+ \right\} e^{i(\omega+\omega_{nn'})t} \\ &\quad + \left\{ \sum_\beta \left[P_{\alpha\beta}^{nn'}(\omega)\right]^* \mathscr{E}_\beta^- \right\} e^{-i(\omega+\omega_{nn'})t} \\ &= m_\alpha^+ e^{i(\omega+\omega_{nn'})t} + m_\alpha^- e^{-i(\omega+\omega_{nn'})t}, \end{aligned} \tag{12.13.1}$$

where

$$m_\alpha^+ = \sum_\beta P_{\alpha\beta}^{nn'}(\omega)\mathscr{E}_\beta^+, \tag{12.13.2}$$

$$m_\alpha^- = \sum_\beta \left[P_{\alpha\beta}^{nn'}(\omega)\right]^* \mathscr{E}_\beta^-, \tag{12.13.3}$$

$$\omega_{nn'} = (\varepsilon_{on} - \varepsilon_{on'}/\hbar). \tag{12.13.4}$$

The oscillating dipole $\underset{\sim}{m}(\varepsilon)$ produces at large distance $\underset{\sim}{L}$ electric and magnetic fields given by[17]

$$\underset{\sim}{\mathscr{E}}\left(t + \frac{L}{c}\right) = -\frac{(\omega+\omega_{nn'})^2}{Lc^2}\{\hat{\underset{\sim}{n}} \times [\hat{\underset{\sim}{n}} \times \underset{\sim}{m}(t)]\}, \tag{12.13.5}$$

$$\underset{\sim}{\mathscr{K}}\left(t + \frac{L}{c}\right) = -\frac{(\omega+\omega_{nn'})^2}{Lc^2}[\hat{\underset{\sim}{n}} \times \underset{\sim}{m}(t)], \tag{12.13.6}$$

where $\hat{n}$ is the unit vector in the director of $\underset{\sim}{L}$. The magnitude of the Poynting vector at $\underset{\sim}{L}$ is given by

$$\begin{aligned} S &= \frac{c}{4\pi}|\underset{\sim}{\mathscr{E}} \times \underset{\sim}{\mathscr{K}}| = \frac{c}{4\pi}\mathscr{E}^2 \\ &= \frac{(\omega+\omega_{nn'})^4}{4\pi L^2c^3}|\hat{\underset{\sim}{n}} \times [\hat{\underset{\sim}{n}} \times \underset{\sim}{m}(t)]\,|^2. \end{aligned} \tag{12.13.7}$$

The vector $\hat{\underset{\sim}{n}} \times [\hat{\underset{\sim}{n}} \times \underset{\sim}{m}(t)]$ is the projection of $\underset{\sim}{m}$ on a plane perpendicular to $\underset{\sim}{L}$; we take $\hat{\underset{\sim}{n}}^1$ and $\hat{\underset{\sim}{n}}^2$ as two mutually perpendicular unit vectors lying in such a plane and forming with $\underset{\sim}{\hat{n}}$ a Cartesian system of axes. It will then be

$$
\begin{aligned}
|\underset{\sim}{\hat{n}} \times [\underset{\sim}{\hat{n}} \times \underset{\sim}{m}]|^2 &= (\underset{\sim}{\hat{n}}^1 \cdot \underset{\sim}{m})^2 + (\underset{\sim}{\hat{n}}^2 \cdot \underset{\sim}{m})^2 \\
&= \left(\sum_\alpha n_\alpha^1 m_\alpha\right)^2 + \left(\sum_\alpha n_\alpha^2 m_\alpha\right)^2 \\
&= \sum_\alpha \sum_\gamma n_\alpha^1 n_\gamma^1 m_\alpha m_\gamma + \sum_\alpha \sum_\gamma n_\alpha^2 n_\gamma^2 m_\alpha m_\gamma \\
&= \sum_{i=1}^{2} \sum_{\alpha\gamma} n_\alpha^i m_\alpha m_\gamma .
\end{aligned}
\tag{12.13.8}
$$

Using this result in (12.13.7) we obtain

$$
\begin{aligned}
S = \frac{(\omega + \omega_{nn'})^4}{4\pi L^2 c^3} \sum_{i=1}^{2} \sum_{\alpha\gamma} n_\alpha^i n_\gamma^i \left[m_\alpha^+ e^{i(\omega+\omega_{nn'})t} + m_\alpha^- e^{-i(\omega+\omega_{nn'})t}\right] \\
\times \left[m_\gamma^+ e^{i(\omega+\omega_{nn'})t} + m_\gamma^- e^{-i(\omega+\omega_{nn'})t}\right] .
\end{aligned}
\tag{12.13.9}
$$

Taking the time average of S over a period of the vibration of the oscillating dipole we obtain

$$
\begin{aligned}
\bar{S} &= \frac{(\omega + \omega_{nn'})^4}{2\pi L^2 c^3} \sum_{i=1}^{2} \sum_{\alpha\gamma} n_\alpha^i n_\gamma^i m_\alpha^+ m_\gamma^- \\
&= \frac{(\omega + \omega_{nn'})^4}{2\pi L^2 c^3} \sum_{i=1}^{2} \sum_{\alpha\gamma} \sum_{\beta\delta} n_\alpha^i n_\gamma^i P_{\alpha\beta}^{nn'}(\omega) \left[P_{\gamma\delta}^{nn'}(\omega)\right]^* \mathscr{E}_\beta^+ \mathscr{E}_\delta^- .
\end{aligned}
\tag{12.13.10}
$$

The energy scattered per unit time into the solid ample $d\Omega$ is then given by

$$
\bar{S} L^2 d\Omega = \frac{(\omega + \omega_{nn'})^4}{2\pi c^3} \sum_{i=1}^{2} \sum_{\alpha\gamma} \sum_{\beta\delta} n_\alpha^i n_\gamma^i P_{\alpha\beta}^{nn'}(\omega) \left[P_{\gamma\delta}^{nn'}(\omega)\right]^* \mathscr{E}_\beta^+ \mathscr{E}_\delta^- d\Omega .
\tag{12.13.11}
$$

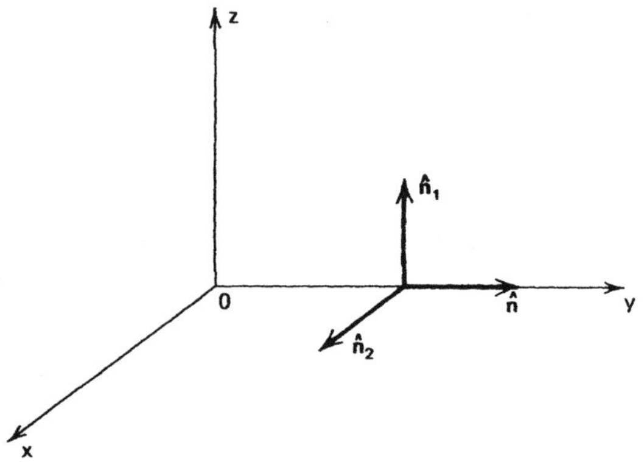

Fig. 12.3. Diagram illustrating a possible Raman scattering experiment.

In order to illustrate this result let us consider the following Raman scattering experiment. The scattering target is located at O in Fig. 12.3, the incoming light of frequency ω is polarized in the z direction and is traveling in the x direction. We wish to know how much light power at frequency $(\omega + \omega_{nn'})$ is scattered into the unit solid angle in the y direction with light polarization in the x direction. In these condition it is

$$\mathscr{E}_x^{\pm} = \mathscr{E}_y^{\pm} = 0, \quad \mathscr{E}_z^{+} = (\mathscr{E}^{-})^* \neq 0,$$
$$\underset{\sim}{\hat{n}}^1 = \underset{\sim}{1}_z; \quad \underset{\sim}{\hat{n}}^2 = \underset{\sim}{1}_x, \quad \underset{\sim}{\hat{n}} = \underset{\sim}{1}_y.$$

The energy scattered per unit time into the solid angle $d\Omega$ in the y direction with light polarization in the x direction is given by

$$\frac{(\omega + \omega_{nn'})^4}{4\pi c^3} P_{xz}^{nn'}(\omega) \left[P_{xz}^{nn'}(\omega)\right]^* (\mathscr{E}_z^+ \mathscr{E}_z^- + \mathscr{E}_z^- \mathscr{E}_z^+)$$
$$= \frac{(\omega + \omega_{nn'})^4}{4\pi c^3} |P_{xz}^{nn'}(\omega)|^2 (\mathscr{E}_z^+ \mathscr{E}_z^- + \mathscr{E}_z^- \mathscr{E}_z^+) d\Omega. \quad (12.13.12)$$

If we divide this quantity by the intensity of the incident light $(c/4\pi)(\mathscr{E}_z^+ \mathscr{E}_z^- + \mathscr{E}_z^- \mathscr{E}_z^+)$ we obtain the scattering cross section:

$$(d\sigma)_{n'x;nz} = \frac{(\omega + \omega_{nn'})^4}{c^4} |P_{xz}^{nn'}(\omega)|^2 d\Omega. \quad (12.13.13)$$

It is interesting to compare the formula above with the quantum-mechanical formula (9.6.42) and to note their equivalence if in the latter formula $\hbar\omega_i$ and $\hbar\omega_f$ are negligible in comparison to $|E_j - E_i|$.

In the usual Raman scattering experiments, the scattered radiation is emitted *spontaneously* by the scatterer. The availability of high densities of incident photons when using laser light sources has made it possible to observe *stimulated Raman scattering.*[18,19] In this case the scattered photons are produced in a time interval short enough and with such high density that they can stimulate the emission of other scattered photons. For treating this effect it is necessary to quantize the radiative field; this leads to the introduction of factors $[n(\omega_s) + 1]$ where $n(\omega_s)$ is the number of scattered photons at frequency $\omega_s = \omega + \omega_{nn'}$ is the initial state of the system. These factors may account for the stimulated emission.

12.14. Selection Rules for Raman Scattering

The transition polarizability tensor is given, according to (12.12.23), by

$$P^{nn'}_{\alpha\beta}(\omega) = \langle n|P_{\alpha\beta}(\omega, \underset{\sim}{X})|n'\rangle, \tag{12.14.1}$$

where

$$P_{\alpha\beta}(\omega, \underset{\sim}{X}) = \sum_{j\neq 0}\left[\frac{M^{\alpha}_{oj}(\underset{\sim}{X})M^{\beta}_{jo}(\underset{\sim}{X})}{\varepsilon_j - \varepsilon_o + \hbar\omega} + \frac{M^{\beta}_{oj}(\underset{\sim}{X})M^{\alpha}_{jo}(\underset{\sim}{X})}{\varepsilon_j - \varepsilon_o - \hbar\omega}\right]. \tag{12.14.2}$$

The tensor $P_{\alpha\beta}(\omega, \underset{\sim}{X})$ can be expanded in a power series of normal coordinates about the nuclear equilibrium configuration,

$$\begin{aligned} P_{\alpha\beta}(\omega, \underset{\sim}{X}) &= P^{(o)}_{\alpha\beta}(\omega, \underset{\sim}{X}) + \sum_{\underset{\sim}{k}j}\left[\frac{\partial}{\partial q^j_{\underset{\sim}{k}}}P_{\alpha\beta}(\omega, \underset{\sim}{X})\right]_o q^j_{\underset{\sim}{k}} \\ &\quad + \frac{1}{2}\left[\sum_{\underset{\sim}{k}j}\sum_{\underset{\sim}{k}'j'}\frac{\partial^2}{\partial q^j_{\underset{\sim}{k}}\partial q^{j'}_{\underset{\sim}{k}'}}P_{\alpha\beta}(\omega, \underset{\sim}{X})\right]_o q^j_{\underset{\sim}{k}} q^{j'}_{\underset{\sim}{k}'} + \cdots. \end{aligned} \tag{12.14.3}$$

Since the normal coordinates can be expressed in terms of creation and annihilation phonon operators, it is clear that the first term in the expansion above corresponds to Rayleigh scattering, the second term to one-phonon Raman scattering, the third term to two-phonon Raman scattering, and so on. In any case the creation of a $(\underset{\sim}{k}, j)$

phonon results always in the appearance of a factor $(n^{j}_{\underset{\sim}{k}} + 1)$; the annihilation of a phonon (k', j') in the appearance of a factor $(n^{j'}_{\underset{\sim}{k'}})^{\frac{1}{2}}$. These factors we introduce will be responsible for the temperature dependence of the scattered light intensity.

In Raman scattering experiments the frequency ω of the incident light is in general such that

$$\hbar\omega \ll (\varepsilon_i - \varepsilon_o). \tag{12.14.4}$$

Therefore it is a good approximation to retain in the denominators in (12.14.2) only the energy differences $(\varepsilon_j - \varepsilon_o)$ and write

$$\begin{aligned} P_{\alpha\beta}(\omega, \underset{\sim}{X}) &\approx P_{\alpha\beta}(\underset{\sim}{X}) \\ &= \sum_{j \neq 0} \left[\frac{M^{\alpha}_{oj}(\underset{\sim}{X}) M^{\beta}_{jo}(\underset{\sim}{X}) + M^{\beta}_{oj}(\underset{\sim}{X}) M^{\alpha}_{jo}(\underset{\sim}{X})}{\varepsilon_o - \varepsilon_j} \right]. \\ &= P_{\beta\alpha}(\underset{\sim}{X}) \end{aligned} \tag{12.14.5}$$

Within this approximation $P_{\alpha\beta}(\underset{\sim}{X})$ is a symmetric tensor.

Since $\underset{\sim}{M}_{oj}(\underset{\sim}{X})$ and $\underset{\sim}{M}_{jo}(\underset{\sim}{X})$ transform as polar vectors under (proper or improper) rotations, the symmetric tensor $P_{\alpha\beta}(\underset{\sim}{X})$ must transform as a second-order rank tensor; that is its component must transform under rotations as second-degree terms in x, y, and z.

Let us consider now the selection rules of the matrix element

$$\langle n | P_{\alpha\beta}(\underset{\sim}{X}) | n' \rangle. \tag{12.14.6}$$

We note that in the expansion of $P_{\alpha\beta}(\omega, \underset{\sim}{X})$ in normal coordinates (12.14.3), only those terms of the expansion that transform as $P_{\alpha\beta}(\omega, \underset{\sim}{X})$ are different from zero. Also, the coefficients of the expansion are quantities calculated in correspondence to the equilibrium nuclear configuration and, as such, are invariant under the operators of the space group G of the crystal. In the approximation (12.14.5), $P_{\alpha\beta}(\omega, \underset{\sim}{X}) \approx P_{\alpha\beta}(\underset{\sim}{X})$ transforms as a second-rank tensor; therefore the coefficients in the first-order terms will be different from zero only for those normal coordinates $q^{j}_{\underset{\sim}{k}}$ $(\underset{\sim}{k} = 0)$ that transform under the operators of the space group G according to the irreducible representations of the crystallographic point group G_o that have as basis functions second-degree terms in x, y, and z. For single-phonon Raman scattering only the normal modes that correspond to these representations can be activated.

At low enough temperatures all the phonon oscillators are in their ground state ($n^j_{\tilde{k}} \approx 0$) and the initial vibrational state $|n\rangle$ transforms as the identity representation. In this case a single-phonon Raman scattering occurs if a normal coordinate for $\tilde{k} = 0$ and one or more of the polarization tensor components $(x^2, y^2, z^2, xy, yz, zx)$ belong to the same irreducible representation of the crystallographic group G_o.

An interesting result can be derived from the fact that the first-order Raman-active modes transform as $(x^2, y^2, z^2, xy, yz, zx)$ whereas the first-order infrared-active modes transform as (x, y, z). If the point group G_o contains the inversion operation, its irreducible representations are either odd or even, and therefore no mode can be at the same time infrared active and Raman active for one-phonon processes.

In two-phonon Raman processes, the two modes participating in the process may have, in general, wave vectors different from zero, and as such do not transform according to irreducible representations of the point group G_o, but rather, according to irreducible representations of the space group G. The product of two normal coordinates $q^j_{\tilde{k}} q^{j'}_{\tilde{k}'}$ transforms as the direct product of the two related space group representations. If this direct product representation, when reduced into irreducible representations of the space group, contains one of the irreducible representations of G_o that have x^2, y^2, z^2, xy, yz, or zx as basis functions, then the corresponding term in the expansion (12.3.3) of the tensor $P_{\alpha\beta}(\tilde{X})$ may have its coefficient different from zero and the two modes $(\tilde{k}, j)$ and $(\tilde{k}', j')$ may participate in the scattering process. Such selection rules, as shown by Cornwell,[10] require the conservation of $\tilde{k}$ vector within a primitive $\tilde{K}$ vector of the reciprocal lattice.

12.15. The Effect of Impurities on Raman Scattering

When impurity ions are present in the crystal the translational invariance is destroyed and phonons of all wave vectors can be active in first-order Raman scattering, instead of only phonons with $\tilde{k} \approx 0$;

local phonon modes may also be active. The first-order Raman spectrum is no longer a line spectrum as in the case of perfect crystal but, rather, is continuous with a structure reflecting the singularities of the phonon spectrum of the perfect crystal and the presence of possible localized modes.[20]

12.16. Brillouin Scattering

As discussed above, Raman scattering involves the absorption or emission of optical phonons. Phonons in the acoustical branch shown in Fig. 6.5 can also be involved in light scattering processes. In the long wavelength limit, the crystal lattice is treated as a continuum for light traveling through the material. Any physical process that produces a local change in the refractive index of the material will cause light to be scattered. For the case of Raman scattering this is best attributed to the local change in polarizability associated with the modes of vibration of neighboring ions (i.e. optical phonons). For the longer wavelength acoustical phonons the change in refractive index is best attributed to a change in density. Light scattering involving the absorption or emission of acoustical phonons is called *Brillouin Scattering*. It can be treated as the scattering of light by density fluctuations traveling at the velocity of sound in the material.

A generic light scattering spectrum for a transparent crystal is shown in Fig. 12.4. In the center of the spectrum is the *Rayleigh line* produced by elastic scattering of photons. Light traveling through the crystal is scattered by random density fluctuations not associated with phonons. To the low energy side of the Rayleigh line are lines associated with inelastic scattering of photons involving the emission of phonons. These are called *Stokes* processes and the shift in frequency from the Rayleigh line is equal to the frequency of the phonon that is created in the process. This is greater for Raman spectral lines associated with high frequency optical phonons than it is for Brillouin spectral lines associated with low frequency acoustical phonons. To the high energy side of the Rayleigh line are the Raman and Brillouin spectral line for *anti-Stokes* processes that involve the absorption of optical and acoustical phonons, respectively. Using the requirements

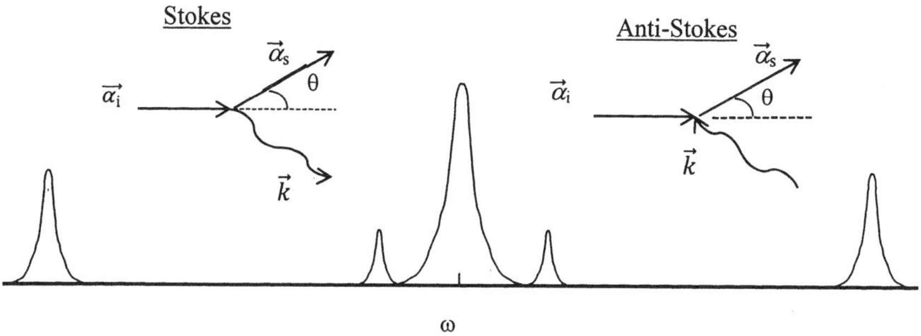

Fig. 12.4. Generic light scattering spectrum.

of conservation of energy and momentum in the scattering processes, information about the phonons involved can be obtained.

Treating Brillouin scattering from a classical physical perspective provides insight into the physics of the process and the fundamental information that can be obtained from Brillouin scattering experiments. In this approach the physical description of the scattering processes is the same as the Bragg diffraction problem discussed in Chapter 5. The geometry for Bragg diffraction of x-rays from a crystal lattice is shown in Fig. 5.6. A similar picture of Brillouin scattering is shown in Fig. 12.5. In both cases the requirement for having a maximum in the intensity of the scattered beam requires constructive interference of the beams scattered from neighboring peaks in the scattering medium. This requirement, along with the requirements of conservation of energy and momentum, to the following equations:

$$\vec{\alpha}_s = \vec{\alpha}_i \mp \vec{k}, \quad \omega_s = \omega_i \mp \omega(\vec{k}), \quad m\lambda/n = 2\Lambda_s \sin(\phi/2). \quad (12.16.1)$$

The main differences between the scattering events shown in Figs. 5.6 and 12.5 are that the first case involves scattering from a static structure with lattice spacing d, while the latter involves scattering from a wave traveling at the speed of sound V_s with a wavelength Λ_s. In Fig. 12.4 the scattering angle ϕ is measured between the incident and scattered directions of the light. The third equation in Eq. (12.16.1) results from the requirement for constructive interference as described in Sec. 5.2. The maximum scattering occurs for m = 1 and $\phi = 180°$. Thus for Brillouin "backscattering"

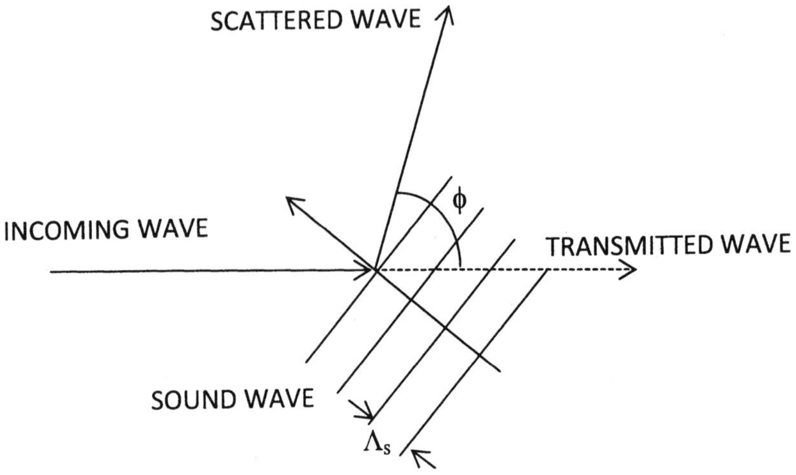

Fig. 12.5. Light scattering from an acoustic wave.

the frequency shift is given by

$$\nu_B = \frac{2n\mathrm{v_a}}{\lambda_{vac}}, \tag{12.16.2}$$

where λ_{vac} is the photon wavelength in a vacuum, n is the refractive index of the material, and $\mathrm{v_a}$ is the speed of sound in the direction of the incident light wave.

Thus measuring the frequency shift of a Brillouin scattering spectral line from the Rayleigh line provides information about the acoustic phonons of the crystal and the speed of sound waves in the crystal. By varying the geometry of the scattering experiment as described Sec. 12.14 Raman scattering, anisotropics of the acoustic properties of a crystal can be elucidated.

The quantum mechanical treatment of the scattering of electromagnetic waves using time-dependent perturbation theory was developed in Sec. 12.12 for Raman scattering. The same formalism applies to Brillouin scattering except that the interaction Hamiltonian in Sec. 12.12 is associated with electrostriction caused by the incident light wave instead of an intrinsic electric dipole moment of the crystal. The steps in the development of the scattering transition probability outlined in Sec. 12.12 will not be repeated here. The major difference in the results is that the Brillouin scattering

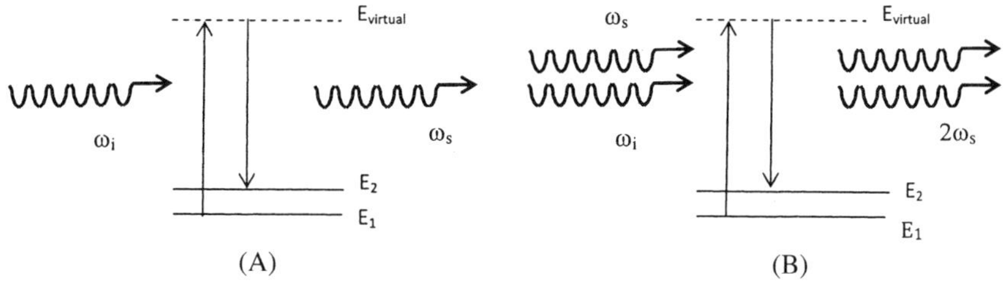

Fig. 12.6. Brillouin Stokes scattering. (A) Spontaneous scattering. (B) Stimulated scattering.

probability and intensity depend on the electrostrictive constant of the material divided by its mean density instead of the intrinsic dipole moment.[21]

With both Brillouin and Raman scattering, if the scattered beam becomes very intense stimulated emission can occur as shown in the schematic energy level diagram in Fig. 12.6 Energy level E_1 is the ground vibrational state of the crystal while E_2 represents a one-phonon excited state. For spontaneous scattering shown in Fig. 12.6(A), the incoming optical photon causes a transition from the ground state to a virtual intermediate electronic state while the scattered photon originates from a transition between the virtual level and the first excited vibrational level. In this Stokes scattering process, some of the incident photon energy is transferred to the lattice to create a phonon. In the case of stimulated scattering shown in Fig. 12.6(B), both an incident photon and a Stokes shifted photon interact with the scattering region. Stimulated emission results in two outgoing Stokes photons. Thus the Stokes beam experiences gain as it travels through the material. For the case of Raman scattering this can be used to produce Raman lasers as discussed in Chapter 13. For the case of Brillouin scattering this can be used to produce phase conjugation to enhance the quality of an optical beam. One of the most important practical effects of Brillouin scattering occurs in optical fibers. The transmission of laser light through small diameter glass fibers with minimum loss in power is important in applications such as communications and medical delivery systems. Stimulated Brillouin backscattering has been identified as one of the major loss mechanisms for optical transmission in a fiber.[22]

Understanding Brillouin scattering is important to minimizing this deleterious effect.

References

1. C. Kittel, *Introduction to Solid State Physics*, Wiley, New York, 1957, p. 113.
2. M. Lax and E. Burstein, *Phys. Rev.* 97, 39 (1955).
3. S. S. Mitra, "Infrared and Raman Spectra Due to Lattice Vibrations," in *Optical Properties of Solids*, edited by S. Nudelman and S. S. Mitra, Plenum, New York, 1969, p. 333.
4. J. T. Houghton and S. D. Smith, *Infra-Red Physics*, Oxford Univ. Press, Iondon, 1966, p. 102.
5. M. Born and K. Huang, *Dynamical Theory of Crystal Lattices*, Clarendon, Oxford, 1954.
6. E. Burstein, "Interaction of Phonons with Photons: Infrared, Raman and Brillouin Spectra," in *Phonons and Phonon Interactions*, edited by T. A. Bak, Benjamin, New York, 1966, p. 276.
7. R. M. Lyddane, R. G. Sachs, and E. Teller, *Phys. Rev.* 59, 673 (1941).
8. W. Cochran and R. A. Cowley, J. *Phys. Chem. Solids* 23, 447 (1962).
9. W. Cochran, *Advan. Phys.* 9, 387 (1960).
10. J. F. Cornwell, *Group Theory and Electronic Energy Bands in Solids*, North-Holland, Amsterdam, 1969, p. 181.
11. J. T. Houghton and S. D. Smith, *Infra-Red Physics*, Oxford Univ. Press, London, 1966, p. 108.
12. L. Genzel, "Impurity-Induced Lattice Absorption," in *Optical Properties of Solids*, edited by S. Nudelman and S. S. Mitra, Plenum, New York, 1969, p. 453.
13. H. R. Zallen, *Phys. Rev.* 173, 824 (1968); I Chen and R. Zallen. *Phys. Rev.* 173, (1968).
14. R. C. Powell, *Symmetry, Group Theory, and the Physical Properties of Crystals*, Springer, New York, (2011).
15. R. Loudon, Advan. Phys. 13, 423 (1964).
16. J. R. Hardy, "Raman Scattering by Phonons," in *Phonons in Perfect Lattices and in Lattices with Point Imperfections*, edited by R. W. H. Stevenson, Plenum, New York, 1966, p. 245.
17. W. Panofsky and M. Phillips, *Classical Electricity and Magnetism*, Addison-Wesley, Reading, Mass., 1962, p. 358.
18. R. Loudon, Proc. Phys. Soc. London 82, 393 (1963).
19. T. R. Gilson and P. J. Hendra, *Laser Raman Spectroscopy*, Wiley, New York, 1970.
20. N. X. Xinh, "Theory of First-Order Raman Scattering by Crystals of the Diamond Structure Containing Substitutional Impurities," in *Localized Excitations in Solids*, edited by R. F. Wallis, Plenum, New York, 1968, p. 167.
21. A. Yeniay, M. M. Delavaux, and J. Toulouse, J. Lightwave Tech. **20** 1425 (2002).
22. D. Cotter, Elec. Lett. **18**, 495 (1982).

Chapter 13

Lattice Vibrations and Lasers

Solid state lasers have become extremely important in many different types of applications and lattice vibrations play a critical role in the performance of these lasers. Typical solid state lasers consist of crystalline or glass host materials doped with optically active rare earth or transition metal ions. The relevant ions are discussed in Sec. 11.4 and an effective Hamiltonian to describe the optically active ions in solid state laser materials was developed. The effects of lattice vibrations are contained in the last term in the Hamiltonian given in Eq. (11.4.3). The strength of the crystal field created by the ligands at the site of the dopant ion determines the relative positions of the electronic energy levels of the ion. The vibrations of the host lattice ions modulate the crystal field strength and symmetry at the site of the optically active ion. This modulation affects the strengths, widths, and positions of spectral lines and causes the occurrence of radiationless and vibronic transitions as described in Chapter 11. All of these effects are crucial in determining the characteristics of laser operation.

Figure 13.1 shows a schematic representation of the electronic energy levels and pumping dynamics of a typical solid state laser material. In general, the energy input for this type of laser comes from optical pumping. The figure shows pump photons of energy $h\nu_p$ exciting an optically active ion from its ground state E_{gs} to a pump band at energy E_p. It is helpful to have a broad energy spread in the pump band to maximize the amount of pump light that is absorbed. The next step in the pumping dynamics is the

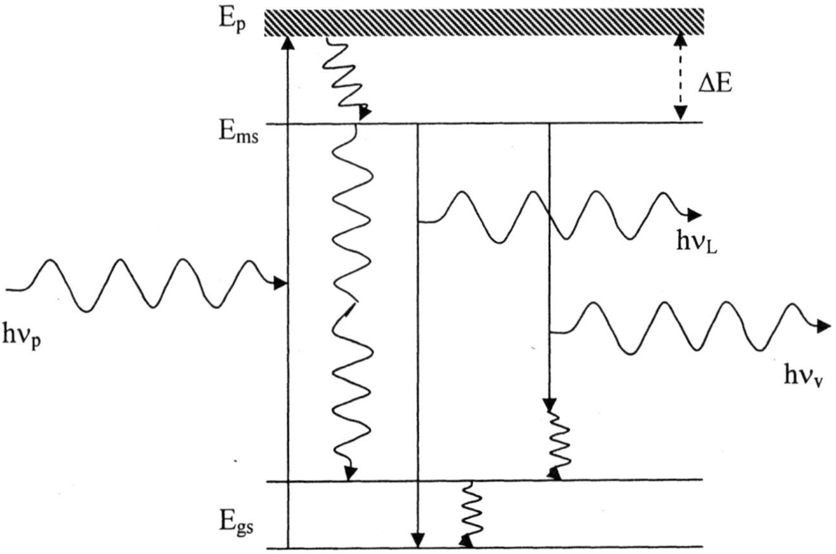

Fig. 13.1. Energy levels and transitions in a typical laser material.

radiationless relaxation of the absorbed energy from the pump band to a long lived metastable state, shown in the figure at energy E_{ms}. For a good laser material, this must be a fast, efficient process so no energy is lost. The metastable state should have a long lifetime and a high radiative quantum efficiency in order to store the energy until stimulated emission occurs. Thus there should be a low probability of radiationless decay processes from this level. The laser emission transition can be either purely radiative giving rise to photons of energy $h\nu_L$ or vibronic with energy $h\nu_v$. If the final state of the laser transition is above the ground state energy, it is important that fast radiationless processes empty this state quickly in order to maintain a population inversion.

The positions of the electronic energy levels shown in Fig. 13.1 depend on the strength of the crystal field created by the neighboring ions in the lattice and the width of the energy levels depends on the modulation of the crystal field through lattice vibrations. The electron-phonon coupling strength determines the importance of the crystal field environment on the spectral properties of the ion. One important parameter determined by the strength of the crystal field is ΔE, the energy separation between the pump band and the metastable state. The effects on laser properties of having large or small values of ΔE are described below.

One important engineering design parameter for solid state lasers is given by the equation[1]

$$\sigma_p \tau_r = \frac{\lambda_L^2}{8\pi n^2 \Delta\nu}, \tag{13.1}$$

where σ_p is the is the peak stimulated emission cross section of the laser transition while λ_L, τ_r and $\Delta\nu$ are the wavelength, radiative lifetime, and line width of the laser transition. Thus, considering the laser wavelength as a fixed design parameter, there are tradeoffs between cross section, lifetime and line width for different materials. For example, for a laser with a low threshold and high gain σ_p and τ_r should be high which requires the laser line width to be small. However, if broadband laser emission is desired for tunability or short mode-locked pulses, the $\sigma_p\tau_r$ product must be small. How the vibrational properties of the host lattice interact with the optically active ion is critical in determining these parameters. This chapter extends the discussion of the optical spectra of ions in crystals given in Chapters 10 and 11 to the specific application to solid state laser materials.

In summary, lattice vibrations affect the spectral properties of laser materials through radiationless and vibronic transitions during the pumping and decay processes, and phonon contributions to the width, position and strength of the laser transition. This chapter discusses the relationship between these effects of lattice vibrations on the spectral dynamics of ions in crystals and the operational characteristics of these materials as solid state lasers. First a more detailed development of phonon emission and absorption processes is given and then the way phonon processes influence the properties of two different types of laser systems is presented. The first type covers lasers with a fixed emission wavelength and the second type covers lasers having wavelength selective emission. The major difference in these two cases is the strength of the crystal field at the site of the optically active ion. This determines the positions of the energy levels of the ions and how these positions are modulated by the dynamic changes of the crystal field introduced by lattice vibrations. A detailed treatment of the properties of solid state lasers can be found in reference one.[1]

13.1. Nonradiative Transitions

Since nonradiative transitions involving the absorption or emission of phonons are so important in the spectral dynamics of solid state lasers as described above, the discussion of these types of processes given in Chapter 10 is extended here to focus on their specific relevance to lasers. The energy of an ion is a crystal consist of the potential well of the electronic state and the vibrational energy levels within this potential well. If an electron is in a specific vibrational energy level in an electronic potential well, it can move to a higher or lower vibrational energy level by absorbing or emitting a phonon. In addition, if two electronic energy levels are separated by an energy gap equal to or less than the phonon cutoff energy for the material, the electron can move from one electronic level to the other by absorbing or emitting phonons. These types of transitions are called "nonradiative" or "radiationless" processes.

As shown in Chapter 10, the probability for a radiationless transition between two vibrational energy levels in the same electronic potential energy well process is proportional to the square of the matrix element

$$M_{nm} = |\langle\phi_m|\mathrm{D}_{ed}^{(2)}|\phi_n\rangle|, \tag{13.1.1}$$

where $\mathrm{D}_{\mathrm{ed}}^{(2)}$ is the second term of the electric dipole moment operator given in Eq. (10.6.11).

In the harmonic approximation the eigenfunctions of the vibrational states are given by

$$\phi_n(x) = \left\{\frac{(\alpha/\pi)}{2^n n!}\right\}^{1/2} \mathrm{e}^{-\xi^2/2} H_n(\xi), \tag{13.1.2}$$

where $\mathrm{H_n}(\xi)$ is the n-th Hermite polynomial with $\xi = \alpha^{1/2}\mathrm{x}$ and the harmonic oscillator constant given by $\alpha = m_0\omega/\hbar$. These are orthonormal functions. Changing variables and integrating Eq. (13.1.1) with the eigenfunctions from Eq. (13.1.2) gives

$$M_{nm} = \begin{cases} 0 & m \neq n \pm 1 \\ \sqrt{(n+1)/2\alpha} & m = n+1 \,. \\ \sqrt{n/2\alpha} & m = n-1 \end{cases} \tag{13.1.3}$$

Thus transitions are only allowed between adjacent vibrational levels.

Along with the matrix element for the transition, the expression for the transition probability has a factor of $2\pi/\hbar$ and the density of final states. The electronic density of states is assumed to be a delta function in energy. Following the development given in Chapter 6, the Debye distribution of phonons is used for the vibrational states. This gives a density of states expressed as

$$\rho(\omega) = \begin{cases} \dfrac{3V\omega^2}{2\pi^2\nu^3} & \text{for } \omega \leq \omega_D \\ 0 & \text{for } \omega > \omega_D \end{cases}, \tag{13.1.4}$$

where ω_D is the Debye cutoff frequency. The expressions for the transition rates of absorption and emission of phonons then become

$$\begin{aligned} W_{nr}^{ab} &= Cn_q \\ W_{nr}^{em} &= C(n_q + 1), \end{aligned} \tag{13.1.5}$$

where C contains a set of constants and n_q represents the number of phonons in the q-th vibrational state.

The radiationless processes are usually very fast (picosecond time scale) so the distribution of electrons in the vibrational states of the electronic potential well reaches thermal equilibrium very quickly. The phonon occupation numbers are given by the Bose-Einstein distribution function

$$n_q = \frac{1}{e^{\hbar\omega_q/k_B T} - 1}, \tag{13.1.6}$$

where k_B is Boltzmann's constant. The rate equations describing the time evolution of the populations of a pair of neighboring vibrational levels in an electronic potential well are

$$\begin{aligned}\frac{\partial n_1}{\partial t} &= -n_1 W_{12} + n_2 W_{21} \\ \frac{\partial n_2}{\partial t} &= n_1 W_{12} - n_2 W_{21}.\end{aligned} \tag{13.1.7}$$

In thermal equilibrium the time derivatives are zero and the solution to these equations give the steady state populations of the vibrational levels as

$$\begin{aligned}n_2 &= \frac{1}{1 + e^{\hbar\omega_q/k_B T}} \\ n_1 &= \frac{e^{\hbar\omega_q/k_B T}}{1 + e^{\hbar\omega_q/k_B T}},\end{aligned} \tag{13.1.8}$$

and the ratio of the populations is given by

$$n_2/n_1 = \mathrm{e}^{-\hbar\omega_q/k_B T}. \tag{13.1.9}$$

If an electron is excited to an upper vibrational level of an electronic potential well, it relaxes to the next lower level by emission of a phonon and continues to cascade downward until it reaches the lowest level consistent with a thermal equilibrium population distribution.

When the electronic energy levels of an ion in a solid are closely spaced, it is possible for an electron to move between these levels by absorbing or emitting phonons. The mathematical treatment of this situation is the same as that discussed above except that it is the first term in the matrix element given before in Eq. (10.6.14) that is relevant instead of the second term. If there is a manifold of closely spaced electronic energy levels the population of electrons in each of the levels again quickly reaches thermal equilibrium as described by the Boltzmann distribution function.

For electronic energy levels with spacing greater than the Debye cutoff energy multiphonon processes can be effective in bridging the energy gap. Higher order perturbation theory can be used to derive an expression for two-phonon or higher processes. The extension to second order perturbation theory for two-phonon processes can be extrapolated to obtain an expression for the transition rate of a p-phonon emission processes[2]

$$W_{nr}^{p}(emission) = W_o(p)(n_{\omega_0+1})^p, \tag{13.1.10}$$

where ω_0 is the "effective phonon frequency" and the factor $W_0(p)$ contains the matrix element for the transition and other constants raised to the 2p power as well as the effective phonon frequency. The assumption is made that the distribution of the p phonons involved in the process is sharply peaked about this frequency.

The temperature dependence of the multi-phonon emission rate is contained in the phonon occupation number. Using Eq. (13.1.6) in Eq. (13.1.10) gives

$$W_{nr}^{p}(emission) = W_o(p)\left[\frac{e^{\hbar\omega_0/k_B T}}{e^{\hbar\omega_0/k_B T}-1}\right]^p . \tag{13.1.11}$$

For a very weak electron-phonon interaction, as it is the case for rare earth ions in crystals, we can define a small number ε as the ratio of the transition rate for a p-order process to that for a (p − 1)-order process. In such a case

$$\begin{aligned} W_{nr}^{p}(emission) = W_{nr}^{p-1}(emission)\varepsilon &= W_{nr}^{p-2}(emission)\varepsilon^2 \\ &= \cdots W_{nr}^{o}(emission)\varepsilon^p \\ &= W_{nr}^{o}e^{p\ln(\varepsilon)} \end{aligned}$$

This can be written as

$$W_{nr}^{p}(emission) = W_{nr}^{o}e^{(\Delta E/\hbar\omega_0)\ln(\varepsilon)}, \tag{13.1.12}$$

where the assumption has been made that p phonons of energy $\hbar\omega_0$ cross an energy gap of ΔE. This "energy gap law" will be discussed further below.

The transition rates for nonradiative processes derived above determine how efficiently the metastable state of a laser is pumped and the population of the terminal state of the laser transition is depleted. The nonradiative transition rates also determine the quantum efficiency of the laser transition. In addition, inverse of the decay rates are related to the width of the initial state of an optical transition through the uncertainty principal.[3] Thus these "lifetime broadening" processes add terms to Eq. (11.7.6) describing the temperature dependence of the position of a spectral line and to Eq. (11.7.12) describing the temperature dependence of the width of a spectral line. The temperature dependences of the contributions of these radiationless processes is contained in the expression for the phonon occupation numbers given in Eq. (13.1.6). Examples of the effects of these types of processes in specific laser materials are given in the following sections.

13.2. Single Wavelength Lasers

One important characteristic of lasers is that they provide a monochromatic source of light. However, different types of applications generally require specific wavelengths of light and matching the source with the application has resulted in the development of many different types of lasers with different wavelengths. Some of these operate at one or a few discrete wavelengths while others can provide a variety or continuum of wavelengths. The former type is discussed in this section while the latter type is discussed in the next section.

One common type of laser with a sharp emission line involves trivalent lanthanide rare earth ions in crystal hosts. The optical transitions for these ions occur between the energy levels of electrons in an unfilled inner shell that have other electrons in outer shells. These outer shell electrons provide a shield for the inner shell electrons so the latter do not feel the full strength of the crystal field created by the ligands at the site of the optically active ion. Thus the affects of the weak crystal field and the modulation of its strength and symmetry by lattice vibrations on the spectral properties are minimum, resulting in sharp absorption and emission lines.

A typical example of a laser with a narrow emission line is Nd-YAG. This commercially important laser system consists of optically active trivalent neodymium ions (Nd^{3+}) as dopants in an yttrium aluminium garnet host crystal ($Y_3A1_5O_{12}$). The neodymium ions replace about 1% of the yttrium ions at lattice sites in the garnet host structure with D_2 symmetry. For this rare earth ion there are three optically active electrons in an unfilled 4f shell that are shielded by two 5s electrons and six 5p electrons. The shielding by the outer shell electrons minimizes the strength of the crystal field at the site of the optically active electrons and thus the modulation of the crystal field by host lattice vibrations is a weak perturbation. Therefore the optical absorption and emission transitions appear as a series of sharp spectral lines.

This laser material has a plethora of energy levels throughout the visible region of the spectrum that can be effective in optical pumping from the ground state through radiative absorption transitions. In general, electrons excited into these high energy levels cascade down to the metastable state in the near infrared region of the spectra through radiationless decay processes. The metastable state is designated $^4F_{3/2}$. The crystal field splits this into two closely spaced levels ($\Delta E = 85\,cm^{-1}$) and at room temperature the populations of these to levels are kept in thermal equilibrium through fast radiationless processes. This system of coupled levels has a radiative lifetime of about 250 μs with a radiative quantum efficiency that can be close to 100%. If the host crystal has impurities such as hydroxyl ions or high concentrations of Nd^{3+} ions, nonradiative cross relaxation can lower this quantum efficiency. The radiative emission from the metastable state occurs as a series of lines terminating on various components of the four multiplets of the ground state term, $^4I_{9/2}$, $^4I_{11/2}$, $^4I_{13/2}$, and $^4I_{15/2}$. More than half of the fluorescence is channeled into the main $^4F_{3/2} \rightarrow ^4 I_{11/2}$ transition at 1.064 μm which is the most important laser transition for Nd-YAG. From this energy level, fast radiationless processes take the electron back to the lowest level of the ground state manifold. Lattice vibrations obviously play an important part in the laser dynamics of this type of system. Vibrational absorption and scattering transitions seen in the infrared and Raman spectra of YAG crystals range between about 120 and $920\,cm^{-1}$. Phonons contributing to radiationless transitions and thermal broadening and shifting of optical spectral lines should lie within this energy range.

13.2.1. *Optical Transitions in Rare Earth Ion Lasers*

As discussed in Chapter 11, the relevant optical transitions for rare earth ions go between different single electron levels of the unfilled f shell. These f-to-f transitions are parity forbidden. For them to occur requires either higher order electric field operators such as magnetic dipole or electric quadruple, or forced electric dipole transitions. The latter require admixing of the energy levels either through odd components of the crystal field perturbation or odd components of the lattice vibrations.[4,5] Forced electric dipole transitions are the most important type of radiative transitions for laser dynamics. The initial and final states of a forced electric dipole are expressed as

$$|\psi_i\rangle = |4f^n\alpha LSJM_J\rangle + \sum_{\beta} \frac{\langle\beta\alpha'' L'' S'' J'' M_J''|\langle 4f^n\alpha LSJM_J|\, V_{\rm int}|\beta\alpha'' L'' S'' J'' M''\rangle}{E_{4f} - E_{\beta}}. \tag{13.2.1}$$

The first term on the right side of the equation is one of the free ion multiplets of the $4f^n$ configuration which has odd parity since f = 3. The L, S, J, and M_J quantum numbers designate different components of angular momentum as usual and α contains all required quantum numbers not shown explicitly. The second term on the right describes how the odd parity perturbation interaction given by V_{int} admixes this state with multiplets of a higher lying even parity state designated as β. Because of the energy denominator, the lowest energy empty state generally dominates the sum. For $4f^3$ ions such as trivalent neodymium this would be the 5d level. This admixture contributes an even parity component to the wave function of the state.

Using the formalism for electric dipole radiative transitions developed in Chapters 9–11 with the initial and final state wave functions given by Eq. (13.2.1) and the dipole moment operator given by Eq. (10.7.2), the radiative transition rate can be determined for absorption and emission transitions of rare earth ions like Nd^{3+} in crystals. The transition rate can be used to determine the important laser parameters in Eq. (13.1) such as the peak emission cross section for the laser transition and the radiative lifetime of the metastable state. These parameters are important in determining the threshold and gain of a solid state laser.

The interaction Hamiltonian causing the admixing of even and odd parity energy levels can either be associated with the odd terms in the static crystal field or odd parity lattice vibrations modulating the crystal field. For cases where this dynamic electric-phonon coupling is important, the interaction Hamiltonian is given by Eq. (11.5.11) and lattice vibrations play a critical role in determining the strength of the optical transitions. This can be treated in a way similar to the treatment of vibronic transitions discussed in Chapters 10 and 11. The electron-phonon coupling in these materials is so weak that vibronic transitions are not generally observed. In general, the important transitions associated with rare earth ion lasers can be attributed to static crystal field admixing so this topic is not discussed further. In Nd-YAG this results in the main laser line having a stimulated emission cross section of $6.5 \times 10^{-19}\,\mathrm{cm}^2$.

Along with the strength of a spectral line discussed above, its width and position are also important for laser operation as seen in Eq. (13.1). Chapter 11 describes the contributions to the width of a spectral line due the Raman scattering of phonons and Sec. 13.1 describes the contributions due to direct phonon absorption and emission processes. Due to the close spacing of the electronic energy levels of rare earth ions in crystals, both types of processes can make significant contributions to the widths of their spectral lines and the variation of these widths with temperature.

As an example, the main laser line in Nd-YAG has a width of about $5\,\mathrm{cm}^1$ at room temperature. At very low temperatures the width of this line is about $1\,\mathrm{cm}^{-1}$ which is due to contributions from both inhomogeneous broadening and radiationless decay of the terminal level. Using Eq. (11.7.12) to fit the results of measurements of the temperature dependence of this transition reveals that the Raman scattering of phonons is responsible for the major part of the linewidth at room temperature.[6] The effective Debye temperature obtained from this analysis is 500°K which gives an energy of about $348\,\mathrm{cm}^{-1}$ that is well within the phonon spectrum of the garnet host crystal.

The main laser line in Nd-YAG also exhibits a shift in position toward longer wavelengths with increasing temperature. Near room temperature this shift is about $0.04\,\mathrm{cm}^{-1}/°\mathrm{K}$. The two-phonon scattering processes described by Eq. (11.7.5) are responsible for most of the thermal lineshift with direct phonon processes making a smaller contribution.[6] For other spectral lines direct phonon absorption and emission processes play a larger role.

13.2.2. *Radiationless Decay Processes in Rare Earth Ion Lasers*

The radiationless decay processes that are involved in the laser pumping dynamics of rare earth doped lasers can be described by the formalism developed in Chapters 10 and 11 and Sec. 13.1. The Hamiltonian for the electron-phonon interaction given in Eq. (11.6.2) can be written in terms of phonon creation and annihilation operators and the vibrational states of the system can be expressed as phonon occupation numbers.[3] The transition rate for phonon decay processes in this formalism depends exponentially on the energy gap between the two states of the transition as discussed above. The temperature dependence of the transition rate can be measured and the results fit using Eq. (13.1.11). This gives the number of phonons involved in the process and the effective phonon energy.

For the cascading radiationless processes describing the relaxation from the higher lying pump bands to the metastable state, the overall relaxation rate is controlled by the process with the largest energy gap. For pumping Nd-YAG the gaps between levels in the visible spectral region range from about 100 to 1200 cm^{-1}. For excitation in the middle of the visible spectral region,[7] the largest gap is around 1148 cm^{-1} and there is no loss of energy through radiative transitions. Theoretical fits to the data give an overall relaxation rate of $2\times10^6\mathrm{s}^{-1}$ and an effective phonon energy of 700 cm^{-1}. If pumping occurs at higher energies, energy gaps as large as 2145 cm^{-1} must be crossed and this results in decay rates as low as $5 \times 10^5\mathrm{s}^{-1}$ which creates a bottleneck in the relaxation from pump band to the metastable state. Some radiative transition loss occurs at these high pump energies.

The energy gap law predicts a nonradiative decay rate from the metastable state in Nd-YAG that is about five orders of magnitude smaller than the radiative decay rate. This is why radiationless emission processes are not important for the laser transition in this material except for the cases of concentration or impurity quenching.[1] This results in a very high radiative quantum efficiency for the $^4\mathrm{F}_{3/2}$ level which helps make Nd-YAG such a good laser material.

Having fast radiationless decay processes deplete the population of the terminal state of the laser transition is critical for allowing Nd-YAG to operate as a four-level level system.[1] The energy gaps between the various components of the $^4I_{11/2}$ level and the $^4I_{9/2}$ level range[7] from 1150 to 1300 cm^{-1} giving decay rates of the order of $10^{-5}\,s^{-1}$. This results in a decay of the $^4I_{9/2}$ level that is fast enough to maintain a population inversion with the $^4F_{3/2}$ level.

Trivalent neodymium ions have been made into lasers in many types of oxide and fluoride crystal hosts other than YAG.[1] The width of the laser transition ranges from slightly smaller than Nd-YAG to about 17 times greater. The cross sections for the laser transition range from an order of magnitude smaller to one and a half times larger. The lifetime of the metastable state ranges from about one third the value for Nd-YAG to almost two and a half times the value. These variations are due to the strength and symmetry of the crystal field at the site of the Nd^{3+} ions and how this is modulated by lattice vibrations in the different host crystals. In general, the spectrum of lattice vibrations cuts off at lower energies in fluoride crystals than in oxide crystals so the effects of electron-phonon interactions are less in fluoride host lasers than in oxides.

In addition to crystalline hosts, Nd^{3+} -doped into oxide and fluoride glass hosts have also been made into lasers.[8] Because of the lack of long range order of the lattice, the dopant ions have a range of sites with slightly different crystal fields which results in strong inhomogeneous broadening of the spectral lines. The atomic vibrations that are important in modulating the crystal field at the site of an ion are local modes that will vary slightly from site to site. Selectively exciting ions in different subsets of sites has shown that properties such as radiative and nonradiative decay rates vary from site to site in glass hosts.[10] This is associated with site-to-site variations of the crystal field strength, symmetry, and vibrational modulation. In general, glasses have lower thermal conductivities than crystals. This can cause problems such as thermal lensing for high average power glass lasers.

All of the lanthanide series rare earth ions discussed in Chapter 11 have been made into lasers in one or more of a wide variety of oxide and fluoride crystal and glass hosts.[1] The effects of lattice vibrations on their laser operational characteristics are similar to those discussed above for the example of Nd^{3+}. One interesting result of having the ability to study a wide variety of combinations of rare earth ions in different hosts is a better understanding of multiphonon radiationless relaxation processes discussed above. Nonradiative decay rates have been measured for transitions having energy gaps ranging from just over 1,000 cm^{-1} to over 4,600 cm^{-1}.[2] In general the results can be explained quite well by Eq. (13.1.12).

One manifestation of the slower radiationless decay rates in fluoride host crystals is "upconversion" pumping of a laser.[9] Low energy photons can cause an absorption transitions to one of the higher energy levels of the ground state term and before the population of this level can relax radiationlessly back to the lowest level a second low energy photon gets absorbed from the intermediate level to the metastable state where laser emission can occur. This can be useful in converting infrared pump light to visible laser light.

Figure 13.2 shows a schematic diagram of the transitions involved in a typical upconversion laser. However there are many variation of accomplishing upconversion depending on the types of ions and pump lasers involved. For an ion with equal spacing of the three relevant electronic levels, two phonons of the same frequency can be absorbed sequentially from one pump laser. If these levels have different energy spacing, the two pump photons must come from two different lasers with different emission wavelengths. In some upconversion lasers two types of ions are involved. The first ion absorbs a

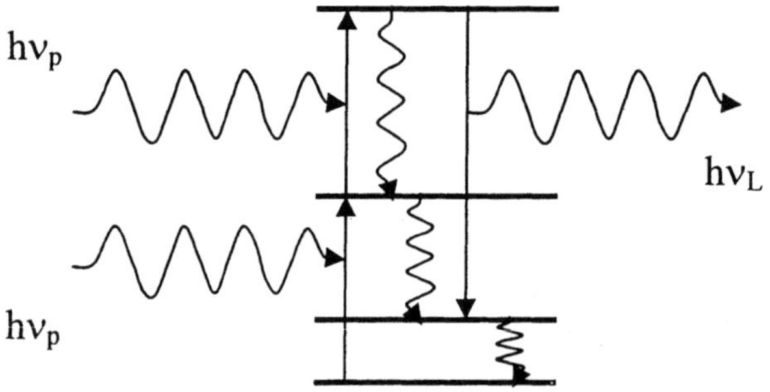

Fig. 13.2. Transitions relevant to upconversion lasers.

photon and transfers the energy to a different type of ion that then absorbs the second photon from an excited state before it emits a laser photon. For any of these different types of configurations the key to efficient upconversion is having very weak phonon decay processes from the intermediate state so ions in this energy level have time to absorb the second photon. Of course for good laser operation the nonradiative decay from the metastable state must also be very weak and the radiationless relaxation of the terminal state of the laser transition (if it is not the ground state) must be very fast.

The most common host crystal for upconversion lasers is $YLiF_4$ but other fluoride crystals and fluoride glass fibers have also been used.[1] Nd^{3+} ions have been used in upconversion lasers but only at low temperatures where the rates of radiationless decay processes are greatly reduced. Trivalent holmium praseodymium, erbium and thulium ions all have been made into upconversion lasers operating at room temperature in fluoride glass fibers. Er^{3+} and Tm^{3+} have also been made into room temperature upconversion lasers in $YLiF_4$ crystals. In addition, energy transfer between Yb^{3+} and Er^{3+} ions or multiple Er^{3+} ions have been made into upconversion lasers. The affects of lattice vibrations are key to determining the laser properties of all of these systems.

13.3. Multiple Wavelength Lasers

First row transition metal ions in a variety of crystalline host materials are successful laser materials.[1] The optically active electrons on these ions are in partially filled d shells that have no outer electrons to shield them. Thus their spectral properties are strongly affected by the local crystal field and its modulation through lattice vibrations. This may result in broad vibronic absorption and emission transitions as well as fast nonradiative processes. The radiationless relaxation processes within an electronic potential well were described in Chapters 10 and 11 and Sec. 13.1. In general these are very fast compared to electronic transitions so the laser properties are determined by the fluorescence emission transition. If the population of thermal levels in the excited electronic state shown in Fig. 11.1 is higher than the energy at which the ground state electronic potential well crosses the excited state potential well, radiationless relaxation can occur from the upper level to the ground state. This would act as a major energy loss factor in the laser dynamics of the material. However, this type of loss is negligible in important tunable laser materials.

Vibronic emission processes were described in Secs. 10.10 and 11.5. Incorporating these types of processes in the expressions for laser cross sections, thresholds, and gain was done by McCumber.[11] His theoretical analysis provides a good explanation for the properties of "phonon terminated" or vibronic lasers.

There are two types of electronic transitions for transition metal ions in crystals. The first type involves all of the optically active electrons remaining in their original electronic configuration but one of them flipping the direction of its spin. This change in spin state does not depend on the local crystal field or its modulation by lattice vibrations so the optical transition involving a spin flip appears as a sharp spectral line. The other type of transition involves an electron moving from one configuration of crystal field states to another. This type of transition is directly related to the strength of the crystal field at the site of the ion that determines the position of the resulting spectral band. The breadth of the absorption or emission band associated with this type of configuration changing transition is determined by the modulation of the crystal field through lattice vibrations. Most transition metal ions have both types of transitions in their optical spectra and their spectral characteristics are generally determined by the relative positions of these different types of transitions. The important parameter demonstrating this in Fig. 13.1 is the energy gap between the pump band and the metastable state, ΔE.

In some materials involving transition metal ions such as ruby ($Al_2O_3 : Cr^{3+}$) the lowest excited state is related to the ground state through a spin-flip transition while the configuration changing transitions are at significantly higher energies giving a large value of ΔE. This results in strong, broad absorption bands in the blue-green spectral region and a weak, sharp line in the red region of the spectrum. The weak electron-phonon coupling associated with the spin-flip transition leads to an emission spectrum that appears as a sharp no-phonon line with weak vibronic sidebands.[3] Energy levels decaying through spin-flip transitions have long lifetimes and therefore make excellent metastable states for laser emission. Ruby lasers operate as single wavelength lasers with properties similar to those discussed in the previous section and described by the schematic diagram in Fig. 13.1.

In other cases such as alexandrite ($BeAl_2O_4:Cr^{3+}$) or Ti-sapphire ($Al_2O_3:Ti^{3+}$) the emission spectra appear as broad vibronic bands. This broad band emission in conjunction with a dispersive element in the cavity allows these solid state lasers to have a continuously tunable output wavelength. This can be critically important in some applications.

Tunable lasers such as Ti-sapphire have only one ground state and one excited state connected through configuration changing absorption or emission transitions. They are best described using a model based on a configuration coordinate diagram such as that shown in Fig. 11.1. The pump transition originates on the lowest vibrational level of the ground electronic state and terminates on an upper vibrational component of the excited electronic state. The energy relaxes radiationlessly to the lowest vibrational state in this potential well as described in Sec. 13.1. The laser emission transition occurs from this state and terminates on an upper vibrational component of the ground state potential well. The energy then relaxes radiationlessly to the bottom of this potential well as discussed above. The radiationless relaxation processes in both the upper and lower electronic potential wells are very fast resulting in a population inversion that allows for good laser operation. Due to the vibronic nature of the emission transition a Ti-sapphire laser can be tuned over 400 nm. This broad laser bandwidth is coupled with a short metastable state lifetime (around 3 ps). The emission bandwidth allows for modelocking to get femtosecond pulses.

Materials such as alexandrite are more complicated since they have more than one excited state involved in the laser pumping dynamics. In this case a model based on the energy level diagram in Fig. 13.1 is more useful. For these types of materials the lowest excited state is still related to the ground state through a spin-flip transition as it is in ruby, but the higher level associated with configuration changing transitions is very close to it in energy. In other words the value of ΔE is very small. The energy gap between the pump band and the lowest excited state is so small that direct phonon absorption and emission process cause the populations of these levels to be in thermal equilibrium at room temperature. The lower level decays through a spin-flip transition that has a weak transition cross section compared to the decay of the upper level configuration changing transition. The coupled system of levels emits fluorescence through strong vibronic transitions to the ground state.

Equation (13.1) shows that the radiative lifetime of the initial level of the laser transition is an important parameter in determining laser properties. For materials such as alexandrite this is a combination of the lifetimes of the pump band and metastable state shown in Fig. 13.1 weighted by the populations of these levels. Since phonon absorption and emission processes keep these levels in thermal equilibrium, the coupled lifetime is given by[1]

$$\tau^{-1} = \tau_{ms}^{-1} + \tau_p^{-1} e^{-\Delta E/(k_B T)}. \tag{13.3.1}$$

Here ΔE is the energy gap between the pump band and the metastable state. For materials such as alexandrite the vibronic transition form the excited state to the ground state has a short intrinsic lifetime and thus a strong radiative decay rate that dominates the laser transition while the longer lived metastable state stores the excited state population to feed into this transition through phonon absorption processes. For situations like this where the active ions are in sites that make the transition to the ground state symmetry forbidden, odd parity lattice vibrations may also be necessary to make the transition occur. Thus fast phonon absorption and emission processes between the pump band and the metastable state are critical in the operational characteristics of an alexandrite laser.

Trivalent chromium ions in fluoride hosts can also be made into tunable vibronic lasers. One important example of this[12] is $LiSrAlF_6:Cr^{3+}$ that has a broad emission band peaking at 825 nm. Vibrational distortions of the local site symmetry play an important role in determining the strength of the laser transition as well as the breadth of the tunable emission band. Chromium in divalent and four-valent states, trivalent titanium, and divalent vanadium, iron, cobalt, and nickel ions have all been made into vibronic lasers. Some of these work only at low temperatures because of nonradiatrive decay processes quenching the fluorescence emission at high temperatures.[1] The lattice vibrations related to both the vibronic and radiationless processes discussed in Chapters 10 and 11 are critical in determining the properties of these lasers.

It is also possible to achieve wavelength selectability from lasers emitting a series of discrete as opposed to the continuum of wavelength obtained from a broadband laser. This can be done using a nonlinear optical material either internal or external to the laser cavity in order to shift the primary frequency. One laser of this type that is based on lattice vibrations is a Raman laser.[13] The Raman scattering of light by phonons was discussed in Chapter 12. Under high intensity pump radiation, stimulated Raman emission may occur and if the material is placed in a resonant cavity under these conditions laser emission may occur at the Raman shifted frequency. A dispersive element can be used to select laser output at the primary laser wavelength or one of the Stokes or anti-Stokes Raman shifted wavelengths.

One successful example of a material for solid state Raman lasers is $Ba(NO_3)_2$. This is a molecular crystal containing tightly bound NO_3 groups as part of the lattice. This subgroup of the lattice has D_{3h} symmetry and its normal modes of vibration can be determined as described in Chapter 5. The strongest Raman transition is associated with the breathing mode of the NO_3 molecule. This lattice vibration acts as a zone center phonon mode that transforms as the A_{1g} irreducible representation. Other molecular crystals such as tungstates, vanadates, niobates and molybdates have similar vibrational modes with strong stimulated Raman emission cross sections that are useful for lasers. The ability to have lasers operating on a Raman emission line provides many additional laser wavelengths that may be needed for specific applications. This class of lasers is enabled by the direct interaction of the pump light with lattice vibrations.

References

1. R. C. Powell, Physics of Solid-State Laser Materials, Springer-Verlag, New York, (1998).
2. C. B. Layne, W. H. Lowdermilk, and M. J. Weber, Phys. Rev. 16, 10 (1977); L. A. Riseberg and M. J. Weber in Progress in Optics, edited by E. Wolf,

North Holland, Amsterdam (1977), Vol. 14, p. 89; L. A. Riseberg and H. W. Moos, Phys. Rev. 174, 429 (1968).
3. B. DiBartolo, Optical Interactions in Solids, John Wiley, New York, (1968).
4. B. R. Judd, Phys. Rev. 127, 750 (1962).
5. G. S. Ofelt, J. Chem. Phys. 37, 511 (1962).
6. T. Kushida, Phys. Rev. 185, 500 (1969).
7. Yu. Perlin, A.A. Kaminskii, M.G. Blazha, and V.N. Enakii, Phys. Status. Solidi B 112, K125 (1982); Yu. Perlin, A.A. Kaminskii, V.N. Enakii, and D.N. Vylegzhanin, Phys. Status. Solidi B 92, 403 (1979); A.A. Kaminskii, and D.N. Vylegzhanin, Dokl. Akad. Nauk SSSR 195, 827 (1970).
8. S.E. Stokowski, in The CRC Handbook of Laser Science and Technology, edited by M. J. Weber, CRC, Boca Raton, FL (1982), Vol. 1, p. 215.
9. R. M. Macfarlane, A. J. Silversmith, F. Tong, and W. Lenth, Appl. Phys. Lett. 52, 1300 (1988).
10. C. Brecher, L.A. Riseberg, and M.J. Weber, Phys. Rev. B 18, 5799 (1978).
11. D. E. McCumber, Phys. Rev. 134, A299 (1964).
12. S. A. Payne, L. L. Chase, and G. D. Wilke, J. Lumin. 44, 167 (1989); H. W. H. Lee, S. A. Payne, and L. L. Chase, Phys. Rev. B 39, 8907 (1989).
13. T. T. Basiev and R.C. Powell in Handbook of Laser Technology and Applications, edited by C. Webb and J. Jones, Institute of Physics Publishing, Bristol (2004) p. 469.

Subject Index